MW01166124

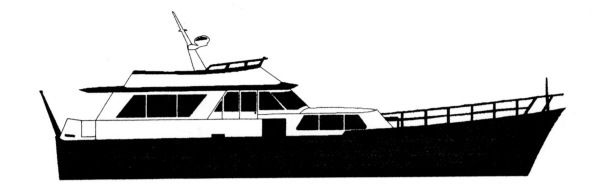

1995
EDITION

McKnew & Parker's
BUYER'S GUIDE TO

MOTOR YACHTS & TRAWLERS

From the Editors of the POWERBOAT GUIDE

Ed McKnew and Mark Parker

International Marine
Camden, Maine

PUBLISHED BY
International Marine
A Division of McGraw-Hill, Inc.
Blue Ridge Summit, PA 17294
1-800-822-8158

COMPILED BY
American Marine Publishing, Inc.
P.O. Box 30577
Palm Beach Gardens, FL 33420
1-800-832-0038

ISBN 0-07-045172-9
Printed and bound in the United States of America.

CONTENTS

INTRODUCTION

This book is written to help buyers sort through the hundreds of imported and domestically produced motor yachts, double cabins, long-range cruisers (LRC), and trawlers currently found on the nation's new and used-boat markets. Over 300 popular models from 27 feet to 83 feet in length, with prices ranging from the affordable to the truly staggering, are featured in these pages. For each, we have included complete factory specifications together with floorplan options, real-world performance data, production history, engine choices, and production updates. Throughout, the authors' opinions are oftentimes freely expressed. Advertising hype notwithstanding, some boats are simply better than others, and those that stand out at either extreme are occasionally noted. No attempt has been made to maliciously abuse a particular model, however, and the comments represent nothing more or less than the opinions of the authors. Needless to say, the services of experienced marine professionals are strongly recommended in the sale or purchase of any boat.

The prices quoted in this book reflect the market conditions projected by our staff for 1995. Those wishing to establish a consistent pattern for depreciation will be disappointed: We know of no such schedule. Rather, we have evaluated each model on its own merits and assigned values based on our own research and experience. *These are real-world prices.* If you expect to purchase a boat in excellent condition for notably less than the published Retail High, you're likely to be very disappointed. In many cases—often depending on location and availability—a well-maintained and equipped example of a given model may sell for 10 to 15 percent higher than the published Retail High figure. Conversely, a boat in particularly poor condition will almost certainly sell for less than the published Retail Low.

McKnew & Parker's Buyer's Guide to Motor Yachts & Trawlers is one of three annual consumer publications written by Ed McKnew and Mark Parker. (The other two are the *McKnew & Parker's Buyer's Guide to Sportfishing Boats* and the *McKnew & Parker's Buyer's Guide to Family & Express Cruisers*.) The series is a spin-off of the hugely successful *PowerBoat Guide*, a now standard industry reference originally compiled for the exclusive use of marine professionals. Future additions to the series will include guides for cruising sailboats, trailerable fishing boats, etc., and under-25-foot pleasure boats.

ACKNOWLEDGMENTS

We wish to thank the following individuals for their support. Without their help we would never have come this far.

Floyd Appling, Jr.

Bill Burgstiner

Steve & Delores Brown

Top & Sandy Cornell

George & Helene Gereke

Edward & Betty Groth

Freddy & Patti Hamlin

HOW TO USE THIS BOOK

For the most part, the contents of this book are straightforward and easily understood. There are, however, a few points that must be kept in mind to use this information to best advantage. Failure to do so may well result in a good deal of confusion and misunderstanding.

Factory Specifications

The specifications listed for each model are self-explanatory, although the following factors are noted:

1. *Clearance* refers to bridge clearance, or the height above the waterline to the highest point on the boat. Note that this is often a highly ambiguous piece of information since the manufacturer may or may not include such things as an arch, hardtop, or mast. Use this figure with caution.

2. *Designer* refers to the designer of the hull only and not necessarily the superstructure and/or the interior.

3. *NA* means that the information is not available.

Performance Data

Whenever possible, performance figures have been obtained from the manufacturer or a reliable dealer or broker. When such information was unavailable (or—as in many cases—manufacturers refused to provide it) the authors have relied upon their own research together with actual hands-on experience. The speeds are estimates and (in most cases) based on boats with average loads of fuel, water, options, and gear.

Speeds are always reported in knots. Those in the Great Lakes or inland waterways may convert knots to miles-per-hour by multiplying a given figure by 1.14.

Cruising Speeds, Gas Engines

Unless otherwise noted, the cruising speed for gas-powered inboard boats is calculated at 3,000–3,200 rpm.

Cruising Speeds, Diesel Engines

The cruising speeds for diesel-powered boats are calculated as follows:

1. Detroit (two-stroke) Diesels: about 250–300 rpm off the top rpm rating.

2. Other (four-stroke) diesels: about 350–400 rpm off the manufacturer's maximum rpm rating.

Floorplans

When there are two or more floorplans, the most recent layout comes last.

Pricing Information

The values provided should not be used after December 31, 1995.

Used-boat prices have been compiled from 1975, the base year for our calculations. Boats whose production runs were previous to that year are noted in the Price Schedule with four asterisks (****).

In the Price Schedule, six asterisks (******) indicate that we have insufficient data to render a value for a particular year.

While diesel engines always add significant value to any boat, there are some cases where the differences in the *type* or *horsepower* of diesel engines installed in a particular model will seriously affect the average resale value. Those cases in which we believe the different diesel options do indeed affect the value of an individual boat have been noted in the Price Schedule.

The *Retail High* is the average selling price of a clean, well-equipped, and well-maintained boat with low-to-moderate engine hours. Boats with an exceptional equipment list or those with unusually low hours will often sell at a figure higher than the published Retail High.

The *Retail Low* is the average selling price of a boat with below-average maintenance, poor equipment, high-time engines, or excessive wear. High-time boats in poor condition may sell for less than the published Retail Low.

Used boats located in the following markets are generally valued at 10–15% higher than published prices:

1. Great Lakes
2. Pacific Northwest
3. Inland Rivers and Lakes

The prices presented in this book reflect our best estimates of new and used boat prices for the model year 1995. They are intended for general use only and are not meant to represent exact market values.

FREQUENTLY ASKED QUESTIONS ABOUT MOTOR YACHTS AND TRAWLERS

In an effort to clear away some of the confusion regarding the purchase of a new or used motor yacht or trawler, we have listed below some of the most common questions asked by potential buyers. The answers to these questions are our own and we welcome responses from others who hold differing views. For the most part, however, we believe the information presented here will address several important issues confronting buyers of boats presented in this publication.

1. I notice that motor yachts are built on modified-V hulls while trawlers are constructed on either a semi-displacement or a full displacement hull form. What's the difference?

In general, motor yachts are built on the same kind of all-purpose modified-V hull forms used in the construction of many sportfishing boats. A modified-V hull—with its moderate transom deadrise and hard chines—provides a combination of performance, comfort, and stability not available with any other configuration. It's a compromise, but a good one.

The difference between a modified-V hull and a semi-displacement hull is oftentimes difficult to distinguish. In general, the semi-displacement design places less emphasis on performance in favor of economical, lower-speed operation. The chines are sometimes softer, and a true semi-displacement hull is often narrower of beam than a modified-V design. These hull forms offer little fuss in the transition from displacement speeds to planing speeds and they can operate efficiently at the transition speeds of a conventional modified-V model. Other semi-displacement designs are trawler-types capable of speeds above hull (displacement) speed but not necessarily in an efficient manner.

Semi-displacement hulls are most often seen on trawler-style boats (Grand Banks, Albin, Marine Trader, etc.) as well as many of the Downeast designs from New England builders. Seldom capable of 20+ knots top, they provide an economical and comfortable ride at less than motor yacht cruising speeds.

Full displacement hulls are seen only in true trawler designs. A displacent hull is characterized by rounded bilges and deep, full-length keels. Note that a true displacement hull cannot exceed its so-called *hull speed,* which is calculated by multiplying the square root of the hull length by 1.13. The Krogen 42 and Nordhavn 46 are examples of true displacement boats.

2. Beam is important to us since we want as much interior volume as possible in a boat. Some models we've been aboard recently seem cavernous inside, and their beams were unusually wide. The question is, how much is too much beam?

Although modern production boats clearly have more beam than their predecessors, many yacht designers still consider one foot of beam for each three feet of length close to the ideal length-to-beam ratio for yachts under fifty-five feet. Having said that, the fact is that most modern production boats have beams that exceed that 3–1 ratio. It's not unusual, for example, to see a 36-footer with a 13 foot beam or, say, a 45-footer with a 16 foot beam.

3. Should I be looking for a boat with diesel engines?

In many cases the answer is yes. All trawlers—regardless of size—come with diesels, and most motor yachts above 40 feet have them as well.

Range is often a factor in a cruising yacht, and diesels can deliver up to 50% more of it than the same boat with gasoline engines. It should be noted, however, that it is not at all uncommon to see gas-powered motor yachts over 40 feet in the Great Lakes and inland waterways, where range is of less importance.

In a smaller aft-cabin cruiser the choice between gas and diesel isn't always so clear. The added expense of diesels in a 35' boat could easily add up up to a quarter of the purchase price. Fortunately, diesels are becoming more affordable as new technology allows manufacturers to reduce both the size and weight of diesel engines while increasing their performance.

In resolving this question, keep in mind that the resale value of a diesel-powered boat will often go a long way toward justifying the added up-front expense.

4. I know that engine hours are important, but what constitutes a lot of hours on a particular set of motors?

This is always a hard question to answer, so we'll just offer some general guidelines. When it comes to gas engines (inboards and I/Os), most dealers and brokers figure those with over 1,000 hours are probably tired. With turbocharged diesels 3,500 hours is a lot of running time, and with naturally aspirated diesels it's not uncommon to pile up 5,000 hours before an overhaul is required.

The tragedy is that many of today's ultra-high-performance diesels never see 2,000 hours before an overhaul. Sometimes this is a manufacturer problem, but premature marine diesel death usually results from improper owner care and maintenance. Lack of use and poor exercise habits may be the number-one killer. Humidity (moisture) on cylinder components can be avoided with regular running and engine heaters. Diesel engines should be run under a load whenever possible. If your mechanical surveyor suggests new oil, fuel, or water hoses, do it. Trying to save money here can be expensive.

Having said that, it's important to note that there are far too many variables to make any buying decisions based on engine hours alone. It's imperative to have the diesels in a used boat surveyed just as you have the boat itself professionally examined before reaching a final decision. It's not quite so critical with gas engines since they cost far less to rebuild than diesels, however it's always worth the small expense of having a compression test done on gas engines just to see what you're getting into.

Determining the actual hours on an engine (or a set of engines) can be difficult. There are generally hour meters installed in boats over 25 feet (that often aren't working) but they're not always found in smaller gas-powered boats. Even if you have access to all the service records, we strongly suggest that you rely on an expert to evaluate the engines in any boat you have a serious interest in owning.

5. I'm in the market for a 40-foot trawler. My broker's telling me to buy a boat with twin engines rather than one for resale purposes, but according to some articles I've read, a single diesel is all I need. Who's right?

With a couple of exceptions, we think your broker is providing some excellent advice. Handling a single-screw 40-footer in a tight marina with the wind blowing hard may not intimidate a seasoned skipper, but it'll severely test the limits of your average weekend family cruiser. In general, trawler-style boats above 36 feet should have twin engines.

It should be noted that a *true* trawler—built on a full-displacement hull with a deep keel—will often have only a single engine due in part to the hull configuration. (Two production models that come to mind are the Krogen 42 and the Nordhavn 46.)

6. What kind of cruising speeds should I look for in a late-model motor yacht?

A modern, properly powered motor yacht (above 45 feet) is capable of cruising in the neighborhood of 16–17 knots. Those that cruise at 20 knots are considered fast and a 25-knot cruising speed is seen only in a very-high-performance model.

Many motor yachts under 45 feet—and that includes a lot of late-model Taiwanese imports—are not able to attain these numbers because they were built with too-small engines. While this tactic allowed importers to sell them at seemingly low prices, the performance of these boats runs a poor second to a properly powered model.

7. I'm looking at a boat with prop pockets. What are the advantages and disadvantages?

Prop pockets are used to reduce shaft angles, which often results in improved fuel economy and engine efficiency at cruising speeds. Manufacturers like Sea Ray and Phoenix have used them for years. A secondary benefit to owning a hull with prop pockets is evident—the ability to operate with less fear of clipping a prop in shallow waters.

One criticism we've often heard of boats with prop pockets is that they don't back down as well since they lack the "bite" that a more exposed propeller can get. Most experienced captains agree that backing a boat with prop pockets into a slip takes a little more finesse (or throttle) than with a conventional hull.

8. Should I reject a boat with bottom blisters?

Generally, no. Blisters can almost always be repaired, although the process can require a fair amount of time and expense. With that in mind, it is rare indeed to see a blistering problem so severe that it actually affects the integrity of the hull.

While some boats tend to re-blister again and again, most bottoms properly dried and protected should remain blister-free for five years or longer.

9. Can I rely on the boat tests that I read in the national magazines?

Yes, they're usually accurate as far as they go. For example, the performance figures—speeds at various rpm's, fuel burn data, etc.—are quite reliable, although it's always wise to keep in mind that these are new boats with light loads and plenty of factory preparation. Don't look for a lot of hard-hitting criticism in these tests, however, because boating magazines depend on boat manufacturers for a major part of their advertising revenues.

We've read a lot of boat tests over the years. In our opinion, the best and most comprehensive are conducted by *Boating* magazine. *Sea* and *Sport Fishing* also have some excellent reviews.

10. How important is a lower helm?

That depends on your location. A lower helm is a great convenience—a luxury, actually—when you're getting an early start on a chilly morning or trying to stay dry on a wet, windy day. For visibility, however (and to avoid seasickness), most skippers prefer the bridge station for heavy weather running in spite of the physical discomforts.

Aside from the added expense (which can be considerable), a lower helm takes up valuable room in the salon that would otherwise be devoted to living space. Not surprisingly, inside helms are most commonly seen in northern climates, especially in the Great Lakes and the Pacific Northwest. On the other hand, a lower helm in Florida (or along the Gulf Coast) is often a hindrance to a boat's resale value since it may be viewed as a useless and unnecessary feature.

Note that lower helms are a standard feature in trawlers and also in most motor yachts over 50 feet in length.

11. Should I hesitate before purchasing a Taiwanese boat? Several dealers and brokers I've met have been very critical of their quality.

Today's Taiwanese imports no longer enjoy the huge price advantage over their U.S. counterparts that they had in the 1970s and 1980s. While it's true that many of the boats imported from Taiwan during that period were cheaply built and suffered from poor quality control, others were surprisingly well built and represented long-term excellent values. The Taiwanese unquestionably had the skills to turn out a first-rate product, but most of their clients were U.S. importers whose only objective was a fat profit back home. Asked to turn out cheap boats in volume, the Taiwanese responded by flooding the market with waves of inexpensive trawlers and sailboats, usually with an abundance of interior and exterior teak woodwork.

The big price advantages Taiwanese products had in the American market began to dry up a few years ago and today their exports must compete on a level playing field with our own domestic yachts. There's been a big shake-out in the boatbuilding business in Taiwan in the past decade and many of the smaller, less productive yards have shut down. Those remaining are experienced boatbuilders with often state-of-the-art technology and highly skilled, well-paid workers. Many of these yards have successfully sold their products into the European market.

Recent history has shown that Asian boats often hold their resale values as well or better than their U.S. counterparts—a good indication of their continuing popularity on the used market. Many of the Taiwanese motor yachts currently being imported into this country are completely modern boats with plenty of high-tech construction and customer appeal, and practically all of today's trawler-style boats are being produced in Taiwan. We can think of no reason to avoid them.

12. Are freshwater boats really worth more?

Sure, no question about it. Salt water is hard on a boat, especially the gelcoat, electronics, paint, metalwork, and engine room components. And while nearly all diesel-powered boats have closed cooling systems, the same is not always true of gas engines. In a saltwater environment it's wise to look for a boat with a closed cooling system since it usually lengthens engine life.

Another reason freshwater boats often bring a premium price has to do with the fact that they generally have fewer engine hours. The boating season in most freshwater regions is shorter than many of the largest saltwater boating areas. Furthermore, the majority of freshwater vessels spend their winters out of the water—many in a protected environment with reduced exposure to the corrosive effects of sun, wind, and rain.

As might be imagined, a well-maintained saltwater vessel is probably a better investment than a poorly maintained freshwater boat. One final factor that equalizes the values between the two is equipment. An East or West Coast saltwater boat is often fitted out with better cruising equipment and more elaborate electronics than a similar Great Lakes or inland waters vessel.

13. Should I avoid a boat if the manufacturer has gone out of business or is currently undergoing hard times?

Emphatically, no. There are plenty of good used boats on the market from manufacturers who couldn't survive the poor economy of the past several years. The parts you will need from time to time are always available from catalog outlets or suppliers. Engine parts, of course, are easily secured from a number of sources. Generally speaking, there are no components used in a production model that cannot be replaced (or repaired) by a good yard.

Note that many of the most popular models on today's brokerage market were built by companies now out of business.

14. For resale, should I consider only a brand of boat with big-name market recognition?

There is no question that certain popular brands have consistently higher resale values. There are, however, many designs from small or regional builders that are highly sought after by knowledgeable boaters. Often, the market for these models is tighter and generally less saturated than the high-production designs—a factor that often works to a seller's advantage.

15. If I decide to buy a used boat, should I use a broker?

If you have plenty of time on your hands you could locate a good boat at a fair price without a broker. Unless you find a boat for sale by its owner, you end up working with a broker anyway—the listing agent.

When choosing a broker consider that you are about to spend a large amount of money. Do your homework and end up with an agent that you feel has your long-term interests at heart. You're not paying for his time and expertise until you purchase a boat through him. Keeping many brokers in competition against one another often results in no one giving you the time and attention that you'll require.

USEFUL TERMS

Abaft—behind
Athwartships—at a right angle to the boat's length
Bulkhead—an upright partition separating compartments in a boat
Bulwark—a raised portion of the deck designed to serve as a barrier
Chine—the point at which the hullsides and the bottom of the boat come together
CID—referring to the cubic inch displacement of an engine, i.e., 454-cid gas engine
Coaming—vertical surface surrounding the cockpit
Cuddy—generally refers to the cabin of a small boat
Deadrise—the angle from the bottom of the hull (not the keel) to the chine
Deep-V Hull—a planing hull form with at least 17.5° of constant deadrise
Displacement Hull—a hull designed to go through the water and not capable of planing speed
Forefoot—the underwater shape of the hull at the bow
Freeboard—the height of the sides of the boat above the waterline
GPH—gallons per hour (of fuel consumption)
Gunwale—(also gunnel) the upper edge of the sheerline
Hull Speed—the maximum practical speed of a displacement hull. To calculate, take the square root of the load waterline length (LWL) and multiply by 1.34.
Knot—one nautical mile per hour. To convert knots to statute mph, multiply by 1.14.
Modified-V Hull—a planing hull form with less than 17.5° of transom deadrise
Nautical Mile—measurement used in salt water. A nautical mile is 6,076 feet.
Planing Speed—the point at which an accelerating hull rises onto the top of the water. To calculate a hull's planing speed, multiply the square root of the waterline length by 2.
Semi-Displacement Hull—a hull designed to operate economically at low speeds while still able to attain efficient planing speed performance
Sheerline—the fore-and-aft line along the top edge of the hull
Sole—a nautical term for floor
Statute Mile—measurement used in fresh water. A statute mile equals 5,280 feet.
Tender—refers to (*a*) a dinghy, or (*b*) lack of stability
WOT—wide open throttle

Directory of Yacht Brokers & Dealers

ALABAMA

A&M Yacht Sales/Mobile Hatteras
5004 Dauphin Island Pkwy.
Mobile, AL 36605
205-471-6949, Fax: 205-479-4625
Hatteras, Viking

Bay Yacht Sales
4960 Dauphin Island Pkwy.
Mobile, AL 36605
205-476-8306, Fax: 205-473-3802

KV Yacht Brokerage
27844 Canal Rd., Sportsman's Marina
Orange Beach, AL 36561
205-981-9600, Fax: 205-981-4304

The Marine Group
Sportsman's Marina, Box 650
Orange Beach, AL 36561
205-981-9200, Fax: 205-981-9137

CALIFORNIA

Ballena Bay Yacht Brokers
1150 Ballena Blvd., #121
Alameda, CA 94501
415-865-8601, Fax: 415-865-5560
Krogen

Bill Gorman Yachts
1070 Marina Village Pky., #100
Alameda, CA 94501
510-865-6151, Fax: 510-865-1220

Cruising World Pacific
2099 Grand St.
Alameda, CA 94501
415-521-1929, Fax: 415-522-6198

Don Trask Yachts
1070 Marina Village Pkwy., #108
Alameda, CA 94501
510-523-8500, Fax: 510-522-0641
Sabreline

Kensington Yacht Brokers
1535 Buena Vista Ave.
Alameda, CA 94501
510-865-1777, Fax: 510-865-8789

Nor-Cal Yachts
2415 Mariner Square Dr.
Alameda, CA 94501
510-523-8773, Fax: 510-865-4383
Ocean Alexander, Cruisers, Riviera, Luhrs

Richard Boland Yacht Sales
1070 Marina Village Pkwy., #107
Alameda, CA 94501
510-521-6213, Fax: 510-521-0118
Viking, Ocean

Newmarks Yacht & Ship Brokerage
3141 Victoria
Channel Islands, CA 93030
805-985-9898, Fax: 805-985-9982

Yachtline International
3150 South Harbor Blvd.
Channel Islands Harbor, CA 93035
805-985-8643, Fax: 805-985-3889
Ocean Alexander

Cays Boat Sales
509 Grand Caribe Isle
Coronado, CA 92118
619-424-4024, Fax: 619-575-7716

Lemest Yacht Sales
24703 Dana Drive
Dana Point, CA 92629
714-496-4933, Fax: 714-240-2398
Mason, Nordhavn

Huntington Harbour Yacht Exchange
16400 Pacific Coast Hwy., #107
Huntington Beach, CA 92649
714-840-2373, Fax: 310-592-9315

Wescal Yachts
16400 Pacific Coast Hwy., #106
Huntington Beach, CA 92649
310-592-4547, Fax: 310-592-2960

Marina Boat Sales
14900 West Highway 12
Lodi, CA 95242
209-367-0111
Carver, Tollycraft, Formula, Larson

Far West Marine
718 W. Anaheim St.
Long Beach, CA 90813
310-437-6461, Fax: 714-673-0733

Flying Cloud Yachts
6400 Marina Dr.
Long Beach, CA 90803
310-594-9716, Fax: 310-594-0710

Long Beach Yacht Sales
6400 E. Pacific Coast Hwy.
Long Beach, CA 90815
213-431-3393, Fax: 213-598-9483

Naples Yacht Sales
5925 Naples Plaza
Long Beach, CA 90803
310-434-7278, Fax: 310-434-0738

Stan Miller Yachts
245 Marina Dr.
Long Beach, CA 90803
310-598-9433, Fax: 310-598-5349
Catalina, Grand Banks, Blackfin

Bob Seldon Yacht Sales
14120 Tahiti Way
Marina del Rey, CA 90292
310-821-5883, Fax: 310-301-1020

Cruising World Pacific
14025 Panay Way
Marina del Rey, CA 90292
310-823-3838, Fax: 310-305-1941

Executive Yacht Management
646-A Venice Blvd.
Marina del Rey, CA 90291
310-306-2555

Purcell Yachts
14000 Palawan Way, #C
Marina del Rey, CA 90292
310-823-2040, Fax: 310-827-1877

Rick Ermshar Yachts
4601 Admiralty Way
Marina del Rey, CA 90292
213-822-4727, Fax: 310-822-6730

Tom Murdock Yachts
13915 Panay Dr.
Marina del Rey, CA 90292
310-822-8333, Fax: 310-822-9404

Yacht Broker & Dealer Directory

Yahama Marina Del Rey
13555 Fiji Way
Marina del Rey, CA 90292
310-823-8964, Fax: 310-821-0569
Tiara, Wellcraft

Thorsen Marine
6038 Shelter Bay Ave.
Mill Valley, CA 94941
415-461-5957, Fax: 415-461-6958

Ardell Yacht & Ship Brokers
2101 W. Coast Hwy.
Newport Beach, CA 92663
714-642-5735, Fax: 714-642-9884

Avion Yacht Sales, Ltd.
177 Riverside Ave., Suite F
Newport Beach, CA 92660
714-642-2827, Fax: 714-642-4127
Princess, Sunseeker

Bayliner Yacht Center
101 Shipyard Way, Cabin G
Newport Beach, CA 92663
714-723-0473, Fax: 714-723-0475
Bayliner

Chuck Hovey Yachts
717 Lido Park Dr. , Ste. A
Newport Beach, CA 92663
714-675-8092, Fax: 714-673-1037
Riva, Fleming, Island Gypsy, Azimuth

Craig Beckwith Yacht Sales
101 Shipyard Way, Suite J
Newport Beach, CA 92663
714-675-9352, Fax: 714-675-2519

Crow's Nest
2801 W. Coast Hwy., #260
Newport Beach, CA 92663
714-574-7600, Fax: 714-574-7610
Bertram, Davis, Hatteras, Tiara, Trojan

Emerald Yacht & Ship Brokers
3300 Irvine Avenue, Ste. 308
Newport Beach, CA 92660
714-553-0695, Fax: 714-752-0462

Falmouth Yachts
510 31st St., Suite D
Newport Beach, CA 92663
714-723-4225, Fax: 714-723-4093

Fraser Yachts
3471 Via Lido, #200
Newport Beach, CA 92263
714-673-5252, Fax: 714-673-8795
Chicago, IL 60614
312-993-7711, Fax: 312772-0891
Tiara

Fredericks Power & Sail
201 Shipyard Way, #A/3
Newport Beach, CA 92663
714-854-2696, Fax: 714-854-4598

H&S Yacht Sales
2001 W. Coast Highway
Newport Beach, CA 92663
714-642-4786, Fax: 714-642-1568
Mainship, Silverton, President

Lido Yacht Brokeage
3412 Via Oporto, #301
Newport Beach, CA 92663
714-675-0915, Fax: 714-675-0805
Rampage

Marine Center,
2200 W. Coast Hwy.
Newport Beach, CA 92663
714-645-3880
Bayliner

Newport Yacht Brokers
Box 5741, 400 S. Bayfront
Newport Beach, CA 92662
714-723-1200, Fax: 714-723-1201

Orange Coast Yachts
201 E. Coast Hwy.
Newport Beach, CA 92660
714-675-3844, Fax: 714-675-3980
Ocean Alexander

Seaward Yacht Sales
101 Shipyard Way, Suite K
Newport Beach, CA 92663
714-673-5950, Fax: 714-673-1058
Vitech, Nordic, Tayana

Venwest Yachts
2505 W. Coast Highway, #201
Newport Beach, CA 92663
714-642-1557, Fax: 714-548-0257
Viking

World Wide Custom Yachts
3412 Via Oporto, #301
Newport Beach, CA 92663
714-675-2179, Fax: 714-675-8210
Symbol

Yacht & Ship Brokers International
2507 W. Coast Hwy. #202
Newport Beach, CA 92663
714-722-7740, Fax: 714-722-8733

D'Anna Yacht Center
11 Embarcadero West, #100
Oakland, CA 94607
510-451-7000, Fax: 510-451-7026
Silverton, Wellcraft, Mainship

Integre Marine Ltd.
1155 Embarcadero
Oakland, CA 94606
415-465-6060, Fax: 415-465-6078

Admiralty Yacht Sales
3600 South Harbor Blvd.
Oxnard, CA 93035
805-985-1686
Wellcraft

Executive Yacht & Ship Brokers
3205 S. Victoria Ave.
Oxnard, CA 93035
805-984-1004, Fax: 805-985-4365

Wright Marine Sales,
3600 S. Harbor Blvd.
Oxnard, CA 93035
800-237-7174, Fax: 805-985-4586
Sea Ray

ACA Marine Yacht Sales
4262 Dauntless
Rancho Palos Verdes, CA 90274
213-541-6186, Fax: 213-541-6053

Pacific Coast Boats
2413 Cormorant Way
Sacramento, CA 95815
916-372-1500
Cruisers

Bower & Kling Yachts
955 Harbor Island Dr., #180
San Diego, CA 92101
619-299-7797, Fax: 619-299-3811

Cabrillo Yacht Sales
2638 Shelter Island Dr.
San Diego, CA 92106
619-523-1745, Fax: 619-523-1746

California Yacht Sales
2040 Harbor Island Dr., #111
San Diego, CA 92101
619-295-9669, Fax: 619-295-9909

Continental Yachts
333 West Harbor Dr.
San Diego, CA 92101
619-696-7400, Fax: 619-696-8029

CR Marine Yacht Sales
PO Box 82838
San Diego, CA 92138
619-295-0305, Fax: 619-298-5738

Crow's Nest
2515 Shelter Island Dr.
San Diego, CA 92106
619-222-1122, Fax: 619-222-3851
Bertram, Davis, Hatteras, Tiara, Trojan

Cruising World Pacific
2323 Shelter Island Dr.
San Diego, CA 92106
619-224-3277, Fax: 619-224-9225

Driscoll Yacht & Ship Brokerage
1050 Anchorage Lane
San Diego, CA 92106
619-222-0325, Fax: 619-222-0326

Fraser Yachts
2353 Shelter Island Dr.
San Diego, CA 92106
619-225-0588, Fax: 619-225-1325

H&S Yacht Sales
955 Harbor Island Dr., #110
San Diego, CA 92101
619-291-2600, Fax: 619-291-2613
Mainship, Silverton, President

Knight & Carver Yacht Sales
1500 Quivira Way
San Diego, CA 92109
619-224-4102, Fax: 619-222-6014

MacDonald Yacht Management
1450 Harbor Island Dr.
San Diego, CA 92101
619-294-4545, Fax: 619-294-8694

Mikelson Yachts
2330 Shelter Island Dr., #202
San Diego, CA 92106
619-222-5007, Fax: 619-223-1194
Mikelson

R.D. Snyder Yacht Sales
1231 Shafter St.
San Diego, CA 92106
619-224-2464, Fax: 619-224-7396

San Diego Yacht Sales
2525 Shelter Island Dr.
San Diego, CA 92106
800-221-8116, Fax: 619-221-0308
Hylas

Shelter Island Yacht Sales
2330 Shelter Island Dr., #200
San Diego, CA 92106
619-222-0515, Fax: 619-222-5283

Suncoast Yachts & Charters
955 Harbor Island Dr., #140
San Diego, CA 92101
619-297-1900, Fax: 619-297-1994
Grand Banks

Sunset Marine
2590 Ingraham St.
San Diego, CA 91209
619-224-3221
Sea Ray

Yachts West
333 West Harbor Dr.
San Diego, CA 92101
619-230-8989

Newmarks Yacht & Ship Brokerage
210 Whalers Walk
San Pedro, CA 90731
310-833-0887, Fax: 310-833-0979

ABC Yachts
One Gate 5 Road
Sausalito, CA 94964
415-332-7245, Fax: 415-332-4580

Fraser Yachts
320 Harbor Dr.
Sausalito, CA 94965
415-332-5311, Fax: 415-332-7036

Lager Yacht Brokerage Corp.
400 Harbor Dr. #C
Sausalito, CA 94965
415-332-9500, Fax: 415-332-9503

Nor-Cal Yachts
400 Harbor Dr., Suite C
Sausalito, CA 94965
415-332-0393
Ocean Alexander, Cruisers, Riviera, Luhrs

Oceanic Yacht Sales
308 Harbor Dr.
Sausalito, CA 94965
415-331-0533, Fax: 415-331-1642
Grand Banks, Midnight Lace

Sausalito Yacht Brokerage
100 Bay St.
Sausalito, CA 94965
415-331-6200, Fax: 415-331-6213

Western California Yacht Sales
6649 Embarcadero
Stockton, CA 95209
209-952-7672, Fax: 209-952-6443

Premier Yacht Sales
1801 Sonoma Blvd., #607
Vallejo, CA 94590
510-652-2109, Fax: 510-658-1635

Larry Dudley Yacht Sales
1559 Spinnaker Dr. #202
Ventura, CA 93001
805-644-9665, Fax: 805-644-9695

Ventura Yacht Sales
1101 Spinnaker Dr.
Ventura, CA 93003
805-644-1888
Grand Banks, Riviera

Newmarks Yacht & Ship Brokers
Berth 204
Wilmington, CA 90744
310-834-2830, Fax: 310-835-7206

CONNECTICUT

Randall Yacht Sales
145 S. Montowese St.
Branford, CT 06405
203-481-3866, Fax: 203-481-8699

Cedar Island Marina
PO Box 181, Riverside Dr.
Clinton, CT 06413
203-669-8681, Fax: 203-669-4157

Coastal Marine
143 River Rd., Box 228
Cos Cob, CT 06807
203-661-5765, Fax: 203-661-6040
Albin

Norwalk Cove Marina
Beach Road
East Norwalk, CT 06855
203-838-2326, Fax: 203-838-9258
Hatteras, Grand Banks, Tollycraft, Cheoy Lee

Boatworks Yacht Sales
PO Box 668
Essex, CT 06426
203-767-3013, Fax: 203-767-7178
Sabreline, Grand Banks

Eastland Yachts
33 Pratt St.
Essex, CT 06426
203-767-8224, Fax: 203-767-9094

Essex Island Yachts
Foot of Ferry St.,
Essex Island Marina
Essex, CT 06426
203-767-8645, Fax: 203-767-0075

Hank Aldrich Yacht Sales,
37 Pratt St., Box 72
Essex, CT 06426
203-767-4988, Fax: 203-767-4998
Ocean

Photo-Boat
PO Box 504
Mystic, CT 06355
203-536-9333, Fax: 203-535-4801

Storm Haven Yachts
PO Box 85
Newtown, CT 06470
203-426-0806

Noank Shipyard
PO Box 9248
Noank, CT 06340
203-536-9651, Fax: 203-572-8140

Don Zak's Shoreline Yacht Sales
54 Ferry Rd.
Old Saybrook, CT 06475
203-395-0866, Fax: 203-395-0877

Northeast Blackfin,
PO Box 429
Portland, CT 06480
203-342-1988, Fax: 203-342-4132

Petzold's Yacht Sales
37 Indian Hill Ave.
Portland, CT 06480
203-342-1196, Fax: 203-342-0462
Silverton, Mainship, Egg Harbor, Stamas, Cruisers

Portland Boat Works,
1 Grove St.
Portland, CT 06480
203-342-1085, Fax: 203-342-0544
Post, Tiara

Yacht Broker & Dealer Directory

Boatworks Yacht Sales
PO Box 265
Rowayton, CT 06853
203-866-0882, Fax: 203-853-4910
Sabreline, Grand Banks

Rex Marine Center
144 Water St.
South Norwalk, CT 06854
203-866-5555, Fax: 203-866-2518
Stamas, Formula, Island Packet

Chan Moser Yachts
6 Woodridge Drive
Stamford, CT 06905
203-322-6668, Fax: 203-322-8288

Brewer Yacht Sales
63 Pilots Point Dr.
Westbrook, CT 06498
203-399-6213, Fax: 203-399-4379

Louis Marine, Ltd.
438 Boston Post Rd.
Westbrook, CT 06498-1722
203-664-4230
Wellcraft, Chris Craft

Sail Westbrook
PO Box 1179
Westbrook, CT 06498
203-399-5515

FLORIDA

Aventura Yacht Sales
20801 Biscayne Blvd.
Adventura, FL 33180
305-933-8285, Fax: 305-933-8287

South Florida Marine Liquidators
4800 N. Federal Hwy., Suite 113B
Boca Raton, FL 33431
407-750-5155, Fax: 407-750-8533

O'Brien Yacht Sales
3010 SW 14th Place
Boynton Beach, FL 33426
407-738-6676, Fax: 407-738-1658

The Boatworks
6921 14th St. W. (U.S. 41)
Bradenton, FL 34207
813-756-1896, Fax: 813-753-9426
Bayliner, Wellcraft

Harbour Yacht Sales
25 Causeway Blvd.
Clearwater Beach, FL 34630
813-446-5617, Fax: 813-441-9173

Hatteras in Miami
2550 S. Bayshore Dr.
Coconut Grove, FL 33133
305-854-1100, Fax: 305-854-1186
Hatteras, Tiara

Reel Deal Yachts
2550 S. Bayshore Dr.
Coconut Grove, FL 33133
305-859-8200, Fax: 305-854-8044
Blackfin, Luhrs, Phoenix, Mainship, Mako

Cozy Cove Marina
300 N Federal Hwy.
Dania, FL 33004
305-921-8800, Fax: 305-922-0173
Blackfin

HMY Yacht Sales
850 NE 3rd. St.
Dania, FL 33004
305-926-0400, Fax: 305-921-2543
Post, Hines-Farley, Viking, Cabo

Intrepid Southeast
850 NE 3rd. St.
Dania, FL 33004
305-922-7544, Fax: 305-922-3858
Intrepid

Oviatt Marine
850 NE 3rd St., Suite 201
Dania, FL 33004
305-925-0065, Fax: 305-925-8822

Daytona Marina & Boatworks
645 S. Beach St.
Daytona Beach, FL 32114
904-253-6266, Fax: 904-253-8174

Eagle Yachts
721 Ballough Rd
Daytona Beach, FL 32114
904-258-7578, Fax: 904-257-5179

Yacht Brokerage USA
3948 S. Peninsula Dr.
Daytona Beach, FL 32127
904-760-9353

Universal Yachts,
1645 SE 3rd Ct., #214
Deerfield Beach, FL 33441
305-786-2911, Fax: 305-786-1937

Yacht Registry
343 Causeway Blvd.
Dunedin, FL 34698
813-733-0334, Fax: 813-733-6754

Alexander Yachts
2150 SE 17th St., Suite 201,
Ft. Lauderdale, FL 33316
305-763-7676, Fax: 305-763-7758

Allied Marine
401 SW First Ave., 2nd Fl.,
Ft. Lauderdale, FL 33301
305-462-7424, Fax: 305-462-0756

American Trading Industries
500 SE 17th St., #220,
Ft. Lauderdale, FL 33316
305-522-4254, Fax: 305-522-4435

Ameriship Corporation
3285 SW 11th Ave.,
Ft. Lauderdale, FL 33315
305-463-7957, Fax: 305-463-3342
Exporter

Ardell Yacht & Ship Brokers
1550 SE 17th Street,
Ft. Lauderdale, FL 33316
305-525-7637, Fax: 305-527-1292

Atlantic Pacific Sailing Yachts
2244 SE 17th St.,
Ft. Lauderdale, FL 33316
305-463-7651, Fax: 305-779-3316

Bollman Yachts
2046 SE 17th St.,
Ft. Lauderdale, FL 33316
305-761-1122, Fax: 305-463-9878

Bradford International
3151 State Road 84,
Ft. Lauderdale, FL 33312
305-791-2600, Fax: 305-791-2655

Broward Yacht Sales
1535 SE 17th St., Suite 202,
Ft. Lauderdale, FL 33316
305-763-8201, Fax: 305-763-9079

Bruce A. Bales Yacht Sales
1635 S. Miami Rd., #2,
Ft. Lauderdale, FL 33316
305-522-3760, Fax: 305-522-4364

Castlemain,
300 SW 2nd St., Suite 4,
Ft. Lauderdale, FL 33312
305-760-4730, Fax: 305-760-4737
Swiftships

Chas. P. Irwin Yacht Brokerage
801 Seabreeze Blvd.
(Bahia Mar Yachting Ctr.),
Ft. Lauderdale, FL 33316
305-463-6302, Fax: 305-523-0056

Colonial Yacht Sales
901 SE 17th St., #203,
Ft. Lauderdale, FL 33316
305-463-0555, Fax: 305-463-8621

Dave D'Onofrio Yacht Sales
1875 SE 17th St. (Marriott Marina),
Ft. Lauderdale, FL 33316
305-527-4848, Fax: 305-462-6817

Dave Pyles Yacht Sales
2596 SW 23rd Terrace,
Ft. Lauderdale, FL 33312
305-583-8104, Fax: 305-797-7669

Emerald Yacht & Ship Brokers
801 Seabreeze Blvd.,
Ft. Lauderdale, FL 33316
305-522-0556, Fax: 305-522-3194

Yacht Broker & Dealer Directory

Everglades Marina
2409 NE 26th Ave.,
Ft. Lauderdale, FL 33305
305-763-3030, Fax: 305-763-3167
Baja, Fountain

Florida Yacht & Ship Brokers
1700 E. Las Olas Blvd.,
Ft. Lauderdale, FL 33316
305-467-1122, Fax: 305-467-0011

Frank Gordon Yacht Sales
801 Seabreeze Blvd.
(Bahia Mar Yachting Center),
Ft. Lauderdale, FL 33316
305-525-8476, Fax: 305-525-6024

Fraser Yachts
2160 SE 17th St.,
Ft. Lauderdale, FL 33316
305-463-0600, Fax: 305-763-1053

Garcia Yacht Sales
1323 SE 17 St, #220,
Ft Lauderdale, FL 33316
305-763-6152, Fax: 305-763-6152

Hal Jones & Co.
1900 SE 15th St.,
Ft. Lauderdale, FL 33316
305-527-1778, Fax: 305-523-5153
Grand Banks

Hatteras of Lauderdale
401 SW 1st Ave.,
Ft. Lauderdale, FL 33301
305-462-5557, Fax: 305-462-0029
Hatteras, Tiara

Helms • Kelly • MacMahon International Yachting
1650 SE 17th St., Suite101,
Ft. Lauderdale, FL 33316
305-525-1441, Fax: 305-525-1110

High-Tech Marine
1535 SE 17th St. Quay,
Ft. Lauderdale, FL 33316
305-524-6911, Fax: 305-524-7107

J. Woods Marine Group
808 NE 20th Ave.,
Ft. Lauderdale, FL 33304
305-764-8770, Fax: 305-764-8771

Jackson Marine Sales
1915 SW 21st Ave.,
Ft. Lauderdale, FL 33312
305-792-4900, Fax: 305-587-8164

Jet Sea Yacht Brokerage
1650 SE 17th St., #204,
Ft. Lauderdale, FL 33316
305-766-2600, Fax: 305-766-2611
Tecnomarine

Luke Brown & Assoc.
1500 Cordova Rd., #200,
Ft. Lauderdale, FL 33316
305-525-6617, Fax: 305-525-6626

Mares Yacht Sales
1535 SE 17th St., #107,
Ft. Lauderdale, FL 33316
305-523-2287, Fax: 305-523-2236
Mares

Marina 84
2698 SW 23rd Ave.,
Ft. Lauderdale, FL 33312
305-581-3313, Fax: 305-797-8986

Merle Wood & Associates
1535 SE 17th St., #201B,
Ft. Lauderdale, FL 33316
305-525-5111, Fax: 305-525-5165

Merritt Yacht Brokers
2040 SE 17th St,
Ft. Lauderdale, FL 33316
305-761-1300, Fax: 305-463-8617

Northrop & Johnson
1901 SW 4th Avenue,
Ft. Lauderdale, FL 33316
305-522-3344, Fax: 305-522-9500

Peter Kehoe & Associates
2150 SE 17th St., #107,
Ft. Lauderdale, FL 33316
305-767-9880, Fax: 305-767-9884

Rex Yacht Sales
2152 SE 17th Street,
Ft. Lauderdale, FL 33316
305-463-8810, Fax: 305-462-3640
Ocean Alexander, Cheoy Lee

Richard Bertram & Co.
651 Seabreeze Blvd.,
Ft. Lauderdale, FL 33316
305-467-8405, Fax: 305-763-2675

Royce Yacht & Ship Brokers
1600 SE 17th St., #418,
Ft. Lauderdale, FL 33316
305-764-0100, Fax: 305-764-0192

Sea Yachts
837 NE 20th Ave.,
Ft. Lauderdale, FL 33304
305-522-0993, Fax: 305-768-9027

Tom Klein Yacht
5200 N. Federal Hwy., Suite 2,
Ft. Lauderdale, FL 33308
305-772-7070, Fax: 305-772-7086

Trans America Yacht Brokers
1535 SE 17th St., Ste. 109,
Ft. Lauderdale, FL 33316
305-462-1177, Fax: 305-462-7858

Trans-Coastal Yacht Brokerage
515 Seabreeze Blvd.,
Ft. Lauderdale, FL 33316
305-767-8830, Fax: 305-767-8942

Walsh Yachts
1900 S.E. 15th Street,
Ft. Lauderdale, FL 33316
305-525-7447, Fax: 305-525-7451

Woods & Oviatt
Pier 66 Marina, 2301 SE 17th St.,
Ft. Lauderdale, FL 33316
305-463-5606, Fax: 305-522-5156

Yacht Brokerage USA
1700 E. Los Olas Blvd., Suite 101,
Ft. Lauderdale, FL 33301
305-463-1255, Fax: 305-463-7733

Yacht Search—The Professional Brokerage
2150 SE 17th St.,
Ft. Lauderdale, FL 33315
305-524-1823, Fax: 305-525-3074

Yacht & Ship Brokers,
2501 S. Federal Hwy.,
Ft. Lauderdale, FL 33316
305-779-7447, Fax: 305-779-3735

Great American Marine
1310 Lee Street,
Ft. Myers, FL 33901
813-334-8622, Fax: 813-334-0207
Grand Banks, Hatteras

Yacht Brokerage USA
1700 Medical Lane,
Ft. Myers, FL 33907
813-936-5595, Fax: 813-936-0544

Yacht-Eng,
13601 McGregor Blvd., #16,
Ft. Myers, FL 33919
813-481-3511, Fax: 813-481-3064

East-West Yachts
10 Avenue A, Ft. Pierce Yacht Center,
Ft. Pierce, FL 34950
407-466-1240, Fax: 407-466-1242

Waterline Yacht Brokerage
2010 Harbortown Dr.,
Ft. Pierce, FL 34946
407-466-5747, Fax: 407-466-5966

Yacht Brokerage USA
PO Box 1552,
Ft. Walton Beach, FL 32549
904-664-1212, Fax: 904-244-1751

Grantour Yachts,
2422 NE 9th St.
Hallandale, FL 33009
305-936-0337, Fax: 305-936-0338
Riviera

Yacht Broker & Dealer Directory

Palm Beach Yacht Center
7848 S. Federal Hwy.
Hypoluxo, FL 33462
407-585-2003, Fax: 407-585-9933
Bayliner

Ortega River Boat Yard
4451 Herschel St.
Jacksonville, FL 32210
904-387-5538, Fax: 904-388-7476
Luhrs

Roger Hansen Yacht Sales
3344 Lake Shore Blvd.
Jacksonville, FL 32210
904-384-3113, Fax: 904-384-6550
Californian, Bertram

Jax Beach Yacht Brokerage
13846 Atlantic Blvd.
Jacksonville Beach, FL 32225
904-246-4975, Fax: 904-246-7573
Halvorsen

North Florida Yacht Sales
2305 Beach Blvd., #105
Jacksonville Beach, FL 32250
904-249-8444, Fax: 904-247-0050

Card Sound Yachts
9 Barracuda Lane
Key Largo, FL 33037
305-367-2727, Fax: 305-367-3962

Perdue Dean,
#2 Fishing Village Dr. ORC
Key Largo, FL 33037
305-367-2661, Fax: 305-367-2128

Marine Unlimited
232 Basin Dr.
Lauderdale-by-the-Sea, FL 33308
305-491-0430, Fax: 305-771-6122

Oceanus Institute,
4332 E. Tradewinds Ave.
Lauderdale-by-the-Sea, FL 33308
305-772-5773

SGK Yacht Sales
218 Commercial
Lauderdale-by-the-Sea, FL 33308
305-776-5525

Donhuser's Yacht Brokerage
3142 N. Federal Hwy.
Lighthouse Point, FL 33064
305-946-9484, Fax: 305-946-9487

Rhodes Yacht Brokers
2901 NE 28th Court
Lighthouse Point, FL 33064
305-941-2404, Fax: 305-941-2507

Waterline Yacht Brokerage
905 N. Harbor City Blvd.
Melbourne, FL 32935
407-254-0452, Fax: 407-254-0516

Cruising Yachts & Ships,
3051 Orange St.
Miami, FL 33133
305-448-3481, Fax: 305-567-9750

Custom Brokerage Yacht Sales
11422 SW 87th Terrace
Miami, FL 33173
305-598-9875, Fax: 305-598-2239

Frank Stanzel Yachts
7350 SW 96th St.
Miami, FL 33156
305-669-0962, Fax: 305-669-0961

Merrill-Stevens Yacht Sales
1270 NW 11th St.
Miami, FL 33125
305-858-5911, Fax: 305-858-5919

Richard Bertram & Co.
3660 NW 21st St.
Miami, FL 33152
305-633-9761, Fax: 305-634-9071

Florida Yacht Charters & Sales
1290 Fifth Street
Miami Beach, FL 33139
305-532-8600, Fax: 305-672-2039

Naples Yacht Brokerage
P.O. Box 882
Naples, FL 33939
813-434-8338, Fax: 813-434-6848

Walker's Yacht Sales
895 10th St. South
Naples, FL 33940
813-262-6500, Fax: 813-262-6693
Parker, Tiara, Formula, Albemarle

Blake Davis Yacht Brokerage
7601 E. Treasure Dr.
North Bay Village, FL 33141
305-866-8329

Gilman Yacht Sales
1212-A U.S. Hwy 1
North Palm Beach, FL 33408
407-626-1790, Fax: 407-626-5870

Camper & Nicholson
450 Royal Palm Way
Palm Beach, FL 33480
407-655-2121, Fax: 407-655-2202

Hatteras in Palm Beach
2410 PGA Blvd., #155
Palm Beach Gardens, FL 33410
407-775-3531, Fax: 407-775-8790
Hatteras, Tiara

Shear Yacht Sales
2385 PGA Blvd., Box 30308
Palm Beach Gardens, FL 33420
407-624-2112, Fax: 407-624-1877
Albin, Island Gypsy, Novatec, DeFever

Singer Island Yacht Sales
11440 U.S. Highway 1
Palm Beach Gardens, FL 33408
407-622-0355, Fax: 407-622-0339

Stella Marine
2385 PGA Blvd.
Palm Beach Gardens, FL 33410
407-624-9950, Fax: 407-624-9949

The Marine Group
2401 PGA Blvd., Suite 104
Palm Beach Gardens, FL 33410
407-627-9500, Fax: 407-627-9503

Coastal Yacht Sales
1496 Treetop Dr.
Palm Harbor, FL 34683
813-787-9300
Ocean

Carson Yacht Brokerage
1035 Riverside Dr.
Palmetto, FL 34221
813-723-1825, Fax: 813-729-8254

K&H Yachts
1055 N. Riverside Dr.
Palmetto, FL 34221
813-729-4449
Albin

Regatta Point Yacht Sales
985 Riverside Dr.
Palmetto, FL 34221
813-722-7755, Fax: 813-722-7757

Grand Lagoon Yacht Brokers
3706 Thomas Dr.
Panama City Beach, FL 32408
904-233-4747, Fax: 904-233-4741

Treasure Island Marina
3605 Thomas Dr.
Panama City Beach, FL 32408
904-234-6533, Fax: 904-235-1299
Sea Ray

Prestige Yachts
600 Barracks St., Suite 102
Pensacola, FL 32501
904-432-6838, Fax: 904-432-8999
Tiara, Silverton

Four Points Yacht & Ship Brokers
101 N. Riverside Dr., Suite 214
Pompano Beach, FL 33062
305-941-5500, Fax: 305-941-5521
Ocean, Jefferson

Ocean Harbor Marine
1500 N. Federal Hwy.
Pompano Beach, FL 33062
305-946-9900, Fax: 305-946-4040
Wellcraft

Yacht Broker & Dealer Directory

Cape Yacht Brokerage
800 Scallop Dr.
Port Canaveral, FL 32920
407-799-4724, Fax: 407-799-0096

Taber Yacht Sales
Pirates Cove Marine, Box 1687
Port Salerno, FL 34992
407-288-7466, Fax: 407-288-7476

Bain Yacht Sales
1200 W. Retta Esplanade
Punta Gorda, FL 33950
813-637-1335, Fax: 813-637-8057
Onset

Yacht Perfection
1601 W. Marion Ave #203 D
Punta Gorda, FL 33950
813-637-8111, Fax: 813-637-9918

Wayne Roman Yachts
207 E. Blue Heron Blvd.
Riviera Beach, FL 33404
407-844-5000, Fax: 407-848-5422

Yacht Brokerage USA
613 Rockledge Dr.
Rockledge, FL 32955
407-636-3600, Fax: 407-636-3606

Great American Marine
1889 N. Tamiami Trail
Sarasota, FL 33580
813-365-1770, Fax: 813-365-1787
Grand Banks, Hatteras

Modern Classic Yachtworks
1666 Main St.
Sarasota, FL 34236
813-955-7733, Fax: 813-957-3132

Sarasota Yacht & Ship Services
1306 Main St.
Sarasota, FL 34236
813-365-9095, Fax: 813-955-1727

First Coast Yacht Sales
103 Yacht Club Dr.
St. Augustine, FL 32095
904-824-7293, Fax: 904-829-6779

Offshore Yacht & Ship Brokers
256-B Riberia St.
St. Augustine, FL 32084
904-829-9224, Fax: 904-825-4292

St. Augustine Yacht Center
3040 Harbor Dr.
St. Augustine, FL 32095
904-829-2294, Fax: 904-829-2298

D.M. Savage Yacht
4326 Central Ave.
St. Petersburg, FL 33711
813-327-1288, Fax: 813-321-0491

Anchor Yachts International
1110 Pinellas Bayway Dr.
St. Petersburg, FL 33715
813-867-8027, Fax: 813-864-1359

Capt. Jack's Yacht Brokerage
101 16th Ave So.
St Petersburg, FL 33701
813-825-0757, Fax: 813-822-6415

Charles Morgan Associates
200 Second Ave. S.
St. Petersburg, FL 33701
813-894-7027, Fax: 813-894-8983

Great American Marine
6810 Gulfport Blvd.
St. Petersburg, FL 33707
813-384-3428, Fax: 813-381-1401
Grand Banks, Hatteras

Mariner Yacht Sales
12022 Gandy Blvd.
St. Petersburg, FL 33702
813-576-3307, Fax: 813-576-4767

Royal Yacht & Ship Brokers
3859 Central Ave.
St. Petersburg, FL 33713
813-327-0900, Fax: 813-327-7797

The Harborage Marina
1110 3rd St. South
St. Petersburg, FL 33701
813-894-7497, Fax: 813-898-2028
Carver, Sport Craft

West Florida Yachts
4880 37th St. South
St. Petersburg, FL 33711
813-864-0310, Fax: 813-867-6860

Yacht Brokerage USA
4401 Central Ave.
St. Petersburg, FL 33713
813-328-1255, Fax: 813-328-1796

Midcoast Yacht Sales
3957 Barcelona St.
Stuart, FL 34997
407-288-4886

Northside Marine Sales
400 NW Alice Ave.
Stuart, FL 34994
407-692-3052, Fax: 407-692-4006
Post, Mainship, Blackfin

Stuart Hatteras
110 N. Federal Hwy.
Stuart, FL 34994
407-692-1122, Fax: 407-692-1341
Hatteras, Tiara

Stuart Yacht
450 SW Salerno Rd.
Stuart, FL 34997
407-283-1947, Fax: 407-286-9800

Flammer Viking Yachts,
650 U.S. 19 North
Tarpon Springs, FL 34689
813-733-9289, Fax: 813-733-8876
Viking, Cabo

Complete Yacht Services
3599 E. Indian River Dr.
Vero Beach, FL 32963
407-231-2111, Fax: 407-231-4465
Grand Banks, Sabreline

Palm Beach Yacht Club Brokerage
800 N. Flagler Dr.
West Palm Beach, FL 33401
407-833-8633, Fax: 407-833-8639

Rybovich-Spencer Group
4200 N. Dixie
West Palm Beach, FL 33407
407-844-4331, Fax: 407-844-8393

GEORGIA

Robert P. Minis
102 McIntosh Dr.
Savannah, GA 31406
912-354-6589

Golden IslesYacht Sales
PO Box 21715
St. Simons Island, GA 31522
912-638-5678, Fax: 912-638-8532

ILLINOIS

Class Sea Yachts
207 N. Hager
Barrington, IL 60010
708-382-2100, Fax: 708-381-1265

Larsen Marine Service
1663 N. Elston Ave.
Chicago, IL 60614
312-993-7711, Fax: 312772-0891
Tiara

Sailboat Sales Co.
2500 S. Corbett St.
Chicago, IL 60608
312-225-2046, Fax: 312-225-6354

Spring Brook Marina
623 W. River Dr., Box 379
Seneca, IL 61360
815-357-8666, Fax: 815-357-8678
Carver, Viking, Harbor Master, Californian

Riverview Marine
515 S. Spaulding St.
Spring Valley, IL 61362
815-663-1000, Fax: 815-663-2628
Fountain. Powerquest

Larsen Marine Service
625 Sea Horse Dr.
Waukegan, IL 60085
708-336-5456, Fax: 708-336-5530
Tiara

Yacht Broker & Dealer Directory

Skipper Bud's at North Point Marina
215 N. Point Dr.
Winthrop Harbor, IL 60096
708-872-3200, Fax: 708-872-3230
Hatteras, Chris Craft, Sea Ray

INDIANA

B&E Marine
Washington Park
Michigan City, IN 46360
219-879-8301, Fax: 219-879-8388
Sea Ray, Rinker

H&M Yacht Brokerage
1 Newport Dr.
Michigan City, IN 46360
219-879-7152

KENTUCKY

Kentuckiana Yacht Sales
Hwy. 641 South
Gilbertville, KY 42044
502-362-8343

LOUISIANA

A&M Yacht Sales/New Orleans Hatteras
126 South Roadway
New Orleans, LA 70124
504-282-6800
Hatteras, Viking

Prestige Yachts
6701 South Shore Harbor Blvd.
New Orleans, LA 70126
504-242-9000, Fax: 504-246-3908
Tiara, Silverton

MAINE

Duffy & Duffy Yacht Sales
HC 63, Box 333
Brooklin, ME 04616
207-359-4658, Fax: 207-359-8948

Camden Harbor Yachts
PO Box 880
Camden, ME 04843
207-236-7112, Fax: 207-236-7113

Casco Bay Yacht Exchange
P.O. Box 413, Freeport, ME 04032
207-865-4016, Fax: 207-865-0759

Steedman & Gray,
PO Box 1094
Kennebunkport, ME 04046
207-967-4211, Fax: 207-967-8428

Indian Point Yachts
HCR 62, Box 63
Mt. Desert, ME 04660
207-288-5258, Fax: 207-288-3093

North Star Yacht Sales
DiMillo's Marina, Long Wharf
Portland, ME 04101
207-879-7678, Fax: 207-879-1471

Robinhood Marine Center
PO Box 460
Robinhood, ME 04530
207-371-2343, Fax: 207-371-2899

Michael Waters Yacht Brokers
112 Beech St.
Rockland, ME 04841
207-594-4234, Fax: 207-596-0726
Lien Hwa Motoryachts

Atlantic Yacht Brokerage
PO Box 2277
South Portland, ME 04016
207-767-3254, Fax: 207-767-5940

The Yacht Connection
Marine East Marina
South Portland, ME 04106
207-799-3600, Fax: 207-767-5937

Hinckley Yacht Brokerage
Box 699, Shore Rd
Southwest Harbor, ME 04679
207-244-5531, Fax: 207-244-9833

Newman Marine
HC 33, Box 5
Southwest Harbor, ME 04679
207-244-5560

Midcoast Yacht Sales
PO Box 221
Wiscasset, ME 04567
207-882-6445, Fax: 207-882-4250

East Coast Yacht Sales
38 Lafayette St., Rt. 88
Yarmouth, ME 04096
207-846-4545, Fax: 207-846-6088
Grand Banks, Sabreline, J Boats

MARYLAND

Annapolis Landing Boat Sales
922 Klakring Rd.
Annapolis, MD 21403
410-263-0090, Fax: 410-626-1857

Annapolis Motor Yachts
P.O. Box 2193
Annapolis, MD 21404
410-268-7171, Fax: 410-268-6921

Annapolis Yacht Sales
7416 Edgewood Rd.
Annapolis, MD 21403
301-267-8181, Fax: 301-267-7409

Atlantic Coast Yacht Sales
Box 5042 - 326 First St.
Annapolis, MD 21403
410-268-5449, Fax: 410-267-6127

Bay Yacht Agency
326 First St.
Annapolis, MD 21403
410-263-2311, Fax: 410-263-2967

Bristol Yacht Sales
623 Sixth St.
Annapolis, MD 21403
410-280-6611, Fax: 410-280-0170
Black Watch, Bristol

Interyacht
7076 Bembe Beach Rd.
Annapolis, MD 21403
410-269-5200, Fax: 410-269-0571

Martin Bird & Associates
326 First St.
Annapolis, MD 21403
410-268-1086, Fax: 410-268-0942

Passport Yachts East
326 First St., #14
Annapolis, MD 21403
410-263-0008, Fax: 410-263-5705

Wilkins Yacht Sales
PO Box 787
Annapolis, MD 21404
410-266-8585, Fax: 410-266-9745
Hatteras

Yacht Net
1912 Forest Dr.
Annapolis, MD 21401
410-263-0993, Fax: 410-267-7967

Fred Quimby's Marine Services
9296 Ocean Gateway
Easton, MD 21601
410-822-8107
Mako, Fountain

Anchor Yacht Basin
1048 Turkey Point Rd.
Edgewater, MD 21037
410-269-6674, Fax: 410-798-6782
Phoenix, Dawson brokerage

Burr Yacht Sales, Inc
1106 Turkey Point Rd
Edgewater, MD 21037
410-798-5900, Fax: 410-798-5911
Fleming, Bertram

Cherry Yachts
2830 Solomons Island Rd.
Edgewater, MD 21037
410-266-3801, Fax: 410-266-3805

Free State Yachts
64 Old River Rd.
Edgewater, MD 21037
410-266-9060, Fax: 410-266-8309

Tidewater Yacht Sales
64A Old South River Rd.
Edgewater, MD 21307
800-899-2799, Fax: 410-224-6919
Bayliner

Yacht Broker & Dealer Directory

Anchor Bay Yacht Sales
202 Nanticoke Rd.
Essex, MD 21221
410-574-0777, Fax: 410-574-8364

Harbour Yacht Agency
PO Box 10
Friendship, MD 20758
410-855-4250, Fax: 410-855-4485

Hartge Yacht Sales
Church Lane
Galesville, MD 20765
410-867-7240, Fax: 410-867-7139

Nautilus Yacht Sales
150 Skipjack Rd., Box 56
Georgetown, MD 21930
410-275-1100, Fax: 410-275-1133

Bayport Yachts
Rt. 50 & Kent Narrows
Grasonville, MD 21638
410-827-5500, Fax: 410-827-5481
Carver

Harrison Yacht Sales
PO Box 98
Grasonville, MD 21638
410-827-6600
Sea Ray, Carver, Viking

Lippincott Marine
Rt. 2, Box 545
Grasonville, MD 21638
410-827-9300, Fax: 410-827-9303

Havre de Grace Yacht Sales
723 Water St.
Havre de Grace, MD 21078
410-939-2161, Fax: 410-939-0220

Tidewater Marine, Foot of Burbon St.
Havre de Grace, MD 21078
410-939-0950, Fax: 410-939-0955

Gunpowder Cove Marina
510 Riviera Dr.
Joppa, MD 21085
410-679-5454
Sea Ray

Jackson Marine Sales
PO Box 483, Hances Point
North East, MD 21901
410-287-9400, Fax: 410-287-9043
Onset

McDaniel Yacht Basin
PO Box E
North East, MD 21901
410-287-8121, Fax: 410-287-8127
Carver

Chesapeake Motoryacht Sales
Tilghman St. @ Town Creek, Box 417
Oxford, MD 21654
410-226-0002, Fax: 410-226-5699
Albin, President

Maryland Yachts
PO Box 216
Oxford, MD 21654
410-226-5571, Fax: 410-226-5080

Oxford Yacht Agency,
317 S. Morris St.
Oxford, MD 21654
410-226-5454, Fax: 410-226-5244
Grand Banks

Arnold C. Gay Yacht Sales
"C" Street, Box 538
Solomons, MD 20688
410-326-2011, Fax: 410-326-2012

Solomons Yacht Brokerage
PO Box 380, 255 "A" Street
Solomons, MD 20688
410-326-6748, Fax: 410-326-2149

William Magness Yachts
301 Pier One Rd., #103
Stevensville, MD 21666
301-643-8434, Fax: 301-643-8437

Warehouse Creek Yacht Sales
301 Pier One Rd.
Stevensville, MD 21666
410-643-7878, Fax: 410-643-7877
Egg Harbor, Cruisers

Clipper Bay Yacht Sales
389 Deale Rd.
Tracys Landing, MD 20779
410-261-5775, Fax: 410-261-5775

Shady Oaks Yacht Sales
846 Shady Oaks Dr.
West River, MD 20778
410-867-7700, Fax: 410-867-1563
Silverton,Tiara

MASSACHUSETTS

Northrop & Johnson
43 Water St.
Beverly, MA 01915
617-569-6900, Fax: 617-569-9247

John G. Alden & Co.
89 Commercial Wharf
Boston, MA 02110
617-227-9480, Fax: 617-523-5465

Onset Bay Yacht Sales
RFD #3, Green St.
Buzzards Bay, MA 02532
508-295-2300, Fax: 508-295-8873
Onset

Carl Bettano Yacht Brokerage
1000 Justin Dr., Admirals Hill Marina
Chelsea, MA 02150
617-889-4849, Fax: 617-889-4814

Danversport Marine
128 Water St.
Danvers, MA 01923
508-777-3822, Fax: 508-777-5478

Norwood Marine
R-24 Ericsson St.
Dorchester, MA 02122
617-288-1000, Fax: 617-282-5728
Silverton, Wellcraft

Yacht Listings
PO Box 465
East Sandwich, MA 02537
508-833-8591, Fax: 508-833-8592

Capt. O'Connell
180 River St., Fall River, MA 02720
508-672-6303, Fax: 508-672-0922
Silverton

Jameson Yacht Sales
PO Box 548, Falmouth, MA 02541
508-540-4750, Fax: 508-564-5940

Eastern Yacht Sales
349 Lincoln St.
Hingham, MA 02043
617-749-8600, Fax: 617-740-4149

Worldwide Yachts
350 Lincoln St., #105
Hingham, MA 02043
617-740-2628, Fax: 617-740-1325

Cape Cod Marine Group
PO Box 220
Hyannis, MA 02601
508-775-6002, Fax: 508-790-1099

Hyannis Marine
21 Arlington St.
Hyannis, MA 02601
508-775-5662, Fax: 508-775-0851

Able Yacht Brokerage
42 Doaks Lane, Little Harbor
Marblehead, MA 01945
617-639-4280, Fax: 617-639-4233

Wells Yachts
91 Front St.
Marblehead, MA 01945
617-631-3003, Fax: 617-639-2503
Luhrs, Phoenix, Pursuit

Powerbrokers International
PO Box 1015
Marion, MA 02738
508-748-3100, Fax: 508-748-3100

Rose Yacht Sales
PO Box 923
Marion, MA 02738
508-748-2211, Fax: 508-748-3773

Bosun's Marine
100 Falmouth Rd./Rte. 28
Mashpee, MA 02649
508-477-4626
Mako

Yacht Broker & Dealer Directory

Cape Yacht Brokers
Box 70, Seabrook Village
Mashpee, MA 02649
508-477-2422, Fax: 508-394-1660

Russo Marine
357 Mystic Ave.
Medford, MA 02155
617-395-0050, Fax: 617-396-8536

Yankee Yacht Sales
23 Low St.
Newburyport, MA 01950
508-462-2781, Fax: 508-465-8847

Dudley Yacht Sales
42 Fiddler Cove Rd.
North Falmouth, MA 02556
508-564-4100, Fax: 508-564-4129

Bay View Yacht Sales
304 Victory Rd.
North Quincy, MA 02171
617-328-1800

Boston Yacht Sales
PO Box 76, Tern Harbor Marina
North Weymouth, MA 02191
617-331-2400, Fax: 617-331-8215
Hatteras, Viking

Nauset Marine
Box 357, Route 6A
Orleans, MA 02653
508-255-0777, Fax: 508-255-0373
Grady-White

Crosby Yacht Yard
72 Crosby Circle
Osterville, MA 02655
508-428-6958, Fax: 508-428-0323

Oyster Harbor Marine
122 Bridge Street
Osterville, MA 02655
508-428-2017, Fax: 508-420-5398
Blackfin, Limestone

Gary Voller's Yacht Sales
3828 Riverside Ave.
Somerset, MA 02726
508-678-0404, Fax: 508-678-0990

Concordia Yacht Sales
South Wharf, Box P203
South Dartmouth, MA 02748
508-999-1381, Fax: 508-992-4682

Buzzards Bay Yacht Sales
PO Box 369
Westport Point, MA 02791
508-636-4010

MICHIGAN

Colony Marine
6509 M-29 Hwy., Box 388
Algonac, MI 48001
313-794-4932, Fax: 313-794-2147
Sea Ray

Bay Harbor Yacht Brokerage
5309 E. Wilder Rd.
Bay City, MI 48706
517-684-3593, Fax: 517-684-5920
Cruisers, Wellcraft, Silverton

Dilworth & Lally Yacht Brokers
100 North Lake Dr.
Boyne City, MI 49712
616-582-6886, Fax: 616-582-6891

Charlevoix Boat Shop
101 E. Mason (Harborside)
Charlevoix, MI 49720
616-547-2710, Fax: 616-547-2444

Irish Boat Shop
1300 Stover Rd.
Charlevoix, MI 49720
616-547-9967, Fax: 616-547-4129
Boston Whaler, Sea Ray

Diamond Boat Sales
28853 N. Gibraltar Rd.
Gibraltar, MI 48173
313-675-3575
Robolo

Grand Isle Marina
1 Grand Isle Dr.
Grand Haven, MI 49417
800-854-2628, Fax: 616-842-8783
Regal

Walstrom Marine
105 Bay Rd.
Harbor Springs, MI 49740
616-526-2141, Fax: 616-526-7527
Hatteras, Tiara

Toledo Beach Yacht Sales
11840 Toledo Beach Rd.
LaSalle, MI 48145
313-243-3830, Fax: 313-243-3815
Hatteras, Silverton

Eldean Boat Sales, Ltd.
2223 South Shore Dr.
Macatawa, MI 49434
616-335-5843, Fax: 616-335-5848
Grand Banks, Bertram, Ocean

Active Marine
31785 S. River Rd.
Mt. Clemens, MI 48045
810-463-7441, Fax: 810-468-7080

John B. Slaven,
Box 864, 31300 N. River Rd.
Mt. Clemens, MI 48046
810-463-0000, Fax: 810-463-4317

McMachen Sea Ray
30099 South River Rd.
Mt. Clemens, MI 48045
810-469-0223, Fax: 810-469-1646
Tiara, Sea Ray

Sun & Sail
31040 N. River Rd.
Mt. Clemens, MI 48045
313-463-4800

Oselka Marina
514 Oselka Dr.
New Buffalo, MI 49117
616-469-2600, Fax: 616-469-0988
Carver

Onekama Marine
Portage Lake
Onekama, MI 49675
616-889-4218, Fax: 616-889-3398
Cruisers, Silverton, Larson

Colony Marine
60 S. Telegraph Rd.
Pontiac, MI 48053
313-683-2500
Sea Ray

Barrett Boat Works
821 W. Savidge St.
Spring Lake, MI 49456
616-842-1202, Fax: 616-842-5735
Zodiac

Keenan Marine
526 Pine St.
Spring Lake, MI 49456
616-846-3830, Fax: 616-846-3821
Chris Craft, Baja

North Shore Marina
18275 Berwyck
Spring Lake, MI 49456
616-842-1488, Fax: 616-842-0143
Silverton, Chaparral

Colony Marine
24530 Jefferson Ave.
St. Clair Shores, MI 48080
810-772-1550
Sea Ray

Jefferson Beach Marina
24400 E. Jefferson
St. Clair Shores, MI 48080
810-778-7600, Fax: 810-778-4766
Viking, Sunseeker, Silverton, Fountain, Cruisers, Formula

Pier 33 Yacht Sales
250 Anchors Way
St. Joseph, MI 49085
616-983-0677
Bayliner, Fountain, Hatteras, Silverton, Wellcraft

Superbrokers of Traverse City
12719 SW Bayshore Dr., #9
Traverse City, MI 49684
616-922-3002, Fax: 616-922-3013

Yacht Broker & Dealer Directory

MINNESOTA

Owens Yacht Sales
371 Canal Park Dr.
Duluth, MN 55802
218-722-9212, Fax: 218-722-4730

MISSISSIPPI

Atlantic Marine Brokers
105 Highway 90
Waveland, MS 39576
800-748-9925, Fax: 601-467-9429
Sea Cat, Answer

NEW JERSEY

Twin Lights
197 Princeton Ave.
Brick, NJ 08724
908-295-3500, Fax: 908-295-0230
Bluewater

Comstock Yacht Sales
704 Princeton Ave
Brick Town, NJ 08724
908-899-2500, Fax: 908-892-3763
Post, Silverton, Blackfin, Regulator

Yacht Quest
197 Princeton Ave.
Brick Town, NJ 08724
908-295-0200, Fax: 908-295-0230

South Jersey Marine Brokers
602 Green Ave.
Brielle, NJ 08730
908-223-2200, Fax: 908-223-0211
Viking

Sportside Marine
201 Union St.
Brielle, NJ 08730
908-223-6677, Fax: 908-223-1215

Clark's Landing Marina
1224 Hwy. 109, Box 2170
Cape May, NJ 08204
609-898-9889, Fax: 609-898-1635
Luhrs, Pursuit, Pro-Line

South Jersey Yacht Sales
PO Box 641
Cape May, NJ 08204
609-884-0880, Fax: 609-884-2995
Viking

Integrity Marine
9401 Amherst Ave.
Margate City, NJ 08402
609-898-0801, Fax: 609-898-0785
Jersey

Olson/Weidman Int'l Yacht Brokers
PO Box 2182
Ocean City, NJ 08226
609-390-2288, Fax: 609-390-1260

Bob Massey Yacht Sales
1668 Beaver Dam Rd.
Pt. Pleasant, NJ 08742
908-295-3700, Fax: 908-892-0649
Jefferson, Onset

Clark's Landing Marina
847 Arnold Ave.
Pt. Pleasant, NJ 08742
908-899-5559, Fax: 908-899-5572
Luhrs, Pursuit, Pro-Line

Northeast Sportfishing
406 Channel Dr.
Pt. Pleasant Beach, NJ 08742
908-899-2600, Fax: 908-899-2701

Catskill Classic Yachts
1 Cherry Lane
Ramsey, NJ 07446
201-327-5000, Fax: 201-327-6848

Sandy Hook Yacht Sales
1246 Ocean Ave.
Sea Bright, NJ 07760
201-530-5500, Fax: 908-530-1323

Cape Island Yacht Sales
Bay & Decatur
Summers Point, NJ 08244
609-927-8886, Fax: 609-927-9707

Caribe Yachts
333 Marc Drive
Toms River, NJ 08753
908-914-1224
Ocean Master

North Sea Yachts
1 Robbins Parkway
Toms River, NJ 08753
908-286-6100, Fax: 908-286-2096
Monk 36

Total Marine
411 Great Bay Blvd.
Tuckerton, NJ 08087
609-294-0480
Phoenix, Powerplay

NEW YORK

City Island Yacht Sales
673 City Island Ave.
City Island, NY 10464
212-885-2300, Fax: 212-885-2385
Egg Harbor

Van Schaick Island Marina
South Delaware Ave.
Cohoes, NY 12047
518-237-2681, Fax: 518-233-8355
Carver, Trojan

Fred Chall Marine
1160 Merrick Rd.
Copaigue, NY 11726
516-842-7777
Hatteras, Wellcraft

Patchogue Shores Marina
28 Cornell Rd.
East Patchogue, NY 11772
516-475-0790, Fax: 516-475-0791

Lakes Yacht Sales,
P.O. Box 512
Freeport, NY 11520
516-378-6070, Fax: 516-546-9457

New York Yacht Corp.
102 Woodcleft Ave.
Freeport, NY 11520
516-546-3377, Fax: 516-223-8393

Star Island Yacht Club
116 Woodcleft Ave.
Freeport, NY 11520
516-623-6256, Fax: 516-868-7332
Tiara

Brewer Yacht Sales
500 Beach Rd.
Greenport, NY 11944
516-477-0770

Higgins Yacht Sales
229 6th St.
Greenport, NY 11944
516-477-2404, Fax: 516-477-8112

Islander Boat Center
555 W. Montauk Hwy.
Lindenhurst, NY 11757
516-957-8721, Fax: 516-957-8744
Bayliner, Regal

Surfside 3 Marina
846 S. Wellwood Ave.
Lindenhurst, NY 11757
516-957-5900, Fax: 516-957-5972
Carver, Sea Ray

McMichael Yacht Brokers
447 E. Boston Post Rd.
Mamaorneck, NY 10543
914-381-5900, Fax: 914-381-5060
Nauticat, J-Boats, Sweden Yachts

McMichael Yacht Brokers
700 Rushmore Ave.
Mamaroneck, NY 10543
914-381-2100, Fax: 914-381-2184
Nauticat, J-Boats, Sweden Yachts

Total Marine, Ltd.
622 Rushmore Ave.
Mamaroneck, NY 10543
914-698-2700, Fax: 914-698-8872
Phoenix, Powerplay

Mattituck Inlet Marina
Mill Road
Mattituck, NY 11952
516-298-4480, Fax: 516-298-4126
Viking, Post

Yacht Broker & Dealer Directory

Ocean Outboard
2976 Whaleneck Dr.
Merrick, NY 11566
516-378-6400, Fax: 516-378-1609
Mako, Fountain, Formula

Star Island Yacht Club
P.O. Box 2180
Montauk Point, NY 11954
516-668-5052, Fax: 516-668-5503
Tiara

Le Comte Company
101 Harbor Lane W.
New Rochelle, NY 10805
914-636-1524, Fax: 914-636-1359

Orange Boat Sales
51-57 Route 9W
New Windsor, NY 12553
914-565-8530, Fax: 914-565-2706
Baja, Thompson, Regal, Rinker, Bayliner, Mainship

Solidmark North America
393 South End Ave.
New York, NY 10280
212-938-0883, Fax: 212-488-6053

Sparkman & Stevens
79 Madison Avenue
New York, NY 10016
212-689-9292, Fax: 212-689-3884

Prime Power Marine
2 Washington St.
Newburgh, NY 12550
914-565-7110, Fax: 914-569-8337
Regal, Pursuit

Smith Boys
278 River Rd.
North Tonawanda, NY 14120
716-695-3472
Sea Ray

Oyster Bay Marine Center
5 Bay Ave.
Oyster Bay, NY 11771
516-922-6331, Fax: 516-922-3542

Ventura Yacht Services
15 Orchard Beach Blvd.
Port Washington, NY 11050
516-944-8415, Fax: 516-944-8415

Shumway Yacht Sales
70 Pattonwood Dr.
Rochester, NY 14617
716-342-3030, Fax: 716-266-4722

Bruce Taite Yacht Sales
PO Box 1928
Sag Harbor, NY 11963
516-725-4222, Fax: 516-725-9886

Maritime Yacht
50 W. Water St., PO Box 1981
Sag Harbor, NY 11963
516-725-2878, Fax: 516-725-2878

White River Marine
5500 Sunrise Hwy.
Sayville, NY 11782
516-589-2502
Grady-White

Krenzer Marine
8495 Greig Street
Sodius Point, NY 14555
315-483-6986, Fax: 315-483-6986
Silverton, Cruisers

Dave Bofill Marine
1810 North Sea Rd.
Southampton, NY 11968
516-283-6736, Fax: 516-283-7041
Robolo, Formula

Port of Egypt Marine
Main Road
Southold, NY 11971
516-765-2445
Albemarle, Grady-White

Canyon Yacht Sales
870 Jericoh Turnpike
St. James, NY 11780
516-724-2424
Bertram, Cabo

Staten Island Boat Sales
222 Mansion Ave.
Staten Island, NY 10308
718-984-7676, Fax: 718-317-8338
Viking

Northeast Yachts,
3451 Burnett Ave.
Syracuse, NY 13206
315-437-1438, Fax: 315-437-2501
Bluewater

Hudson Boat Sales
Beach Road
West Haverstraw, NY 10993
914-429-5100
Bayliner

South Shore Boats of Suffolk
Box 1706, Library Ave.
Westhampton Beach, NY 11978
516-288-2400
Rampage

RCR Yachts
PO Box 399
Youngstown, NY 14174
716-745-3862, Fax: 716-745-9671

NORTH CAROLINA

Causeway Marina
300 Morehead Ave., Box 2366
Atlantic Beach, NC 28512
919-726-6977, Fax: 919-726-7089
Stamas

Beaufort Yacht Sales
328 Front St.
Beaufort, NC 28516
919-728-3155, Fax: 919-728-6715
Viking, Freedom, Valient

Nelson Yacht Sales
103 Hill St., Box 1129
Beaufort, NC 28516
919-728-3663, Fax: 919-728-5333
Island Packet, Cabo Rico

Pembroke Yacht Sales
PO Box 384
Edenton, NC 27932
919-482-5151, Fax: 919-482-8754

New Hope Marine
3717 S. New Hope Rd.
Gastonia, NC 28056
704-824-7653
Grady-White

Slane Marine
PO Box 6687
High Point, NC 27262
919-861-6100, Fax: 919-861-8329

70 West Marina
Highway 70
Morehead City, NC 28557
919-726-5171, Fax: 919-726-9993
Tiara, Jersey, Albemarle

Spooners Creek Yacht Sales
Rt. 2, Lands End Rd.
Morehead City, NC 28557
919-726-8082, Fax: 919-726-9806

Quay & Associates
PO Box 563
Oriental, NC 28571
919-249-1825, Fax: 919-249-2240

Whittaker Creek Yacht Sales
PO Box 357
Oriental, NC 28571
919-249-0666, Fax: 919-249-2222

Baker Marine
3410 River Rd.
Wilmington, NC 28412
910-256-8300, Fax: 910-256-9542
Grand Banks, Hatteras, Tiara

Bennett Brothers Yachts
8118 Market St.
Wilmington, NC 28405
919-686-9535, Fax: 919-686-1332

Carolina Yacht Sales
1322 Airlie Rd.
Wilmington, NC 28403
919-256-9901, Fax: 919-256-8526

Pages Creek Marina
7000 Market St.
Wilmington, NC 28405
919-799-7179, Fax: 919-799-1096
Mako

Harbourside Yachts
PO Drawer 896
Wrightsville Beach, NC 28480
910-350-0660, Fax: 910-350-0506
Carver, Blackfin, Luhrs, Trojan

OHIO

Captain's Cove Marina
4241 Kellogg Ave.
Cincinnati, OH 45226
513-321-1111, Fax: 513-654-2466
Baja, Formula, Fountain

North Shore Boat Brokerage
1787 Merwin
Cleveland, OH 44113
216-241-2237

Anchor Yacht Brokerage
10905 Corduroy Rd.
Curtice, OH 43412
419-836-8985

Bolton's Marine Sales
160 Forest Dr.
Eastlake, OH 44095
216-942-7426, Fax: 216-942-7404

South Shore Yacht Supply
36355 Lakeshore Blvd.
Eastlake, OH 44095
216-946-2266

Lakeside Marine
Erie Beach Rd.
Lakeside, OH 43440
419-798-4406, Fax: 419-798-4089
Tiara

Island Yacht Sales
4236 E. Moore's Dock Rd.
Port Clinton, OH 43452
419-797-9003, Fax: 419-797-6846

Riverdale
1805 W. Lakeshore Dr.
Port Clinton, OH 43452
419-732-2150, Fax: 419-732-8820
Tollycraft

OKLAHOMA

Ugly John's Custom Boats
Route 1, Box 50
Cleveland, OK 74020
918-243-5220, Fax: 918-243-5238
Bluewater, Carver, Celebrity, Fountain, Gibson, Harbor Master

OREGON

Irwin Yacht Sales
865 NE Tomahawk Island Dr.
Portland, OR 97211
503-298-0074, Fax: 503-286-2213
Wellcraft

Oregon Yacht Sales
2305 NW 133rd Place
Portland, OR 97229
503-629-0717
Tollycraft, Tierra

Seaward Yacht Sales
0315 SW Montgomery, #200
Portland, OR 97201
503-224-2628, Fax: 503-224-5210
Grand Banks

RHODE ISLAND

Shaw Yachts
305 Oliphant Lane
Middletown, RI 02840
401-848-2900, Fax: 401-848-2904

Bartram & Brakenhoff
2 Marine Plaza, Goat Island
Newport, RI 02840
401-846-7355, Fax: 401-847-6329

Bass Harbor Marine
49 Americas Cup Ave.
Newport, RI 02840
401-849-0240, Fax: 401-849-0620
Mason, Taswell, Sabre, Nordhavn

Newport Yacht Services
PO Box 149
Newport, RI 02840
401-846-7720, Fax: 401-846-6850

Alden Yacht Brokerage
1909 Alden Landing
Portsmouth, RI 02871
401-683-4285, Fax: 401-683-3668

Alden Yachts, Power & Sail
Brewer's Yacht Sales
222 Narragansett Blvd.
Portsmouth, RI 02871
401-683-3977, Fax: 401-683-0696

Eastern Yacht Sales
One Lagoon Rd.
Portsmouth, RI 02871
401-683-2200, Fax: 401-683-0961

Little Harbor Yacht Sales
One Little Harbor Landing
Portsmouth, RI 02871
401-683-7000, Fax: 401-683-7029
Little Harbor Custom Yachts, Grand Banks, Black Watch

SOUTH CAROLINA

Charleston Yacht Sales
3 Lockwood Dr., #201
Charleston, SC 29402
803-577-5050, Fax: 803-723-9829
Hunter

Pilot Yacht Sales
3 Lockwood Dr., #301
Charleston, SC 29401
803-723-8356, Fax: 803-723-2102

Sea Island Yacht Sales
105 Wappoo Creek Dr., #3A
Charleston, SC 29412
803762-2610, Fax: 803-762-2615

Yacht Brokerage USA
2345 Tall Sail Dr., #E
Charleston, SC 29414
803-763-1224, Fax: 803-763-4215

Ashley Marina Yacht Sales
33 Lockwood Drive
Charleston, SC 29401
803-722-1996, Fax: 803-720-3623

DYB Charters & Yacht Sales
14 New Orleans Rd., Suite 14
Hilton Head Island, SC 29928
803-785-4740, Fax: 803-785-4827
Catalina

Hilton Head Yachts, Ltd.
PO Box 22488
Hilton Head Island, SC 29925
803-686-6860, Fax: 803-681-5093

American Yacht Sales
1880 Andell Buffs Blvd.
Johns Island, SC 29455
800-234-8814, Fax: 803-768-7300
Luhrs, Mainship, Ocean

Berry-Boger Yacht Sales
Box 36, Harbour Place, #101
N. Myrtle Beach, SC 29597
803-249-6167, Fax: 803-249-0105
Marine Trader

Wilkins Boat & Yacht Co.
1 Harbour Place
N. Myrtle Beach, SC 29582
803-249-6032, Fax: 803-249-6523

TENNESSEE

Phil's Marine Sales
4935 Highway 58, Suite C
Chattanooga, TN 37416
615-892-0058, Fax: 615-894-3281

Leader Marine
720 E. College St.
Dickson, TN 37055
615-446-3422, Fax: 615-446-9819
Cruisers

Jim Bennett Yacht Sales
Route 4, Box 532
Iuka, TN 38852
601-423-9999, Fax: 601-423-3339
Bluewater, Carver

Fox Road Marina
1100 Fox Rd.
Knoxville, TN 37922
615-966-9422, Fax: 615-966-9475

Nashville Yacht Brokers
1 Vantage Way, Suite B-100
Nashville, TN 37228
615-259-9444, Fax: 615-259-9481

TEXAS

Coastal Yacht Brokers
715 Holiday Dr. North
Galveston, TX 77550
409-763-3474, Fax: 713-488-8782

HoustonYacht Sales
585 Bradford Ave., Suite B
Kemah, TX 77565
713-334-7094, Fax: 713-334-4936
Hatteras, Marlin, Ocean

Ship and Sail
300 Admiralty Way
Kemah, TX 77565
713-334-0573, Fax: 713-334-2697
Luhrs, Mainship, Carver

Delhomme Marine
2551 S. Shore Blvd., #C
League City, TX 77573
713-334-3335, Fax: 713-334-1402

Fox Yacht Sales
Box 772, Island Moorings Marina
Port Aransas, TX 78373
512-749-4870, Fax: 512-749-4859

Jay Bettis & Company
2509 NASA Road 1
Seabrook, TX 77586
713-474-4101, Fax: 713-532-1305
DeFever, Shamrock, Mainship

VIRGINIA

Doziers Dockyard
PO Box 388
Deltaville, VA 23043
804-776-6711, Fax: 804-776-6998

Norton's Yacht Sales
PO Box 220
Deltaville, VA 23043
804-776-9211, Fax: 804-776-9044
Luhrs, Hunter, Silverton

Commonwealth Yachts
PO Box 1070
Gloucester Point, VA 23062
804-642-2156, Fax: 804-642-4766

Bluewater Yacht Sales
25 Marina Rd.
Hampton, VA 23669
804-723-0793, Fax: 804-723-3320
Hatteras, Viking, Tiara

Casey Marine
1021 W. Mercury Blvd.
Hampton, VA 23666
804-591-1500, Fax: 804-244-7805
Pro-Line, Baja, Robolo, Maxum, Luhrs

Virginia Yacht Brokers
4503 Ericcson Dr.
Hampton, VA 23669
804-722-3500, Fax: 804-722-7909

Norview Yacht Sales
PO Box 740
Hayes, VA 23072
804-776-7233, Fax: 804-776-7940

Bay Harbor Brokerage
1553 Bayville St.
Norfolk, VA 23503
804-480-1073, Fax: 804-587-4612

Prince William Marine Sales
207 Mill St.
Occoquan, VA 22125
703-494-6611
Sea Ray

Tidewater Yacht Sales
10A Crawford Parkway
Portsmouth, VA 23704
804-393-6200, Fax: 804-397-1193
Bayliner

Atlantic Yacht Brokers
932 Laskin St., Suite 200
Virginia Beach, VA 23451
804-428-9000, Fax: 804-491-8632
Ocean, Cabo

Casey Marine
4417 Shore Dr.
Virginia Beach, VA 23455
Pro-Line, Baja, Robolo, Maxum, Luhrs

VERMONT

Bruce Hill Yacht Sales
219 Harbor Rd.
Shelburne, VT 05482
802-985-3336, Fax: 802-985-3337

WASHINGTON

Anacortes Yacht Sales
P.O. Box 855
Anacortes, WA 98221
206-293-0631, Fax: 206-293-0633

Skipper Cress Yacht Sales
1019 Q Ave., Suite B
Anacortes, WA 98221
800-996-9991, Fax: 206-293-7874
Nordic Tugs

Bellingham Yacht Sales
1801 Roeder Ave.
Bellingham, WA 98225
800-671-4244, Fax: 206-671-0992
Sabreline

Padden Creek Marine
809 Harris Ave.
Bellingham, WA 98225
206-733-6248, Fax: 206-733-6251

Edmonds Yacht Sales
300 Admiral Way
Edmonds, WA 98020
206-774-8878, Fax: 206-771-7277
Riviera

Superior Yacht Sales
628 Daley St., Unit 1
Edmonds, WA 98020
206-771-3786, Fax: 206-771-3786
Vitesse

Northwest Yachts
PO Box915, Friday Harbor, WA 98250
206-378-7196, Fax: 206-378-7197

Northwest Yachts
3805 Harborview Dr.
Gig Harbor, WA 98335
206-858-7700, Fax: 206-851-8649
Krogen

Sunset Yacht Sales
2905 Harborview Dr.
Gig Harbor, WA 98335
206-858-8811, Fax: 206-858-7373

Sailboats & Yachts Northwest
11207 101st Ave. NE
Kirkland, WA 98033
206-623-9011, Fax: 206-282-8815

Yacht Doc
8031 NE 112th St.
Kirkland, WA 98034
206-820-9659, Fax: 206-823-8913

La Conner Yacht Sales
612 Dunlap, #C
La Conner, WA 98257
206-466-3300, Fax: 206-466-3533

Adventure Yacht Sales
2400 Westlake Ave. North, #1
Seattle, WA 98109
206-283-3010, Fax: 206-283-8611

Alliance Yacht Sales
2130 Westlake Ave. North
Seattle, WA 98109
206-283-8111
Offshore

American Yacht Sales
2144 Westlake Ave. N.
Seattle, WA 98109
206-284-6354, Fax: 206-285-8772

Brigadoon Yacht Sales
1111 Fairview Ave. North
Seattle, WA 98109
206-282-6500, Fax: 206-282-2410

Cruising Yachts
927 N. Northlake Way
Seattle, WA 98103
206-632-4819, Fax: 206-548-1050
Nordhavn, Lord Nelson

Dave Maples Yacht Sales
1530 Westlake North
Seattle, WA 98109
206-284-0880, Fax: 206-285-7903
Canoe Cove

Elliott Bay Yachting Center
2601 West Marina Pl., #E
Seattle, WA 98199
206-285-9499, Fax: 206-281-7636

Fraser Yachts
1500 Westlake Ave. N.
Seattle, WA 98109
206-282-4943, Fax: 206-285-4956

Intrepid Yacht Sales
2000 Westlake Ave. North
Seattle, WA 98109
206-282-0211
Grand Banks

Lager Yacht Brokerage Corp.
2601 W. Marine Place
Seattle, WA 98199
206-283-6440, Fax: 206-283-4707

Lake Union Yacht Sales
3245 Fairview Ave. E. #103
Seattle, WA 98102
206-323-3505, Fax: 206-323-4751
Island Gypsy, Catalina

Maple Bay Boat Co.
1333 N. Northlake Way
Seattle, WA 98103
206-547-4780

Ocean Alexander Yacht Sales
1001 Fairview Ave. North
Seattle, WA 98109
206-223-0809, Fax: 206-223-0812
Ocean Alexander, Nordlund

Ray Rairdon Yacht Sales
1800 Westlake Ave. N., #101
Seattle, WA 98103
206-284-5527, Fax: 206-284-5537

Tatoosh Marine
809 Fairview Place N., #150
Seattle, WA 98109
206-625-1580, Fax: 206-682-1473

Tradewind Yacht Sales
2470 Westlake Ave. N.
Seattle, WA 98109
206-285-0926

Trans Coastal Yacht
1800 Westlake Avenue North, #201
Seattle, WA 98109
206-284-4547, Fax: 206-284-4337

West Coast Yachts
1836 Westlake Ave. N., #201
Seattle, WA 98109
206-298-3724, Fax: 206-298-0227

Yacht Sales International
1220 Westlake Ave. North
Seattle, WA 98109
206-282-0052, Fax: 206-283-2297
Tollycraft, Tiara, Ocean

River City Marina
E. 6326 Trent
Spokane, WA 99212
509-534-5444, Fax: 509-534-3179
Silverton

Murray Wasson Marine Sales
4224 Marine View Dr.
Tacoma, WA 98422
206-927-9036, Fax: 206-927-9034
Shamrock, Blackfin

Picks Cove Marine Center
1940 East D Street
Tacoma, WA 98421
206-572-3625
Symbol

Preferred Yacht & Ship Brokers
1802 East D Street
Tacoma, WA 98421
206-272-4550, Fax: 206-272-4804

WISCONSIN

Capt. Jim's Yacht Sales
5136 Sheridan Rd.
Kenosha, WI 53140
414-652-8866, Fax: 414-652-5453
Mainship, Silverton

Kewaunee Marina
77 N. Main St., Box 261
Kewaunee, WI 54216
414-388-4550
Pro-Line

Emerald Yacht Ship Mid America
759 N. Milwaukee Street
Milwaukee, WI 53202
414-271-2595, Fax: 414-271-4743

Professional Yacht Brokerage
9501 W. Morgan Ave.
Milwaukee, WI 53228
414-321-8880, Fax: 414-321-7411

Fox River Marina
P.O. Box 1006
Oshkosh, WI 54902
414-235-2340
Wellcraft, Cruisers

Lakeside Marina
902 Taft Ave.
Oshkosh, WI 54901
414-231-4321
Bayliner, Carver, Doral, Chaparral

Professional Yacht Sales
451 S. Second St.
Prescott, WI 54021
715-262-5762, Fax: 715-262-5658

Palmer Johnson Yacht Sales
811 Ontario St.
Racine, WI 53402
414-633-8883, Fax: 414-633-4681

Cal Marine
1024 Bay Shore Dr.
Sister Bay, WI 54234
414-854-4521, Fax: 414-854-5137
Tiara, Powerquest

Sturgeon Bay Yacht Harbor
306 Nautical Drive
Sturgeon Bay, WI 54235
414-743-3311, Fax: 414-743-4298
Alexander, Formula, Mako

Marine Surveyor Directory

Professional Surveyor Associations

ABYC..American Boat & Yacht Council
ASA..American Society of Appraisors
AIMS ....American Institute of Marine Surveyors
MTAM..........................Marine Trade Association of Maryland
NAMI ....National Association of Marine Investigators
NAMS.....................National Association of Marine Surveyors
NFPA ....National Fire Protection Association
SAMS..........................Society of Accredited Marine Surveyors
SNAME........Society of Naval Architects & Marine Surveyors

ALABAMA

Donald Smith
Port City Marine Services
PO Box 190321
Mobile, AL 36619
205-661-5426
SAMS, ABYC

Michael Schiehl
M.J. Schiehl & Assoc.
PO Box 1990
Orange Beach, AL 36561-1990
NAMS

ALASKA

George Sepel
Sepel & Son Marine Surveying
PO Box 32223
Juneau, AK 99803
907-790-2628
SAMS

ARKANSAS

Angus Rankin
Marine Surveyor
PO Box 264
Maynard, AR 72444
501-892-8300
SAMS

CALIFORNIA

Kurt Holland
R.J. Whitfield & Assoc.
One Pacific Marina, Apt. 807
Alameda, CA 94501
800-344-1838
SAMS

Stanley Wild
Stan Wild & Associates
1912 Stanford
Alameda, CA 94501
510-521-8527
NAMS

Mare Colomb
Colomb Yacht Surveyor
2619 Willow Lane
Costa Mesa, CA 92627
714-646-7807
SAMS

Ronald Grant
Grant Marine Surveys
25611-114 Quail Run
Dana Point, CA 92629
714-240-8353
SAMS

Michael Whitfield
R.J. Whitfield & Associates
4471 Amador Rd.
Fremont, CA 94538-1201
800-344-1838
SAMS

Douglas Malin
Malin Marine Surveyors
5942 Edinger Ave., #113
Huntington Beach, CA 92649
714-897-6769
SAMS

Hans Anderson
Anderson Int'l. Marine Surveyors
433 North H St., Ste. G
Lompoc, CA 93436
805-737-3770
SAMS

Clark Barthol, CMS
Clark Barthol Marine Surveyors
27 Buccaneer St.
Marina del Rey, CA 90292
310-823-3350
NAMS, ABYC

William Butler
Marine Surveyor
PO Box 11914
Marina Del Rey, CA 90295
310-396-1791
SAMS

Terrence O'Herren
Marine Surveyor
2021 Ashton Ave.
Menlo Park, CA 94025
415-854-8380
SAMS

Donald Young
Donru Marine Surveyors & Adjusters
32 Cannery Row
Monterey, CA 93940
408-372-8604
SAMS, ABYC

Richard Christopher
Marine Surveyor
14705 Watsonville
Morgan Hill, CA 95037
415-368-8711
SAMS, ABYC

James Wood
Marine Surveyor
PO Box 968
Morro Bay, CA 93443
805-772-0110
SAMS

Marine Surveyor Directory

John Kelly
Kelly & Associates
PO Box 1031
Napa, CA 94581
707-641-1061
SAMS, ABYC

Bill Beck
Marine Surveyor
444-A N. Newport Blvd.
Newport Beach, CA 92663
714-642-6673
SAMS

Don Parish
Marine Surveyor
4140 Oceanside Blvd., #159-320
Oceanside, CA 92056
619-721-9410
ABYC

Richard Quinn
Oceanside Marine Surveyors
425 Calle Corazon
Oceanside, CA 92057
619-757-5586
SAMS

Mike Pierce
Mike Pierce Industries
1811 Diego Way
Oxnard, CA 93030
805-657-9490
SAMS

Skip Riley
Maritime Surveyors
3203 S. Victoria Ave., Ste. B
Oxnard, CA 93035
805-984-8889
NAMS, SAMS

Donald Brandmeyer
Brandmeyer International
2447 Sparta Dr.
Rancho Palos Verdes, CA 90274-6538
310-519-1979
NAMS

Marine Survey Group
1310 Rosecrans St., #K
San Diego, CA 92106
619-224-2944
NAMS

Anthony Tillett
A.N. Tillett & Associates
663 Switzer St.
San Diego, CA 92101
619-235-0766
NAMS, SAMS

Charles Driscoll
Frank Wyatt Marine Surveyors
1967 Shaffer St.
San Diego, CA 92106
619-223-8167
NAMS

Leroy Lester
Lester & Lester Marine Survey
1310 Rosecrans St., #K
San Diego, CA 92106
619-224-2944
SAMS

Marvin Henderson
Marvin Henderson Marine Surveyors
2727 Shelter Island Dr., #C
San Diego,CA 92106
619-224-3164
NAMS, ABYC, NFPA, SNAME

Todd Schwede
Todd & Associates
2390 Shelter Island Dr., #220
San Diego, CA 92106
619-226-1895
SAMS

Bruce Sherburne
Sherburne & Associates
6130 Monterey Rd., #23
San Jose, CA 95138
800-882-7124
SAMS

Jack Mackinnon
Marine Surveyor
PO Box 335
San Lorenzo, CA 94580-0335
415-276-4351
SAMS

Mike Pyzel
Marine Surveyor
PO Box 4217
Santa Barbara, CA 93140
805-640-0900
SAMS

Thomas Bell
Thomas Bell & Associates
1323 Berkeley Street
Santa Monica, CA 90404
310-306-1895
SAMS, ABYC

Archibald Campbell
Campbell's Marine Survey
340 Countryside Drive
Santa Rosa, CA 95401
707-542-8812
SAMS, ABYC, ASME, SNAME

Michael Wilson
Marine Surveyor
1001 Bridgeway, Suite 722
Sausalito, CA 94965
415-332-8928
SAMS

Peggy Feakes
R.J. Whitfield & Associates
7011 Bridgeport Circle
Stockton, CA 95207-2357
209-956-8488
SAMS

Rod Whitfield
R.J. Whitfield & Associates
7011 Bridgeport Circle
Stockton, CA 95207
209-956-8488
SAMS

Robert Downing
Marine Surveyor
PO Box 4154
Vallejo, CA 94590
707-642-6346
SAMS

Basil Dalseme
Marine Surveyor
PO Box 24353
Ventura, CA 93001
805-643-6407
SAMS

CONNECTICUT

J. Mitchell DePalma
Connecticut Yacht Survey Corp.
PO Box 842
Branford, CT 06405
203-488-0265
SAMS, ABYC

Richard Tudan
Bosun's Yacht Survey
26 Hickory Hill Lane
Branford, CT 06405
203-481-5099
SAMS

Grant Westerson
New England Marine Surveyors
PO Box 533, 19 Commerce St.
Clinton, CT 06413
203-669-4018
SAMS

William Robbins
New England Marine Surveyors
19 Commerce St., Box 533
Clinton, CT 06413
203-669-4018
SAMS

Harry Hartzell
Hartzell Marine Surveys
92 Willard Dr.
Enfield, CT 06082
203-292-7179
SAMS

Albert Truslow
Truslow Marine Surveying
PO Box 9185
Forestville, CT 06011-9185
203-583-6503
SAMS

Marine Surveyor Directory

George Stafford
Aetna Casualty
One Civic Ctr. Plaza, Box 2954
Hartford, CT 06143
203-240-6765
SAMS

Welles Worthen
Marine Surveyors, Inc.
102 Milford Point Rd.
Milford, CT 06460
203-874-2445
NAMS

Marine Surveyors Bureau
1440 Whalley Ave, #128
New Haven, CT 06515
203-323-0225
NAMS, SAMS, ABYC, NFPA

Robert Keaney, Jr.
Marine Surveyor
7 Candlewood Heights
New Milford, CT 06776
203-354-1372
SAMS

Kenneth Johnson
Johnson Marine Survey
2 Haley Farm Lane
Noank, CT 06340
203-444-8576
SAMS

Robert Krauss
Connecticut Compass Service
3 Anchorage Lane
Old Saybrook, CT 06475
203-388-2019
SAMS

William Stadel
Marine Surveyor
1088 Shippan Ave.
Stamford, CT 06902
203-324-2610
SAMS

Thomas Greaves
Greaves Yacht Service
30 Toby Hill Road
Westbrook, CT 06498
203-399-6966
SAMS

David Robotham
Robotham Marine Surveyors
PO Box 2143
Westport, CT 06880
203-227-9640
NAMS, SAMS

FLORIDA

Larry Vanscoy
Accredited Marine Surveys
PO Box 331162
Atlantic Beach, FL 32233
904-636-4382
SAMS

Dean Greger
Coastal Marine Surveyors
23 Winston Dr.
Belleair, FL 34616
813-581-0914

Kermit Naylor
Southern Yacht Surveyors
2895 Del Rio Dr.
Belleair Bluffs, FL 34640
813-585-8949
NAMS

Dick Williamson
Professional Marine Surveys, Inc.
7491-C5 N. Federal Hwy, #232
Boca Raton, FL 33487
407-272-1053
SAMS, ABYC, NFPA

John Greeley
Marine Surveyor
22177 Thomas Terrace
Boca Raton, FL 33433
305-360-3330
SAMS

Jeff Brown
JGB Corporation
8716 54th Ave. West
Bradenton, FL 34210
813-794-3998
SAMS

Stephen Fredrick
Preferred Claims Adjusters
PO Box 10265
Bradenton, FL 34282
813-794-3552
SAMS

James Hughes
Capt. J. R. Hughes Marine Services
368 Harbor Drive
Cape Canaveral, FL 32920
407-783-3832
SAMS

Lawrence O'Pezio
Canaveral Marine Consultants
677 George King Blvd., #112
Cape Canaveral, FL 32920
407-783-1771
NAMS, ABYC, NFPA

Ralph Strauss
Island Marine Service
PO Box 10636
Clearwater, FL 34617
813-581-0942
SAMS

Channing Chapman
Clyde Eaton & Assoc., Inc.
PO Box 231862
Cocoa, FL 32923-1862
407-633-0860
ABYC, NFPA

Anthony Pavlo
Alp's Marine Surveying, Inc.
281 NW 42nd Ave.
Coconut Creek, FL 33066
305-973-1135
SAMS

Jeffrey Turner
Turner & Associates
801 NE 3rd Street
Dania, FL 33004
305-922-3333
SAMS

Stephen Rhodes
Boating Services
6 Sunset Terrace
Daytona Beach, FL 32118
904-257-5112
SAMS

H. Jack MacDonald
H. Jack MacDonald, Inc.
23 NW 18th Street
Delray Beach, FL 33444
407-731-0471
SAMS, ABYC, NFPA

James Sanislo
C&J Marine Surveyors
4163 Frances Dr.
Delray Beach, FL 33445
407-495-4920
SAMS

Rollie Tallman
American Boat Brokerage
2548 Alton Rd.
Deltona, FL 32738
904-789-0971
SAMS

John Marrocco
Marine Surveyor
PO Box 891
Edgewater, FL 32132
904-426-0368
SAMS

Marine Surveyor Directory

Capt. Michael McGhee
Black Pearl Marine Specialities
6695 NW 25th Terrace
Ft. Lauderdale, FL 33309
305-970-8305
SAMS

Drew Kwederas
Global Adventure Marine Associates
4120 NE 26th Ave.
Ft. Lauderdale, FL 33308
305-566-4800
ABYC, NFPA, SNAME

Edward Rowe
Ed Rowe & Associates
1821 SW 22nd Ave.
Ft. Lauderdale, FL 33312
305-792-6062
SAMS

Gene Thornton
Gene Thornton Diesel Survey
4564 NE 11th Ave.
Ft. Lauderdale, FL 33334
305-776-7242
SAMS (Diesel Specialist)

Gerald Slakoff
Slakoff & Associates
1525 S. Andrews Ave.
Ft. Lauderdale, FL 33316
305-525-7930
SAMS

Gregory Mitchell
Marine Surveyor
1007 N. Federal Hwy., #84
Ft. Lauderdale, FL 33304
407-286-3924
SAMS

Gregory Newton
Marine Evaluation Service
1323 SE 17th St., #119
Ft. Lauderdale, FL 33316
305-763-9562
NAMS

Jerome Cramer
Gerald Slakoff & Associates
1524 S. Andrews Ave.
Ft. Lauderdale, FL 33316
305-525-7930
SAMS

Junko Pascoe
Marine Surveyor
501 SW 14th St.
Ft. Lauderdale, FL 33315
305-527-5741
SAMS

Kurt Merolla
Merolla Marine Surveyors & Consultants
4761 NE 29th Ave.
Ft. Lauderdale, FL 33308
305-772-8090
NAMS, SAMS, ASA SNAME, ABYC, NFPA

Marc Slakoff
Slakoff & Associates
1525 S. Andrews Ave.
Ft. Lauderdale, FL 33316
305-525-7930
SAMS

Norman Schreiber II
Transtech—Marine Division
PO Box 350247
Ft. Lauderdale, FL 33335
305-537-1423
NAMS, SNAME, ABYC, NFPA

Randal Roden
The Marine Surveyors
PO Box 100145
Ft. Lauderdale, FL 33310
800-522-5119
SAMS

Robert Heekin
The Marine Surveyors
PO Box 100145
Ft. Lauderdal, FL 33310
800-522-5119
SAMS

William Casey
SCS & Associates
3215 NW 10th Terrace, #209
Ft. Lauderdale, FL 33315
305-563-6900
SAMS

Steven Berlin
Independent Marine Surveyors
18400 San Carlos Blvd.
Ft. Myers Beach, FL 33931
813-466-4544
SAMS

Richard Cain
Independent Marine Surveyors
144 Bay Mar Dr.
Ft. Myers Beach, FL 33931
813-466-4544
SAMS

Donna Summerlin
Summerlin's Marine Survey
200 Naco Rd., Suite C
Ft. Pierce, FL 34949
407-461-3244
SAMS

John McCulley
McCulley Marine Services
101 Sea Way Dr., Suite A
Ft. Pierce, FL 34950
407-489-6069
SAMS

Thomas Price
Price Marine Services, Inc.
9418 Sharon St. SE
Hobe Sound, FL 33455
407-546-0928
SAMS, ABYC, NFPA

William King
Atlantic Marine Survey
6201 SE Monticello Terrace
Hobe Sound, FL 33455-7383
407-545-0011
SAMS

James Macefield
Macefield Marine Services, Inc.
3389 Sheridan St., #178
Hollywood, FL 33021
305-784-9188

Capt. Larry C. Dukehart
Marine Surveyor & Consultant
PO Box 1172
Islamorada, FL 33036-1172
305-664-9452
SAMS, ABYC, NFPA, NAMI

Downing Nightingale, Jr.
North Florida Marine Services
3360 Lake Shore Blvd.
Jacksonville, FL 32210-5348
904-384-4356
SAMS, AMS

Mickey Strocchi
Strocchi & Company
PO Box 16541
Jacksonville, FL 32245-6541
904-398-1862
SAMS

Ted Willandt
Marine Network
2771-25 Monument Rd., Box 210
Jacksonville, FL 32225-3547
904-641-3334
SAMS, ABYC, NAMI

Ted Stevens
Stevens & Stevens, Ltd.
3250 Candice Ave., #132
Jensen Beach, FL 34957
407-229-6394
SAMS

Michael Bennett
Marine Surveyor
12129 181st Court North
Jupiter, FL 33478
407-744-0213
SAMS

Robert Camuccio
Master Marine of South Florida
101425 Overseas Highway, Ste. 710
Key Largo, FL 33037
305-662-6644
SAMS

Edwin Crusoe
Key West Marine Services
PO Box 4854
Key West, FL 33040
305-872-9073
NAMS

Marine Surveyor Directory

George Stuck
Marine Surveyor
PO Box 5481
Key West, FL 33045
305-294-4959

Mark Perkins
Marine Surveyor
901 Fleming Street
Key West, FL 33040
305-294-7635
SAMS

William Colby
Florida Keys Community College
5901 W. Junior College Rd.
Key West, FL 33040
305-296-9081
SAMS

Ed Stanton
Rhodes Marine Surveyors
4701 N. Federal Hwy., Ste. 340, Box C-8
Lighthouse Point, FL 33064-6563
305-9466779
SAMS, ABYC, NFPA

Mark Rhodes
Rhodes Marine Surveyors
4701 N. Federal Hwy., Ste. 340, Box C-8
Lighthouse Point, FL 33064-6563
305-9466779
SAMS, ABYC, NFPA

Melvin Wamsley
Accurate Marine Surveying
2130 NE 42nd St., Apt. #3
Lighthouse Point, FL 4802133064
305-942-9206
SAMS

Mike Rhodes
Rhodes Marine Surveyors
4701 N. Federal Hwy., Ste. 340, Box C-8
Lighthouse Point, FL 33064-6563
305-9466779
SAMS, ABYC, NFPA

L. Frank Hamlin
L.F. Hamlin, Inc.
14085 E. Parsley Dr.
Madeira Beach, FL 33708
813-393-1905
NAMS, SNAME

Dewey Acker
Acker Marine Surveyors
551 61st St. Gulf
Marathon, FL 33050
305-743-3434
SAMS, ABYC

Brough Treffer
Treffer Marine Survey, Inc.
2865 S. Tropical Trail
Merritt Island, FL 32952
407-453-6046
SAMS, NFPA, ABYC

James Robbins
Marine Surveyor
7701 Pine Lake Dr.
Merritt Island, FL 32953
407-459-1196
SAMS

Joanna Bailey
Brevard Marine Service
150 E. Merritt Island Causeway
Merritt Island, FL 32952
407-452-8250
SAMS

Dave Alter
Dave Alter & Associates
6500 SW 129th Terrace
Miami, FL 33156
305-667-0326
SAMS

William Ballard
Ballard & Associates
18845 SW 93rd Ave.
Miami, FL 33157
305-378-9674
SAMS

Brett Carlson
Carlson Marine Surveyors & Adjusters
1002 NE 105th St.
Miami Shores, FL 33138
305-891-0445
SAMS

Allen Perry
Ocean Adventures
453 Spinnaker Dr.
Naples, FL 33940
813-261-5466
SAMS, ABYC, NAMS

Donald Walwer
D&G Marine Company
58 Ocean Blvd.
Naples, FL 33942
813-643-0028
SAMS

Eugene Sipe, Jr.
Nautical Services Technologies
424 Production Blvd., #70
Naples, FL 33940
813-434-7445
SAMS

Veronica Lawson
Veronica M. Lawson & Associates
PO Box 1201
Naples, FL 33939
813-434-6960
NAMS

Vikki Hughes
Marine Surveyor
236 Polk Place
Naples, FL 33942
813-643-5101
SAMS

Bobby Crawford
Professional Marine Surveys
6823 Tidewater Dr.
Navarre, FL 32566
904-939-1848
SAMS

Richard Jacobs
Gulf Coast Yacht Service
3444 Marinatown Lane, #8
North Ft. Myers, FL 33903
813-997-8822
SAMS

Richard Koogle
Marine Surveyor
5849 Millay Ct.
North Ft. Myers, FL 33903
813-997-5146
SAMS

Ronald Silvera
R.E. Silvera & Associates
1904 SW 86th Ave.
North Lauderdale, FL 33068
305-720-8660
SAMS, SNAME, ABYC

Tom Drennan
Continental Marine Consultants, Inc.
700 North U.S. Highway 1
North Palm Beach, FL 33408
305-844-6111
NAMS

Tony Uselis
Maritime Yachting Services
30 Turtle Creek Circle
Oldsmar, FL 34677
813-789-4226
SAMS

Jerry Wheeler
Wheeler Marine Surveying
60 Canterbury Court
Orange Park, FL 32065
904-269-2171
SAMS

Charles Akers
Marine Surveyor
2816 Ahern Dr.
Orlando, FL 32817
407-658-0622
SAMS

William Streeter
B&S Marine Inc.
PO Box 690082
Orlando, FL 32869-0082

Russell Thomas
Thomas Marine Surveyors
737 Bywood Dr., NE
Palm Bay, FL 32905
800-352-6287
NAMS, SAMS, SNAME

Marine Surveyor Directory

Richard Thompson
Lakes/Coastal Marine Surveys
235 E. Tall Oaks Circle
Palm Beach Gardens, FL 33410
407-622-9283
SAMS

William Slattum
The Marine Surveyors
3450 Northlake Blvd., #207
Palm Beach Gardens, FL 33403
407-627-4639
SAMS

David Wyman
Marine Surveyor
798 Wood Ave.
Panama City, FL 32401
904-769-6280
SAMS

Doug Wagner
Wagner & Associates
7231 LaFitte Reef
Pensacola, FL 32507
904-492-3475
SAMS

Eugene Briggs
Gene Briggs & Associates
505 Decatur Ave.
Pensacola, FL 32507
904-456-4968
SAMS

Richard Everett
Marine Surveyor
PO Box 13512
Pensacola, FL 32591
904-435-9026
SAMS, ABYC, NFPA

Elaine Miranda
Marine Surveyor
9400 Mainlands Blvd. W.
Pinellas Park, FL 34666
813-577-4128
SAMS

Robert Zimmerman
Integrity Marine Surveying
PO Box 543
Placida, FL 33946
813-697-4799
SAMS

Arthur Buchman, Jr.
Marine Surveyor
12118 Chancellor Blvd.
Port Charlotte, FL 33953
813-743-2198

Pat Guckian
Aquarius Marine Systems
160 SE Duxbury Ave.
Port St. Lucie, FL34983
407-871-0364
SAMS

Adrian Volney
Ardian J. Volney & Co.
5806 Whistlewood Circle
Sarasota, FL 34232
813-371-8781
SAMS

Roy Bowen
Bowen Enterprises
1114 Sylvan Dr.
Sarasota, FL 34234
813-350-3123
SAMS

Roy Pesta
Darling & Co., Marine Surveyors
6336 Brentwood Ave.
Sarasota, FL 34231
813-922-5341
SAMS

William Willien
W.F. Willien Associates
15 Crossroads, Suite 250
Sarasota, FL 34239
813-951-6138
SAMS

Gary Flack
Flack Marine Survey Service
9276 Elm Circle
Seminole, FL 34646
813-398-2267
SAMS

Chris Ramsdell
Marine Consultants, Inc.
13060 Gandy Blvd.
St. Petersburg, FL 33702
813-577-2033
SAMS

Alvin Kushner
Yacht Services
275 SE Salerno Rd.
Stuart, FL 34997
407-286-7961
SAMS

Charles Corder
Chapman School of Seamanship
4343 SE St. Lucie Blvd.
Stuart, FL 34997
407-283-8130
SAMS

Douglas Newbigin
Stuart Yacht Design
450 SW Salerno Rd.
Stuart, FL 34997
407-283-1947
SAMS

Marty Merolla
Independent Marine Surveyor
4300 SE St. Lucie Blvd., #128
Stuart, FL 34997
407-286-4880
NAMS

T. Richard Garlington
Garlington Marine Services
1083 SE St. Lucie Blvd.
Stuart, FL 34996
407-283-5102
SAMS

Tom Fexas
Tom Fexas Yacht Design
333 Tressler Dr., Suites B&C
Stuart, FL 34994
407-287-6558
NAMS

James Garrett
Garrett Marine Survey & Consultants
PO Box 333
Summerland Key, FL 33042
305-745-9989
SAMS

John Reynolds
Jack Reynolds International
11172 NW 35th St.
Sunrise, FL 33351
800-833-9698
SAMS

Sidney Kaufman
Surfside Harbor Associates
PO Box 54-6514
Surfside, FL 33154
305-358-1414
ABYC

Capt. E. Bay Hansen
Capt. E. Bay Hansen, Inc.
1302 N. 19th St., #101
Tampa, FL 33605
813-248-6897
NAMS

Charles Harden
Harden Marine Associates, Inc.
P.O. Box 13256
Tampa, FL 33681-3256
813-254-4273
SAMS

Henry Pickersgill
Henry W. Pickersgill & Co., Inc.
4118 W. Euclid Ave.
Tampa, FL 33629
800-348-8105
NAMS

Robert Buckles
USCG Marine Safety Office
155 Columbia Dr.
Tampa, FL 33606-3598
813-228-2196

Melvin Allen
Allen's Boat Surveying & Consulting
638 North U.S. Hwy. 1, Suite 207
Tequesta, FL 33469-2397
407-747-2433
SAMS, ABYC, NFPA

Marine Surveyor Directory

Omar Sultan
Maro Marine
610 Cheney Highway
Titusville, FL 32780
407-268-2655
SAMS

Richard Fortin
Marine Surveyor & Consultant
1405 19th St., SW
Vero Beach, FL 32962
407-567-9286
ABYC, SNAME

Capt. A.T. Kyle
Marine Consultants & Surveyors
6428 Heather Way
West Palm Beach, FL 33406
407-964-6189
SAMS

Robert Despres
Despres & Associates
332 Pine St.
West Palm Beach, FL 33407
407-820-9290
SAMS

GEORGIA

John Woodside III
Woodside Surveys
15 Howell Mill Plantation
Atlanta, GA 30327
404-355-3732
SAMS

Wilbur Wennersten
Wentek Associates
2137 Tully Wren
Marietta, GA 30066
404-516-5623
SAMS

HAWAII

Dennis Smith
Marine Surveyors & Consultants
677 Ala Moana Blvd., Ste. 812
Honolulu, HI 96813
808-545-1333
SAMS

E.H. "Chip" Gunther
All Ship & Cargo Surveys, Ltd.
965-A2 Nimitz Highway
Honolulu, HI 96817
808-538-3260
NAMS

John Mihlbauer
All Ship & Cargo Surveys, Ltd.
965-A2 Nimitz Highway
Honolulu, HI 96817
808-538-3260
NAMS

Michael Doyle
Mike Doyle, Ltd.
575 Cooke St., #B
Honolulu, HI 96813
808-521-9881
NAMS

IDAHO

Kirk Marshall
Accurate Marine Surveys
1906 North 9th Street
Coeur D'Alene, ID 83814
208-667-2610
SAMS

ILLINOIS

John Boltz
Inland Surveyors, Inc.
307 N. Michigan Ave., #1008
Chicago, IL 60601
312-329-9881
NAMS

Lee H. Asbridge
Marine Surveyor
480 N. McClurg Ct., #1002
Chicago, IL 60611
312-527-4860
SAMS

James Singer
Marine Surveyor
1854 York Lane
Highland Park, IL 60035
708-831-9157
SAMS, ABYC

Kenneth Martin
Professional Development Assoc.
PO Box 712
Mt. Prospect, IL 60056
708-476-7321
SAMS

Paul Petersen
Great Lakes Marine Surveys
710 E. Camp McDonald Rd.
Prospect Heights, IL 60070
708-253-3102
SAMS

INDIANA

Tim Kleihege
Great Lakes Marine Surveying, Inc.
2831 Lakewood Trail
Porter Beach, IN 46304
312-663-2503
SAMS, ABYC

Robert Craig
Marine Surveyor
323 West Main St.
Richmond, IN 47374-4161
317-966-9807
SAMS

KENTUCKY

Jim Hill
RR 1, Box 306
Gilbertsville, KY 42044-9801

Gregory Weeter
Riverlands Marine Surveyors
935 Riverside Dr.
Louisville, KY 40207
502-897-9900
NAMS, ABYC

Robert Urso
Urso Marine Surveying
PO Box 765
Prospect, KY 40059-0765
502-426-3997
SAMS, ABYC

LOUISIANA

Brendan O'Connor
Celtic Marine Corp.
357 Dunstan Circle
Baton Rouge, LA 70815
504-275-5320
SAMS

Kenneth Firm
Celtic Marine Corp.
357 Dunstan Circle
Baton Rouge, LA 70815
504-275-5320
SAMS

Michael O'Connor
Celtic Marine Corp.
357 Dunstan Circle
Baton Rouge, LA 70815
504-275-5320
SAMS

Roger Cheek
Celtic Marine Corp.
357 Dunstan Circle
Baton Rouge, LA 70815
504-275-5320
SAMS

John Illg
Summit Design Services
6444 Jefferson Hwy.
Harahan, LA 70123
504-737-3267
NAMS

Frank Basile
Entech & Associates
PO Box 1470
Houma, LA 70361
504-868-5524
SAMS

Cesar Lurati
Argos Marine Surveyors
PO Box 640191
Kenner, LA 70062
504-466-7333
SAMS

Marine Surveyor Directory

Alfred Cutno
Celtic Marine Corp.
1532 Natchez Lane
LaPlace, LA 70068
504-275-5320
SAMS

Curtley Boudreaux
Marine Surveyor
PO Box 321
Lockport, LA 70374
504-787-2391
SAMS

Chander Gorowara
Maritech Commercial, Inc.
4605 Alexander Dr.
Metairie, LA 70003-2809
504-455-7372
NAMS, SAMS

Pete Peters
Bachrach & Wood/Peters Assoc.
PO Box 7415
Metairie, LA 70010-7415
504-454-0001
SAMS

Andre Chauvin
Chauvin & Associates
PO Box 788
Morgan City, LA 70381
504-385-1043
SAMS

J. Anthony Brown
A.B. Marine Consulting
1397 E. Stephensville Rd.
Morgan City, LA 70380
504-384-5184
SAMS

Douglass Westgate
Westocean Marine, Inc.
PO Box 57446
New Orleans, LA 70157
504-895-7388
SAMS

Hjalmer Breit
Breit Marine Surveying
1311 Leonidas St.
New Orleans, LA 70118
504-866-1814
NAMS, SAMS

Sewell "Si" Williams
Arthur H. Terry & Co.
101 W. Robt. E. Lee Blvd., #200
New Orleans, LA 70124
504-283-1514
NAMS, SAMS, ASA, SNAME

Hubert Gallagher
Marine Surveyor
52246 Highway 90, Apt. 2
Slidell, LA 70461
504-641-2921

MAINE

Marvin Curtis
Marine Surveyor
HC 64, Box 355
Blue Hill, ME 04614
207-374-5342
SAMS

Bob Cartwright
North American Marine Surveying, Ltd.
PO Box 205
Boothbay, ME 04537
207-633-5062
NAMS, SNAME

Jeffrey Johnson
Marine Surveyor
PO Box 1305, Suite 130
Brunswick, ME 04011
207-729-6711
SAMS

Malcolm Harriman
Marine Surveyor
8 Country Club Rd.
Manchester, ME 04351
207-622-2049
SAMS

Jesus Artiaga
North Atlantic Marine
65 W. Commercial St.
Portland, ME 04530
207-775-7317
SAMS

William Leavitt
Chase, Leavitt & Company
10 Dana St.
Portland, ME 04112
207-772-3751
NAMS

MARYLAND

Ernie Leeger
Independent Marine Surveyor
2506 Buckingham Court
Abingdon, MD 21009
410-515-0155
SAMS, ABYC

Hartoft Marine Survey
PO, Box 3188
Annapolis, MD 21403
410-263-3609
ABYC, NAMS, MTAM

C. Robert Skord, Jr.
Skord & Company
400 Forest Beach Rd.
Annapolis, MD 21401
410-757-7454
NAMS, SAMS

Clyde Eaton
Clyde Eaton & Assoc., Inc.
PO Box 4609
Annapolis, MD 21403
800-347-7331
ABYC, NFPA

Frederick Hecklinger
Frederick E. Hecklinger, Inc.
17 Hull Ave.
Annapolis, MD 21403
410-268-3018
NAMS

Michael Kaufman III
Kaufman Design, Inc.
222 Severn Ave., Box 4219
Annapolis, MD 21403
410-263-8900
NAMS

Patricia Kearns
Marine Associates
PO Box 3441, 2 Leeward Ct.
Annapolis, MD 21403
410-263-2419
NAMS, SNAME

Terence Fitzsimmons
Kaufman Design, Inc.
222 Severn Ave., Box 4219
Annapolis, MD 21403
410-263-8900
NAMS

Anthony Eversmier
Marine Surveyor
8110 Woodhaven Rd.
Baltimore, MD 21237
410-391-4200
SAMS

Michael Wright
KIS Marine
5830 Hudson Wharf Rd.
Cambridge, MD 21613
410-228-1448
SAMS

Don Miller
Beacon Marine Surveys
2916 Cox Neck Rd. E.
Chester, MD 21619
410-643-4390
SAMS

Roy Beers
Full Circle Marine Surveyors
PO Box 835
Chestertown, MD 21620
410-778-0247
SAMS

Woodrow Loller
Woodrow W. Loller, Inc.
204 Washington Ave.
Chestertown, MD 21620
410-778-5357
NAMS, ABYC

Marine Surveyor Directory

Steve Sanders
East Coast Surveying
6375 Genoa Rd.
Dunkirk, MD 20754
410-257-3134
SAMS

John Griffiths
John R. Griffiths, Inc.
785 Knight Island Rd.
Earleville (Eastern Shore), MD 21919
410-275-8750
NAMS, ABYC

Kenneth Henry
McHenry Marine Services
38 Oak Hill Lane
Elkton, MD 21921
410-287-2028
SAMS

Charles Wilson
C.R. Wilson, Inc.
3912 Earon Drive
Jarrettsville, MD 21084-1314
410-692-6718
SAMS

Catherine C. McLaughlin
Marine Surveyor
29142 Belchester Rd.
Kennedyville, MD 21645
410-348-5188
SAMS

Marvin Dawson
Chesapeake Marine Surveyors
PO Box 322
Mayo, MD 21106-0322
301-798-5077
NAMS, NFPA

Rick Hall
Marine Surveyor
272 Hance Point Rd.
North East, MD 21901
410-287-2516
SAMS

Thomas Lucke
Oxford Marine Survey
4383 Holly Harbor Rd.
Oxford, MD 21654
410-226-5616
NAMS, SAMS

Marc Cruder
Marine Surveyor
514 Heavitree Garth
Severna Park, MD 21146
202-267-1055
SAMS

Harry Langley
Marine Surveyor
PO Box 220
Solomons, MD 20688
410-326-2001
SAMS

William Thomte
Atlantic Marine Surveyors
PO Box 299
St. Michaels, MD 21663
410-745-3080
NAMS, SNAME

MASSACHUSETTS

Edwin Boice
Robert N. Kershaw, Inc.
25 Garden Park, Box 285
Braintree, MA 02184
617-843-4550
SAMS

Robert Kershaw
Robert N. Kershaw, Inc.
PO Box 285
Braintree, MA 02184
617-843-4550
NAMS, SAMS

Allen Perry
Ocean Adventures
419 Sippewissett Rd.
Falmouth, MA 02540
508-540-5395
SAMS, ABYC, NAMS

Tom Hill
Atlantic & Pacific Marine Surveyors
27 Ferry St.
Gloucester, MA 01930
508-283-7006
SAMS, ABYC, NFPA

Norman Schreiber II
Transtech—Marine Division
140 Wendward Way
Hyannis, MA 02601
508-776-1670
NAMS, SNAME, ABYC, NFPA

Ralph Merrill
Certified Marine Surveyors
48 19th Street
Lowell, MA 01850
508-459-3082
ABYC

William Kirby
Coastal Associates
51 Bowman St.
Malden, MA 02148
617-322-5458
SAMS

Joseph Lombardi
Manchester Yacht Survey
PO Box 1576
Manchester, MA 01944
508-526-1894
SAMS

Capt. Guilford Full
Capt. G.W. Full & Associates
46 Cedar St.
Marblehead, MA 01945
617-631-4902
NAMS

Arnold Cestari
Metropolitan Property & Casualty
PO Box 821
Mattapoisett, MA 02739
800-634-9740
SAMS

Donald Walwer
D&G Marine Company
PO Box 635
North Eastham, MA 02651
508-255-2406
SAMS

Chris Leahy
Leahy Associates
PO Box 6313
North Plymouth, MA 02362
508-846-1725
SAMS

Donald Linde
Marine Surveyor
Old Centre St.
Pembroke, MA 02359
617-294-1919
SAMS

Capt. Norman LeBlanc
Marine Surveyor
23 Congress St.
Salem, MA 01970
508-744-8289
SAMS

Donald Pray
Marine Surveyor
91 Blanchard Rd.
South Weymouth, MA 02190
617-335-3033
SAMS, ABYC, NFPA

Morris Johnson
Marine Surveyor
PO Box 531
West Yarmouth, MA 02673
508-771-8054
SAMS

Wayne Robinson
Admiralty Consulting & Surveying
50 Dunster Lane
Winchester, MA 01890
617-721-7307
SAMS

Marine Surveyor Directory

MICHIGAN

Jeff Amesbury
Independant Marine Surveyor
213 Franklin St.
Boyne City, MI 49712
616-582-7329

Harry Canoles
Marine Appraisal Survey Service, Inc.
4888 Sherman Church Ave. SW
Canton, MI 44706-3966
216-484-0144

John M. Dionne
South Arm International
508 North Lake St.
East Jordan, MI 49727
616-536-7343
SAMS, ABYC

Jim Cukrowicz
Personal Marine Services
52671 CR 388
Grand Junction, MI 49056
616-434-6396
SAMS

Steve Zinner
Zinner Marine Services
38400 Elmite
Mt. Clemens, MI 48045
313-465-4898
SAMS

Capt. A. John Lobbezoo
Great Lakes Marine Surveyors, Inc.
Box 466, 16100 Highland Dr.
Spring Lake, MI 49456-0466
616-842-9400
SAMS, ABYC, NFPA

Donald Rzeppa
Rzeppa Brothers, Inc.
22418 LaVon Rd.
St. Clair Shores, MI 48081
313-778-0123
SAMS

Michael Koch
R.M. Jay Marine
3665 E. 11 Mile Rd.
Warren, MI 48092
313-573-2563
SAMS

MINNESOTA

Paul Liedl
Croix Marine Consultants
531 Mariner Dr.
Bayport, MN 55003
612-439-7748
SAMS, ABYC

John Rantala, Jr.
Rantala Marine Surveys & Services
1671 10th Ave, #2
Newport, MN 55055
612-458-5842
SAMS

A. William Fredell
Marine Surveyor
408 Quarry Lane
Stillwater, MN 55082
612-439-5795
SAMS

MISSISSIPPI

Rush Andre, AMS
Marine Surveyor & Consultant
414 McGuire Circle
Gulfport, MS 39507
601-863-5962
SAMS, ABYC, NFPA, NAMI

Clarence Hamilton
Hamilton, Inc.
PO Box 378
Ocean Springs, MS 39564
601-875-5800
MSPG, IAAI, ASA

Robert Payne
Marine Management, Inc.
PO Box 1803
Ocean Springs, MS 39564
601-872-2846
NAMS, ABYC

MISSOURI

Richard Thompson
Gay & Taylor—THK Marine & Aviation
1721 W. Elfindale St., Ste. 109
Springfield, MO 65807-8400
417-883-7053
SAMS

NEW HAMPSHIRE

Gerald Poliskey
Independent Marine Surveyors & Adjusters
819 2nd St., #B281
Manchester, NH 03102
603-644-4545
NAMS, SAMS

Capt. David Page
Associated Marine Services
2456 LaFayette Rd.
Portsmouth, NH 03801
603-433-1568; 603-431-6150
SAMS, ABYC, NAMI, NFPA

Patrick Enright
Vessel Management Group
PO Box 6579
Portsmouth, NH 03802
603-433-8914
SAMS

NEW JERSEY

John Klose
Bayview Associates
PO Box 368
Barnegat Light, NJ 08006
609-494-7450
SAMS

Dennis Kelly
Teal Yacht Services
668 Main Ave.
Bay Head, NJ 08742
908-295-8225
SAMS

A. William Gross III
Mid Atlantic Marine Consulting
39 Waterford/Blue Anchor Rd.
Blue Anchor, NJ 08037
609-694-6099
SAMS

Richard Thompson
R. T. Marine Associates
275 Shepherd Avenue
Bound Brook, NJ 08805
908-563-0615
SAMS

Robert Gibble
Robert Gibble, Inc.
25 Black Oak Dr.
Ocean View, NJ 08230
609-390-3708
NAMS, SAMS

William Campbell
W.J. Campbell, Marine Surveyor
9 Gate Rd.
Tabernacle, NJ 08088
609-268-7476
NAMS, SNAME

Terry Randolph
East Coast Marine Surveying
126 Emerald Ave.
West Cape May, NJ 08204
609-884-4668
SAMS

NEW YORK

Walter Lawrence
Lawrence Marine Services
PO Box 219
Alton, NY 14413
315-483-6680
SAMS

John Robertson
Fire Traders, Inc.
One Washington Place
Amityville, NY 11701
516-598-2824
NAMS

Marine Surveyor Directory

Joseph Connelly
All Points Marine
417 E. 2nd Street
Brooklyn, NY 11218-3905
718-851-0736

Thomas Crowley
Upstate Marine Consultants, Inc.
8840 New Country Dr.
Cicero, NY 13039
315-699-0024
ABYC, NFPA

Frederic Hamburg
Marine Surveyor
65 Buckley Street
City Island, NY 10464
212-885-1866
SAMS

Marine Surveyors Bureau
30 S. Ocean Ave.
Freeport, NY 11520
516-683-1199
NAMS, SAMS, ABYC, NFPA

Manuel Rebelo
Keel to Rafter
PO Box 1025
Greenwood Lake, NY 10925
914-477-9422
SAMS

Gerald Van Wart
Marine Surveyor
PO Box 795
Hampton Bays, NY 11946
516-728-5706
SAMS

Steve Maddick
Clyde Eaton & Assoc.
PO Box 796
Hampton Bays, NY 11946-0701
516-728-7970
ABYC, NFPA

Chris Garvey
Garvey & Scott Marine
15 Trail Rd.
Hampton Bays, L.I., NY 11946
516-728-5429
SAMS

Shawn Bartnett
Bartnett Marine Services, Inc.
52 Ontario St.
Honeoye Falls (Rochester), NY 14472
716-624-1380
NAMS, ABYC, NFPA, SNAME

Edward Viola
Edward J. Viola Marine Surveying
PO Box 430
Mattituck, NY 11952
516-298-9518
NAMS

Arnold Gaba
Gaba Marine Survey
PO Box 727
Merrick, NY 11566
516-868-1266
SAMS

Melvin Black
Black Marine Enterprise
753 Webster Ave.
New Rochelle, NY 10804
914-633-5499
SAMS

David McClay
Quality Boat Carpentry
57 Maple Ave.
Northport, NY 11768
516-757-9415
SAMS

Paul Robinson
Marifax Marine Services
21 Swanview Dr.
Patchogu, NY 11772
516-654-3300
SAMS, ABYC

Capt. Henry Olsen
Marine Surveyor
PO Box 283
Port Jefferson, NY 11777
516-928-0711
SNAME

Victor Baum
Marine Surveyor
1540 Middle Neck Rd.
Port Washington, NY 11050
516-358-3489
SAMS

Jerry Masters, Jr.
Marine Surveyor
PO Box 727
Poughkeepsie, NY 12602
800-982-6466
SAMS

Herbert Andrews
Marine Surveyor
308 Riverheights Circle
Rochester, NY 14612
716-663-2342
SAMS

James Gambino
Marine Surveyor
66 Browns Blvd.
Ronkonkoma, NY 11779
516-588-5308
SAMS, ABYC, NFPA, NAMI

Gerald LaMarque
LaMarque Marine Services
6 Red Oak Dr.
Rye, NY 10580
914-967-7731
NAMS, ABYC, NFPA

Long Island Marine Surveyor, Inc.
PO Box 542
Sayville, NY 11782
516-589-6154
ABYC, NFPA

Donald Cunningham
McGroder Marine Surveyors
Box 405, 228 Central Ave.
Silver Creek, NY 14136
716934-7848
NAMS, ABYC, NFPA

John Fitzgibbon
McGroder Marine Surveyors
Box 405, 228 Central Ave.
Silver Creek, NY 14136
716-934-7848
NAMS, ABYC, NFPA

Edwin Fleming
Marine Surveyor
21 Coolidge Ave.
Spencerport, NY 14559
716-352-8832
SAMS

Joseph Gaigal
Suffolk Marine Surveying
RFD 1, Box 174G
St. James, NY 11780
516-584-6297
SAMS, ABYC, NFPA, NAMI

William Foster
Marine Surveyor
185 Harrison Place
Staten Island, NY 10310
718-816-0588
SAMS

James Olsen
Marine Surveyor
101 Atlantic Ave.
West Sayville, NY 11796
516-563-8160
SAMS

William Matthews
Admiralty Marine Surveyors & Adjusters
PO Box 183
Westhampton, NY 11977-0183
516-288-3263
NAMS, SNAME. ABYC

Capt. Jim Dias, CMS
Marine Surveyors Bureau
221 Central Ave.
White Plains, NY 10606
914-684-9889
NAMS, SAMS, ABYC, NFPA

Kenneth Weinbrecht
Ocean Bay Marine Services
PO Box 668
Yaphank, NY 11980
516-924-4362
SAMS

Marine Surveyor Directory

NORTH CAROLINA

Carl Foxworth
Industrial Marine Claims Service
9805 White Cascade Dr.
Charlotte, NC 28269
704-536-7511
SAMS, ABYC, NAMI, NFPA

Michael Burns
Industrial Marine Claims
PO Box 1873
Davidson, NC 28036
704-536-7511
SAMS

Bert Quay
Quay Carolina Marine Surveys
PO Box 809
Oriental, NC 28571
919-249-2275
SAMS

Ron Reeves
Atlantic Maritime Services
PO Box 344
Oriental, NC 28571
919-249-1830
SAMS

W. Thomas Suggs
Marine Surveyor & Consultant
PO Box 400
Oriental, NC 28571
919-249-0374
NAMS, SAMS

Lloyd Moore
Moore Marine Surveying
11516 Hardwick Ct.
Raleigh, NC 27614
919-847-1786
SAMS

T. Fred Wright
M.B. Ward & Son, Inc.
PO Box 3632
Wilmington, NC 28406
919-392-1425
NAMS, ABYC

OHIO

Donald Blum
Neptune Marine Surveys
6603 Gracely Dr.
Cincinnati, OH 45233
513-941-4700
SAMS

George Jeffords
Davis & Company
4367 Rocky River Dr., #5
Cleveland, OH 44135
216-671-5181
SAMS

Ray McLeod
Douglas & McLeod, Inc.
209 River St., Box 398
Grand River, OH 44045
216-352-6156
SAMS, ABYC

Capt. Darrell Walton
West Sister Marine Survey
2260 S. Harris Salem Rd.
Oak Harbor, OH 43449-9339
419-898-1118
SAMS

Leroy Wenger
Wenger Enterprises
526 46th Street
Sandusky, OH 44870
419-626-3103
SAMS

Robert Walsh
Marine Survey Professionals
14532 Pearl Rd., #102
Strongsville, OH 44136
216-572-0866
SAMS

Ted Polgar
Marine Surveyor & Consultant
2745 Pine Knoll Dr.
Toledo, OH 43617
419-841-3600
ABYC

Lawrence Imhoff
Ideal Watercraft
PO Box 8027
Toledo,OH 43605
419-691-1600
SAMS

OREGON

Charles Thompson
Marine Surveyor
450 W. Lexington Ave.
Astoria, OR 97103
503-325-4062
SAMS

Peter Kelleher
Marine Surveyor
450 West Lexington
Astoria, OR 97103
503-325-4062
SAMS

PENNSYLVANIA

William Major
Bristol Yacht Services, Inc.
110 Mill St.
Bristol, PA 19007
215-788-0870
SAMS, ABYC

A.S. Impagliazzo
Chesapeake Marine Services
PO Box 218
New London, PA 19360
215-255-4411
SAMS

Charles Limbruner
Anchor's Away Marine Appraisals
3412 Harrisburg St.
Pittsburgh, PA 15204
412-922-3340
SAMS

PUERTO RICO

Julian Ducat
Octagon Marine Services
PO Box 3209, Old San Juan Station
San Juan, PR 00902-3209
809-722-8785

RHODE ISLAND

Charles Morvillo
Star Marine Surveyors
1700 Smith St.
N. Providence, RI 02911
401-353-1960
SAMS

Robert Daigle
Marine Surveyor
141 Plain Road
North Kingstown, RI 02852
401-295-8061
SAMS

Steve Dolloff
Marine Surveyor
38 Dorr Avenue
Riverside, RI 02915
401-433-4155
SAMS

SOUTH CAROLINA

George Lee
Independent Marine Surveyor
PO Box 30040
Charleston, SC 29417
803-571-2526
SAMS

George Barth
Barth Canvass
755 River Rd.
Columbia, SC 29212
803-781-0031
SAMS

John Peeples
Marine Surveyor
614 Regatta Road
Columbia, SC 29212
803-781-2250
SAMS

Marine Surveyor Directory

Mason Draper
Industrial Marine Claims
Rt. 2, Box 542
Marion, SC 29571
803-423-7624
SAMS

TENNESSEE

David Timpani
Marine Surveyor
PO Box 948
Goodlettsville, TN 37072
615-851-9456
SAMS

John Walker
Walker Marine Services
2845 Lebanon Rd.
Nashville, TN 37214
615-859-2337
SAMS

TEXAS

James Merritt
Tangent Development Co.
1715 Harlequin Run
Austin, TX 78758
512-266-9248
SAMS

Robert Hanson
Gulf Coast Surveyors
PO Box 5267
Beaumont, TX 77726-5267
409-866-4403
NAMS, SAMS

Marc McAllister
McAllister Marine Surveying Co.
PO Box 6375
Corpus Christi, TX 78466-6375
512-992-6633
NAMS, ABYC, NFPA

Peter Davidson
Able Seaman Marine Surveyors
341 Melrose Ave.
Corpus Christi, TX 78404
512-884-7245
SAMS

Richard Frenzel
Dixieland Marine, Inc.
PO Box 2408
Corpus Christi, TX 78403
512-946-5566
SAMS, ABYC

Kurtis Samples
Marine Surveyors of North Texas
3401 St. Johns
Dallas, TX 75205
903-786-6082
SAMS, ABYC, NFPA

J.K. Martens
J.K.M. Consulting
Route 1, Box 674
Dickinson, TX 77539
713-339-1267
SAMS

J.B. Oliveros
J.B. Oliveros, Inc.
127 Marlin St.
Galveston, TX 77550
409-763-3123
NAMS, ABYC

Charles Harrison
Russell Brierly & Assoc.
1712 Mercury Dr.
Houston, TX 77029
713-671-2163
SAMS

Fred Struben
The Dutchman Co.
604 Pebbleshire Dr.
Houston, TX 77062
713-480-7096
NAMS, SNAME, ASA

John Kingston
John L. Kingston & Associates
14425 Torry Chase Blvd., #240
Houston, TX 77014
713-537-7770
NAMS, SAMS

Lee Pearson
Pearson Enterprises
PO Box 301169
Houston, TX 77030-1169
713-622-8802
SAMS, ABYC, SNAME

Robert Cwalenski
Russell Brierly & Assoc.
1712 Mercury Dr.
Houston, TX 77029
713-671-2163
SAMS

Ron Ridgeway
Ridgeway Marine Survey
329 Piper Dr., Box 826
Port Aransas, TX 78373
SAMS

Drake Epple
Perry's Marine Survey Co.
1902 Bayport Blvd., #109
Seabrook, TX 77586
713-474-5273
NAMS, ABYC

Michael Firestone
Newberry & Associates
PO Box 998
Seabrook, TX 77586
713-326-6672
SAMS

Roy Newberry
Newberry & Associates
PO Box 998
Seabrook, TX 77586
713-326-6672
SAMS

Terry Moore
Newberry & Associates
PO Box 998
Seabrook, TX 77586
713-326-6672
SAMS

David Boyd
Marine Surveyor
P.O. Box 1416
Victoria, TX 77901
512-578-2708
SAMS

VIRGINIA

Timothy Warren
Bay Yacht Survey
PO Box 300
Carrollton, VA 23314
804-238-3833
SAMS

Edward Harbour
Harbour Marine Services
217 Silver Maple Dr.
Chesapeake, VA 23320
804-482-9119
SAMS

George Zahn, Jr.
Ware River Associates
Rt. 3, Box 1050
Gloucester, VA 23061
804-693-4329
SAMS, SNAME, ABYC, NFPA

Bill Coker
Entre Nous Marine Services
15 Marina Road
Hampton, VA 23669
804723-2883
ABYC, NFPA, SAMS

Gary Naigle
American Yacht Surveys, Inc.
PO Box 3214
Norfolk, VA 23514
804-622-7859
SAMS

Richard Radius
RHR Computer Services
105 Rens Road, #13
Poquoson, VA 23662
804-868-7355
SAMS

Marine Surveyor Directory

Steven Knox
Knox Marine Consultants
355 Crawford St., #601
Portsmouth, VA 23704
804-393-9788
NAMS

Ralph Brown
Marine Surveyor
11337 Orchard Lane
Reston, VA 22090-4431
703-435-1258
SAMS

Richard Geisel
Hoffman Geisel Surveying
8800 Three Chopt Rd., #301
Richmond, VA 23229
804-257-4140
SAMS

WASHINGTON

Matthew Harris, CMS
Reisner, McEwen & Harris
1333 Lincoln St., #323
Bellingham, WA 98226
206-647-6966
NAMS, SAMS, ABYC

Joe Stevens
Sound Surveyors
9651 South 206th Place
Kent, WA 98031
206-854-4375
SAMS

Steve Belzer
North Latitude Marine
26833 Border Way
Kingston, WA 98346
206-282-8806
SAMS

Kenneth Rider
Rider Associates
338 E. Cascade Place
Oak Harbor, WA 98277
206-675-8475

Barrie Arnett
Arnett & Berg Marine Surveyors
PO Box 70424
Seattle, WA 98107-0424
206-283-8884
SAMS

Carl Anderson
Carl A. Anderson,Inc.
8048 9th Ave. NW
Seattle, WA 98117
206-789-2315
SAMS, ABYC

David Berg
Arnett & Berg Marine Surveyors
PO Box 70424
Seattle, WA 98107-0424
206-283-8884
SAMS

Dennis Johnson
Dennis C. Johnson Marine Surveyor, Inc.
15734 Greenwood Ave. North
Seattle, WA 98133
206-365-6591

Robert McEwen
Reisner, McEwen & Associates
2500 Westlake Avenue North, Suite D
Seattle, WA 98109
206-285-8194
NAMS, ABYC

Ronald Reisner
Reisner, McEwen & Associates
2500 Westlake Ave. N., Suite D
Seattle, WA 98109
206-285-8194
NAMS, SAMS

William L. Hockett
Marine Surveyor
3415 NW 66th
Seattle, WA 98117
206-783-7617
NAMS

Martin Braune
NOR-PAC Marine Surveyors
East 570 Strong Rd.
Shelton, WA 98584
206-426-9118
SAMS

Doug McNeill
Tillikum Marine Services
7305 24th Street West
Tacoma, WA 98466
206-566-0737
SAMS

WISCONSIN

Christopher Kelly
Professional Yacht Services
2132–89th Street
Kenosha, WI 53143
414-694-6603
SAMS

Earl Shaw
Rice Adjustment Company
11422 N. Port Washington Rd.
Milwaukee, WI 53217-0529
414-241-6060
NAMS

CANADA

Ivan Herbert
Universal Marine Consultants, Ltd.
5 Carriageway Ct.
Bedford, Nova Scotia
Canada B4A 3V4
902-835-2283
NAMS

Peter Larkins
Larkins Marine Surveyors
6570 68th St.
Delta, B.C.
Canada V4K 4E2
604-940-1221

Wallace Nisbet
Nisbet Marine Surveyors
345 Lakeshore Rd., #301
Oakville, Ontario
Canada L6J 1J5
416-844-6670
NAMS, ASA, ABYC

Geoffrey Gould
Quality Marine Surveyors, Ltd.
PO Box 1105
Prince Ruppert, B.C.
Canada V8J 4H6
604-624-4138

Kelvin Colbourne
Kelvin Colbourne & Associates
PO Box 24 FP
Washago, Ontario
Canada L0K 2B0
705-689-8820
ABYC

Chris Small
Chris Small Marine Surveyors
15219 Royal Ave.
White Rock, B.C.
Canada V4B 1M4
604-681-8825
NAMS

ABOUT THE AUTHORS

Ed McKnew has been a yacht broker and powerboat design enthusiast for many years. He holds a business degree from Oakland University in Rochester, Michigan, and worked for several years in the solid waste business before becoming a yacht broker in 1977. Moving to the Houston area in 1984, he operated a yacht brokerage office in Clear Lake, Texas, before leaving the business in 1987 to work on the manuscript for the original *PowerBoat Guide.* Ed currently works full time on several publishing projects while spending his spare time pursuing his interest in Civil War literature. He is single and lives in Palm Beach Gardens, Florida.

Mark Parker has been a powerboat enthusiast since before he can remember. A graduate of Southwest Texas State University with a business degree in marketing, Mark also holds a Master's license and has captained several large sportfishing boats. He is a native Texan and worked as a broker in both Texas and Florida for twelve years. Mark currently works full time with American Marine Publishing. He and his wife, Sherri, reside in Palm Beach Gardens, Florida.

ALBIN 33 TRAWLER

SPECIFICATIONS

Length	32'6"	Water	150 gals.
Beam	11'5"	Fuel	330 gals.
Draft	3'7"	Hull Type	Semi-Disp.
Weight	16,800#	Designer	Albin
Clearance	NA	Production	1979–80

Although she enjoyed only a brief production run (1979–80), used Albin 33s seem to show up regularly on the used-boat markets. She's basically a scaled-down version of Albin's popular 36 Trawler—a traditional double-cabin design with a deckhouse galley and two private heads. The profile of the 33 is pleasing in spite of her short LOA. She's built of fiberglass on a semi-displacement hull with a sharp entry and generous flare at the bow, hard aft chines, and a deep keel for directional stability and prop protection. The interiors of Albin 33s were finished with either teak or mahogany. A U-shaped dinette is abaft the lower helm, and there are two well-placed overhead grab-rails in the salon. The galley is on the small side but still adequate for basic food prep. Port and starboard sliding salon doors provide easy access to the (teak) walkaround decks with their raised bulwarks. A solid and efficient family cruiser, the Albin 33 burns about 3 gph at her 7-knot hull speed with a single diesel. ❑

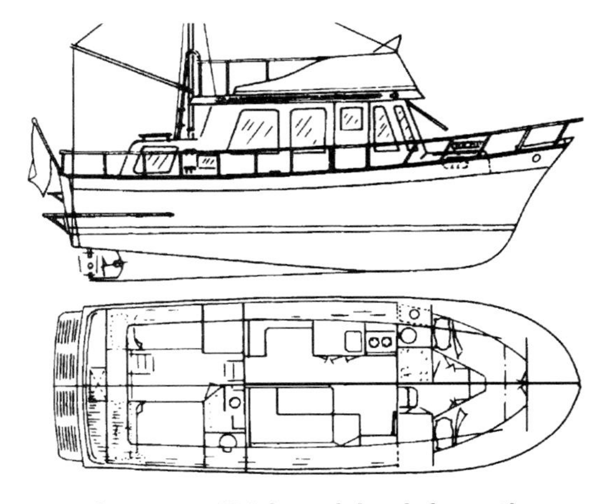

See page 231 for pricing information.

ALBIN 36 TRAWLER

SPECIFICATIONS

Length	35'9"	Clearance	12'4"
LWL	31'3"	Water	220 gals.
Beam	13'2"	Fuel	350 gals.
Draft	3'6"	Hull Type	Semi-Disp.
Weight	18,500#	Production	1978–Current

The Albin 36 remains a popular boat with those who enjoy the look and feel of a traditional trawler-style yacht and the economy of single-screw propulsion. A handsome design, she features a conventional double-cabin layout with a full teak interior. An in-line galley is located in the salon along with a lower helm, deck access door, and an L-shaped convertible settee. Both fore and aft staterooms have private heads, and there's a tub/shower combination aft. Construction is solid fiberglass (although a few early models had glass-over-plywood superstructures), and a full-length skeg protects the running gear. Her flybridge will seat eight and comes with a mast and boom. Finished with an abundance of exterior brightwork, the teak decks, rails, cabintop, doors, and hatches all add much to her appearance but require considerable maintenance. A standard 135-hp Lehman diesel will cruise the Albin 36 Trawler at 7.5 knots for 750–800 miles while burning only 3 gph. Newer models with optional 210-hp Cummins will cruise around 10–11 knots. ❑

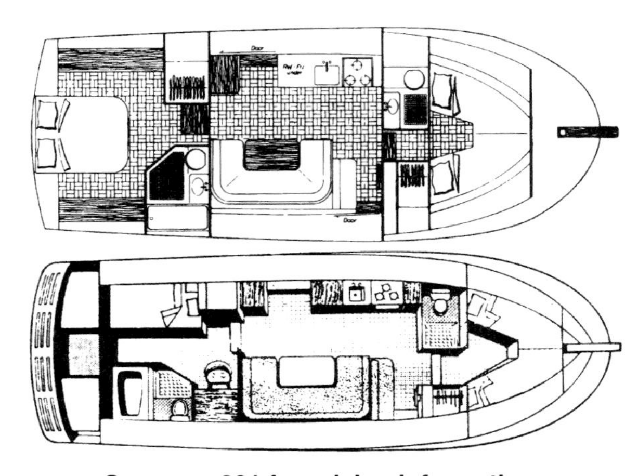

See page 231 for pricing information.

ALBIN 40 TRAWLER

SPECIFICATIONS

Length	39'5"	Clearance	NA
LWL	36'6"	Water	220 gals.
Beam	13'2"	Fuel	400 gals.
Draft	3'6"	Hull Type	Semi-Disp.
Weight	23,500#	Production	1987–Current

The Albin 40 is a traditional trawler-style cruiser with walkaround sidedecks, a simulated lapstrake hull and plenty of exterior teak trim. Construction is solid fiberglass on a semi-displacement hull with moderate beam, a long prop-protecting keel, and hard chines aft. The standard galley-down floorplan of the Albin 40 has remained unchanged since she was introduced in 1987. The salon dimensions are about average for a 40-footer, and the full teak interior is a model of Asian woodworking. A lower helm is standard, and deck access doors are located port and starboard in the salon. The aft cabin has a queen berth on the centerline with a tub/shower in the adjoining head. Outside, the sidedecks are comfortably wide and protected by raised bulwarks all around. A mast and boom are standard. The Albin 40 Trawler has been offered with a variety of engines. Twin 135-hp Lehman diesels will cruise at 7–8 knots (10 knots top), and 210-hp Cummins will cruise at 12 knots (15 top). ❑

See page 231 for pricing information.

ALBIN 40 SUNDECK

SPECIFICATIONS

Length	39'5"	Clearance	NA
LWL	36'6"	Water	220 gals.
Beam	13'2"	Fuel	400 gals.
Draft	3'6"	Hull Type	Semi-Disp.
Weight	23,500#	Production	1987–Current

The Albin 40 Sundeck is a traditional Taiwan-trawler design built on a conventional semi-displacement, hard chine hull with generous flare at the bow, and an attractive sheerline. (Note that her sistership, the Albin 40 Trawler, is the same boat but with a walkaround aft deck and slightly smaller owner's cabin dimensions). Both models share the same roomy two-stateroom interior layout with the galley down and heads fore and aft. The abundant teak woodwork results in a rather dark interior but one most traditionalists will probably find to their liking. The joinerwork is very good throughout. Outside, the flybridge will accommodate six, and there's room on the full-width aft deck for several deck chairs. While a single 135-hp Lehman diesel was once standard, most Albin 40s have been sold with twin 135-hp Lehmans (10 knots cruise/13 knots top) or—more recently—210-hp Cummins (12 knots cruise/15 top). Note that the props and rudders are well protected from grounding by the deep skeg. Moderately priced and inexpensive to operate, the Albin 40 is a good-looking family cruiser with a comfortable ride and good range. ❑

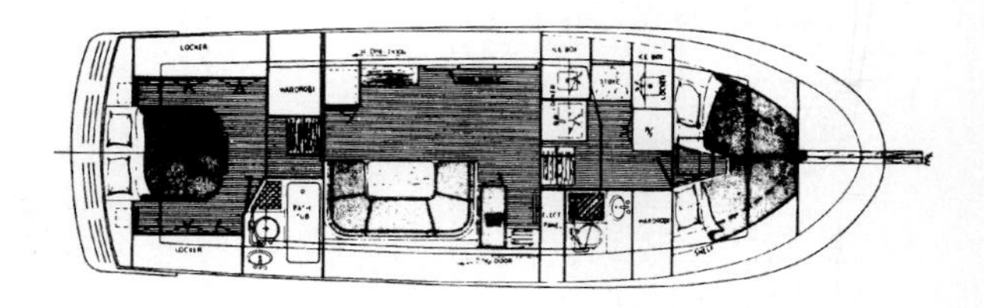

See page 231 for pricing information.

ALBIN 43 TRAWLER

SPECIFICATIONS

Length	42'6"	Clearance	NA
LWL	37'11"	Water	300 gals.
Beam	14'6"	Fuel	500 gals.
Draft	4'1"	Hull Type	Semi-Disp.
Weight	30,000#	Production	1979–Current

Built in Taiwan, the Albin 43 Trawler is a classic trunk cabin design with full walkaround sidedecks and a distinctive trawler profile. Inside, her galley-down layout insures a spacious salon and particularly generous aft cabin accommodations. Storage, interior lighting, helm visibility, and deck access are all excellent. Construction is solid glass, and a deep keel protects the running gear. Underway, the Albin 43 handles like the heavy displacement yacht that she is, with a comfortable ride in most sea conditions. The extensive use of interior and exterior teak woodwork is typical of Asian boats, and in the Albin 43 the joinerwork is impressive indeed. A single diesel was standard until recently, however most Albin 43s have been delivered with twin 120/135-hp Lehman engines (8 knots cruise/11 knots top). Newer models have been fitted with 210-hp Cummins for a 12–13 knot cruising speed. The average cruising range is an impressive 650–700 miles. Long a popular boat, used Albin 43s are commonly found in most East Coast boating markets. ❑

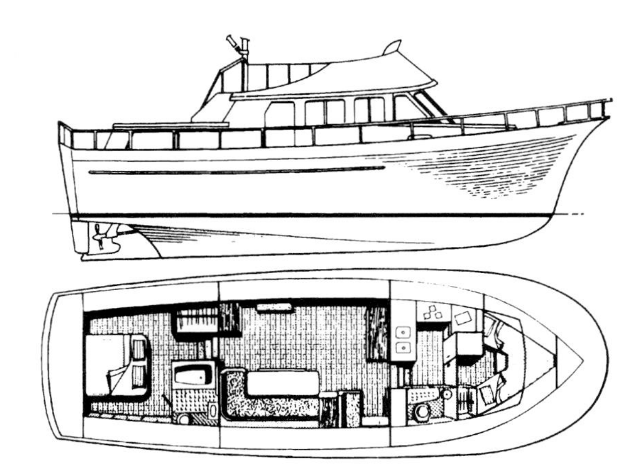

See page 231 for pricing information.

ALBIN 43 SUNDECK

SPECIFICATIONS

Length	42'6"	Clearance	NA
LWL	37'11"	Water	300 gals.
Beam	14'6"	Fuel	500 gals.
Draft	4'1"	Hull Type	Semi-Disp.
Weight	30,000#	Production	1981–Current

Based upon the popular Albin 43 Trawler, the 43 Sundeck features a full-width aft deck (as opposed to the trunk cabin profile of the 43 Trawler) and slightly larger aft cabin dimensions than her sistership. In other respects, the two boats are about the same. The hull is constructed of solid fiberglass, and a deep keel provides grounding protection for the running gear. Albin has offered a choice of three interior floorplans in the 43 Sundeck including a three-stateroom, deckhouse galley arrangement. Well-crafted teak interior paneling and cabinetry abound, and large wraparound windows allow for plenty of natural salon lighting. A lower helm is standard, and deck access doors are fitted to port and starboard in the salon. Unlike most sundeck designs from other manufacturers, the Albin 43 lacks a salon access door from the aft deck—a genuine shortcoming. Twin 135-hp Lehmans will cruise at 8–9 knots, and the now-standard 210-hp Cummins cruise at 12–13 knots. Note that the Albin 49 Cockpit (1979–current) is the same boat with a cockpit extension. ❑

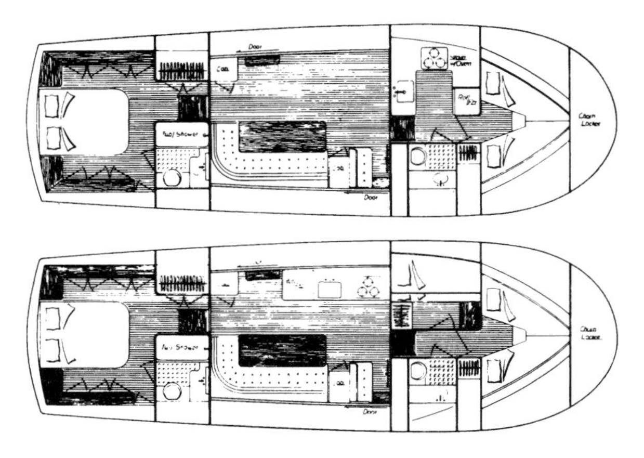

See page 232 for pricing information.

ALBIN 48 CUTTER

SPECIFICATIONS

Length	47'9"	Water	300 gals.
LWL	43'0"	Fuel	600 gals.
Beam	14'0"	Hull Type	Deep-V
Draft	4'0"	Deadrise Aft	24°
Weight	31,000#	Designer	Hunt Assoc.
Clearance	NA	Production	1983–89

Originally called the Albin Palm Beach 48 when she was introduced in 1983, the 48 Cutter features a Hunt-designed deep-V (24° deadrise) Airex cored hull together with a relatively narrow beam—characteristics that can be expected to result in good heavy-weather performance. Albin 48 Cutters were built in Taiwan and came with the full teak interior typical of most Asian boats. Two floorplans were offered, both with cockpit access from the aft cabin. The three-stateroom version has stacked single berths in the guest cabin, V-berths forward, and the galley aft in the salon. The two-stateroom layout features a much larger salon with the galley-down and an offset double berth forward. In both, the owner's cabin has a queen-size berth with a tub/shower in the head. The large cockpit is fitted with a teak sole, transom door, swim platform, and hideaway shower. Volvo 307-hp diesels were standard in later models (17 knots cruise/21 top), and Cat 375-hp diesels were optional (21 knots cruise/25 knots top). ❑

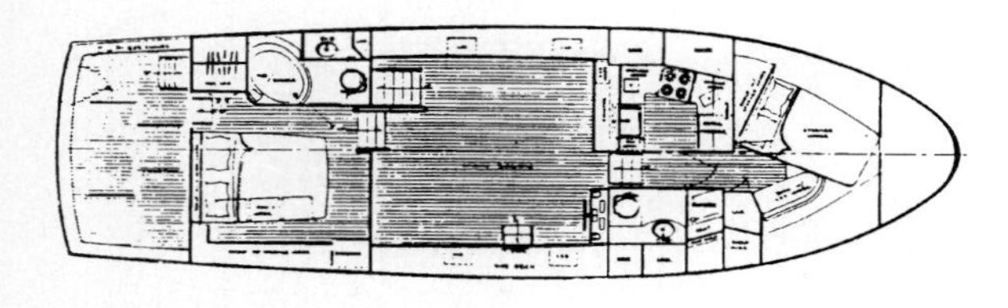

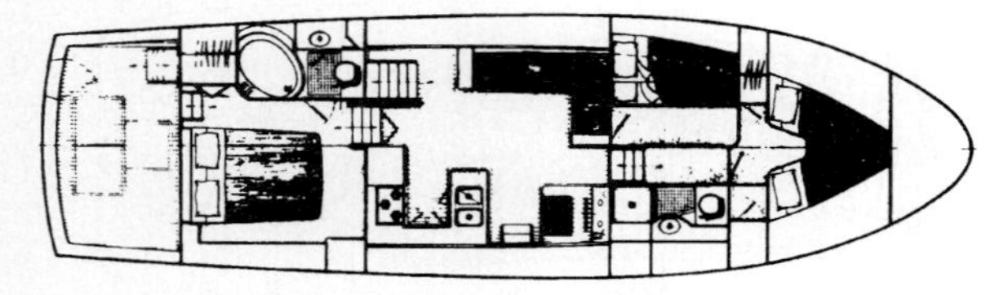

See page 232 for pricing information.

ALBIN 49 COCKPIT TRAWLER

SPECIFICATIONS

Length	48'4"	Water	300 gals.
LWL	43'0"	Fuel	675 gals.
Beam	14'6"	Hull Type	Semi-Disp.
Draft	4'1"	Designer	G. Stadel
Weight	38,500#	Production	1979–Current

Those familiar with Albins will recognize the 49 Cockpit Trawler pictured above as an Albin 43 Sundeck model with a cockpit. Heavily built on a semi-displacement hull with hard chines, flat aftersections, and a full-length keel, the graceful sheer and well-flared bow of the Albin 49 give her a business-like and particularly salty appearance. Inside, the two-stateroom floorplan (finished with teak woodwork and cabinetry throughout) features an oversize master stateroom with excellent storage and a tub in the adjoining head. A lower helm is standard, and there are port and starboard deck doors in the salon. The Albin 49's full-width aft deck is quite spacious, and the cockpit extension allows for extra fuel and storage capacity in addition to providing a platform for swimming and diving activities. Additional features include a hardtop, radar arch, and plenty of guest seating on the bridge. Note that the props are keel-protected. Twin 250-hp Cummins diesels will cruise the Albin 49 Cockpit Trawler efficiently at 10–11 knots and deliver a top speed of about 16 knots. ❑

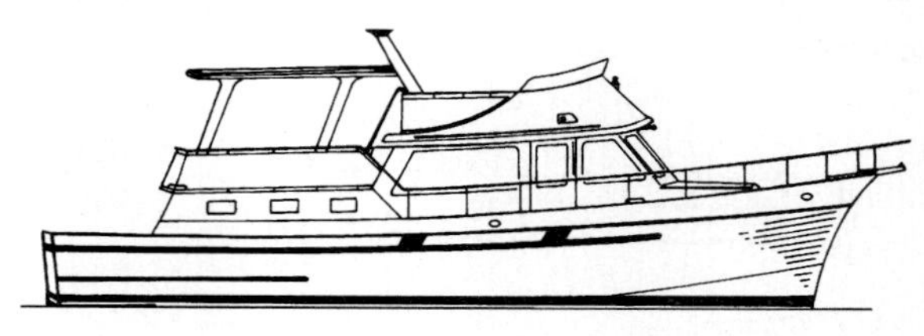

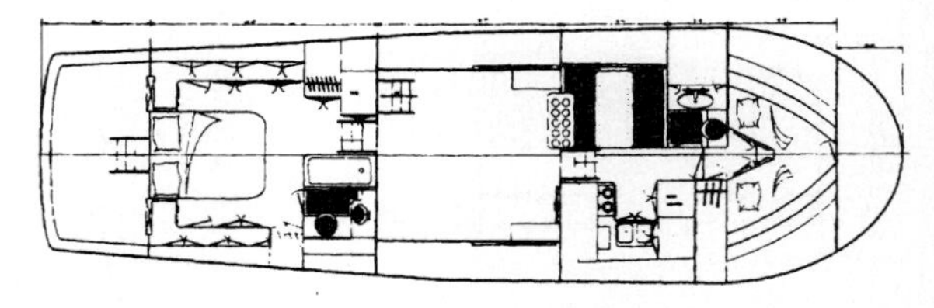

See page 232 for pricing information.

ALBIN 49 TRI CABIN

SPECIFICATIONS

Length	48'4"	Cockpit	NA
LWL	43'0"	Water	320 gals.
Beam	15'1"	Fuel	620 gals.
Draft	3'8"	Hull Type	Semi-Disp.
Weight	39,050#	Designer	G. Stadel
Clearance	13'6"	Production	1979–Current

The Albin 49 Tri Cabin is a heavily built trunk cabin trawler with full walkaround sidedecks and generous interior accommodations. Constructed on a semi-displacement hull with a fine entry, moderate beam, and a relatively deep keel, this is a comfortable and economical long range cruising yacht. While the Albin 49 doesn't have the raised, full-width sundeck of most newer aft-cabin yachts, her full sidedecks, traditional interior, and protective overhangs are practical and attractive cruising features. The "Tri Cabin" designation refers to the number of staterooms: there are three, each with a double berth (a slide-out affair in the forward stateroom). Notably, there are bathtubs in both heads. Access to the flybridge is via a ladder in the wheelhouse. The interior is completely finished with teak paneling and cabinetry, and the joinerwork is excellent throughout. Several engine options have been offered over the years. Among them, twin 120/135-hp Lehman diesels will cruise the Albin 49 Tri Cabin at an economical 8–9 knots with a cruising range of about 800 miles. ❑

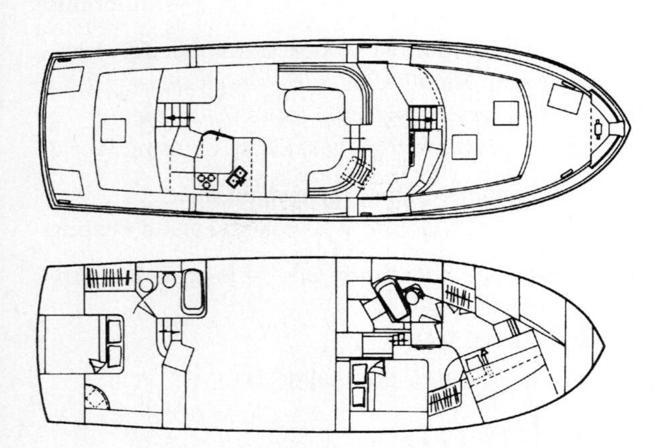

See page 232 for pricing information.

AQUARIUS 41 MOTOR YACHT

SPECIFICATIONS

Length	40'6"	Clearance	NA
LWL	37'0"	Water	200 gals.
Beam	14'10"	Fuel	400 gals.
Draft	3'0"	Hull Type	Modified-V
Weight	26,000#	Production	1987–Current

With her excessively outlined salon windows and prominent reverse arch, the unusual styling characteristics of the Aquarius 41 MY make her an easy boat to spot in a crowd. She's built in Taiwan on a solid fiberglass hull with a wide beam and a relatively long keel. Her two-stateroom, galley-down floorplan features an expansive salon comparable in size to what might be expected in a larger boat. A dinette is opposite the galley, and a built-in entertainment center is to port in the salon. The generous salon dimensions come at a price since both staterooms are a little tight—especially the forward cabin which is probably more useful for storage than adult sleeping quarters. Not surprisingly, teak woodwork is applied extensively throughout the interior. Outside, the aft deck is small with limited entertainment capabilities. The flybridge, however, is quite roomy with plenty of guest seating. Twin Perkins 200-hp (or the more recent Cummins 210-hp) diesels will cruise the Aquarius economically at 12–13 knots. Optional 375-hp Cats will cruise around 21 knots. ❑

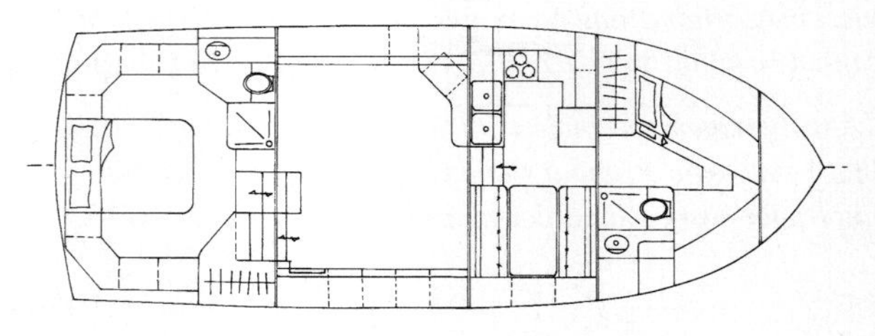

See page 232 for pricing information.

ATLANTIC 37 DOUBLE CABIN

SPECIFICATIONS

Length	36'7"	Water	200 gals.
LWL	33'0"	Fuel	350 gals.
Beam	13'9'	Hull Type	Semi-Disp.
Draft	3'3"	Deadrise Aft	NA
Weight	22,000#	Designer	Jack Hargrave
Clearance	14'0"	Production	1982–92

Boating enthusiasts will recognize the Atlantic 37 Double Cabin as the old Prairie 36 Coastal Cruiser, an attractive Hargrave design built on a semi-displacement hull with soft chines and a full-length keel. Prairie produced the 36 model from 1979 until early 1982 when the company closed. Meanwhile, the 36 had been well-received, and she quickly gained a reputation as a good-quality, long-range family cruiser, albeit at a somewhat hefty price. Following Prairie's demise, Atlantic Yachts acquired the tooling and molds, and reintroduced the boat in 1982 as the Atlantic 37 DC. Prairie's original galley-up floorplan was rearranged in the new model by moving the galley down, adding a stall shower in the forward head, and installing a queen berth in the aft cabin. Note that Atlantic uses more interior teak and fewer laminates than in the previous Prairies. Those powered with twin 120/135-hp Lehman diesels will cruise at up to 10 knots, and the twin 250-hp Cummins (or GM) diesels cruise at 14 knots and reach a top speed of 18 knots. ❑

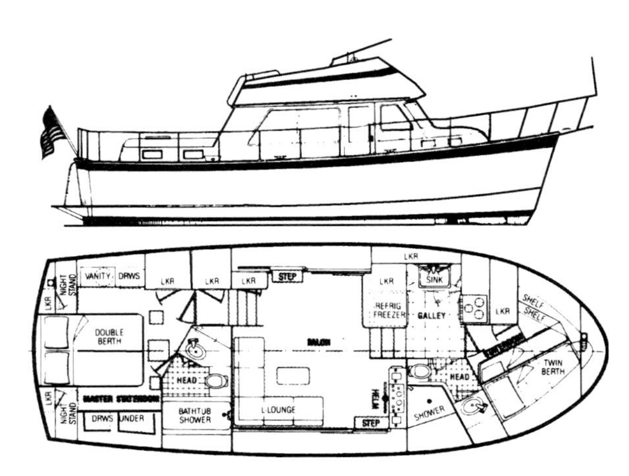

See page 232 for pricing information.

ATLANTIC 44 MOTOR YACHT

SPECIFICATIONS

Length	43'8"	Water	240 gals.
LWL	38'6"	Fuel, Std.	320 gals.
Beam	14'0"	Fuel, Opt.	620 gals.
Draft	3'5"	Hull Type	Modified-V
Weight	30,000#	Designer	Jack Hargrave
Clearance	13'8"	Production	1977–92

The Atlantic 44 MY is a traditional Hargrave-designed cruising yacht with one of the more attractive profiles in her class. A popular model, she remained in production for fifteen years without major alterations to her appearance. Several two-stateroom floorplans were offered over the years. The latest versions had the galley and dinette down a few steps from the salon with a choice of lower helm or entertainment center. (Note that early models have twin single berths in aft stateroom.) The interior is mostly finished with well-crafted teak paneling and cabinetry. Outside, the afterdeck (a full 12' x 14') is fairly large for a 44-foot yacht. Designed with long-range cruising in mind, the Atlantic 44 has been powered with a variety of twin-diesel installations from trawler-speed 135-hp Lehmans up to the more powerful 375-hp Cats. At her 8.3-knot hull speed, she'll cruise economically for 800–900 miles with optional fuel. Planing speeds of 18–19 knots are possible with the Caterpillar diesels. ❑

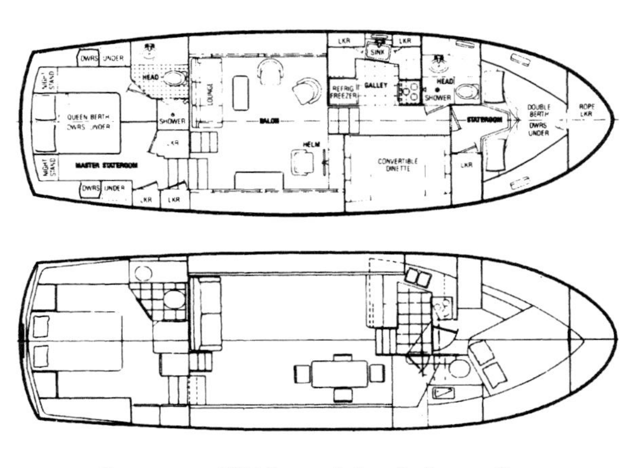

See page 232 for pricing information.

ATLANTIC 47 MOTOR YACHT

SPECIFICATIONS

Length	46'9"	Water	400 gals.
LWL	42'6"	Fuel, Std.	400 gals.
Beam	16'0"	Fuel, Opt.	800 gals.
Draft	3'9"	Hull Type	Modified-V
Weight	41,000#	Designer	Jack Hargrave
Clearance	18'0"	Production	1982–92

The Atlantic 47 MY began life back in 1981 as the Prairie 46 LRC. Atlantic Yachts purchased the tooling in 1982 when Prairie went out of business. Although designed more than a decade ago, the Atlantic 47 remains a stylish and good-looking yacht today. Once aboard, her most striking feature is the huge afterdeck. This is the result of locating two of the three staterooms aft of the salon rather than forward. (The third stateroom is actually a den with a desk, berth, and bookcases.) An inside helm was a popular option. Teak paneling and woodwork are used extensively throughout the interior, and a light oak interior became available in later years. Note that a hard aft enclosure (optional) turns the afterdeck into an air-conditioned, carpeted, and fully paneled second salon. Standard 375-hp Cats cruise the Atlantic 47 MY at a sluggish 14–15 knots. Optional 435-hp 6V71 (or 450-hp 6V92) diesels cruise at 17 knots (20 knots top), and the 550-hp 6V92s will cruise at a steady 19–20 knots. ❑

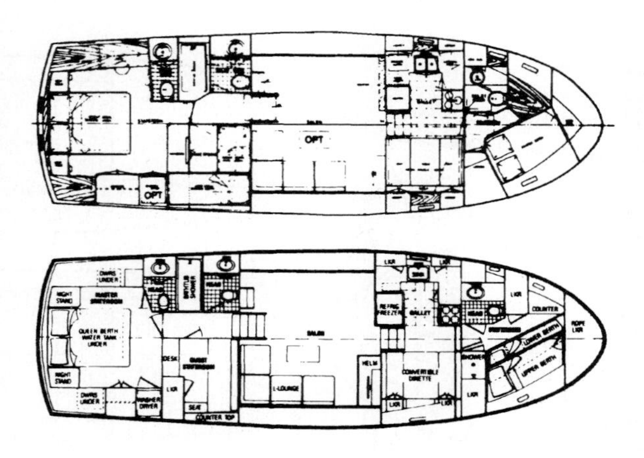

See page 232 for pricing information.

BAYLINER 4387 AFT CABIN MY

SPECIFICATIONS

Length	43'1"	Water	100 gals.
LWL	NA	Fuel	300 gals.
Beam	14'3"	Hull Type	Modified-V
Draft	3'0"	Deadrise Aft	14°
Weight	20,000#	Designer	Bayliner
Clearance	13'6"	Production	1990–93

Probably the most affordable boat in her class, the Bayliner 4387 is a fairly straightforward double cabin cruiser with a modern profile and a practical interior layout. Although the salon dimensions are not notably generous for a 43-footer, the large wraparound cabin windows and light oak interior trim create a surprisingly spacious impression. A lower helm is standard, and the galley and dinette are down from the salon level. Outside, the 4387 features a large flybridge with seating for six, a foredeck sun pad, and an integral transom platform with molded boarding steps. The aft deck comes with a wet bar and hardtop. A light boat for her size (the hull is fully cored and incorporates prop pockets), standard 330-hp gas engines will cruise at 16–17 knots and reach 26 knots top. Optional 250-hp Hino diesels cruise at 18–19 knots and reach about 24 knots top. The fuel capacity is small for a boat this size, especially with gas engines. Note that the current Bayliner 4587 is basically the same boat with a cockpit. ❑

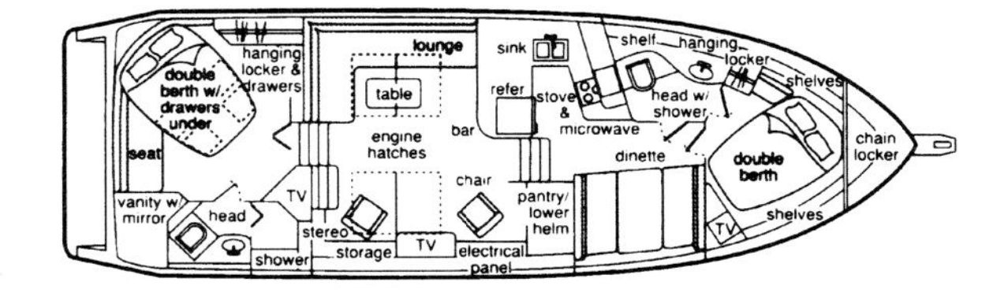

See page 233 for pricing information.

BAYLINER 4587 COCKPIT MY

SPECIFICATIONS

Length	45'1"	Water	100 gals.
Beam	14'3"	Fuel	300 gals.
Draft	3'0"	Hull Type	Modified-V
Weight	22,000#	Deadrise Aft	14°
Clearance	NA	Production	1994–Current

The Bayliner 4587 MY is basically a now-discontinued 4387 Aft Cabin MY with a cockpit extension and a slightly modified bottom. The prop pockets used on the 4387 to reduce shaft angles have been eliminated in the 4587 (the extra hull length makes them unnecessary), and a spray rail has been added at the chine for a drier ride. Aside from the added versatility and convenience provided by a cockpit, the additional hull length adds considerably to the boat's appearance. The two-stateroom, galley-down floorplan is unchanged and features a dinette opposite the galley and a lower helm in the salon. (Note that the woodwork is teak and not light oak of the earlier model.) The forward stateroom in this layout is small, and the adjoining head lacks a stall shower. Outside, the cockpit comes with a transom door, and there's room beneath the cockpit sole for the generator. Standard 250-hp Hino diesels will cruise the Bayliner 4587 at 19–20 knots and reach a top speed of around 26 knots. The fuel capacity is limited. ❑

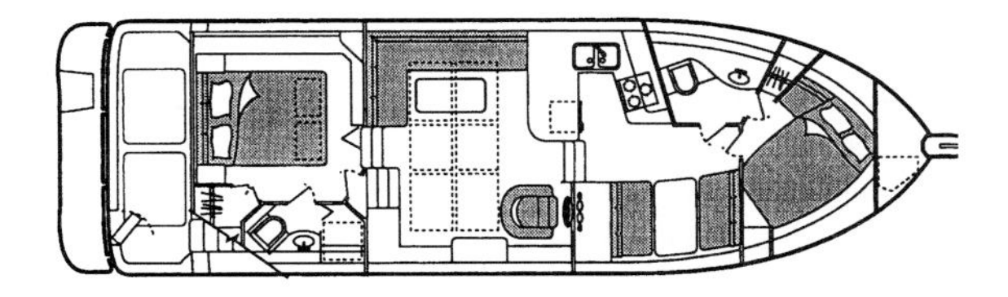

See page 233 for pricing information.

BAYLINER 4588 PILOTHOUSE MY

SPECIFICATIONS

Length	45'4"	Fuel	500 gals.
Beam	14'11"	Hull Type	Modified-V
Draft	3'0"	Deadrise Aft	6°
Weight	28,000#	Designer	Bayliner
Clearance	15'6"	Production	1984–93
Water	200 gals.		

Long the flagship of Bayliner's fleet, this handsome pilothouse yacht has become recognized as one of the best big-boat values in the industry. A careful blend of Pacific Northwest styling and comfortable interior accommodations result in a long-range cruising yacht with instant eye appeal. (Note that she was called the 4550 MY through 1988.) Pilothouse yachts make excellent cruisers, and they offer obvious advantages in cold climates. While she can sleep three couples, this is really a two-couple boat with the third stateroom used as a den/study. Outside, a bridge overhang provides weather protection in the cockpit, and the flybridge will accommodate a dinghy. With standard 220-hp diesels (250-hp since 1991), she'll cruise at 16–17 knots at 1 mpg—excellent economy in a 45-foot yacht. The hull is fully cored to reduce weight, and a sharp entry and 15" keel provide good seakeeping qualities. The engine room is a tight fit. With over 350 built during her production, the 4588 MY has always been considered a lot of boat for not a lot of money. ❑

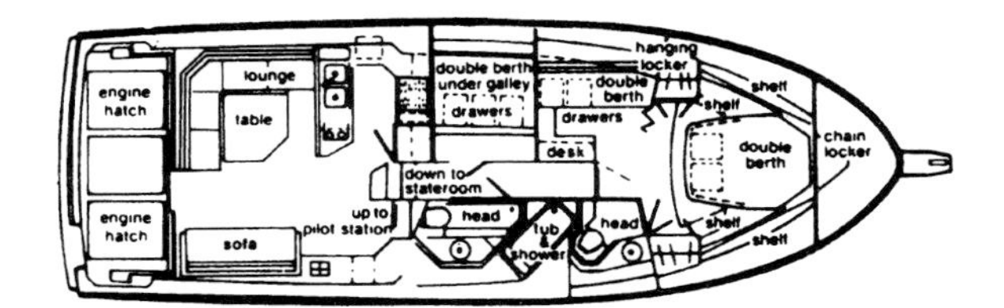

See page 233 for pricing information.

BAYLINER 4788 PILOTHOUSE MY

SPECIFICATIONS

Length	47'4"	Water	200 gals.
Beam	14'11"	Fuel	444 gals.
Draft	3'0"	Hull Type	Modified-V
Weight	30,000#	Deadrise Aft	6°
Clearance	18'2"	Production	1993–Current

The Bayliner 4788 is a slightly restyled and lengthened version of the very popular 4588 Pilothouse model (which many industry professionals still consider one of the best big-boat values in the business). Aside from the additional two feet of hull length, Bayliner designers reversed the radar arch on the new 4788 and softened the flybridge profile for a sleeker and more modern appearance. Inside, the superb pilothouse floorplan of the original 4588 is continued in the new model with few changes. The L-shaped lounge has been moved to starboard, and the increased hull length of the 4788 can be seen in the enlarged galley and salon dimensions. Note that the forward guest stateroom is actually an office with a convertible settee, and the mid-stateroom is arranged with the berth extending below in-galley. Visibility from the raised pilothouse is excellent. On the downside, the engine room is a tight fit, and the overall quality of the boat is marginal. Optional 310-hp Hino diesels will cruise at 17 knots and deliver a top speed of about 21 knots. ❑

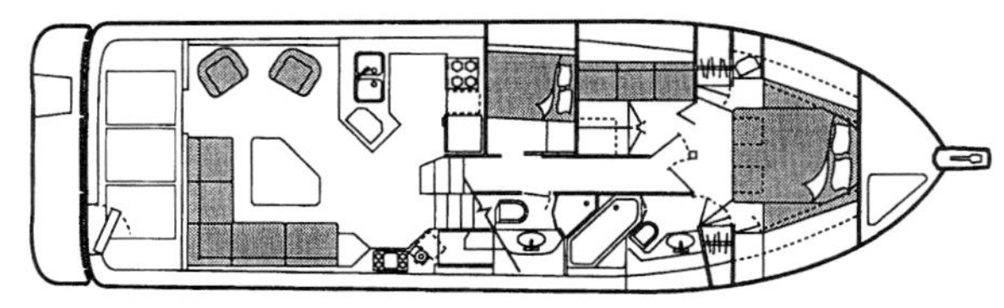

See page 233 for pricing information.

BERTRAM 42 MOTOR YACHT

SPECIFICATIONS

Length	42'6"	Water	150 gals.
Beam	14'10"	Fuel	406 gals.
Draft	4'0"	Hull Type	Deep-V
Weight	39,000#	Deadrise Aft	17°
Clearance	17'11"	Production	1973–87

A distinctive flush deck profile gives the Bertram 42 MY the stately appearance still admired by many motor yacht traditionalists. Designed on a deep-V hull with 17° of deadrise aft, this is the same hull used in the production of the Bertram 42 Convertible. The interior layout is certainly the most notable feature of the 42 MY. By placing the forward salon companionway to starboard, the galley space was enlarged, and the salon took on impressive visual proportions. A new teak interior became available in 1983, and a queen berth in the master stateroom was standard beginning in 1986. With her deep-V hull, the 42 MY can handle the kind of sea conditions that keep other motor yachts her size in protected waters. Most were powered with 335-hp 6-71 diesels. With these engines the range is about 250 miles at a 18-knot cruising speed. Later models using the 435-hp 6V71 diesels cruise around 21 knots. Figures for the standard 330-hp gas engines (thankfully rare) are 15 knots at cruise and 22–23 knots top. ❑

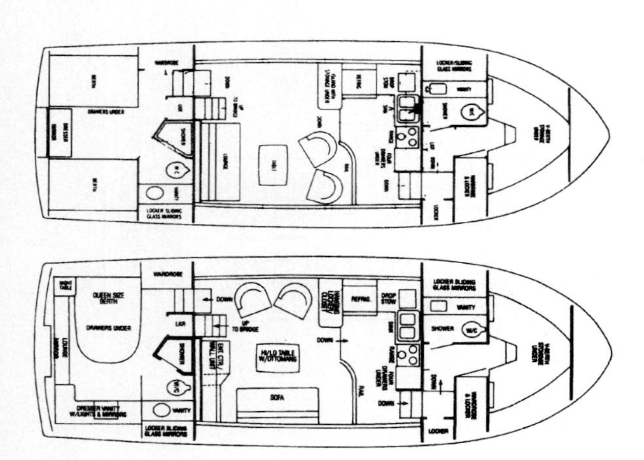

See page 233 for pricing information.

BERTRAM 46 MOTOR YACHT

SPECIFICATIONS

Length	46'6"	Water	230 gals.
Beam	16'0"	Fuel	615 gals.
Draft	4'8"	Hull Type	Deep-V
Weight	45,600#	Deadrise Aft	19°
Clearance	18'8"	Production	1973–87

A rugged cruiser with spacious interior accommodations and traditional flush-deck styling, the Bertram 46 MY was designed for those who place a premium on solid construction and exceptional seakeeping qualities. She was built on the same beamy deep-V hull used in the production of the Bertram 46 Convertible. Below, a conventional interior layout includes a spacious salon with a mid-level galley and dinette and two roomy staterooms. A tub/shower is fitted in the aft head compartment. The large aft deck provides an excellent entertainment platform with space for several pieces of deck furniture. Design changes in the Bertram 46 MY were few. A stall shower was added to the forward head in 1977; an updated teak interior became standard in 1982; and a queen berth (previously optional) replaced twin single beds in the master stateroom in 1986. GM 8V71TI diesels will cruise the Bertram 46 Motor Yacht at 19–20 knots with a range of about 300 miles. Optional 570-hp 8V92s offered after 1980 will cruise around 21 knots and reach a top speed of 24 knots. ❑

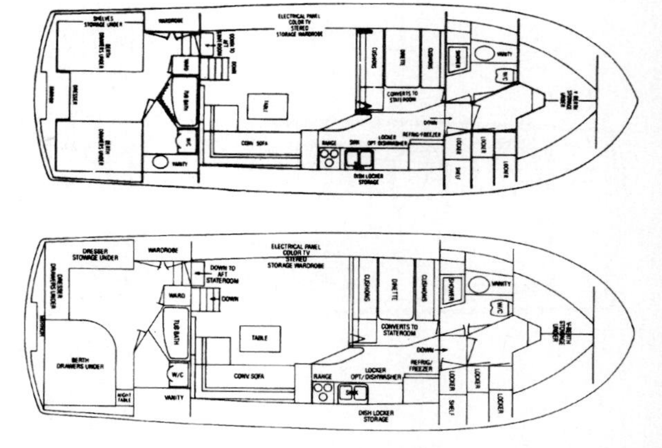

See page 233 for pricing information.

BERTRAM 58 MOTOR YACHT

SPECIFICATIONS

Length	58'3"	Water	275 gals.
Beam	17'11"	Fuel	1,250 gals.
Draft	5'4"	Hull Type	Modified-V
Weight	87,500#	Deadrise Aft	15°
Clearance	18'0"	Production	1976–86

The largest of Bertram's production motor yachts, the 58 MY incorporates the traditional flush deck styling of the smaller Bertram 42 and 46 models and features a lavish three-stateroom interior of spacious dimensions. She shares the same hull as the Bertram 58 Convertible and enjoys a reputation for being a very comfortable offshore cruiser. Her flybridge is small by today's standards, but the sidedecks are wide, and the covered aft deck features a protected lower helm and plenty of entertainment space. Inside, the main salon and cabins are a tasteful blend of designer fabrics and grain-matched teak cabinetry and paneling. The huge owner's stateroom is particularly impressive and comes with a walkaround king-size bed and a tub/shower in the head. GM 12V71 diesels (650-hp/675-hp) will cruise the Bertram 58 around 18 knots and reach a top speed of 21–22 knots. The cruising range is an excellent 350–400 miles. Note that several 58 MYs were fitted with 10' cockpit extensions adding great versatility while taking nothing away from her performance. ❑

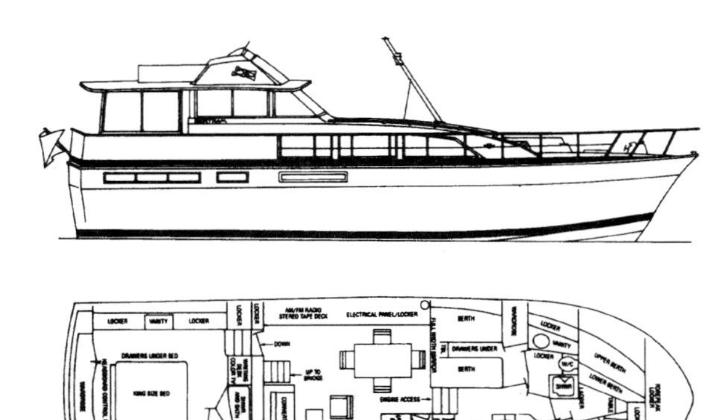

See page 233 for pricing information.

BLUEWATER 44 COASTAL CRUISER

SPECIFICATIONS

Length	47'10"	Water	116 gals.
LWL	36'2"	Fuel	240 gals.
Beam	14'0"	Hull Type	Modified-V
Draft	1'11"	Deadrise Aft	7°
Weight	24,000#	Designer	Bluewater
Clearance	11'7"	Production	1987–89

The Bluewater 44 Coastal Cruiser is a comfortable cruising yacht designed for inland and coastal waters. She was built on the same narrow-beam, shallow-draft hull used in the construction of all Coastal Cruiser models with solid fiberglass construction and prop pockets at the transom. In the Bluewater 44 one can expect the following: an oversized interior layout, big wraparound cabin windows for a magnificent outside view, a completely outrageous party-sized sundeck, reasonable performance, and (thanks to her prop pockets) the unique ability to cruise in only two feet of water. For those who enjoy exploring islands and out-of-the-way inlets the Bluewater 44 offers tremendous close-in flexibility. Actually, the 44 Coastal Cruiser is an enlarged and restyled version of the popular Bluewater 42 Coastal Cruiser (1984–89) featuring a molded swim step, sleeker foredeck lines, and more rake at the bow. The same floorplans were offered in both boats. Relatively small 270-hp Crusader gas engines will cruise the Bluewater 44 Coastal Cruiser at 14–15 knots with a top speed of around 22 knots. ❑

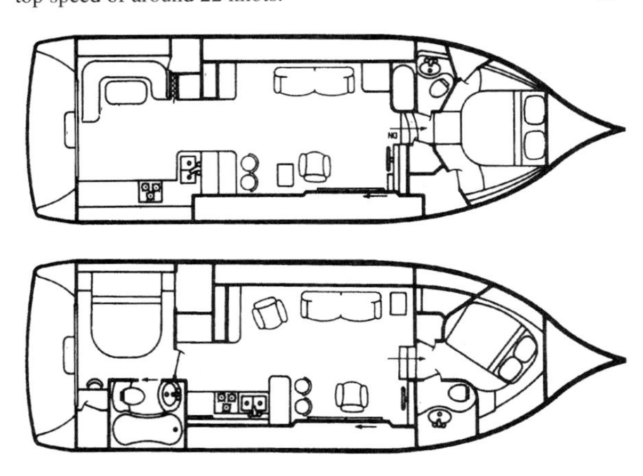

See page 233 for pricing information.

BLUEWATER 462

SPECIFICATIONS

Length	49'7"	Water	116 gals.
LWL	42'7"	Fuel	375 gals.
Beam	14'0"	Hull Type	Modified-V
Draft	1'11"	Deadrise Aft	7°
Weight	30,300#	Designer	Bluewater
Clearance	14'4"	Production	1992–Current

The Bluewater 462 (originally called the Bluewater 45 Motor Yacht) is built on the same shallow-draft hull used in the production of all Bluewater yachts including the 60 MY. A stylish boat with plenty of eye appeal, construction is solid fiberglass, and her modest keel and prop pockets result in a draft of less than two feet—an important sales feature for those who cruise in shallow waters. Unlike most other Bluewater models, the floorplan is arranged with the master stateroom forward. The salon layout was changed in 1993 with the addition of a built-in dinette. A lower helm and deck-access door are standard, and a sliding door aft opens to the integral swim deck. This is a particularly inviting layout, and the 462 contains more interior living and entertaining space than any other boat in her class. And that's not counting her bridge which is huge. A sluggish performer with standard 454-cid gas engines, the Bluewater 462 cruises at just 13–14 knots and reaches a top speed of about 19 knots. ❑

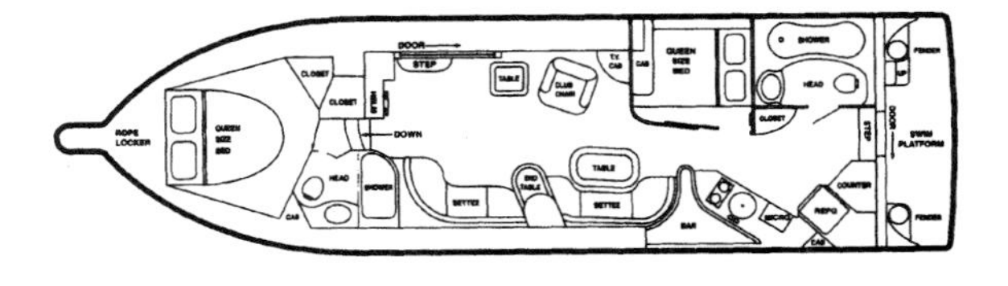

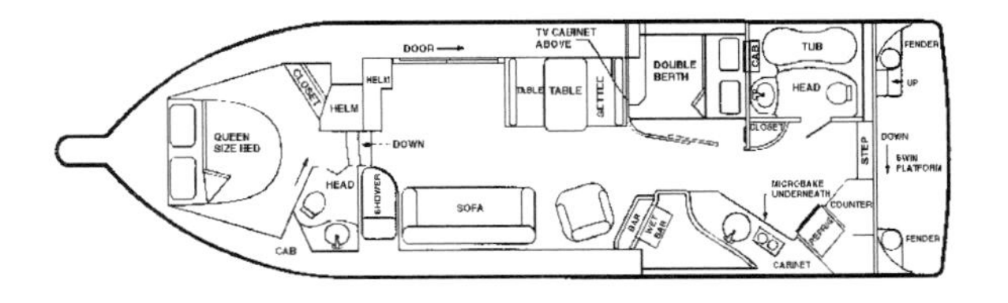

See page 233 for pricing information.

BLUEWATER 48 COASTAL CRUISER

SPECIFICATIONS

Length	48'8"	Water	116 gals.
LWL	NA	Fuel	375 gals.
Beam	14'0"	Hull Type	Modified-V
Draft	1'11"	Deadrise Aft	7°
Weight	25,000#	Designer	Bluewater
Clearance	11'2"	Production	1990–92

The Bluewater 48 Coastal Cruiser features the attractive lines and sculptured, low-freeboard profile common to all of the newer Bluewater models. Having made the Coastal Cruiser name synonymous with waterborne entertainment, the 48 retains the vast bridgedeck and popular single-level interior layout that Bluewater enthusiasts have come to love. A shallow two-foot draft allows close-in running, and the props are protected in a grounding. This ability to explore shallow waters is unique in a motor yacht, and Bluewater actually encourages owners to beach their boats rather than using a dinghy to get ashore. Two floorplans were offered: a two-stateroom layout or a more open single-stateroom arrangement with the dinette and galley aft in place of the guest cabin. The single-level, apartment-style interior is finished with attractive light oak woodwork, white laminates, and oversized cabin windows for plenty of natural lighting. With standard 454-cid Crusader gas engines the Bluewater 48 Coastal Cruiser will cruise at a modest 13–14 knots and reach a top speed of around 20 knots. ❑

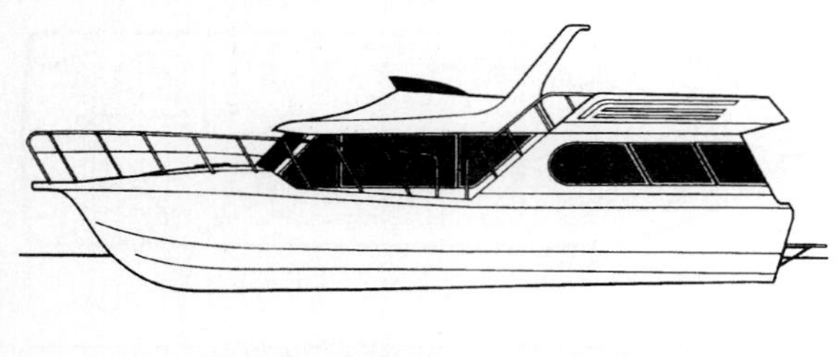

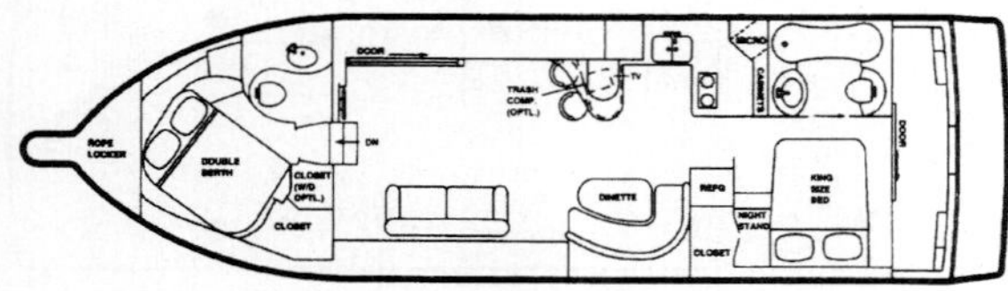

See page 233 for pricing information.

BLUEWATER 51 COASTAL CRUISER

SPECIFICATIONS

Length	51'0"	Water	130 gals.
LWL	49'5"	Fuel	320 gals.
Beam	14'0"	Hull Type	Modified-V
Draft	1'11"	Deadrise Aft	7°
Weight	26,000#	Designer	Bluewater
Clearance	11'7"	Production	1984–89

The Bluewater 51 Coastal Cruiser is a light-displacement cruising yacht designed primarily for inland and coastal waters. She combines spacious, almost extravagant houseboat-style accommodations with the versatile shallow-draft hull found in all Bluewater Coastal Cruisers. With only fourteen feet of beam, the 51 is a narrow boat, but the interior is surprisingly expansive and open. Her standard floorplan is somewhat unconventional when compared with most other modern motor yachts. Rather than dividing the fore and aft cabins with companionways and bulkheads in the normal motor yacht fashion, the Bluewater 51 has the living areas essentially on a single level. Two- and three-stateroom accommodation plans were offered in the 51 Coastal Cruiser, together with a U.S. Coast Guard-certified "Party" version with zero staterooms, his and hers heads forward, and the rest of the boat turned over to entertainment for up to 49 guests. Not a fast boat, Crusader 454-cid gas engines will cruise at 13–14 knots and reach about 19 knots wide open. ❑

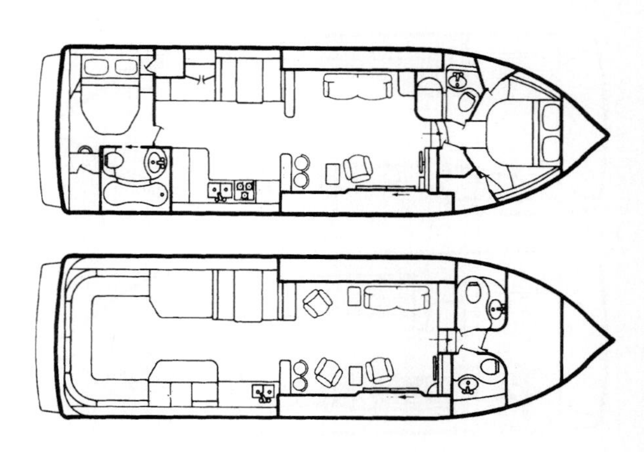

See page 233 for pricing information.

BLUEWATER 543

SPECIFICATIONS

Length	54'7"	Water	116 gals.
LWL	44'3"	Fuel	375 gals.
Beam	14'0"	Hull Type	Modified-V
Draft	1'11"	Deadrise Aft	7°
Weight	34,800#	Designer	Bluewater
Clearance	14'5"	Production	1991–Current

Originally called the Bluewater 53 Coastal Cruiser, the Bluewater 543 is built on the same narrow-beam hull as the rest of the fleet with a shallow keel and prop pockets. (Other manufacturers must envy the ease with which Bluewater uses this standard hull mold to produce so many models.) Like her sisterships, the 543 has a wide-open floorplan arranged on a single level from the stern all the way forward to the helm. Her three-stateroom interior is attractively finished with teak trim and Formica cabinets and countertops. All of the appliances are house-size, and there's more living area in the Bluewater 543 than many larger boats. The wide integrated swim platform/deck is accessed from the master stateroom *or* the flybridge via wide molded steps. The spacious flybridge is a real party platform. Engine access is, well—unique: one is under the berth in the midships stateroom, and the other is beneath a swing-away galley counter. Standard 454-cid gas engines will cruise at 12–13 knots (about 20 knots top). ❑

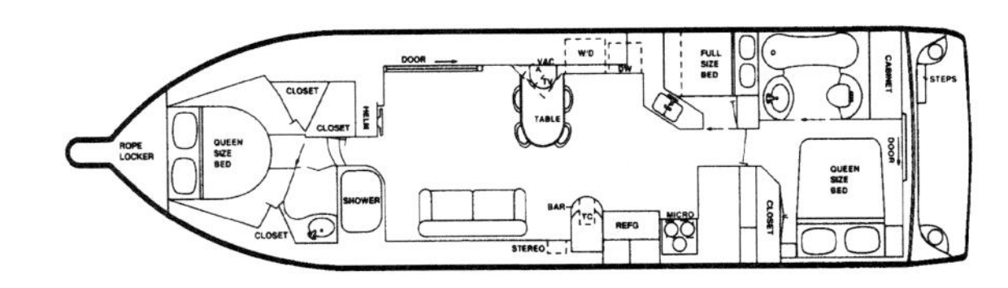

See page 233 for pricing information.

BLUEWATER 55 COASTAL CRUISER

SPECIFICATIONS

Length	55'0"	Water	140 gals.
LWL	NA	Fuel	375 gals.
Beam	14'0"	Hull Type	Modified-V
Draft	1'11"	Deadrise Aft	7°
Weight	29,000#	Designer	Bluewater
Clearance	14'9"	Production	1987–89

The Bluewater 55 Coastal Cruiser is a restyled version of the popular Bluewater 51 with a more rakish profile and a new integral swim platform. If expansive interior dimensions and party-size sundecks are priorities, the 55 Coastal Cruiser is the only answer short of buying a true houseboat. Every inch of the 55's relatively narrow beam is used inside, and the living quarters from the stern to the forward stateroom are arranged on a single level. The visual effect is that of a much larger and more expensive yacht. The home-style interior decor is contemporary, and the comfort level is high indeed. Note that the lower helm station was standard. The 55 Coastal Cruiser was available in either a two- or three-stateroom layout in addition to a zero-stateroom Coast Guard-certified "Party" version for commercial or charter service. As far as the vast upper-level sundeck is concerned, it must be seen to appreciate fully its entertainment potential. With 454-cid gas engines, she'll cruise at 13–14 knots and reach about 20 knots wide open. ❑

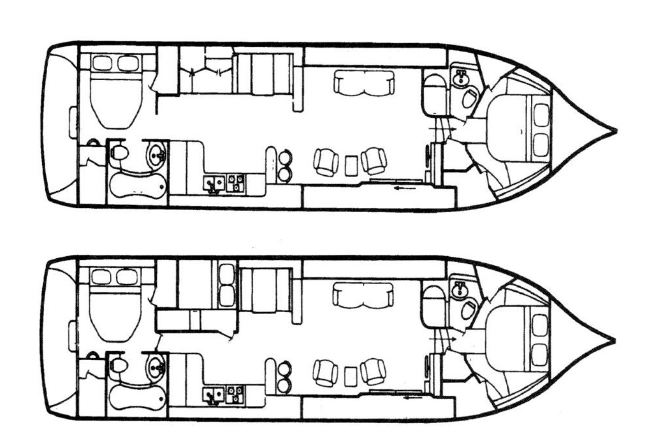

See page 234 for pricing information.

BLUEWATER 55 YACHT

SPECIFICATIONS

Length	59'0"	Water	140 gals.
LWL	NA	Fuel	600 gals.
Beam	14'0"	Hull Type	Modified-V
Draft	1'11"	Deadrise Aft	7°
Weight	29,000#	Designer	Bluewater
Clearance	11'6"	Production	1990–92

A popular model, the Bluewater 55 Yacht is a good-looking boat with a notably rakish, low-slung profile. Built on a narrow shallow-draft hull, she's basically a warmed-over version of the previous Bluewater 55 Coastal Cruiser (1987–89) with a slightly modified interior and a sleeker profile. Like all recent Bluewaters, she's designed to be beached, and the bow is reinforced with Kevlar to handle the extra loads. The props are recessed into tunnels at the transom, and a long 10" keel provides grounding protection. Company officials claim that the 55 Yacht is at home in offshore waters, but she's really more of an inland boat with coastal cruising potential. Her three-stateroom layout includes two full heads and a lower helm with a hideaway console. Port and starboard stairways lead to the huge full-width bridgedeck where guests can take advantage of more sundeck space than anything short of an aircraft carrier. Standard 360-hp (502-cid) gas engines cruise the Bluewater 55 at 14–15 knots and deliver a top speed of around 19 knots. ❑

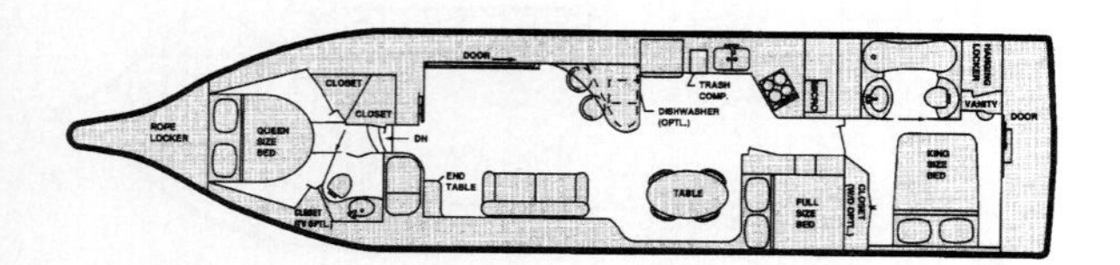

See page 234 for pricing information.

BLUEWATER 622C YACHT

SPECIFICATIONS

Length	64'0"	Water	200 gals.
LWL	54'5"	Fuel	600 gals.
Beam	14'0"	Hull Type	Modified-V
Draft	1'11"	Deadrise Aft	7°
Weight	47,700#	Designer	Bluewater
Clearance	14'9"	Production	1990–Current

A good-looking boat with a modern profile and a wide-open layout, the Bluewater 622C (originally called the Bluewater 60 Yacht) is an extremely narrow cockpit cruiser designed for inland and coastal waters. She's constructed on the same 14-foot-wide hull used in other Bluewater yachts with a shallow keel and relatively flat 7° of transom deadrise. Like all current Bluewater models, her two-stateroom, two-head floorplan is arranged on a single level from the cockpit forward to the guest stateroom bulkhead. This is a roomy interior arrangement in spite of her too-narrow beam with home-size furnishing and appliances and a centerline lower helm station. Oversize cabin windows and good headroom add to the impression of space inside. Not surprisingly, the large cockpit (with refrigerator and wet bar) and the enormous party-time sundeck (with retractable instrument panels) are the real focal points of the boat. Standard 400-hp Cummins diesels will cruise the Bluewater 622C at 19–20 knots and reach a top speed of about 22 knots. ❑

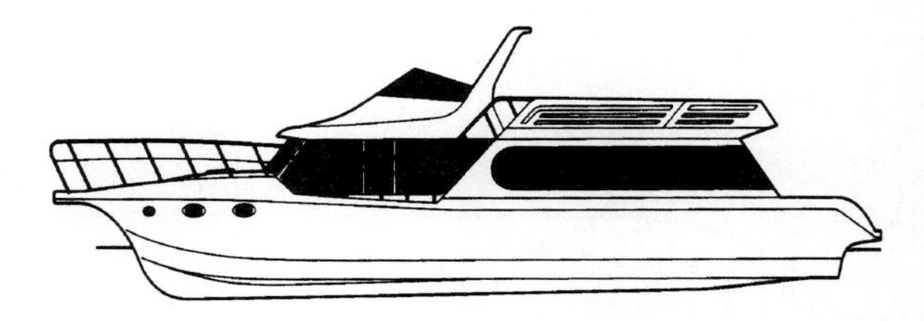

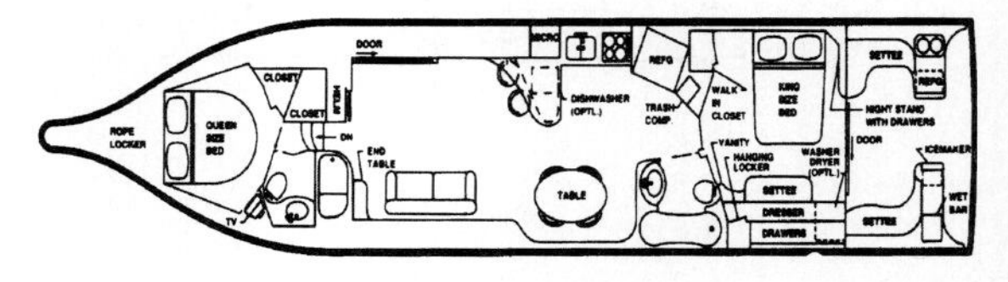

See page 234 for pricing information.

BLUEWATER 623

SPECIFICATIONS

Length	64'0"	Water	200 gals.
LWL	54'5"	Fuel	600 gals.
Beam	14'0"	Hull Type	Modified-V
Draft	1'11"	Deadrise Aft	7°
Weight	48,000#	Designer	Bluewater
Clearance	14'9"	Production	1990–Current

To paraphrase *Boating* magazine, the Bluewater 623 is a party boat of *major* proportions. This is an extremely narrow boat, and her interior accommodations, while open, are considerably less than one might expect in a boat of this size. Her three-stateroom, three-head layout is arranged on a single level from the helm aft with a small swim platform added at the stern. The desk/office area in the master stateroom is a thoughtful touch as is the tub in the en-suite head. Each stateroom has a double berth, and the apartment-size galley is huge. The feature that draws the most attention, however, is the enormous sundeck with its built-in dinette, sunpad, and lounge seating for a crowd. (Note that the instrument panels at the helm are retractable.) Indeed, with some 300 sq. ft. of platform few boats can approach her open-air entertainment potential. Now-standard 400-hp Cummins diesels will cruise the Bluewater 623 at 19–20 knots and reach a top speed of about 22 knots. ❑

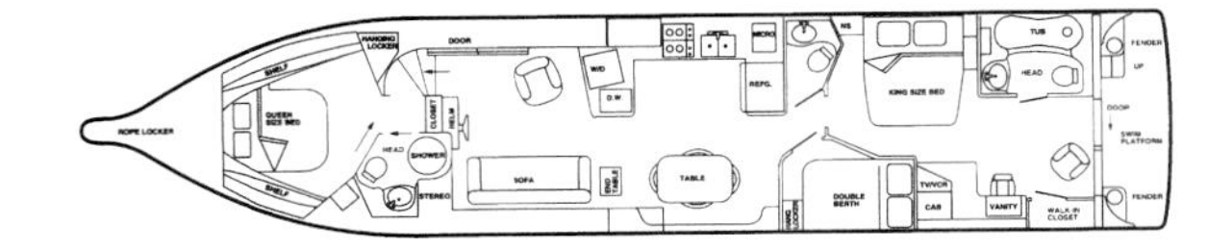

See page 234 for pricing information.

CALIFORNIAN 34 LRC

SPECIFICATIONS

Length	34'6"	Fuel	250 gals.
Beam	12'4"	Cockpit	70 sq. ft.
Draft	3'2"	Hull Type	Modified-V
Weight	18,000#	Deadrise Aft	NA
Clearance	10'8"	Designer	J. Marshall
Water	75 gals.	Production	1979–82

Traditionally styled and showing a distinct trawler profile, the Californian 34 is a still-popular family cruiser with exceptional low-speed economy and (when equipped with the right engines) true planing-speed performance. The 34's modified-V hull has a fine entry forward and gradually levels out into nearly flat aftersections for stability and greater speed. While most trawler-style boats have been imported from Asia, the Californians were among the few to have been built in the U.S. Wide sidedecks give her a rather narrow salon although the two staterooms below come as a surprise on such a small boat. Dark mahogany woodwork is applied throughout the interior. Outside, the aft deck area is large enough for a few deck chairs, and there's seating for everyone on the roomy flybridge. Cruising speed is about 7 knots with standard 85-hp Perkins diesels. Among several engine options, a pair of 210-hp Cat (or 200-hp Perkins) diesels will cruise the Californian 34 efficiently around 16–17 knots while reaching 20+ knots wide open. ❑

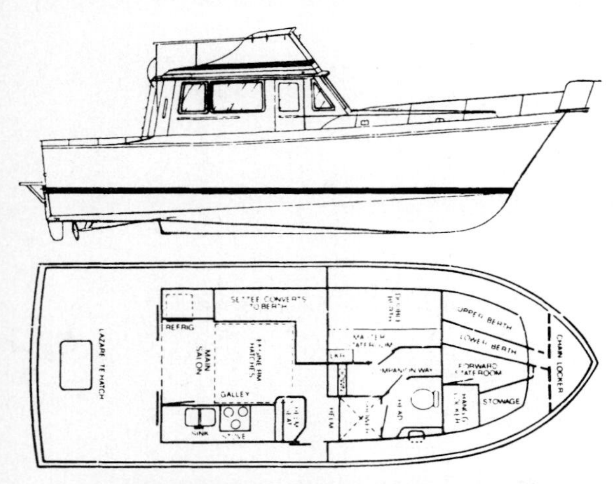

See page 234 for pricing information.

CALIFORNIAN 35 MOTOR YACHT

SPECIFICATIONS

Length	34'11"	Water	75 gals.
LWL	NA	Fuel	270 gals.
Beam	12'4"	Hull Type	Modified-V
Draft	3'2"	Deadrise Aft	NA
Weight	19,000#	Designer	B. Collier
Clearance	NA	Production	1985–87

Building a double-cabin yacht on a 35-foot hull is no easy trick, and the results are generally less than satisfactory from the standpoint of styling. The Californian 35 MY is clearly an exception. Indeed, her profile is surprisingly easy on the eye (unlike her Sea Ray and Carver counterparts). Perhaps this is due to the window treatment, but in any case the Californian 35's lines are actually attractive considering her small size. A popular boat, her features include an all-teak interior, an optional lower helm, a small but well-designed master stateroom, overnight berths for six, and (surprise) stall showers in *both* heads. This is a lot of interior for only 35 feet, and the Californian makes the most of it with room for the entire family. The hardtop was a popular option, and the aft deck area is roomy enough for a few deck chairs. Gas engines were standard, but many Californian 35 MYs were sold with the optional 210-hp Caterpillar diesels for a comfortable 16 knots at cruise and 20 knots wide open. ❑

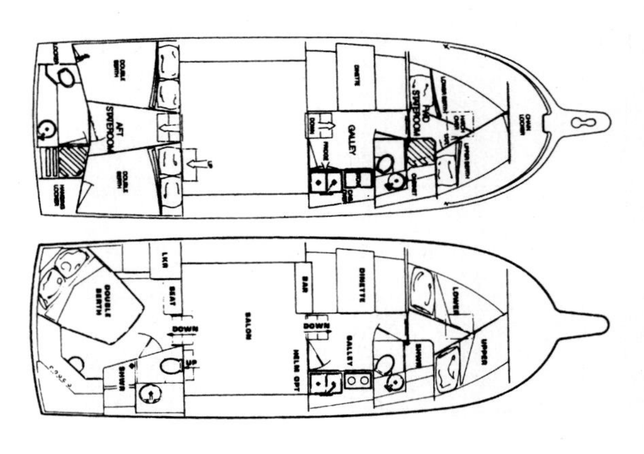

See page 234 for pricing information.

CALIFORNIAN 38 LRC

SPECIFICATIONS

Length	37'8"	Water	100 gals.
LWL	36'6"	Fuel	400 gals.
Beam	13'0"	Cockpit	NA
Draft	3'6"	Hull Type	Modified-V
Weight	28,000#	Designer	J. Marshall
Clearance	14'6"	Production	1980–84

The Californian 38 Long Range Cruiser has the same trawler-like profile and modified-V hull form as the smaller Californian 34. Available with either a single- or twin-stateroom interior layout, the 38 LRC might best be described as a comfortable sedan trawler with surprisingly un-trawlerlike performance. Mahogany interiors were a prominent feature in the early Californian models, and the woodwork in the 38 is well-crafted. As a cruising boat the Californian will disappoint no one with her comfortable salon, roomy cockpit, and wide, well-protected sidedecks. A lower helm was standard, and large wraparound cabin windows provide enough outside lighting to partially offset the dark woodwork. Twin sliding doors open into the cockpit, where there's room for a couple of deck chairs and even some light tackle fishing. Notably, exterior teak trim is kept to a minimum. The helm is set well forward on the bridge with seating for as many as six. A good performer with optional 300-hp Cats, she'll cruise easily at 19–20 knots and reach about 23 knots top. ❑

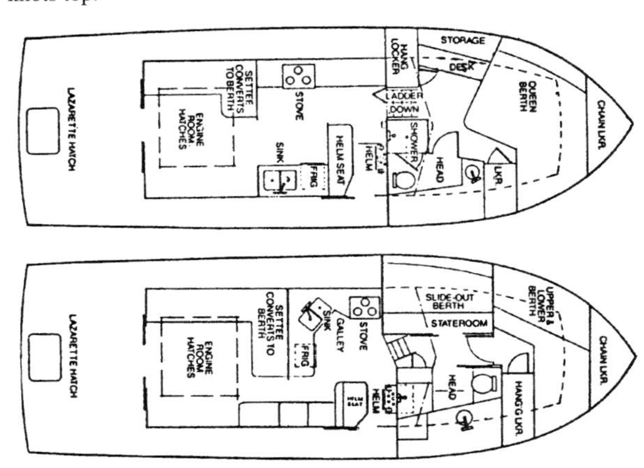

See page 234 for pricing information.

CALIFORNIAN 38 MOTOR YACHT

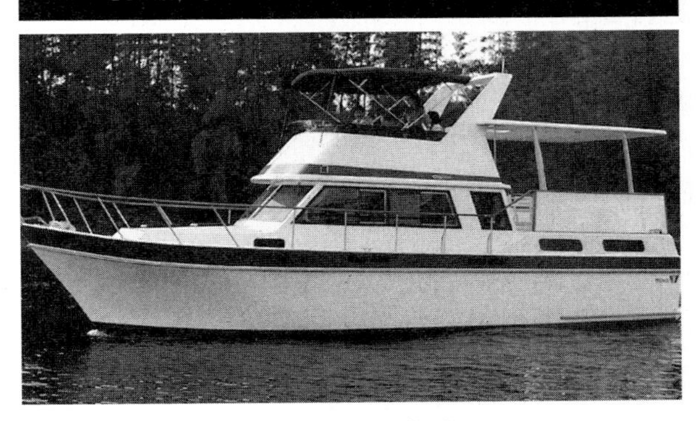

SPECIFICATIONS

Length	37'8"	Water	100 gals.
LWL	36'6"	Fuel	365 gals.
Beam	13'3"	Hull Type	Modified-V
Draft	3'6"	Deadrise Aft	NA
Weight	28,000#	Designer	J. Marshall
Clearance	14'6"	Production	1983–87

A good-looking small yacht with a distinctive profile, the Californian 38 MY is built on a conservative low-deadrise hull with moderate beam, a three-quarter-length skeg keel, and relatively flat aftersections for efficient planing performance. Hull construction is solid fiberglass. Her floorplan is arranged with double berths fore and aft and both heads are fitted with separate stall showers. If the salon dimensions are somewhat limited, the master stateroom is quite roomy for a 38-footer. A sliding deck access door is adjacent to the lower helm station—a worthwhile feature. Two variations of this layout were available: one with the galley down and the other with the galley in the main salon. Hinged steps in the forward companionway provide access to a particularly well-designed engine room with good working space outboard of the motors. Standard 210-hp Cat diesels will cruise the 38 MY around 14 knots, and the 300-hp turbo-Cats will cruise at 18–19 knots. Note that the Californian 43 Cockpit MY is the same boat with a 5-foot cockpit extension. ❑

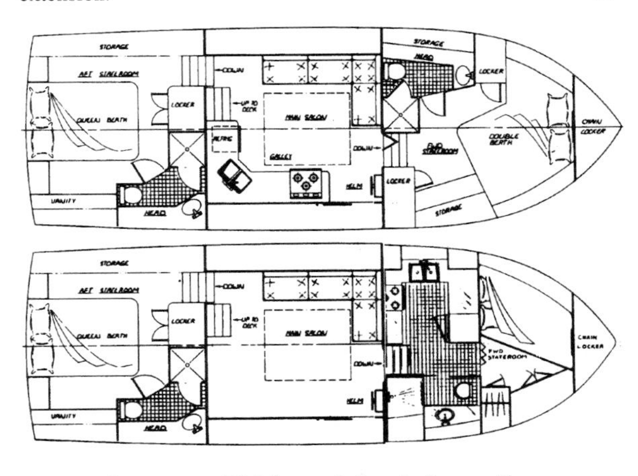

See page 234 for pricing information.

CALIFORNIAN 42 LRC

SPECIFICATIONS

Length	41'8"	Water	175 gals.
Beam	13'8"	Fuel	500 gals.
Draft	3'4"	Hull Type	Semi-Disp.
Weight	31,000#	Designer	J. Marshall
Clearance	NA	Production	1979–84

A popular boat, the Californian 42 is a traditional trawler-style design with a prominent bow and a distinctive appearance. For her length, she's not an especially roomy boat inside (her narrow beam and wide sidedecks take their toll in the salon dimensions), but the layout is still practical, and the mahogany woodwork is appealing and well-crafted. The forward stateroom configuration is somewhat unusual with stacked single berths to starboard just opposite the head—a seemingly practical layout for cruising and one that may well appeal to experienced cruisers. A stall shower is fitted in the forward head, and a tub/shower is aft. There are port and starboard deck doors in the salon as well as direct access to the cockpit from the aft stateroom. (Note that there were three different interiors offered for the 42 LRC, but we've only been able to find the one pictured below.) Twin 210-hp Caterpillar diesels were standard. With a light load and trim tabs, it's possible to get the Californian 42 on plane and cruising at 12–13 knots. ❑

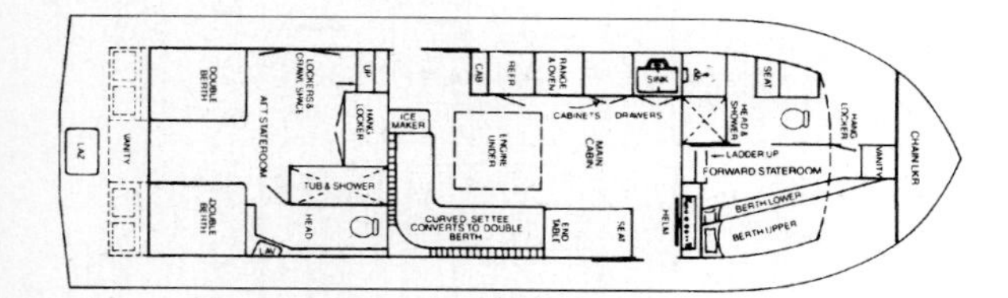

See page 234 for pricing information.

CALIFORNIAN 43 COCKPIT MY

SPECIFICATIONS

Length	43'8"	Water	140 gals.
LWL	NA	Fuel	400 gals.
Beam	13'3"	Hull Type	Modified-V
Draft	3'6"	Deadrise Aft	NA
Weight	32,000#	Designer	J. Marshall
Clearance	14'6"	Production	1983–87

Take a good-looking 38-foot motor yacht, add a cockpit, and you end up with a more versatile and perhaps better-looking boat. A popular model, the Californian 43 CMY is, of course, a Californian 38 MY with a five-foot cockpit extension. While her profile shows a distinctive trawler-style bow, the 43 CMY is built on a conventional modified-V hull with hard chines aft and a long skeg below. The standard layout has the galley in the salon and a huge forward stateroom. The optional galley-down floorplan opens up the salon considerably at the expense of a smaller guest stateroom—probably the preferred arrangement. The rich walnut woodwork is impressive and a pleasant departure from teak in spite of its dark color. The aft deck is particularly roomy, and there's space in the cockpit for a couple of anglers. A comfortable seaboat with an easy ride, Cat 210-hp diesels were standard in the Californian 43. She'll cruise economically at 13–14 knots and reach 17 knots top. Larger Cats or GM diesels were optional. ❑

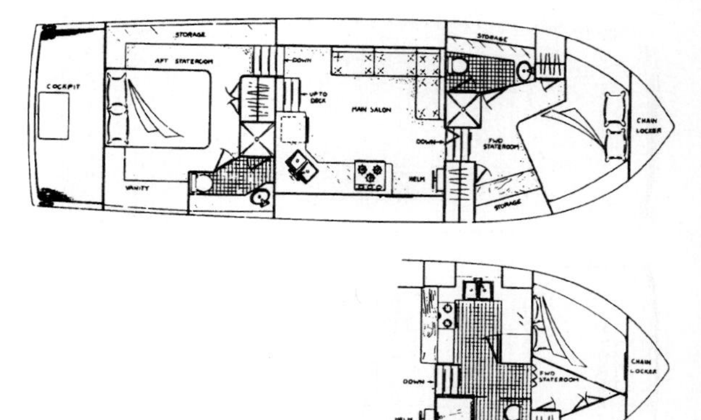

See page 234 for pricing information.

CALIFORNIAN 45 MOTOR YACHT

SPECIFICATIONS

Length	45'0"	Water	190 gals.
LWL	36'8"	Fuel	400 gals.
Beam	15'2"	Hull Type	Modified-V
Draft	4'0"	Deadrise Aft	15°
Weight	40,000#	Designer	B. Collier
Clearance	17'3"	Production	1988–91

The Californian 45 MY is a good quality cruising yacht with a handsome profile and comfortable accommodations. She's built using the same sea kindly hull as other boats in the Californian series with moderate beam and a relatively steep 15° of transom deadrise. Her two-stateroom interior is arranged with the galley and dinette down resulting in a spacious and open salon. The most recent layout features a centerline double berth in both staterooms, and each head has a separate stall shower. Hand-rubbed teak cabinetry and woodwork are used extensively throughout, and large cabin windows provide plenty of natural lighting in the salon and galley. The aft deck hardtop was standard. Her wide sidedecks allow secure passage fore and aft. No lightweight, the Californian 45 MY is generally recognized as a good performer in a chop. Powered with 375-hp Cat diesels, she'll cruise at 15–16 knots and reach a top speed of about 21 knots. GM 485-hp 6-71s are optional. Note that the Californian 52 Cockpit MY is the same boat with a cockpit extension. ❑

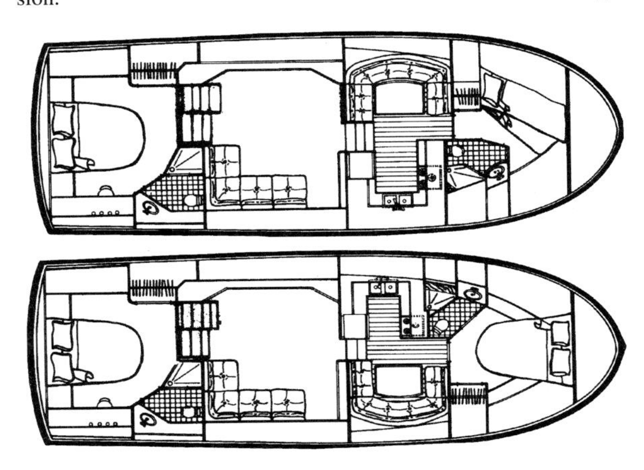

See page 234 for pricing information.

CALIFORNIAN 48 MOTOR YACHT

SPECIFICATIONS

Length	48'5"	Water	210 gals.
LWL	40'11"	Fuel	560 gals.
Beam	15'2"	Hull Type	Modified-V
Draft	4'8"	Deadrise Aft	15°
Weight	43,000#	Designer	B. Collier
Clearance	17'3"	Production	1985–91

Originally built on a solid fiberglass hull, the 48 Motor Yacht underwent significant changes in 1987 following Californian's purchase by Carver Boats. Foam coring was introduced in the hullsides above the waterline, and the resulting weight reduction is said to have added to the 48's performance. With her clean and well-proportioned profile, the Californian 48 is a particularly handsome boat with the added advantage of a huge aft deck. Below, the interior is tastefully finished with teak cabinetry and woodwork throughout (walnut in pre-'88 models). Walkaround double berths are located in the fore and aft staterooms, and a slide-out settee in the small guest stateroom converts into a double. Notably, all three heads are fitted with separate stall showers. A modest performer with standard 375-hp Cats, she'll cruise at 15 knots and reach a top speed of about 19 knots. The optional 485-hp 6-71s will cruise at 18–19 knots and reach 22 knots wide open. Note that the Californian 55 Cockpit MY is the same boat with a cockpit. ❑

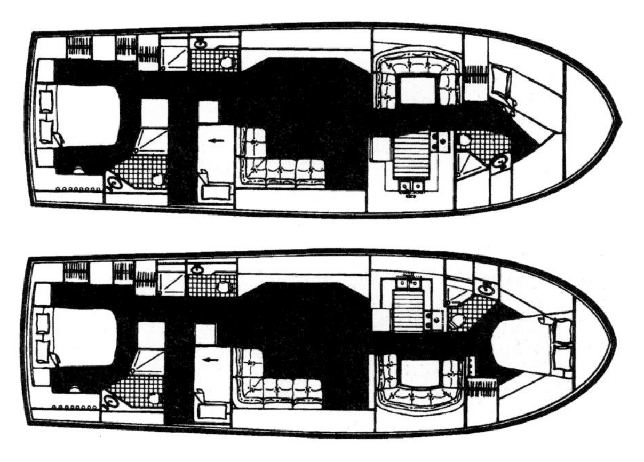

See page 234 for pricing information.

CALIFORNIAN 48 COCKPIT MY

SPECIFICATIONS

Length	48'5"	Water	190 gals.
LWL	40'11"	Fuel	500 gals.
Beam	15'2"	Hull Type	Modified-V
Draft	4'8"	Deadrise Aft	15°
Weight	41,000#	Designer	B. Collier
Clearance	16'3"	Production	1986–89

Cockpit motor yachts have become increasingly popular in recent years, and the Californian 48 CMY is one of the more stylish to be found in her size range. She was introduced in 1986 as a cockpit version of the then-existing Californian 42 MY (1985–87). A comfortable cruising boat with an attractive profile, the 48 CMY has a conventional two-stateroom interior layout with the galley and dinette down from the salon level. Both staterooms have double berths, and stall showers are included in each head. One of the more appealing features of the Californian 48 is the grain-matched teak cabinetry and woodwork. (Those built prior to 1988 have equally attractive dark walnut interiors.) The hardtop was standard. Although the cockpit is too small for any serious fishing ambitions, it is equipped with a transom door, and its existence certainly makes boarding a more civilized procedure. A good sea boat, standard 375-hp Cat diesels will cruise the Californian 48 Cockpit MY at 16 knots and reach a top speed of about 20 knots. ❑

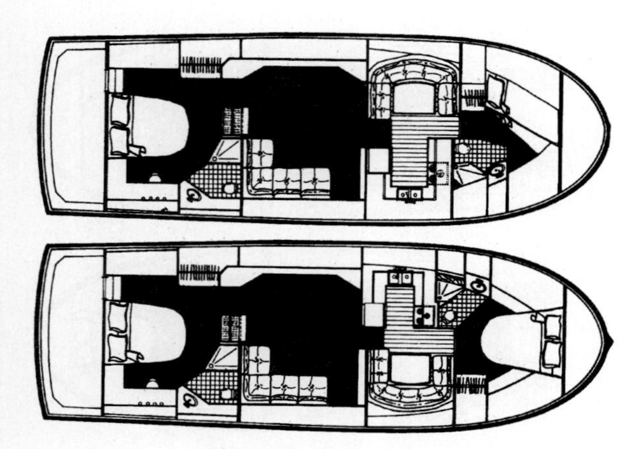

See page 234 for pricing information.

CALIFORNIAN 52 COCKPIT MY

SPECIFICATIONS

Length	51'11"	Water	185 gals.
LWL	NA	Fuel	740 gals.
Beam	15'2"	Hull Type	Modified-V
Draft	4'5"	Deadrise Aft	15°
Weight	43,500#	Designer	B. Collier
Clearance	12'10"	Production	1990–91

The 52 Cockpit MY evolved from the Californian 45 MY, and they share the same deckhouse profile and two-stateroom floorplan. She's constructed on an efficient modified-V hull with cored hullsides, a shallow keel, and a solid fiberglass bottom. The transom deadrise is a fairly steep 15°, and like all Californian hulls she's a sea kindly boat offshore with good handling characteristics. With a generous 70 sq. ft., the cockpit is large enough for some light tackle fishing, and a sliding glass door provides direct access from the master stateroom—a practical and convenient feature. Both staterooms include walkaround double berths, and there's a stall shower in the forward head. The full teak interior is lush, and the sensation of quality and craftsmanship is no illusion. The afterdeck (flush with the sidedecks, not raised) is big enough for a crowd, and the flybridge is arranged with the helm forward and guest seating aft. A good-looking yacht, optional 425-hp Cats will cruise the Californian 52 CMY at 16 knots and deliver 20 knots wide open. ❑

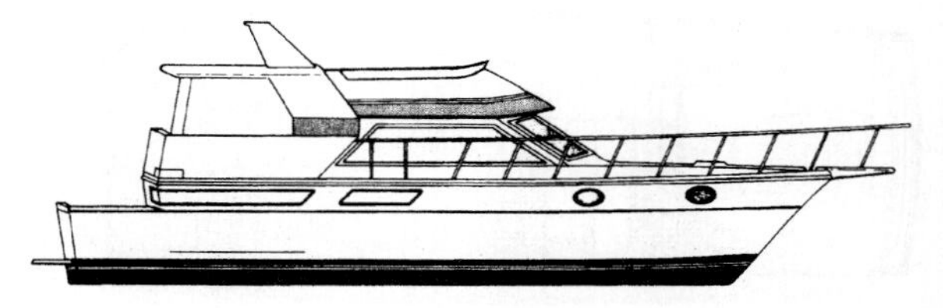

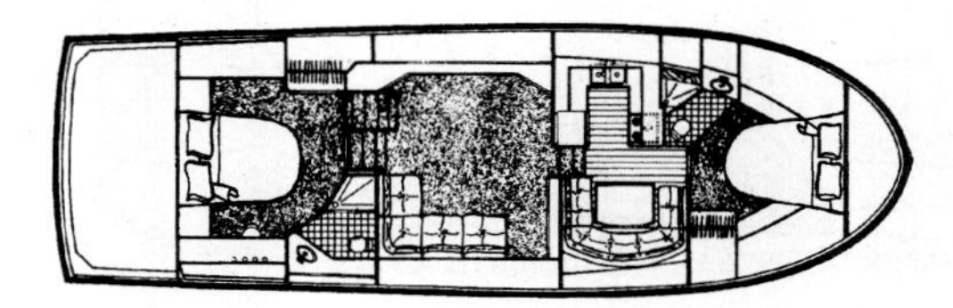

See page 234 for pricing information.

CALIFORNIAN 55 COCKPIT MY

SPECIFICATIONS

Length	54'6"	Water	210 gals.
LWL	NA	Fuel	650 gals.
Beam	15'2"	Hull Type	Modified-V
Draft	4'3"	Deadrise Aft	15°
Weight	46,200#	Designer	B. Collier
Clearance	17'3"	Production	1988–91

The Californian 55 CMY is one of the better-looking cockpit motor yachts on the market. Basically a Californian 48 MY with a big 7' cockpit, one of the most prominent features of the 55 CMY is her enormous afterdeck—one of the largest to be found in any boat of this type. The flybridge is also quite large, and the wrap-around helm console provides space for flush-mounting most electronics. Inside, the three-stateroom floorplan (identical to the 48 MY) has the second guest stateroom *aft* of the salon, and all three heads have separate stall showers. Teak paneling and cabinetry replaced the original dark walnut woodwork in 1989. The upscale interior is lavishly decorated with rich fabrics, contemporary furnishings, and top-quality hardware and appliances throughout. A popular boat (about 40 were built), 375-hp Cat diesels were standard in the 55 CMY, but most were equipped with the optional 485-hp 6-71s (around 19 knots cruise/22 knots top) or—better yet—550-hp 6V92s which cruise at a respectable 21 knots and reach 24 knots top. ❑

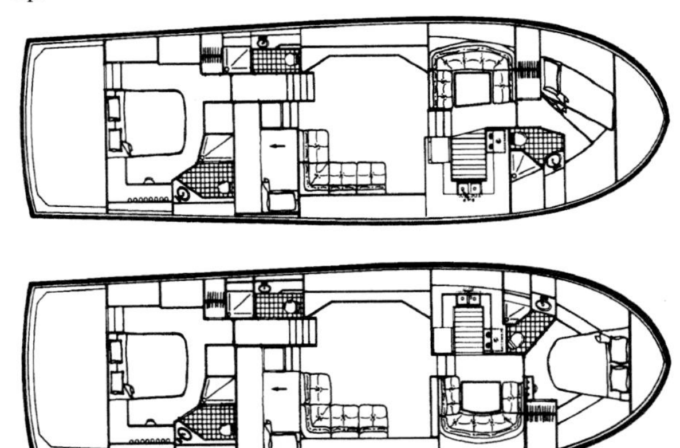

See page 234 for pricing information.

CAMARGUE 48 YACHTFISHER

SPECIFICATIONS

Length	48'0"	Water	210 gals.
LWL	NA	Fuel	600 gals.
Beam	15'5"	Hull Type	Modified-V
Draft	3'3"	Deadrise Aft	NA
Weight	38,000#	Designer	Camargue
Clearance	NA	Production	1987–93

The Camargue 48 YF stands out among Asian imports for her modern profile, excellent glasswork, and above-average fit and finish. Introduced in 1987, she's basically a Camargue 42 Sundeck with a cockpit extension. (About fifty of the 48s were built during her production run.) The hull is a modified-V affair with moderate beam and relatively flat aftersections, and the hullsides are Airex-cored above the waterline. She has a very practical two-stateroom floorplan and manages to provide a lot of salon living-space without skimping in any of the stateroom dimensions. The portside lower helm is cleverly concealed within an exquisite roll-top console—very nice. The teak interior woodwork is very impressive. The extra-wide sidedecks are also worth noting as is the well-organized engine room and comfortable bridge. A handsome boat with attractive lines, standard 375-hp Caterpillar diesels will cruise the Camargue 48 YF around 18–19 knots, and the optional 485-hp 6-71s will provide a brisk 22–23 knots at cruise and 27 knots wide open. ❑

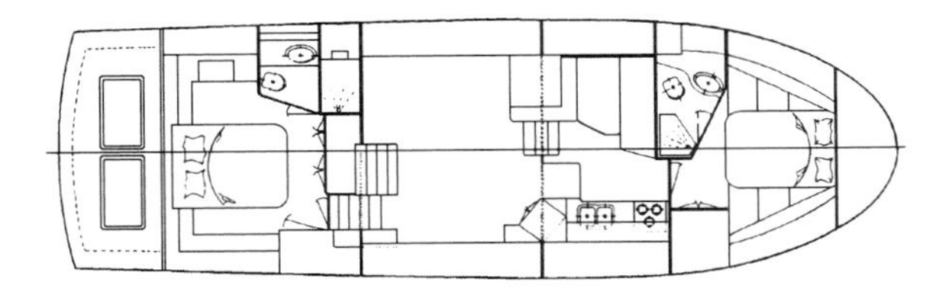

See page 234 for pricing information.

CARVER 32 AFT CABIN

SPECIFICATIONS

Length	32'0"	Water	84 gals.
LWL	28'1"	Fuel	182 gals.
Beam	11'7"	Hull Type	Modified-V
Draft	2'10"	Deadrise Aft	10°
Weight	12,000#	Designer	Carver
Clearance	11'6"	Production	1983–90

The fact that the Carver 32 Aft Cabin managed to remain in production for so many years says a lot about her popularity with the public. Her somewhat boxy exterior appearance masks a roomy double-cabin interior with surprisingly comfortable cruising accommodations. Sharing the same low-deadrise solid fiberglass hull as the 32 Convertible, the conventional aft-cabin floorplan of the Carver 32 is designed with generous salon dimensions, a roomy galley (with an upright refrigerator no less), two heads (one with a separate stall shower), and two economical staterooms. A lower helm station was standard, and there's sufficient storage space for undemanding weekend cruisers. Outside, the aft deck platform can accommodate a couple of folding chairs (a hard top was not an option for this boat), and the flybridge will seat five. Unlike the 32 Convertible, the engines in the Aft Cabin are located below the salon sole with a conventional straight-drive installation. A stiff ride in a chop, the cruising speed with standard 270-hp Crusader gas engines is 16 knots, and the top speed is 25–26 knots. ❑

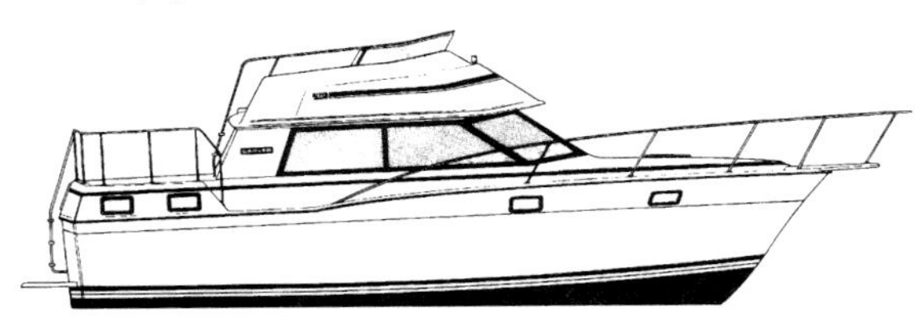

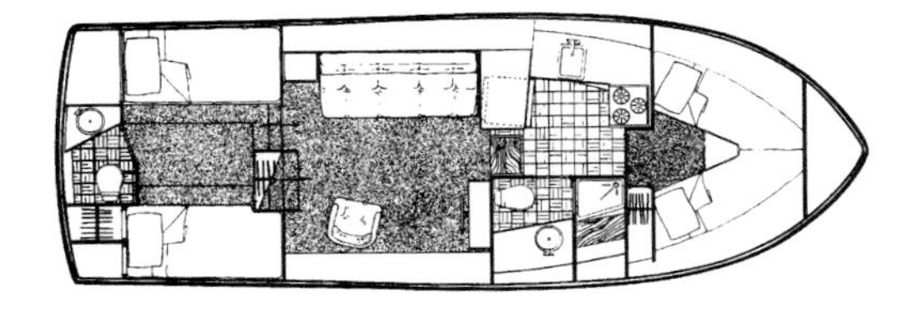

See page 234 for pricing information.

CARVER 350 AFT CABIN

SPECIFICATIONS

Length	33'10"	Water	81 gals.
LOA (w/platform)	36'0"	Fuel	220 gals.
Beam	13'3"	Hull Type	Modified-V
Draft	2'7"	Deadrise Aft	11°
Weight	16,600#	Designer	Carver
Clearance	15'0"	Production	1991–Current

Building an aft cabin design on a less-than-40' hull involves some compromises, and the results are often disappointing. With a hull length of just 33 feet (she was introduced in 1991 as the Carver 33 Aft Cabin), the 350 Aft Cabin uses her super-wide beam and very high freeboard to pack a lot of living space into an unusually small package. This maxi-volume interior, however, results in a boxy, topheavy exterior appearance common to nearly all small double cabin designs. Inside, the 350's galley-down floorplan is cleverly arranged to include double berths in both staterooms, two heads (one with a stall shower!), *and* a dinette—very impressive in spite of the compact salon dimensions and tiny staterooms. Storage space is at a premium. Additional features include flybridge seating for six, decent sidedecks, and good engine-room access. Standard 454-cid gas engines will cruise at 19 knots (26–27 top), and optional 200-hp Volvo diesels cruise economically at 18 knots (22 knots top). The small fuel tanks limit range, especially with gas engines. ❑

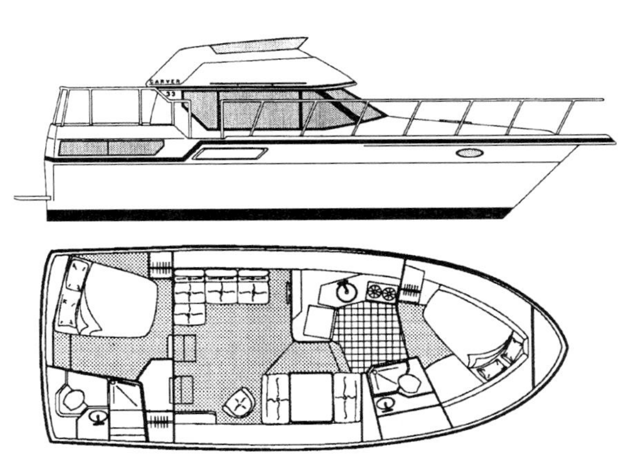

See page 235 for pricing information.

CARVER 36 AFT CABIN

SPECIFICATIONS

Length	35'7"	Water	109 gals.
LWL	31'4"	Fuel	240 gals.
Beam	12'6"	Hull Type	Modified-V
Draft	3'2"	Deadrise Aft	8°
Weight	18,500#	Designer	Carver
Clearance	11'9"	Production	1982–89

When the Carver 36 Aft Cabin was introduced in 1982 she was the biggest boat ever built by Carver. Immediately successful (due in part to her affordable price), she was built on a "dual mode" trawler-style hull, meaning that it can run efficiently (and very economically) at slow 7–8 knot displacement speeds and still plane out when the throttles are fully applied. (The same hull was later used in the production of the 36 Mariner.) While the 36 isn't a fast boat, her interior accommodations are roomy and well-suited to the needs of family cruisers. Her conventional double-cabin floorplan features a large master stateroom aft with a tub/shower in the adjoining head compartment. A lower helm was standard, and the galley and dinette are both down from the salon level. Outside, the raised aft deck platform is large enough for a few folding chairs, and the bridge is only three steps removed with additional guest seating. Standard 350-hp gas engines will cruise the Carver 36 Aft Cabin around 16 knots and reach a top speed of approximately 26 knots. ❑

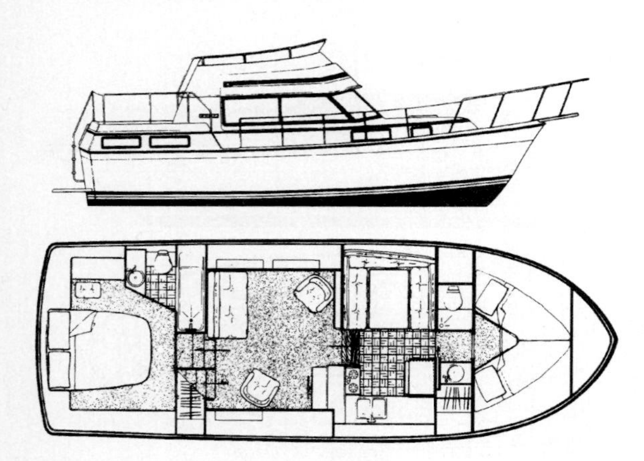

See page 235 for pricing information.

CARVER 370 AFT CABIN MY

SPECIFICATIONS

Length	35'9"	Water	80 gals.
LWL	NA	Fuel	240 gals.
Beam	13'10"	Hull Type	Deep-V
Draft	3'1"	Deadrise Aft	19°
Weight	18,500#	Designer	Carver
Clearance	15'0"	Production	1990–Current

Introduced in 1990 as the Carver 36 Aft Cabin MY, the re-named 370 Aft Cabin is a beamy double-cabin family cruiser with a surprisingly spacious interior and a somewhat boxy profile. This is a much more open boat than Carver's earlier 36 Aft Cabin model (1982–89) thanks primarily to a wider beam and greater freeboard. Notably, she's built on a deep-V hull which accounts for her soft and comfortable ride. The hullsides are cored with balsa, and the bottom is solid fiberglass. Inside, Carver engineers managed to pack double berths into both staterooms as well as providing two head compartments, a big U-shaped dinette, and reasonably spacious main salon dimensions—not bad for just a thirty-six foot hull. A lower helm station is a popular option. While the aft deck is necessarily small, and the sidedecks are narrow, the flybridge is very roomy with lounge seating for six. No racehorse, standard 454-cid gas engines will cruise the Carver 370 at 15 knots and deliver a top speed of about 24 knots. ❑

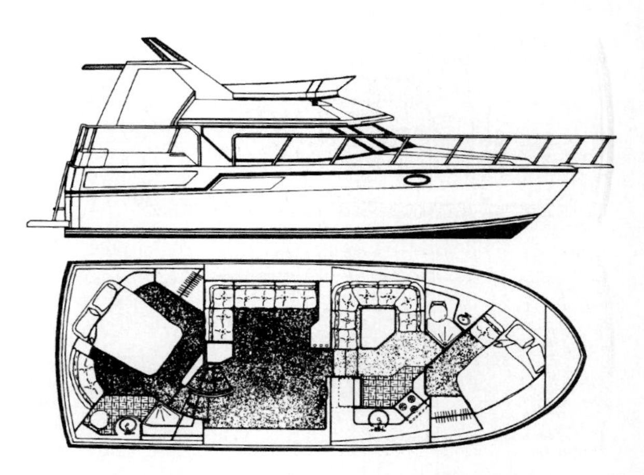

See page 235 for pricing information.

CARVER 390 AFT CABIN

SPECIFICATIONS

LOA (w/platform)...38'10"	Water...91 gals.
Length...37'6"	Fuel...280 gals.
Beam...14'0"	Hull Type...Modified-V
Draft...3'4"	Deadrise Aft...12°
Weight...22,750#	Designer...Carver
Clearance...15'4"	Production...1987–Current

A popular boat since her introduction in 1987, the 390 Aft Cabin (better known as the Carver 38 Aft Cabin) is built on a conventional low-deadrise hull with balsa-cored hullsides and a solid fiberglass bottom. She's a good-looking boat for her length with a small aft deck, a large flybridge, and a remarkably complete interior plan. There are double berths in both staterooms; both heads have separate stall showers; and a full-size dinette is opposite the galley a few steps down from the salon level. Note that a slightly revised floorplan in 1993 eliminated the standard lower helm. The aft deck is small because the master stateroom is small, but there's room enough for some deck chairs. The flybridge is well-arranged with bench seating forward of the helm. With her relatively heavy displacement, she's a handful for gas engines. Standard 454-cid Crusaders will cruise the Carver 390 at a modest 14–15 knots with a top speed of around 24 knots. Optional 375-hp Cat diesels cruise at 21 knots and reach 25 knots wide open. ❑

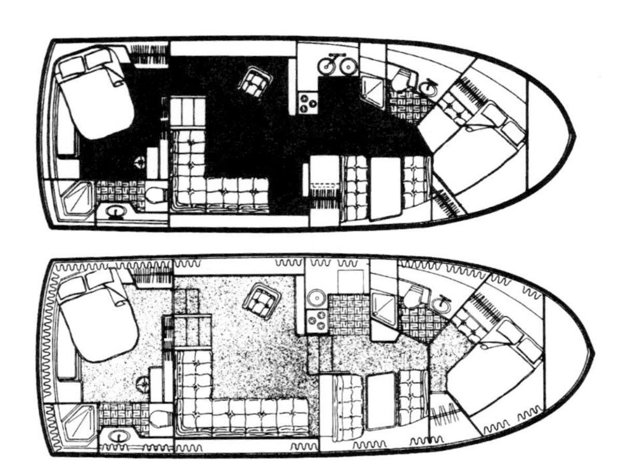

See page 235 for pricing information.

CARVER 390 COCKPIT MY

SPECIFICATIONS

LOA (w/platform)...40'2"	Fuel...336 gals.
Beam...13'3"	Hull Type...Modified-V
Draft...2'7"	Deadrise Aft...11°
Weight...17,400#	Designer...Carver
Clearance...15'0"	Production...1993–Current
Water...81 gals.	

The Carver 390 is actually a Carver 350 Motor Yacht with a cockpit extension. A few benefits accrue with the addition of a cockpit, especially in a small boat. Aside from the versatility that a cockpit provides (they make great dive or fishing platforms, and boarding is a lot less hassle), there is no doubt that a cockpit's extra waterline length adds much to the profile of any boat. (The 390 is indeed a better-looking boat than the 350 MY.) The interior in the 390 is surprisingly complete and includes two heads, a roomy galley, full dinette, and double berths in both staterooms. The staterooms are small, and storage space is at a premium (especially in the aft cabin), but the layout certainly delivers on its promise of big-boat accommodations. The optional lower helm is a popular feature, and the salon is open and quite spacious. Standard 454-cid gas engines will cruise the 390 Cockpit MY at 18 knots (26–27 knots top), and optional 200-hp Volvo diesels cruise around 18 knots as well (22 knots top). ❑

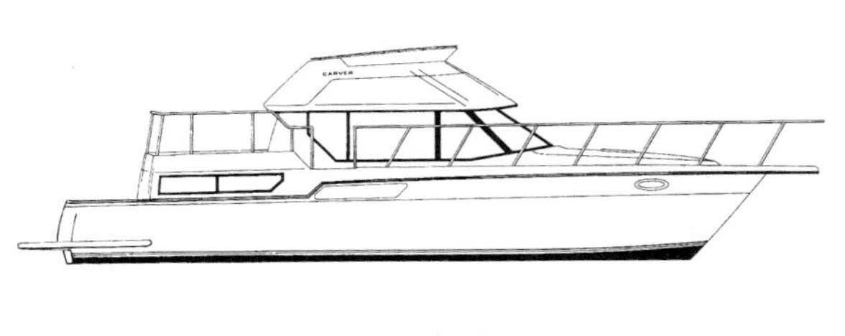

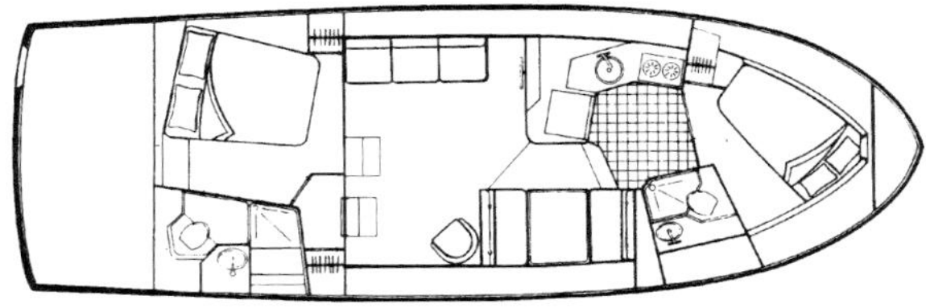

See page 235 for pricing information.

CARVER 42 MOTOR YACHT

SPECIFICATIONS

Length	42'0"	Fuel	400 gals.
Beam	15'0"	Hull Type	Modified-V
Draft	3'6"	Deadrise Aft	12°
Weight	28,500#	Designer	Carver
Clearance	20'0"	Production	1985–91
Water	170 gals.		

A distinctive boat with her stylish cabin windows, aggressive bridge overhang, and reversed radar arch, the 42 Motor Yacht was Carver's first attempt at the big-boat market. She sold well during her production years, and used Carver 42s remain popular today thanks to a roomy, well-arranged floorplan and generally attractive aftermarket prices. She was built on a conventional modified-V hull with balsa coring in the hullsides and a solid fiberglass bottom. Two floorplans were offered: one with a dinette and the other without. The salon seems big for a 42-footer, and both heads are fitted with stall showers. The aft deck has room for a table and chairs, and the bridge will seat six comfortably. Most were sold with a lower helm and hardtop, both optional. While diesels are certainly preferred in a boat this size, many Carver 42 MYs were sold with the standard 454-cid gas engines (just 13–14 knots cruise/23 knots top). Cat 375-hp diesels were a popular (and very desirable) option, since they'll cruise the boat an honest 20 knots. ❑

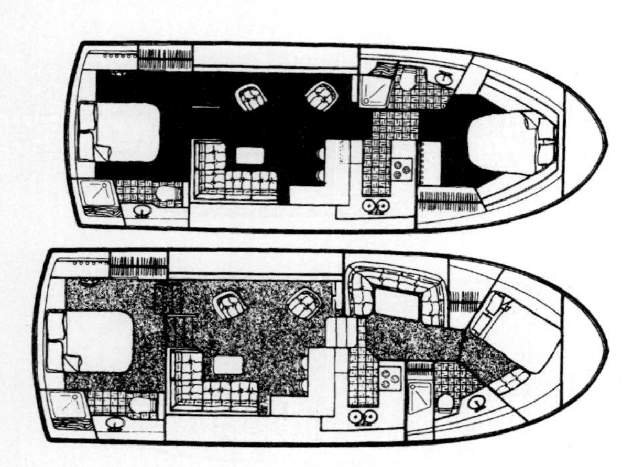

See page 235 for pricing information.

CARVER 430 COCKPIT MY

SPECIFICATIONS

LOA (w/platform)	44'3"	Fuel, Std.	280 gals.
Beam	14'0"	Fuel, Opt.	394 gals.
Draft	3'4"	Hull Type	Modified-V
Weight	25,620#	Deadrise Aft	11°
Clearance	15'4"	Designer	Carver
Water	91 gals.	Production	1991–Current

The 430 Cockpit MY is basically a Carver 390 MY with a stretched hull. The extra length of the cockpit gives the 430 a much-improved profile from the 390's stocky motor-yacht appearance. Carver packs a lot of living space into the interior of the 430 with double berths in both staterooms, a full dinette, a roomy salon area, and stall showers in both heads. (A lower helm station is optional.) A sliding door in the master stateroom leads directly out to the cockpit—a practical feature and a genuine convenience. The foredeck can be fitted with sun pads, and wide sidedecks make getting around the deckhouse easy and safe. While the cockpit dimensions aren't notably spacious, a transom door and swim platform are standard, and there's enough room in the cockpit for some diving gear or even a wet bike. The poor performance with standard 454-cid gas engines (14–15 knots cruise) is to be expected. Optional 375-hp Cat diesels will cruise around 21 knots and reach a top speed of 24–25 knots. ❑

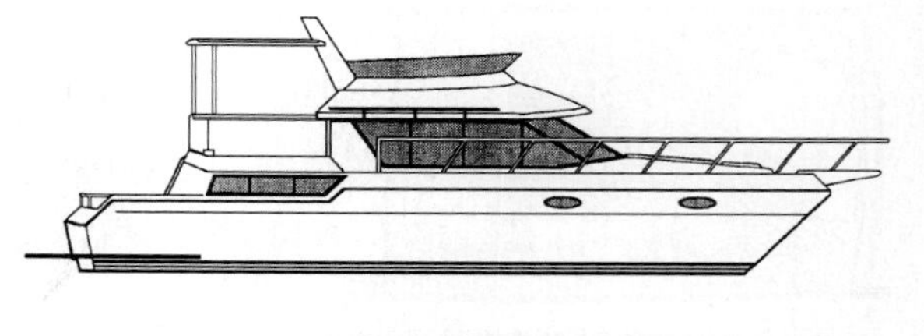

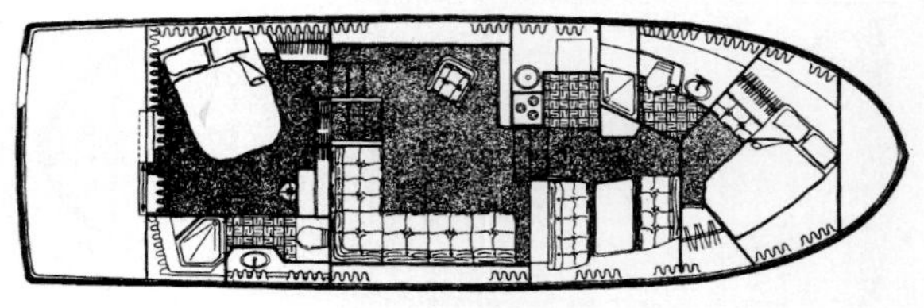

See page 235 for pricing information.

CHEOY LEE 32 TRAWLER

SPECIFICATIONS

Length	31'11"	Water	200 gals.
LWL	29'6"	Fuel	360 gals.
Beam	12'0"	Cockpit	NA
Draft	4'5"	Hull Type	Semi-Disp.
Weight	19,000#	Designer	Cheoy Lee
Clearance	NA	Production	1977–86

After more than a decade in production, the Cheoy Lee 32 Trawler earned a reputation as a durable and economical cruiser with a large cockpit and basic (but certainly adequate considering her size) interior accommodations. Her solid fiberglass hull incorporates a deep forefoot and a full-length skeg for low-speed stability and prop protection. The cockpit, decks, and cabintop are teak-over-fiberglass, and the window frames, doors, and caprail are also teak. Wide sidedecks make passage around the Cheoy Lee 32's deckhouse especially easy. Inside, the salon is narrow but functional. Visibility from the lower helm is very good thanks to large deckhouse windows. (Note the curved corner windows up front.) Traditional teak paneling and cabinetry are found throughout the interior. Outside, a spacious cockpit area makes the Cheoy Lee 32 an excellent utility-type boat for fishing, diving, or coastal cruising. (Fresh water should never be in short supply with 200 gallons aboard.) A single Lehman 6-cylinder diesel provides economical 7-knot performance with a cruising range of nearly 1,000 miles. ❑

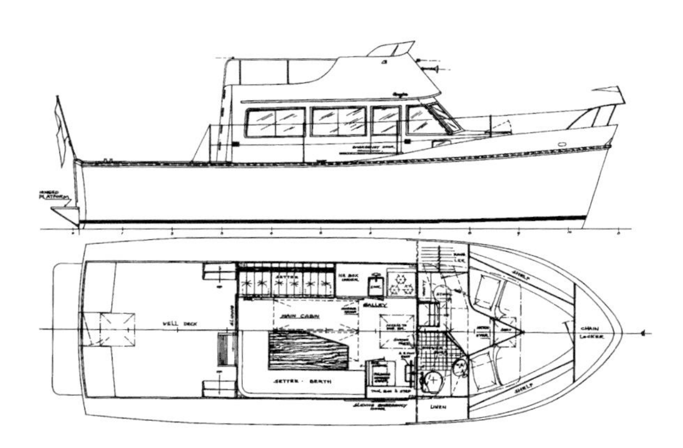

See page 235 for pricing information.

CHEOY LEE 35 TRAWLER

SPECIFICATIONS

Length	34'11"	Water	210 gals.
LWL	32'6"	Fuel	650 gals.
Beam	12'0"	Cockpit	NA
Draft	3'7"	Hull Type	Semi-Disp.
Weight	21,000#	Designer	Cheoy Lee
Clearance	NA	Production	1979–86

Boasting an impressive 1,200-mile cruising range at 7 knots, the Cheoy Lee 35 Trawler will require only a single load of fuel annually to meet the cruising needs of the average family. Her sedan profile is accented with teak trim and includes an extended hardtop for dingy storage and weather protection for the cockpit. At 21,000 lbs., the Cheoy Lee 35 is a heavy boat for her size, and the ride is predictably soft and comfortable. The full-length skeg protects the props and running gear in shallow water while adding stability offshore. The hull construction is solid fiberglass with a teak overlay on the decks and cabintop. Wide walkways and sturdy rails on both sides of the house provide secure access to the foredeck. Inside, the cabin arrangements will accommodate four with the salon settee converted in the normal fashion. The forward stateroom is fitted with a double berth as well as a dressing table and settee. Visibility from the lower helm is good in nearly all directions, and salon access to the single Lehman diesel is excellent. ❑

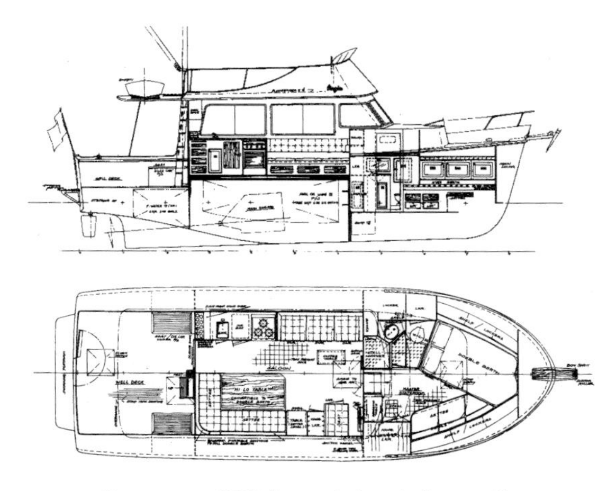

See page 235 for pricing information.

CHEOY LEE 40 LRC

SPECIFICATIONS

Length	40'0"	Water	250 gals.
LWL	35'8"	Fuel	650/710 gals.
Beam	14'6"	Hull Type	Semi-Disp.
Draft	4'8"	Designer	Cheoy Lee
Weight	38,000#	Production	1973–86

The Cheoy Lee 40 Long Range Cruiser is a traditional trunk cabin trawler with only her rounded window frames and radar mast disturbing her otherwise plain-Jane profile. Hull construction is solid fiberglass, and a teak overlay is applied on the decks and cabintop. A deep full-length keel protects the props and running gear and gives the owner a valuable margin of safety in shallow waters. Her all-teak interior originally had twin single berths in the aft stateroom, but later models were fitted with a double berth. Visibility from the lower helm station is very good. With the salon settee converted, overnight accommodations are provided for up to six. The great appeal of owning a trawler, of course, is the ability to cruise economically for long distances. With nearly a thousand nautical miles of range, the Cheoy Lee 40 Long Range Cruiser is easily capable of serious bluewater passages. Her reliable twin 6-cylinder Lehman diesels burn only 6 gph at a steady 8-knot displacement speed. Top speed is about 11–12 knots. ❑

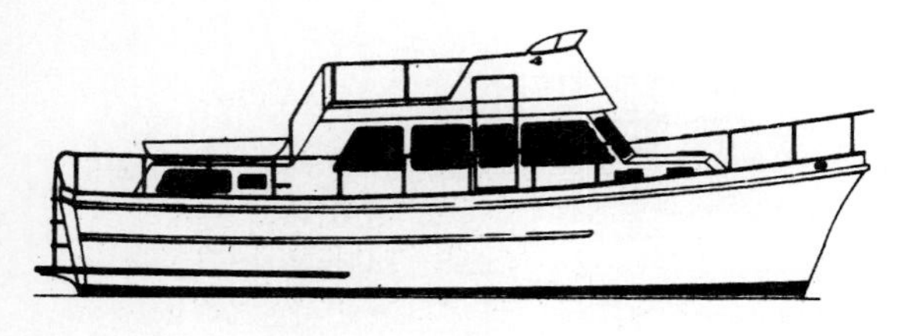

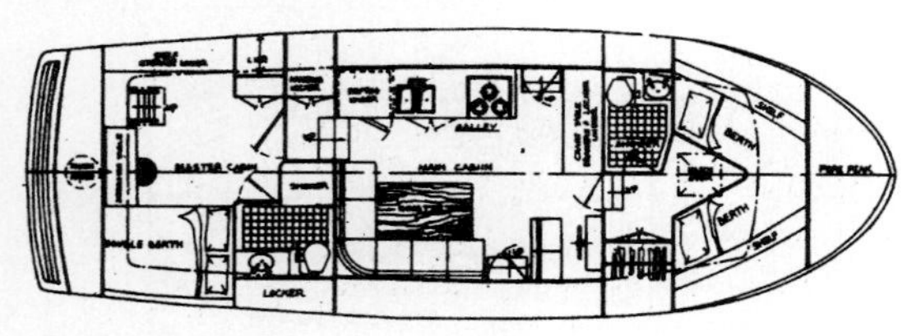

See page 235 for pricing information.

CHEOY LEE 46 TRAWLER

SPECIFICATIONS

Length	45'11"	Water	510 gals.
LWL	42'0"	Fuel	820 gals.
Beam	14'8"	Hull Type	Disp.
Draft	4'8"	Designer	Cheoy Lee
Weight	49,200#	Production	1978–81

The Cheoy Lee 46 Trawler has the distinction of being the smallest double-deck production yacht ever built. A sturdy long-range cruiser with an upright profile, she's constructed on a solid fiberglass displacement hull with a keel deep enough to protect the running gear. Her floorplan is arranged with the galley on the deckhouse level and three staterooms below. Because the beam is moderate and the sidedecks are wide, the salon dimensions are fairly compact for a 46-foot yacht. Aft in the salon is a staircase leading down to the master stateroom with its centerline queen berth and direct access to the stand-up engine room. The large midships VIP stateroom is nearly equal to the master stateroom in size and comforts but lacks an en-suite head. Other features include a full teak interior, day berth in the wheelhouse, teak decks, and a semi-protected aft deck. The only engines offered were 120-hp Lehmans. She'll cruise economically at a 7–8 knots (burning about 6 gph) with a range of 800–850 miles. ❑

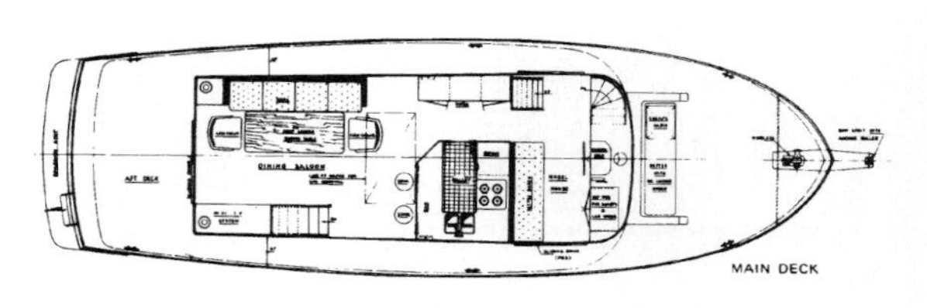

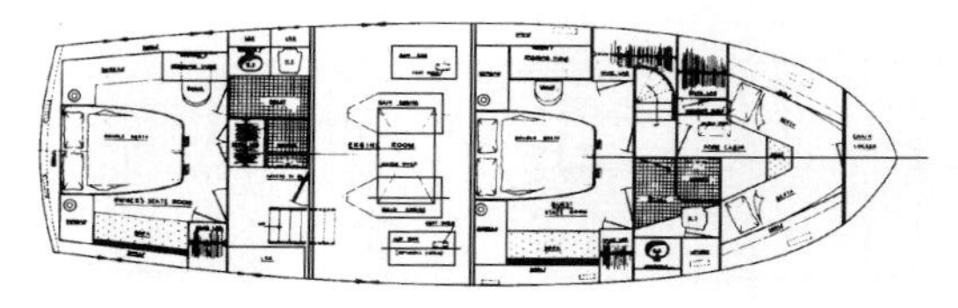

See page 235 for pricing information.

CHEOY LEE 50 TRAWLER

SPECIFICATIONS

Length (w/Pulpit)	51'6"	Water	1,200 gals.
LWL	NA	Fuel	1,600 gals.
Beam	15'7"	Hull Type	Disp.
Draft	5'7"	Designer	Cheoy Lee
Weight	67,000#	Production	1974–80

The Cheoy Lee 50 Trawler is another in the series of long range yachts that Cheoy Lee used to build for the American market. Originally a raised pilothouse sedan, in 1979 the Tri Cabin model (pictured above) featured an aft cabin below an elevated main deck. The Cheoy Lee 50 is an all-fiberglass boat built on a heavy displacement hull with a deep keel and protected props. The aft cabin floorplan comes as a surprise considering her sedan-style appearance. Reached from a stairway in the salon, this roomy stateroom comes with a centerline double berth, private head, and a writing desk. With her very wide sidedecks the salon dimensions are relatively narrow. Note the bridge overhang around the decks and cockpit. Obviously built for those seeking tradition over glitz, the interior is finished with teak cabinetry and woodwork throughout. There were several engine choices offered in the Cheoy Lee 50 including twin 120-hp Lehmans and 210-hp Cats. With 1,600 gallons of fuel her cruising range can exceed 2,000 miles at 7 knots. ❑

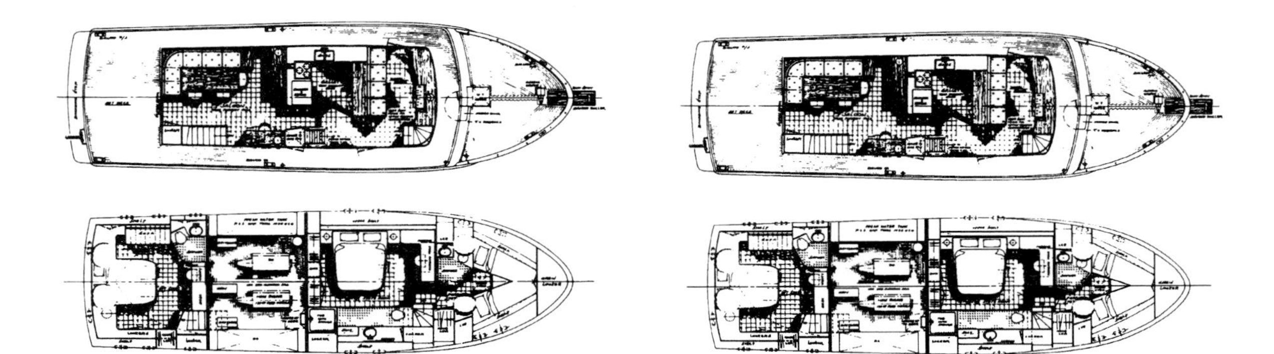

See page 235 for pricing information.

CHEOY LEE 52 EFFICIENT MY

SPECIFICATIONS

Length	51'11"	Water	450 gals.
LWL	47'6"	Fuel	1,000 gals.
Beam	15'6"	Hull Type	Semi-Disp.
Draft	3'10"	Designer	Cheoy Lee
Weight	51,500#	Production	1984–Current
Clearance	16'1"		

The 52 Efficient MY is an unusually distinctive design with a somewhat plain-Jane appearance to go with her economical operation. She's designed with the emphasis on interior accommodations and long-range comfort. Featuring a spacious wide-body salon and a deckhouse galley, the 52 EMY's teak-paneled, three-stateroom floorplan includes a huge master stateroom aft and two guest cabins forward. The latest layout featuring a full-beam amidships guest cabin was introduced in 1987. The salon is open to the galley, and a stairway from the wheelhouse leads to the flybridge. The covered aft deck area has space for a couple of deck chairs, and a door in the cockpit provides direct access to the master stateroom. At her 9-knot hull speed, the Cheoy Lee 52 has a range of nearly 1,500 miles. Twin GM 8.2 diesels are standard and deliver cruising speeds in the neighborhood of 13–14 knots, and the optional 375-hp Cats will cruise around 18–19 knots. Note that the Cheoy Lee 47 Efficient MY is the same boat without the cockpit. ❑

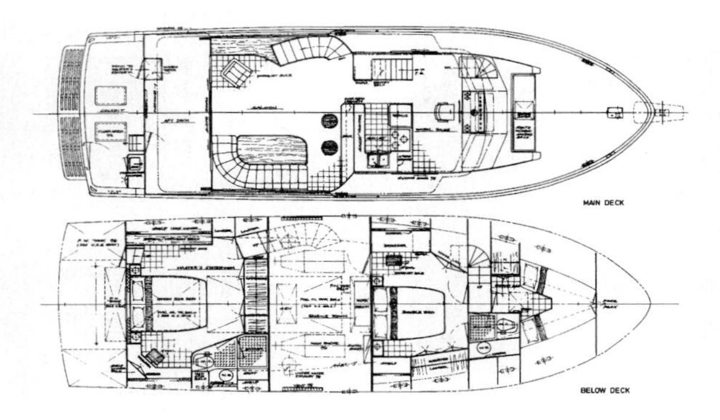

See page 235 for pricing information.

CHEOY LEE 55 LONG RANGE MY

Cheoy Lee 55 Walkaround

Cheoy Lee 55 Wide Body

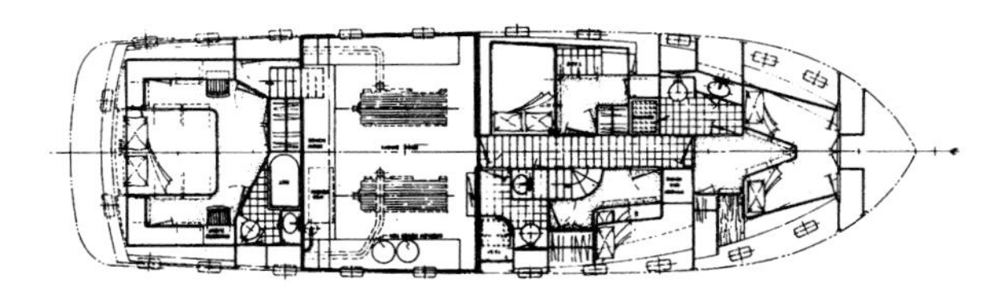

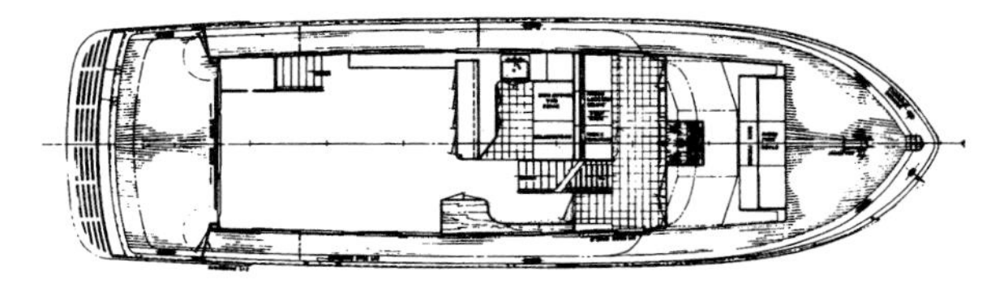

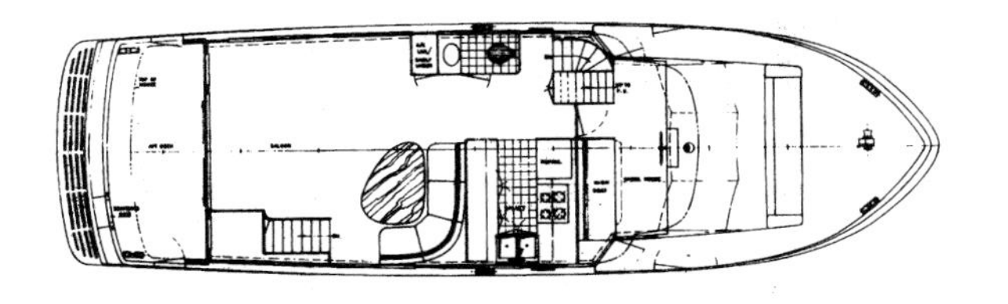

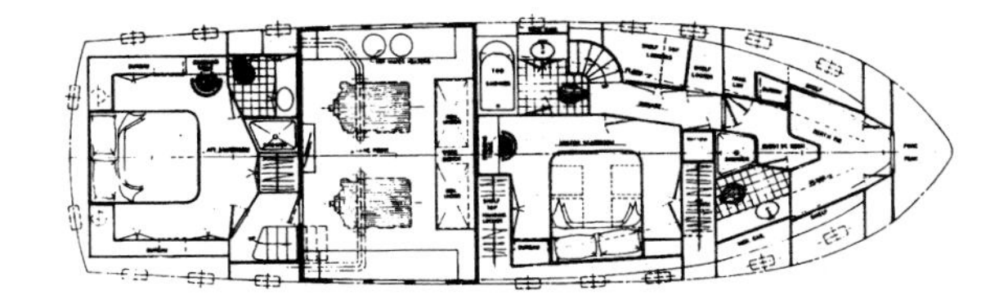

SPECIFICATIONS

Length	54'10"	Water	450 gals.
LWL	50'0"	Fuel	2,500 gals.
Beam	17'2"	Hull Type	Semi-Disp.
Draft	5'4"	Designer	C. Wittholz
Weight	80,000#	Production	1981–Current

It doesn't take a brain surgeon to recognize that the Cheoy Lee 55 Long Range MY is a serious bluewater passagemaker. With 2,500 gallons of fuel she has genuine transatlantic capabilities. Her solid fiberglass hull has the same full displacement underbody as the earlier Cheoy Lee 55 Trawler (1975–80), but an improved deck and superstructure profile have given her a more modern appearance. Two versions of this boat are available: the standard four-stateroom model with full sidedecks, and the Wide Body introduced in 1987 with an enlarged salon and one less stateroom below. (Significantly, only a couple have been built in the widebody configuration.) The lower deck living areas in both versions are separated fore and aft by a spacious walk-in engine room. Topside, there's room on the bridge for a dinghy, and a bridge overhang protects the decks all around the house. Standard 210-hp Caterpillar diesels will cruise efficiently at 9.5 knots burning approximately 7 gph. Note that the Cheoy Lee 61 Cockpit MY is the same boat with a 6-foot cockpit extension. ❑

See page 236 for pricing information.

CHEOY LEE 66 LONG RANGE MY

SPECIFICATIONS

Length	65'6"	Water	700 gals.
LWL	59'0"	Fuel	2,300 gals.
Beam	18'0"	Hull Type	Disp.
Draft	5'3"	Designer	C. Wittholz
Weight	87,000#	Production	1983–Current

A popular and good-selling yacht, the Cheoy Lee 66 Long Range MY is an updated version of the earlier Cheoy Lee 66 Trawler first imported back in 1978. Restyled outside and updated with a fresh interior, the new model was introduced in 1983 and aimed at the high end of the luxury trawler market. The Cheoy Lee 66 will likely satisfy the most demanding owners with her sturdy appearance and extravagant cruising accommodations. The popular five-stateroom, galley-up layout provides overnight berths for up to twelve people. A four-stateroom, galley-down layout is also available. The sidedecks are very wide and protected by a bridge overhang, and the flybridge is immense. The Wide Body model, with a full-width salon and enlarged wheelhouse, was introduced in 1985 and currently accounts for about half of the 66 MY production. A comfortable offshore passagemaker, twin 350-hp 8V71 (or 320-hp Cat) diesels cruise the Cheoy Lee 66 at 10 knots with a top speed of 12–13 knots. The cruising range is approximately 2,000 nautical miles. ❑

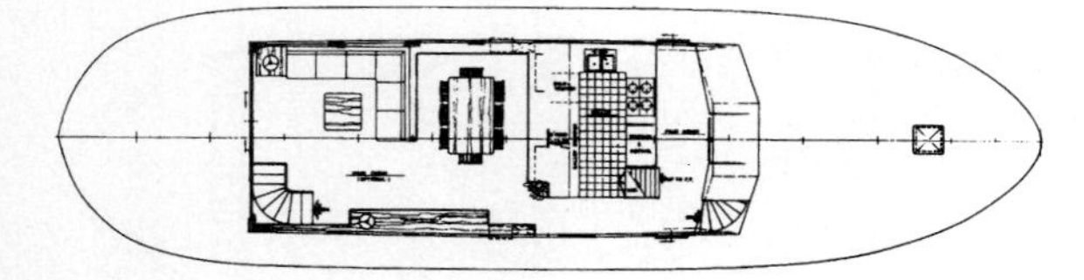

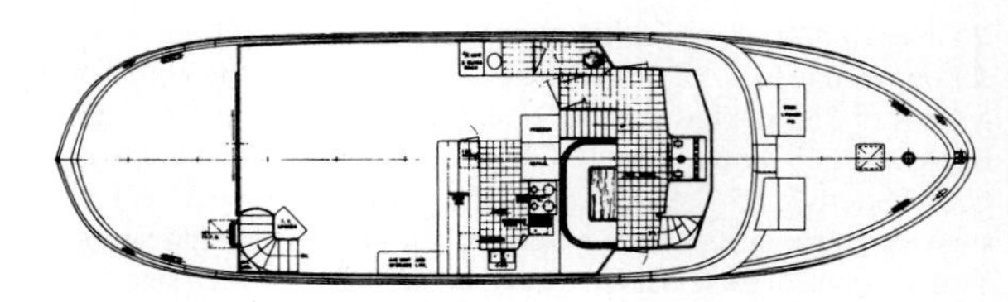

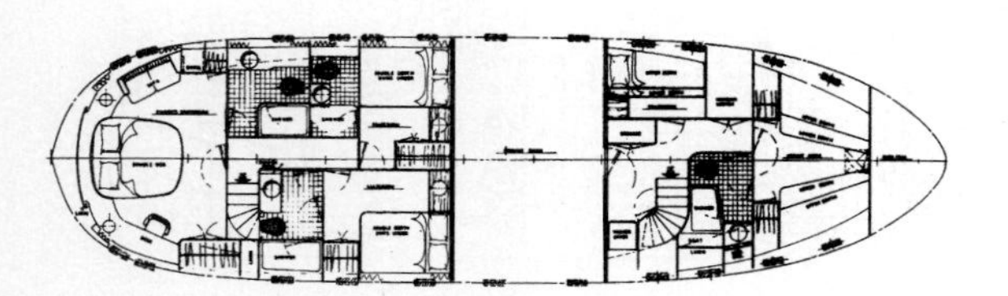

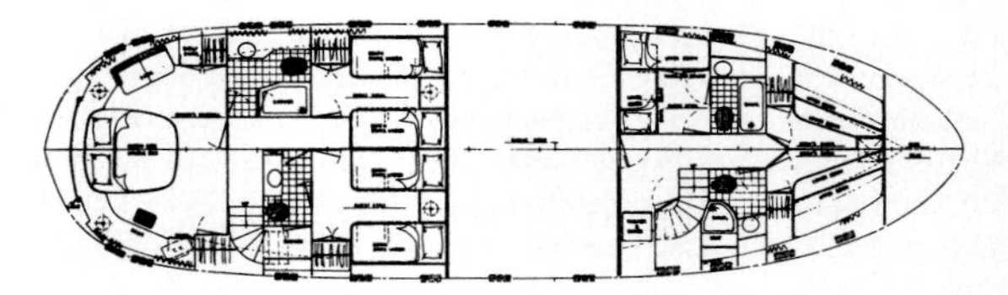

See page 236 for pricing information.

CHEOY LEE 66 FAST MY

SPECIFICATIONS

Length	66'0"	Cockpit	NA
LWL	57'0"	Water	370 gals.
Beam	19'0"	Fuel	1,670 gals.
Draft	4'6"	Hull Type	Modified-V
Weight	83,500#	Designer	Tom Fexas
Clearance	19'10"	Production	1984–1987

When she was introduced in 1984 the Fexas-designed 66 Fast Motor Yacht represented a radical departure from normal motor yacht standards. Even today her dramatic European profile, high-tech construction, and innovative floorplan continue to make this yacht the subject of interest among designers and boating enthusiasts. As production models go, the Cheoy Lee 66 Fast MY was not a notable sales success since only four were constructed. She was built on the same lightweight Airex-cored hull used in the production of the Cheoy Lee 66 Sport Yacht. By increasing the shaft angles to the maximum possible, Fexas was able to place the motors well aft in the hull and free up some interior space forward. The five-stateroom floorplan is unique—the owner's stateroom is amidships, and the aft stateroom serves as guest quarters. The engine room is a tight fit. A superb performer for her size, standard 12V71 diesels will cruise the 66 Motor Yacht at a surprising 21 knots and deliver a top speed of 23–24 knots. ❑

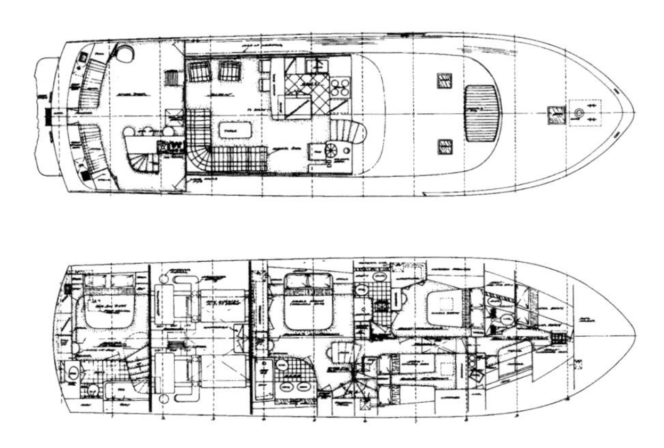

See page 236 for pricing information.

CHEOY LEE 83 COCKPIT MY

SPECIFICATIONS

Length	82'11"	Water	700 gals.
LWL	72'10"	Fuel	3,000 gals.
Beam	21'1"	Hull Type	Modified-V
Draft	5'4"	Deadrise Aft	NA
Weight	143,000#	Designer	Tom Fexas
Clearance	17'10"	Production	1987–Current

A few years ago the concept of a production 80-foot fiberglass motor yacht was just that—an idea whose time had not yet arrived. Today, it's quite a different story, and the Cheoy Lee 83 Cockpit MY is just one of several production mega-yachts affluent buyers can choose from. Among them, the Cheoy Lee 83 has been one of the most successful. She's another Tom Fexas design with the sleek profile and lightweight fully cored construction for which he and Cheoy Lee have become famous. A semi-custom yacht, her standard floorplan includes two huge master staterooms, two guest staterooms, and two more staterooms forward for the crew. Separate engine rooms allow a corridor to connect the staterooms, and pump rooms forward house the generators. Standard 900-hp (12V71) diesels will deliver a surprising 20-knot cruising speed, and the larger 1,200-hp Cats will cruise at about 23 knots. At hull speed (11 knots), the range is 2,500+ miles. Note that the Cheoy Lee 77 MY is the same boat without a cockpit. ❑

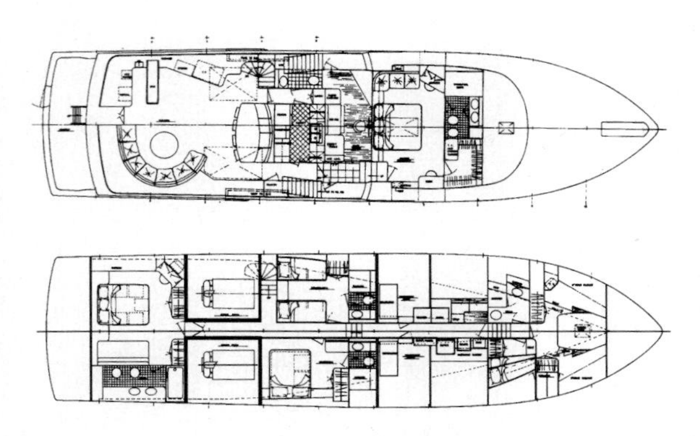

See page 236 for pricing information.

CHRIS CRAFT 350 CATALINA

SPECIFICATIONS

Length	35'1"	Water	55/100 gals.
LWL	NA	Fuel	180/250 gals.
Beam	13'1"	Hull Type	Modified-V
Draft	2'10"	Designer	Chris Craft
Weight	17,229#	Production	1974–87
Clearance	10'8"		

The Chris 350 Catalina is one of the most popular family cruising boats produced in recent years. Her topheavy profile is certainly less than inspiring, but the interior accommodations are those of a much larger boat. She's built in a solid fiberglass hull with moderate beam, high freeboard, and a relatively flat bottom. There have been a total of four different floorplan arrangements offered over the years—all with two staterooms, two heads, and the galley up. Twin berths were standard in the aft stateroom in early models, although an optional double bed was always available. Outside, the helm is positioned just forward of the aft deck in a manner that allows the captain and guests to enjoy the ride together. Wide sidedecks make passage around the house easy and safe. Additional features include foredeck seating, a stall shower aft, and a reasonably spacious engine room. Among several engine options, twin 235-hp Crusaders will cruise at 14–15 knots and reach 24 knots top. Note that the fuel capacity was increased to 250 gallons in 1983. ❑

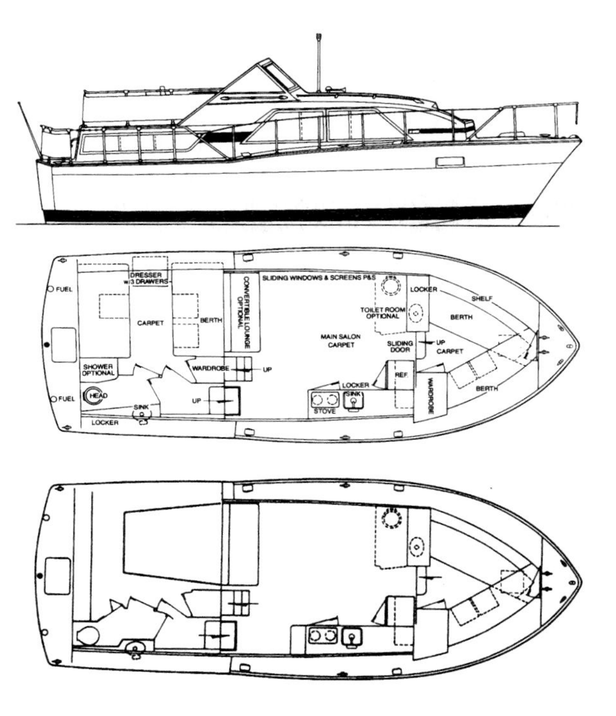

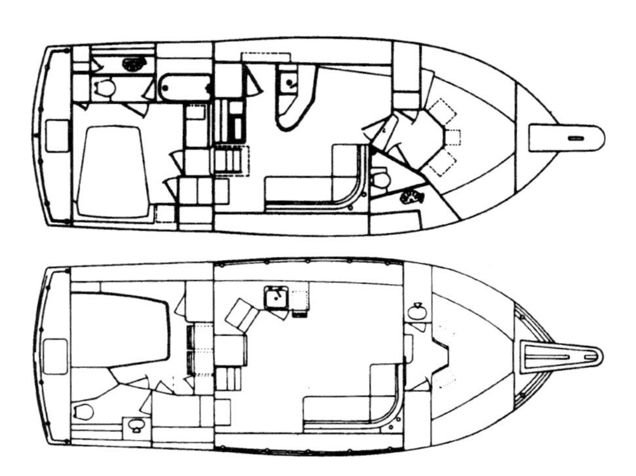

See page 236 for pricing information.

CHRIS CRAFT 380 CORINTHIAN

SPECIFICATIONS

Length	38'0"	Water	65 gals.
LWL	33'6"	Fuel	400 gals.
Beam	14'0"	Hull Type	Modified-V
Draft	3'0"	Deadrise Aft	NA
Weight	22,500#	Designer	Chris Craft
Clearance	12'2"	Production	1978–86

The versatile Chris 380 Corinthian has long been a favorite with family cruisers because of her excellent indoor and outdoor accommodations and affordable price. True, she has the slab-sided lines of a houseboat, and she gets blown around by the slightest breeze, but her multi-level layout includes everything required for enjoyable cruising activities—including a cockpit. There were several floorplan modifications and upgrades made to the 380 Corinthian over the years, and the two shown below were among the most recent. (A queen berth in the master stateroom became standard in 1981.) Notable design features include a full-size cockpit door to the master stateroom, a tub/shower in the aft head, an extended flybridge with seating for a crowd, and a sliding deck door at the lower helm. The sidedecks are wide, and a transom door and swim platform were standard. Not the softest ride in a chop, the 380 Corinthian still performs better than she looks. Standard 454-cid gas engines will cruise at 16 knots and reach a top speed of 26–27 knots. ❑

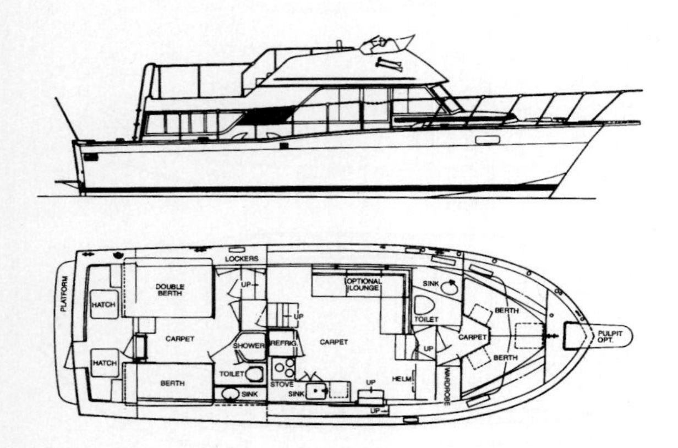

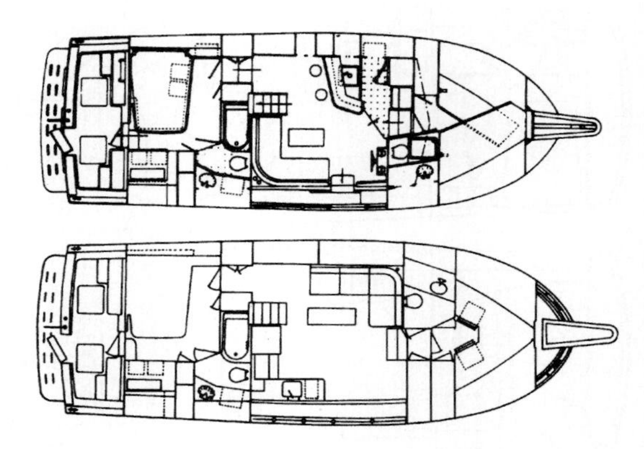

See page 236 for pricing information.

CHRIS CRAFT 381 CATALINA

SPECIFICATIONS

Length	38'0"	Water	65 gals.
LWL	NA	Fuel	410 gals.
Beam	14'0"	Hull Type	Modified-V
Draft	3'0"	Deadrise Aft	NA
Weight	21,600#	Designer	Chris Craft
Clearance	11'7"	Production	1980–89

The 381 Catalina Double Cabin is an enlarged version of the popular Chris 350 Catalina introduced a few years earlier. Built on a solid-fiberglass, low-deadrise hull with a relatively wide beam, the 381 has an enormous interior for her size as well as a small-but-entertainment-friendly helm/aft deck platform. She may not be easy on the eye from the outside (the comparison to a beached whale has been made), but her interior accommodations include two roomy staterooms, stall shower enclosures in both heads (a tub/shower aft, actually) and full walkaround decks—not bad for a 38' cruiser. A lower helm was never offered, because the front windshield area is fiberglassed-over to provide foredeck seating. Note that a walkaround queen bed in the master stateroom became standard in 1983. Topside, the helm is level with the aft deck, so the helmsman isn't separated from the passengers. The 381 Catalina will cruise at 17 knots with standard 330-hp gas engines and reach a top speed of 26–27 knots. ❑

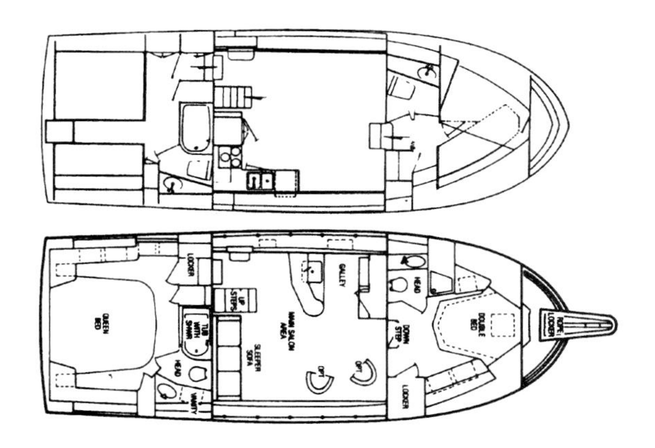

See page 236 for pricing information.

CHRIS CRAFT 410 MOTOR YACHT

SPECIFICATIONS

Length	41'0"	Water	100 gals.
Beam	14'0"	Fuel	350 gals.
Draft	3'3"	Hull Type	Modified-V
Weight	26,565#	Designer	Chris Craft
Clearance	15'10"	Production	1972–86

It's a shame that Chris Craft never credits the designers of their boats in any company literature, because whoever drew the plans for the original Chris Craft 41 MY had his finger right on the public pulse. This popular double-cabin motor yacht had a 15-year production run—the longest of any modern motor yacht design except the Hatteras 53 MY. Offering a spacious interior layout, the Chris 410's flush deck profile makes her appear larger than she actually is. While her lines are conservative by today's standards, the addition of a restyled and enlarged flybridge in 1980 enhanced her profile considerably. Inside, the accommodations are comfortable and well-arranged, especially in the master stateroom where a queen berth (optional in earlier models) became standard in 1981. Various decor schemes have been used over the years—some good and others not so good. Big-block 454-cid gas engines were standard and provide a cruising speed of 15–16 knots and 24 knots top. With just 350 gallons of fuel, the 410's range is limited to about 200 miles. ❑

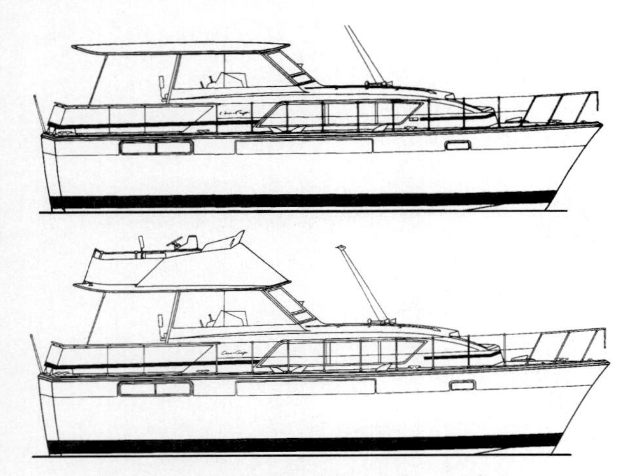

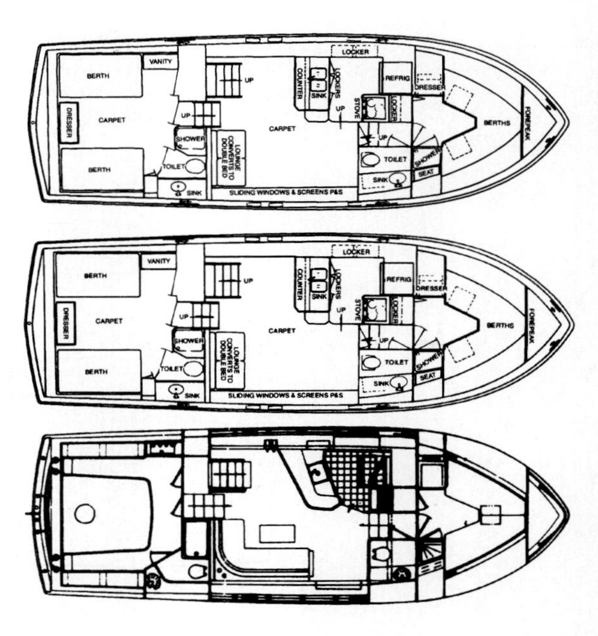

See page 236 for pricing information.

CHRIS CRAFT 426/427 CATALINA

SPECIFICATIONS

Length	42'0"	Water	160 gals.
LWL	37'7"	Fuel	400 gals.
Beam	14'9"	Hull Type	Modified-V
Draft	3'6"	Deadrise Aft	13°
Weight	33,000#	Designer	A. Nordtvedt
Clearance	12'10"	Production	1985–90

The Chris 427 Catalina pictured above began life in 1971 as the Uniflite 42 Double Cabin. She turned out to be one of Uniflite's most popular boats and earned a reputation for sturdy construction and a practical, well-arranged interior. In 1984 Uniflite ceased operations. Chris Craft bought the molds (as well as the old Uniflite production facility in Bellingham, WA) and resumed production of this model the following year, calling her the Chris 425 Catalina Double Cabin. While she looked exactly like the old Uniflite, the new Catalina had a fresh interior layout with walkaround queen berths fore and aft—something the Uniflite 42 never had. The 425 designation lasted only a year; in 1986 she was called the 426 Catalina and benefited from a restyled and enlarged flybridge. The 427 designation came in 1987, but there were no further updates of significance. With standard 454-cid gas engines the cruising speed of the Catalina Double Cabin is 17–18 knots with about 27 knots at full throttle. The optional 375-hp Cat diesels will cruise at 21+ knots. ❑

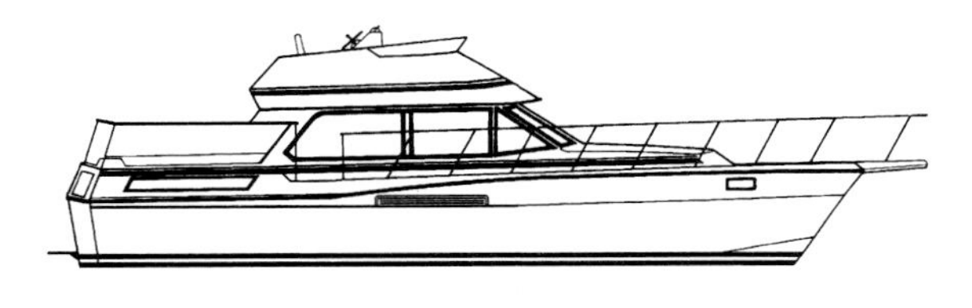

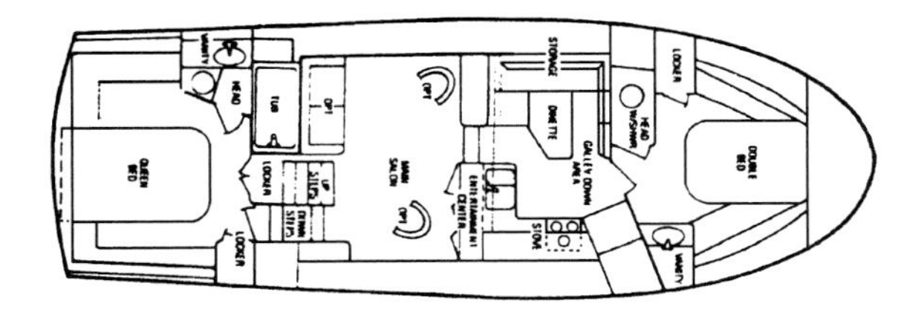

See page 236 for pricing information.

CHRIS CRAFT 45 COMMANDER MY

SPECIFICATIONS

Length	45'0"	Water	200 gals.
LWL	NA	Fuel	500 gals.
Beam	15'0"	Hull Type	Modified-V
Draft	4'2"	Deadrise Aft	NA
Weight	39,400#	Designer	Chris Craft
Clearance	16'3"	Production	1972–81

The Chris Craft 45 Commander MY is a traditional flush-deck design with the classic Chris Craft motor yacht appearance. As a weekend-retreat or coastal-cruising yacht the accommodations are more than adequate with a huge main salon dominating the floorplan. Early models were built with roomy two-stateroom interior layouts and are remembered today for a unique, full-width forward stateroom with the head fitted in the forepeak. A popular three-stateroom arrangement with a deckhouse galley became available in 1977, and a more conventional two-stateroom interior with the galley down followed the next year. Interestingly, a dinette was never offered in the 45 Commander MY—a rare omission in a boat this size. The flybridge proved a popular option, although a few were sold as hardtops. A comfortable yacht with a reputation for good handling characteristics, many 45 Commander MYs were powered with 325-hp GM 8V71 diesels for an economical cruising speed of 16 knots and a top speed of about 19 knots. The cruising range is around 225–250 nautical miles. ❑

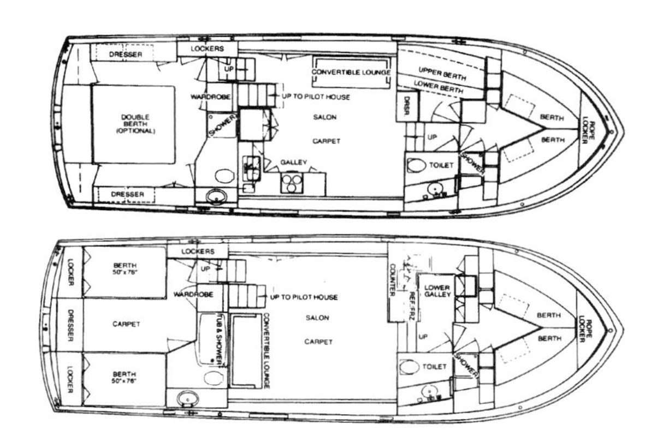

See page 237 for pricing information.

CHRIS CRAFT 47 COMMANDER MY

SPECIFICATIONS

Length	47'0"	Water	80 gals.
Beam	15'0"	Fuel	350 gals.
Draft	4'2"	Hull Type	Modified-V
Weight	41,500#	Deadrise Aft	NA
Clearance	13'0"	Production	1966–76

The Chris Craft 47 Commander MY may be dated by today's standards, but many dealers and brokers have acquired from her a firsthand appreciation of the way Chris Craft used to build boats when that company's name stood foremost in the industry. The massive solid-fiberglass hull of the 47 Commander is probably bulletproof. The gelcoat is long-lasting, and all of the deck hardware was custom-crafted at the Chris Craft factory in Algonac, MI. A comfortable cruising yacht, the three-stateroom interior layout proved very popular and will be seen on nearly all used models. In 1973 the model designation was changed to Flush Deck, but no major changes were made in design. One common criticism is the small fuel and water tankage, which seriously limits her effective cruising range. Various diesel engine options were offered over the years with the GM 8V53s probably the most popular. They provide a very sedate cruising speed of 12–13 knots depending on load. It is widely known that the 47 Commander's handling and seakeeping qualities are very good. ❑

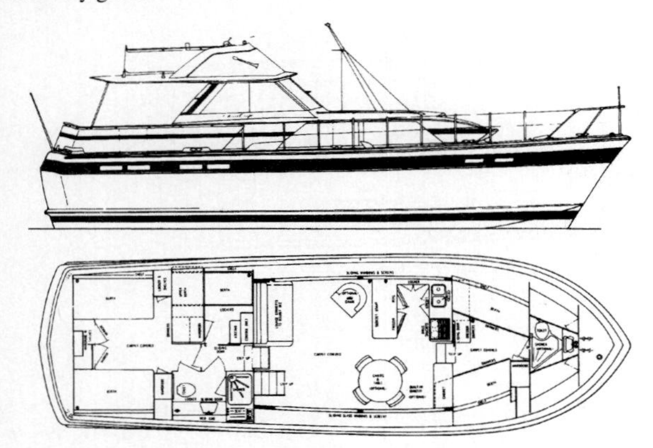

See page 237 for pricing information.

CHRIS CRAFT 480 CATALINA

SPECIFICATIONS

Length	48'0"	Fuel	590 gals.
Beam	14'9"	Cockpit	NA
Draft	3'6"	Hull Type	Modified-V
Weight	34,000#	Deadrise Aft	13°
Clearance	12'10"	Designer	A. Nordtvedt
Water	160 gals.	Production	1985–89

Called the 480 Corinthian when she was introduced in 1985, this is actually the old Uniflite 48 Yacht Fisherman that Chris Craft acquired when they took over the assets of Uniflite in 1984. The big change was her new floorplan with walkaround double berths in both staterooms—something the Uniflite 48 YF never had. In 1987 Chris Craft began calling her the 480 Catalina—a name that stuck until she went out of production in 1989. A popular boat, her interior accommodations are roomy and well-arranged, and there's plenty of entertaining space on the large aft deck. Keep in mind that this is basically a Chris 426 Double Cabin with a six-foot cockpit extension and extra fuel. The additional hull length makes her a better-looking boat and might even add a knot to her performance. The popular 375-hp Cats will cruise the Chris 480 Catalina around 21 knots with a top speed of 25–26 knots—impressive speeds for such modest power and reflective of a good, fast-bottom hull design. ❑

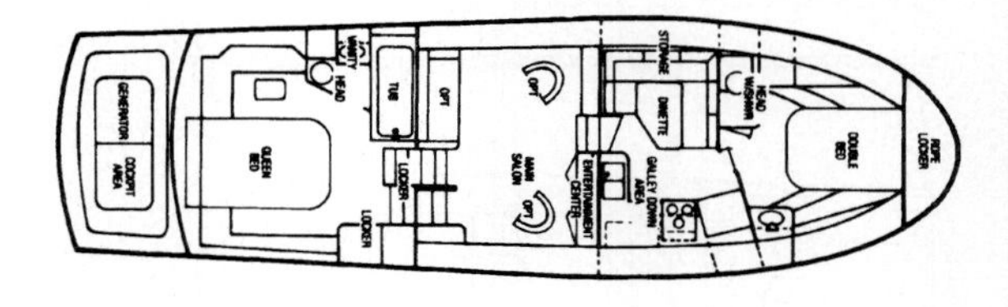

See page 237 for pricing information.

CHRIS CRAFT 500 MOTOR YACHT

SPECIFICATIONS

Length	50'6"	Fuel	600 gals.
Beam	15'3"	Hull Type	Modified-V
Draft	4'4"	Deadrise Aft	4°
Weight	54,000#	Designer	D. Martin
Clearance	17'1"	Production	1985–89
Water	160 gals.		

This Chris 500 MY (pictured above) has a fairly complex history. She was introduced in 1977 as the Pacemaker 46 MY, a classic David Martin design with handsome lines. She remained in production until Pacemaker went out of business in 1980. Uniflite then acquired the tooling, and production resumed in 1981 as the Uniflite 46 MY. By late 1983 Uniflite designers had stretched the mold four feet, and the Uniflite 50 Motor Yacht was introduced. It was a short production run, because Uniflite closed in 1984. Chris Craft acquired the molds and resumed production of both boats in 1985 calling them the 460 and 500 Constellations. (The 460 ended production in 1988.) For those who admire the traditional styling of a flush-deck layout, the Chris 500 has much to offer. A tremendous amount of living space is available in her three-stateroom floorplan, including a fully enclosed and paneled aft deck that serves as a second salon. The Chris 500 MY has a cruising speed of 18 knots with standard 550-hp 6V92s and a top speed of 21 knots. ❏

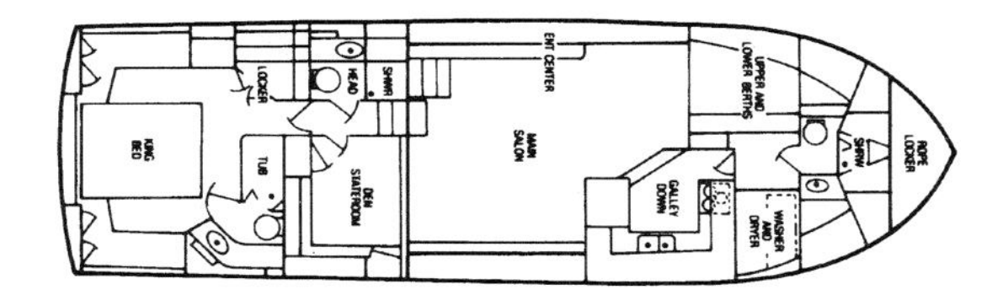

See page 237 for pricing information.

CHRIS CRAFT 501 MOTOR YACHT

SPECIFICATIONS

Length	50'8"	Fuel	778 gals.
Beam	15'5"	Hull Type	Modified-V
Draft	4'6"	Deadrise Aft	4°
Weight	49,000#	Designer	Chris Craft
Clearance	NA	Production	1987–90
Water	260 gals.		

The Chris Craft 501 MY (called the 50 Constellation in 1990) is yet another refinement of the original Pacemaker 46 hull. A handsome and good-selling yacht, her extended-deckhouse layout is aimed at those who prefer the increased entertainment capabilities of a full-width salon. While the 501 was built on a stretched mold, the deck and superstructure were new and designed by Chris Craft. By moving the pilothouse forward and relocating the engines further aft, the 501 gains an expansive salon, genuine privacy in the aft cabin, and an engine room with standing headroom. A spiral staircase leads from the salon down to a roomy master stateroom where a tub is located in the adjoining head compartment. A new galley-up arrangement joined the original galley-down layout in 1990. Since the walkaround sidedecks are eliminated in this wide-body configuration, Chris Craft designers thoughtfully added a small aft deck suitable for line-handling duties. Cruising speed of the Chris 501 MY with 550-hp GM 6V92TA diesels is 17 knots, and the top speed is 18–19 knots. ❑

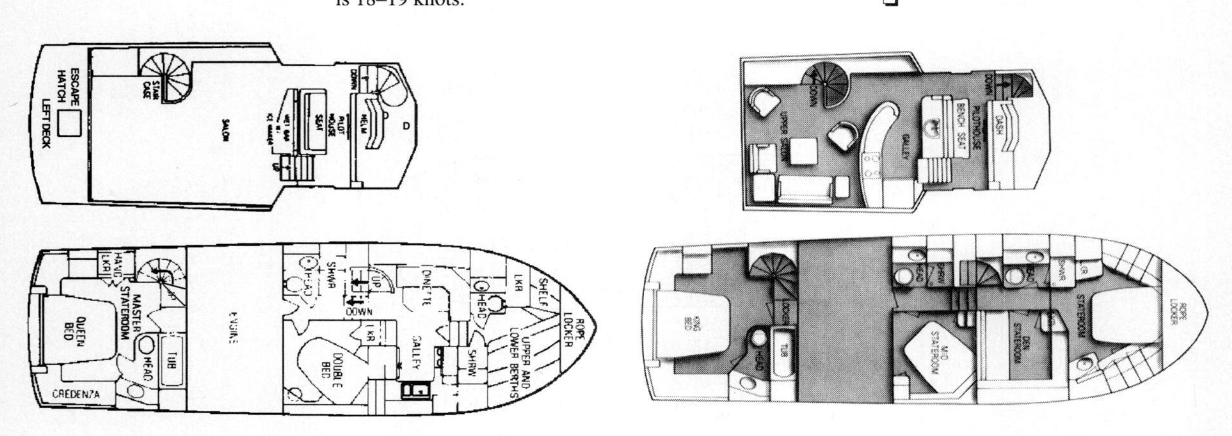

See page 237 for pricing information.

CHRIS CRAFT 55 FLUSH DECK MY

SPECIFICATIONS

Length	55'0"	Water	162 gals.
Beam	16'6"	Fuel	570 gals.
Draft	4'0"	Hull Type	Modified-V
Weight	57,800#	Designer	Chris Craft
Clearance	16'9"	Production	1970–77

When the 55 Flush Deck was introduced in 1970, she was the largest fiberglass yacht in the Chris Craft fleet. Like many of the larger motor yachts of her era, the 55 was designed to be operated as a crewed vessel. She was built using a unique three-piece hull—the bottom is one mold and the hullsides are two separate pieces. A molded spray rail runs around the hull where the bottom and sides are joined, and a long keel below adds directional stability. The first 55s were hardtop models with a covered afterdeck and semi-enclosed helm. In 1975 the Enclosed Flush Deck model turned the aft deck into a huge full-width and fully paneled salon. The interior is arranged with the owner's stateroom and guest cabin aft, a mid-level dining salon and galley, and a guest stateroom at the bow. Most of the 55 Flush Deck yachts were powered with 425-hp 8V71 diesels which will cruise at 15 knots (18 knots top). Due to her limited fuel capacity the cruising range is less than 200 miles. ❑

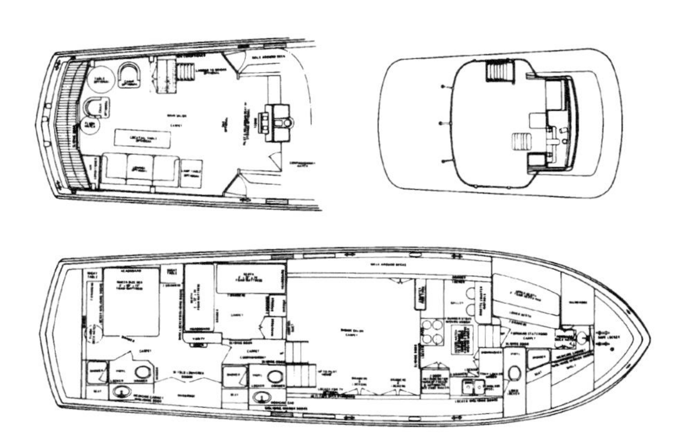

See page 237 for pricing information.

CRUISERS 3850 AFT CABIN

SPECIFICATIONS

Length	39'0"	Fuel	400 gals.
Beam	14'0"	Hull Type	Modified-V
Draft	3'4"	Deadrise Aft	16°
Weight	20,000#	Designer	Cruisers
Clearance	17'3"	Production	1991–Current
Water	100 gals.		

Face it, there are no really good-looking double cabin boats under 40 feet. What the Cruisers 3850 Aft Cabin lacks in exterior styling, however, she more than makes up for with a truly spacious interior with the accommodations of a much larger boat. Packed into this beamy, fully cored hull (with prop pockets) are two staterooms with double berths, two full heads with a tub aft, a roomy U-shape galley with excellent storage *and* a washer/dryer, and a comfortable salon with curved companionway steps. Elegantly furnished and decorated, this is an impressive interior layout for a 38-foot boat. The aft deck is on the small side, but the innovative reverse transom with its molded boarding steps is a practical addition on a boat with this much freeboard. There's plenty of guest seating on the bridge, and a sun pad can be fitted on the foredeck. A thirsty boat with big-block 502-cid gas engines, the 3850 cruises at just 15–16 knots (at 33 gph) and reaches 25 knots wide open. ❑

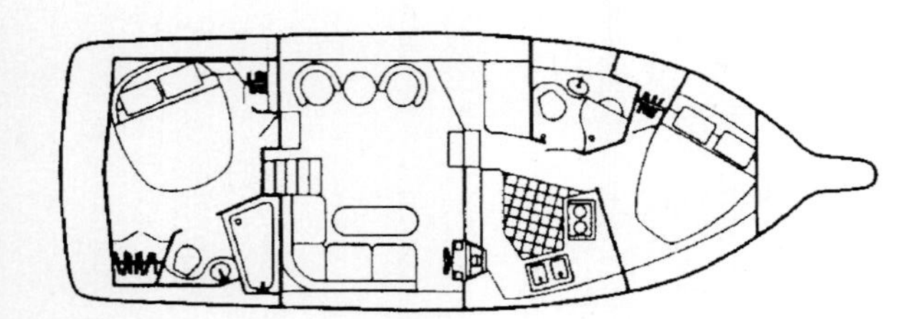

See page 237 for pricing information.

CRUISERS 4280/4285 EXPRESS BRIDGE

SPECIFICATIONS

Length	42'0"	Fuel	400 gals.
Beam	14'6"	Cockpit	91 sq. ft.
Draft	3'6"	Hull Type	Modified-V
Weight	27,000#	Deadrise Aft	16°
Clearance	13'3"	Designer	Jim Wynne
Water	160 gals.	Production	1988–Current

For those who enjoy their socializing on a grand scale, the Express Bridge is loaded with features not usually expected in a mid-size cruiser. Restyled on the outside for 1992, she's still no award winner when it comes to beauty on the outside. Inside, however, the expansive full-width interior is laid out on a single-level, providing unusually comfortable living accommodations. The 4280 model has a two-stateroom layout (with two heads beginning in 1992) with the galley in the salon. The 4285 model (introduced in 1990) has only a single-stateroom but a much larger salon, a huge head, and a separate galley. Topside, there's seating for a crowd on the massive bridge. Molded steps on both sides of the cockpit lead up to the bridge—very convenient. Her modified-V hull is fully cored and includes prop pockets for reduced shaft angles. Gas engines are standard (13 knots cruise/20 top). With optional 375-hp Cats or 400-hp 6V53s the Express Bridge will cruise at 20 knots and reach a top speed of around 24 knots. ❑

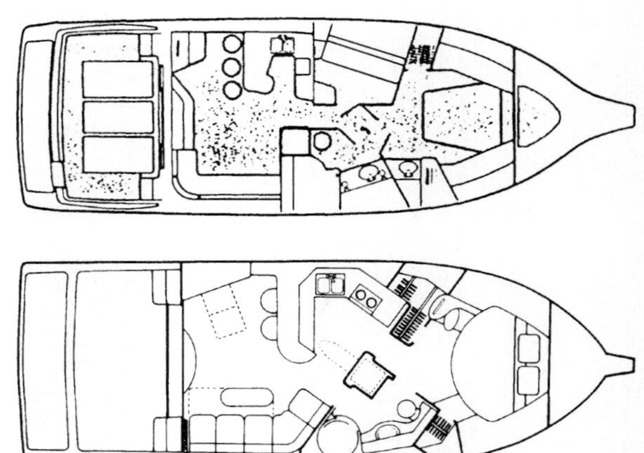

See page 237 for pricing information.

CT 35 TRAWLER

SPECIFICATIONS

Length	34'11"	Water	200 gals.
LWL	30'5"	Fuel	300 gals.
Beam	12'0"	Hull Type	Semi-Disp.
Draft	3'5"	Designer	Ta Chiao
Weight	19,800#	Production	1977–Current
Clearance	NA		

Still in production, the CT 35 is a traditional trunk cabin trawler with a smart, business-like profile and a roomy double-cabin layout. She's built by Ta Chiao in Taiwan on a solid fiberglass semi-displacement hull with hard chines and a full-length, prop-protecting keel. Inside, the all-teak floorplan features a compact galley and convertible dinette in the salon and a cockpit entryway to the aft cabin. With a double berth in the master stateroom, a big engine room, and excellent storage, the CT 35 packs a good deal of living space into a relatively short LOA. Additional features include wide sidedecks, teak window frames (aluminum in more recent models), a functional mast & boom, Sampson post, and swim platform. Most CT 35s have been powered with a single 120/135-hp Lehman diesel (7.5 knots cruise/9–10 knots top). Larger engines will deliver speeds to 15 knots. Note that the CT 35 Sundeck model is the same boat with a full-width aft deck and enlarged master stateroom. ❑

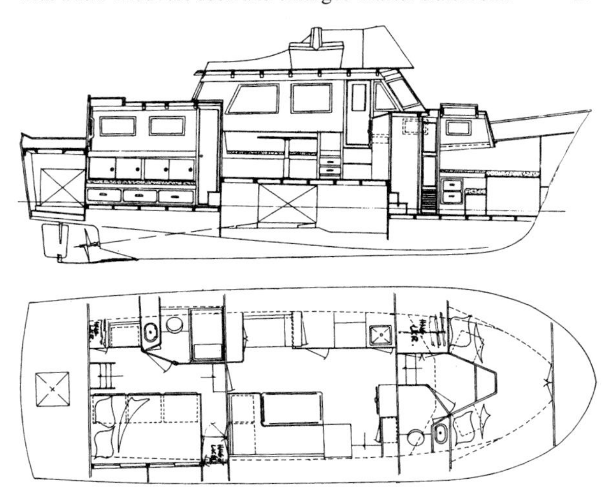

See page 237 for pricing information.

DEFEVER 40 OFFSHORE CRUISER

SPECIFICATIONS

Length	40'0"	Water	200 gals.
LWL	35'6"	Fuel	400 gals.
Beam	14'6"	Hull Type	Semi-Disp.
Draft	3'10"	Designer	DeFever
Weight	29,000#	Production	1985–91
Clearance	NA		

The DeFever 40 is an unusual trunk-cabin cruiser with a flat sheerline—something seldom seen in a trawler-style design. She's built by CTF in Taiwan on a solid glass hull with hard aft chines and a deep keel for prop protection. Her galley-up floorplan includes a large aft cabin, although it comes at the expense of the salon dimensions. The U-shaped galley is aft in the salon—not the normal galley location but a satisfactory arrangement in the eyes of many. A lower helm is standard, and there are port and starboard deck doors in the salon. Teak paneling and cabinetry is used throughout the interior in the traditional Asian mode. Additional features of the DeFever 40 include full walkaround decks, a well-arranged flybridge with guest seating around the helm, radar arch, and a large engine room with good access to the motors. Standard 240-hp Perkins diesels cruise at a comfortable 12–13 knots (about 15 knots top). Optional 320-hp Cats will cruise around 15 knots and reach 18 knots wide open. ❑

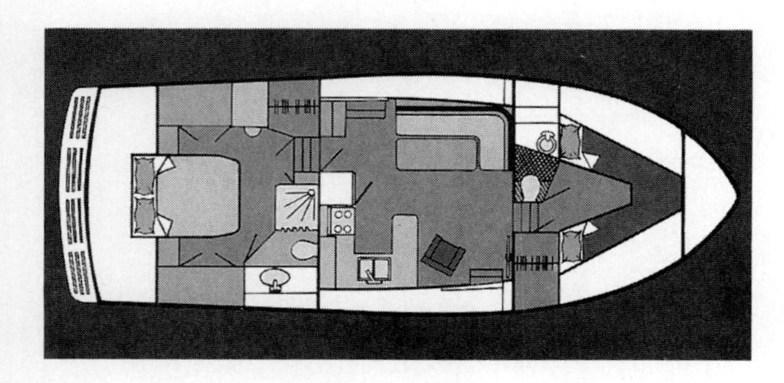

See page 237 for pricing information.

DEFEVER 41 TRAWLER

SPECIFICATIONS

Length	40'7"	Water	250 gals.
LWL	34'7"	Fuel	400/483 gals.
Beam	14'2"	Hull Type	Semi-Disp.
Draft	4'0"	Designer	DeFever
Weight	33,000#	Production	1983–88

This traditional double-cabin trawler was built in Taiwan by Bluewater Marine and still enjoys a good deal of owner popularity. She's constructed on a solid fiberglass hull with a deep forefoot and a long keel, which fully protects the running gear in the event of grounding. Her aft cabin floorplan consists of a very spacious master stateroom (with direct access to the aft deck), a compact salon with a lower helm and deck access doors port and starboard, and V-berths forward. The varnished teak interior woodwork—and there's plenty of it—is quite impressive. The house is set well forward on the deck, and there's adequate space on the trunk cabin aft of the bridge for dinghy storage. Additional features include teak overlaid decks and cabintop, wide sidedecks with raised bulwarks all around, and tub/shower in the aft head. With so much exterior teak, this boat is a maintenance nightmare. Most DeFever 41s were fitted with single (or twin) 135-hp Lehman diesels. She'll cruise at a leisurely 7–8 knots burning 4–5 gph. ❑

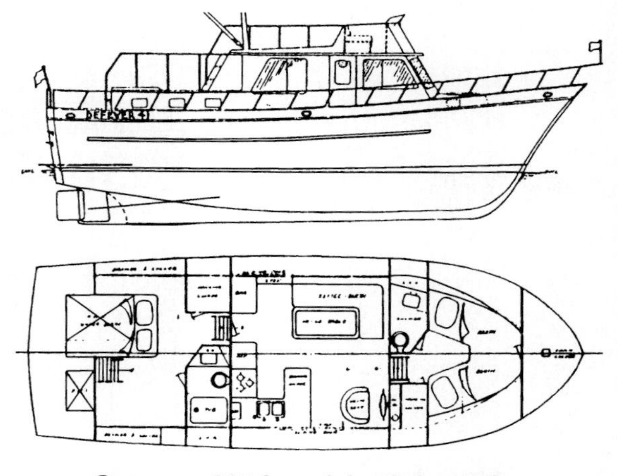

See page 237 for pricing information.

DEFEVER 43 TRAWLER

SPECIFICATIONS

Length	42'2"	Water	500 gals.
LWL	36'7"	Fuel	1,072 gals.
Beam	14'0"	Hull	Disp.
Draft	4'5"	Designer	DeFever
Weight	40,836#	Production	1978–85

A true displacement trawler with a distinctive flush deck profile, the DeFever 43 was built in Taiwan of solid fiberglass with a deep keel, moderate beam, and simulated lapstrake hullsides. Note the 40,000 lbs. displacement and tremendous fuel and water tankage—this is a serious offshore cruiser with an honest 1,500-mile range. The unusual flush-deck layout of the DeFever 43 results in a very large afterdeck (with room for a dinghy) while still maintaining adequate headroom in the master stateroom below. It also provides for a spacious engine room with near-standing headroom and excellent outboard engine access. Her two-stateroom floorplan is arranged with the U-shaped galley aft in the salon—an ideal layout in a cruising boat of this size. Teak paneling and teak parquet floors are used throughout the interior. Outside, the decks, doors, pulpit, handrails, and swim platform are teak—not exactly a low-maintenance boat. The flybridge has guest seating abaft the helm. Twin 120-hp Lehman diesels will cruise the DeFever 43 Trawler effortlessly at 7–8 knots. ❑

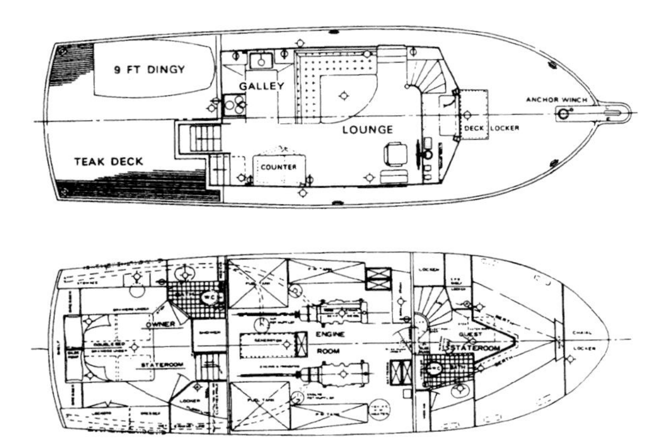

See page 237 for pricing information.

DEFEVER 44 DIESEL CRUISER

SPECIFICATIONS

Length	43'9"	Water	364 gals.
LWL	38'6"	Fuel	900 gals.
Beam	14'9"	Hull Type	Semi-Disp.
Draft	4'7"	Designer	DeFever
Weight	44,000#	Production	1980–Current
Clearance	NA		

Few trawler-style boats enjoy the continued popularity of the DeFever 44. Heavily built by the CTF yard in Taiwan and designed for long range cruising, the absence of too much exterior teak continues to attract those who are seeking a traditional trawler design without the usual maintenance headaches. She's heavily built on a solid fiberglass hull with hard chines aft and a full-length keel. Below, her teak interior is arranged with the galley aft and to port in the salon—a pleasant departure from the normal galley-down layout found in most trawlers. There are two sliding deck access doors in the salon and visibility from the lower helm is very good. A notable feature of the DeFever 44 is her excellent engine room (reached via a door in the forward stateroom) with nearly standing headroom. With her rugged and seakindly hull, the DeFever 44 is considered a serious bluewater cruiser with better-than-average resale values. Most were sold with twin 120/135-hp Lehman diesels. She'll cruise efficiently at 8–9 knots with a range about 1,200 miles . ❑

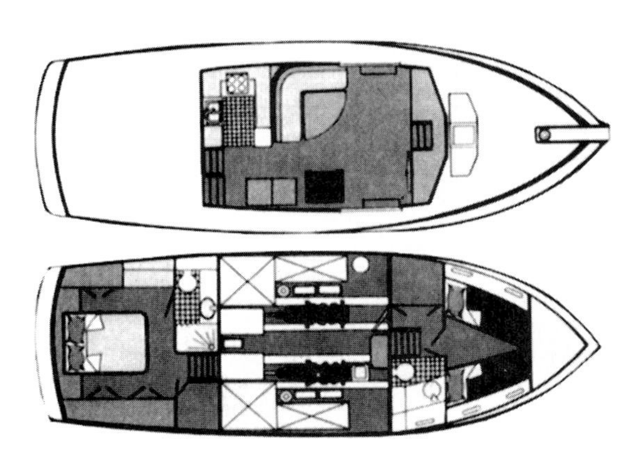

See page 237 for pricing information.

DEFEVER 48 DIESEL CRUISER

SPECIFICATIONS

Length	47'3"	Water	500 gals.
LWL	40'10"	Fuel	950 gals.
Beam	15'0"	Hull Type	Semi-Disp.
Draft	4'9"	Designer	A. DeFever
Weight	50,000#	Production	1978–Current

A handsome design, the DeFever 48 is a heavy displacement long-range cruiser built by the CTF yard in Taiwan. She's constructed on a solid fiberglass semi-displacement hull with a moderately flared bow, a fine entry and hard aft chines. She shares several design features with the smaller DeFever 44, namely the galley arrangement (a U-shaped affair located aft in the salon) and the remarkably spacious engine room. Inside, the DeFever is finished with teak paneling and cabinetry throughout—all done to high Asian standards. The floorplan includes three staterooms, two full heads, and a reasonably roomy salon with a lower helm and sliding deck doors port and starboard. Considered a trawler by most, the DeFever 48 is capable of planing speeds with engines larger than the standard twin 135-hp Lehman diesels. The optional Cat 375-hp diesels (generally found in later models) provide an almost-brisk cruising speed of 14–15 knots. A combination of good looks and heavy construction have earned the DeFever 48 a good reputation on the resale market. ❑

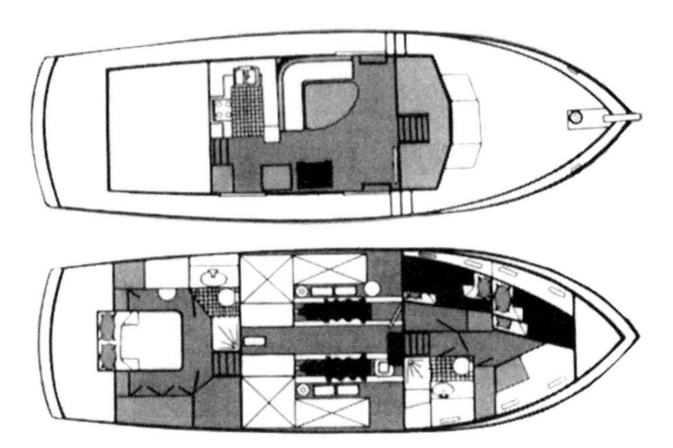

See page 237 for pricing information.

DEFEVER 49 PILOTHOUSE

SPECIFICATIONS

Length	49'10"	Water	400 gals.
LWL	42'0"	Fuel	800 gals.
Beam	15'0"	Hull Type	Semi-Disp.
Draft	4'6"	Designer	DeFever
Weight	50,000#	Production	1977–Current
Clearance	17'0"		

Arthur DeFever first designed a series of popular pilothouse trawlers for American Marine (Grand Banks) back in the late 1960s. Called Alaskans, these sturdy wooden passagemakers soon attracted a following of serious cruisers, and it was only a matter of time before similar designs became available in fiberglass. The DeFever 49 Pilothouse Trawler appeared in 1977. She was built by the Sen Koh and CTF yards in Taiwan on a solid fiberglass hull with teak decks, teak window frames, doors, rails, etc. A notably handsome boat with a go-anywhere appearance and enough bulk to deal with heavy seas, the DeFever 49's floorplan locates the master stateroom directly below the raised pilothouse with the guest stateroom forward. The interior is total teak. In recent years the DeFever 49 has been redesigned to eliminate some of the maintenance (less exterior teak), and the original soft-chined hull has been given flatter aft sections to reduce rolling. Powered with twin 135-hp Lehman diesels, the DeFever 49 is economical to operate and offers genuine long-range cruising potential. ❑

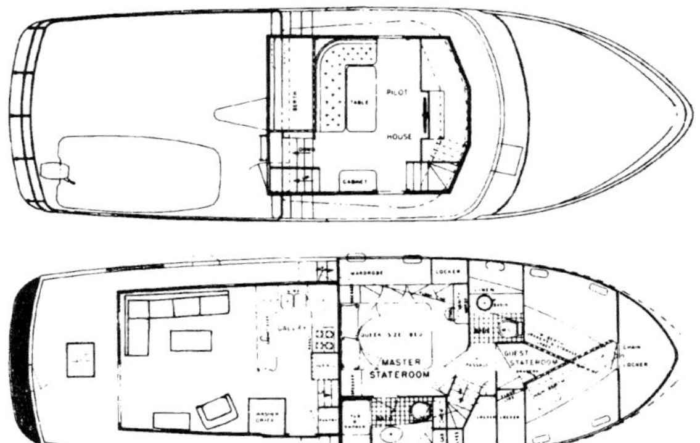

See page 238 for pricing information.

DEFEVER 52 MOTOR YACHT

SPECIFICATIONS

Length	51'7"	Water	500 gals.
LWL	45'5"	Fuel	1,500 gals.
Beam	16'8"	Hull	Disp.
Draft	4'9"	Designer	DeFever
Weight	77,000#	Production	1980–Current

The DeFever 52 MY is a sturdy offshore passagemaker with a distinctive trawler profile and a superb cruising layout. She's built by CTF in Taiwan on a solid fiberglass displacement hull with a long, deep keel and soft chines. The double-deck layout is arranged with the galley on the deckhouse level, where it's wide open to the salon and cockpit. Note the small day head opposite the galley—a great convenience for guests. The pilothouse comes with a watch berth, deck doors, and direct bridge access, and two of the three staterooms below are huge. All of the woodwork is teak, and there's plenty of it. Additional features include a walk-in engine room, stall showers in each head, optional teak decks, and a large flybridge with seating for a crowd. Standard 210-hp Cat diesels will cruise economically at 8–9 knots with a range of up to 2,000 miles. A Widebody version with a full-width salon (at the expense of the sidedecks—a poor trade) is also available. ❏

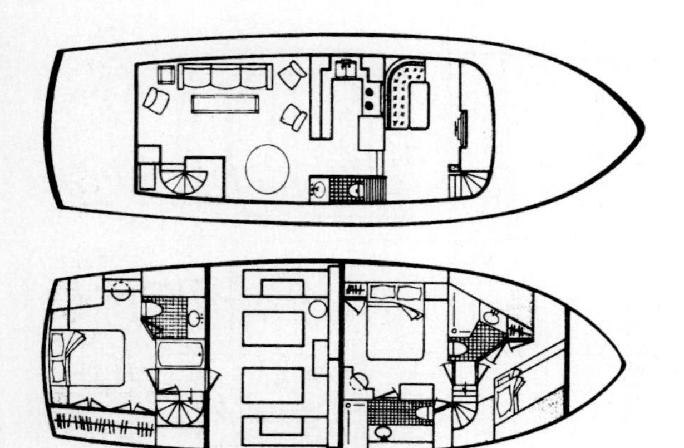

See page 238 for pricing information.

DEFEVER 53 MOTOR YACHT

SPECIFICATIONS

Length	52'7"	Water	400 gals.
Beam	16'6"	Fuel	800 gals.
Draft	4'8"	Hull Type	Semi-Disp.
Weight	55,000#	Designer	DeFever
Clearance	18'5"	Production	1986–92

The DeFever 53 Performance Offshore Cruiser (POC for short) was built in Taiwan with modern glass laminates and Divinycell foam coring in the hullsides. The first impression of the DeFever 53 is that of a much larger boat. The house is set well forward on the deck—a good indication of the spacious accommodations found below. The 53 is also distinguished by her wide and well-protected sidedecks (a disappearing feature in many modern motor yachts), which run completely around the house. (Note that a Widebody version was also available—a boxy and very unattractive boat.) The all-teak interior is laid out on two levels with three staterooms and three full heads. The floorplan on the deckhouse level is fairly conventional with the pilothouse separated from the galley and main salon. The walk-in engine room is especially spacious and well-designed. The popular 375-hp Caterpillar diesels will cruise the DeFever 53 at 13–14 knots and reach a top speed of about 17 knots. A good-selling boat, over 70 were built. ❏

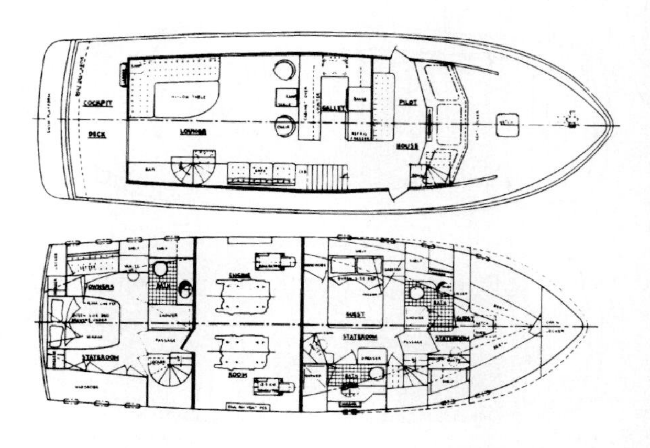

See page 238 for pricing information.

DEFEVER 60 OFFSHORE CRUISER

SPECIFICATIONS

Length	59'3"	Water	750 gals.
Beam	17'8"	Fuel	2,500 gals.
Draft	5'6"	Hull	Semi-Disp.
Weight	84,000#	Designer	DeFever
Clearance	NA	Production	1978–Current

The DeFever 60 Offshore is a good-looking flush-deck cruiser with a classic trawler profile and long-range capability. Built by CTF in Taiwan, the hullsides are Airex-cored, and a deep keel protects the props and running gear. Her standard floorplan is arranged with four staterooms and three heads on the lower level with private salon access to the master suite. The galley is open to the salon, and twin doors open to the large, protected afterdeck. While the deckhouse dimensions aren't extravagant compared with other yachts her size, the spacious aft deck and walkaround decks are desirable characteristics of any serious cruising yacht. Additional features include a fake stack on the flybridge, weather-protected sidedecks, port and starboard wheelhouse deck doors, wing doors, and a big engine room with access from the master stateroom. Standard 290-hp Cat 3306 diesels will cruise the DeFever 60 at 9–10 knots with about 2,000 miles of range. Note that the hull of this boat has been stretched to accommodate models up to 72 feet in length. ❑

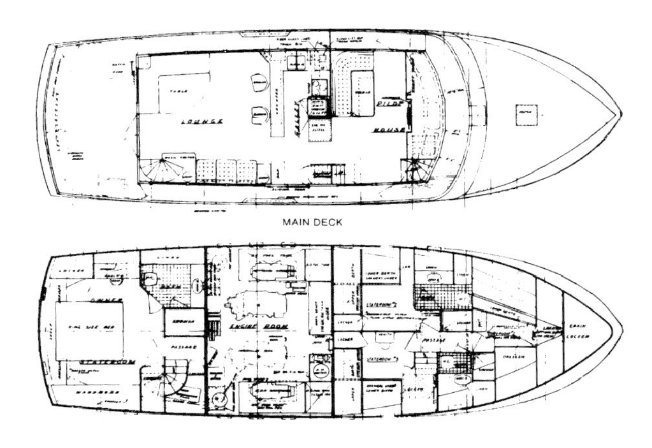

See page 238 for pricing information.

DYNA 53 COCKPIT MY

SPECIFICATIONS

Length	52'6"	Fuel	650/750 gals.
Beam	15'6"	Cockpit	78 sq. ft.
Draft	3'6"	Hull Type	Modified-V
Weight	38,000#	Deadrise Aft	NA
Clearance	15'6"	Designer	Terry Yen
Water	250 gals.	Production	1990–Current

Equally well-known in marine circles as the Vantare 53, the Dyna 53 Cockpit MY is a modern, low-profile yacht with a streamlined appearance, moderate beam, and an unusually large cockpit. The hullsides are cored from the chines up, and there's a shallow keel below for stability. Designed as a two-stateroom yacht, a lower helm is optional, and the galley and dinette are down in the conventional manner. Note that both heads have shower stalls—definitely a plus—and both staterooms have walkaround double berths. Like most Taiwan imports, the salon furnishings are "built-ins"—*not* a plus, but that's the way they come. The interiors in early models were teak, but most now have the more contemporary light oak woodwork. The Dyna is a well-finished yacht, and the detailing is often above average. Additional features include wing doors, dinghy storage on the hardtop, side exhausts, and wide sidedecks around the house. The engine room is a tight fit. The most popular engine option (among many) for the Dyna 53 CMY is the 485-hp GM 6-71s which will cruise at 18–19 knots and turn 22 knots wide open. ❑

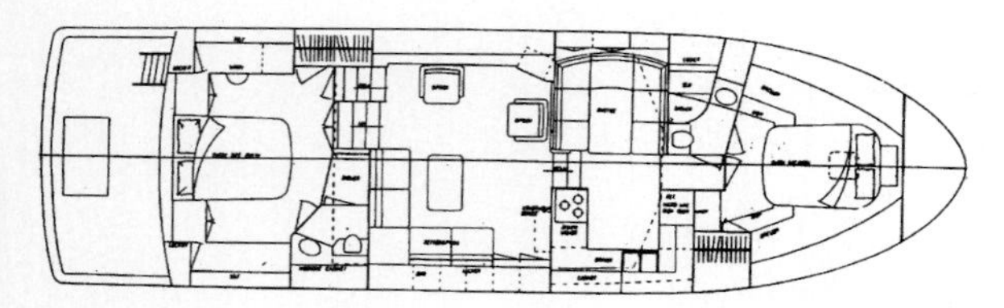

See page 238 for pricing information.

EAGLE 32 TRAWLER

SPECIFICATIONS

Length	32'0"	Water	100 gals.
LWL	28'0"	Fuel	150 gals.
Beam	11'6"	Hull Type	Semi-Disp.
Draft	3'4"	Designer	K. Hankinson
Weight	16,000#	Production	1985–Current

The salty lines and workboat profile of the Eagle 32 portray a sense of confidence unique to tugboat-style designs. Indeed, the Eagle looks like she was born for the sea. Built in Taiwan by Transpacific, her semi-displacement hull is constructed of solid glass with a full-length skeg for prop and rudder protection. Her slightly rounded transom moderates the effects of a following sea, while a hard chine and long keel provide good stability. The teak interior of the Eagle is well-finished and detailed. Visibility from the pilothouse is excellent, and a door closes off the salon for glare-free nighttime running. The stateroom is a little tight, but there's a stall shower in the head, and the salon and aft deck areas are comfortable indeed. Note the upper helm hidden within the false stack on the bridge. A single Lehman 90-hp diesel will cruise at 7 knots (about 1 gph) resulting in an impressive 750–800 mile range with just 150 gallons of fuel! The optional 135-hp Lehman diesel will cruise at 9–10 knots. ❑

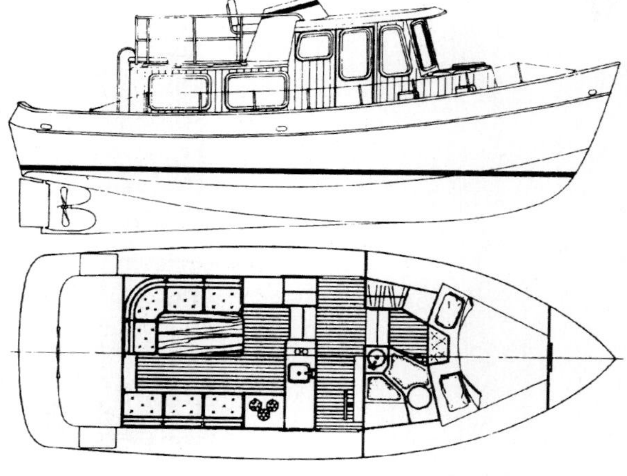

See page 238 for pricing information.

EGG HARBOR 40 MOTOR YACHT

SPECIFICATIONS

Length	40'0"	Fuel	300 gals.
Beam	14'1"	Hull Type	Modified-V
Draft	2'11"	Deadrise Aft	8°
Weight	30,000#	Designer	D. Martin
Clearance	16'2"	Production	1982–86
Water	100 gals.		

The Egg Harbor 40 MY was one of the last flush-deck motor yacht designs introduced by any major U.S. manufacturer. Ending production in 1986, she was built on the reworked hull of the earlier Pacemaker 40 MY whose molds and tooling Egg Harbor had acquired after Pacemaker closed down in 1980. A new superstructure and interior were added along with a traditional teak interior decor. A practical and straightforward cruising yacht, her conventional two-stateroom layout is centered around a salon that appears larger than her dimensions might suggest. An enormous U-shaped galley is down from the salon level, and only the aft head has a separate stall shower. Elsewhere, a built-in washer/dryer is fitted under the wet bar, and the teak woodwork is all of furniture quality. The aft stateroom is large for a 40-foot yacht and includes a home-size closet with real wardrobe space. Gas engines were standard (15-knots cruise/24 top). Optional 410-hp 6-71TIs will cruise at a brisk 20 knots and reach 24 knots top. ❑

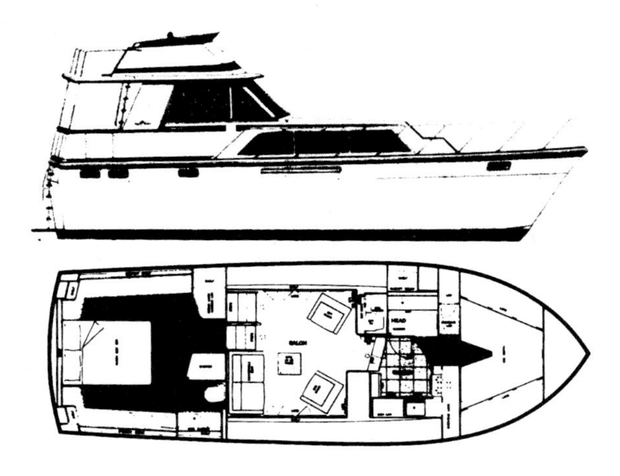

See page 238 for pricing information.

EGG HARBOR 41 MOTOR YACHT

SPECIFICATIONS

Length	40'7"	Fuel	300 gals.
Beam	14'1"	Hull Type	Modified-V
Draft	2'11"	Deadrise Aft	6°
Weight	22,000#	Designer	Egg Harbor
Clearance	NA	Production	1975–77
Water	100 gals.		

The Egg Harbor 41 Motor Yacht is a traditional flush-deck design built on a fiberglass hull with a glassed-over mahogany superstructure. In terms of popularity, the 41 suffered from her wood construction during a period when most other builders had long since made the switch to all-fiberglass designs. Not unexpectedly, the production run was short-lived. Two interior floorplans were available (galley-up or galley-down versions), and an optional queen berth aft was offered from the beginning. Varnished mahogany joinerwork is found in every cabin. Topside, the 41 has an attractive mahogany control station and good helm visibility in all directions. The open aft deck is large enough for dockside entertaining with plenty of room for several furniture arrangements. The optional flybridge—small by today's standards—greatly improves the resale value of used models. Those powered with the popular 310-hp J&T 6-71 diesels will cruise around 16–17 knots. On balance, the Egg Harbor 41 is a comfortable boat but obviously dated by today's motor yacht standards. ❑

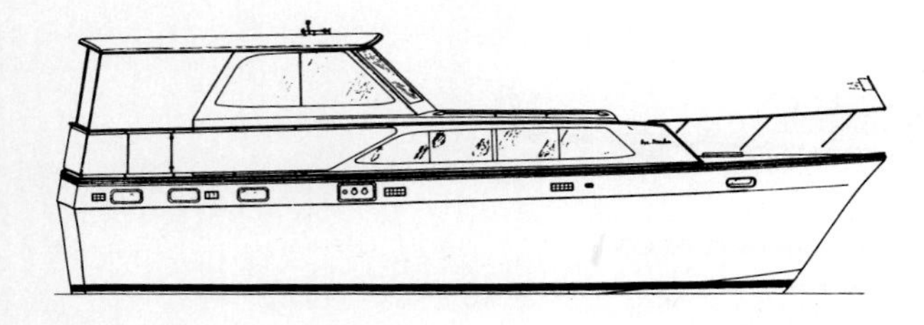

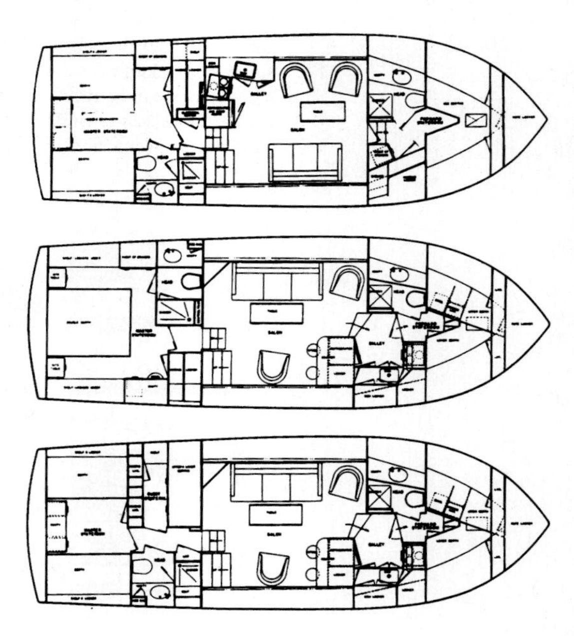

See page 238 for pricing information.

EMBASSY 444 SUNDECK MY

SPECIFICATIONS

Length	44'4"	Water	200 gals.
LWL	NA	Fuel	500 gals.
Beam	15'0"	Hull Type	Modified-V
Draft	3'6"	Deadrise Aft	17°
Weight	26,000#	Designer	Nova
Clearance	NA	Production	1987–Current

Built by Nova Marine in Taiwan, the Embassy 444 is an attractive and popular double-cabin design with a modern profile to go with her roomy interior accommodations. The hull is cored with balsa above the waterline, and the 15-foot beam is about average for a boat this size. Inside, the two-stateroom floorplan is arranged with the galley and dinette down, and both heads are fitted with shower stalls. Note the angled companionway steps leading from the salon down to the galley. Most 444s have been delivered with a traditional teak interior, although light ash or oak woodwork is available. The sidedecks are wide enough for secure passage around the house, and the raised aft deck is roomy enough for several pieces of deck furniture. Topside, the Embassy's boxy, plain-Jane helm console provides space for flush mounting some electronics, and there's plenty of bench seating for guests. With the popular 375-hp Cat diesels, the Embassy 444 will cruise economically at 21 knots and reach a top speed of around 26 knots. ❑

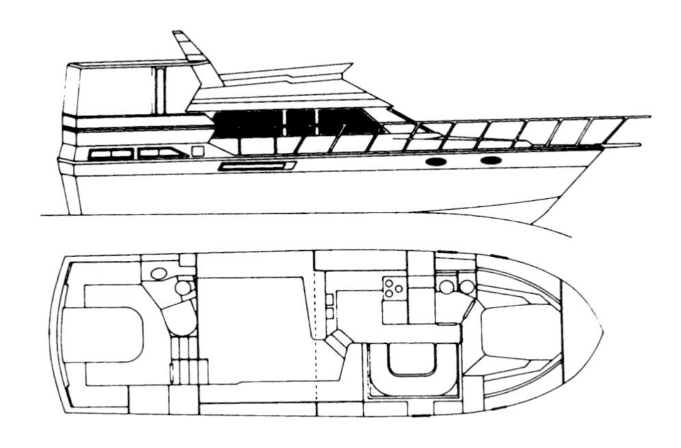

See page 238 for pricing information.

FLEMING 55 PILOTHOUSE MY

SPECIFICATIONS

Length	50'9"	Water	300 gals.
LWL	45'10"	Fuel	1,000 gals.
Beam	16'0"	Hull Type	Semi-Disp.
Draft	5'0"	Designer	L. Drake
Weight	60,000#	Production	1987–Current
Clearance	NA		

A good-selling boat, the Fleming 55 is a modern-day version of the old Alaskan 49, a DeFever-designed pilothouse trawler built of wood by Grand Banks in the 1960s. The Fleming's beam is wider for added stability and interior volume, and the enlarged flybridge has more lounge seating, but otherwise the lines are nearly identical. Flemings are built by Tung Hwa in Taiwan on an easily driven semi-displacement hull with simulated lapstrake hullsides and a prop-protecting keel. Inside, the look is total teak, and the joinery is flawless. The standard two-stateroom floorplan has the galley in the salon and a big dinette on the pilothouse level opposite the lower helm. The master stateroom is huge, and both heads are fitted with stall showers. (A three-stateroom floorplan is optional.) The engine room is a little tight, but the lazerette storage is excellent. A thoroughly impressive, very high-quality boat, Cat 210-hp diesels will cruise at 10 knots (12 knots top), and optional 425-hp Cats cruise at 14 knots and reach 17–18 knots top. ❑

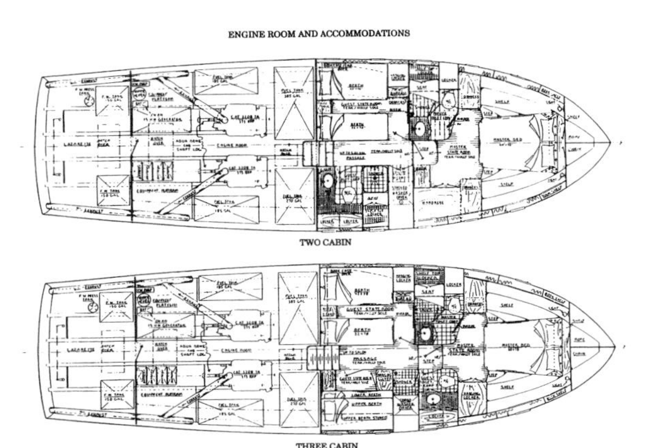

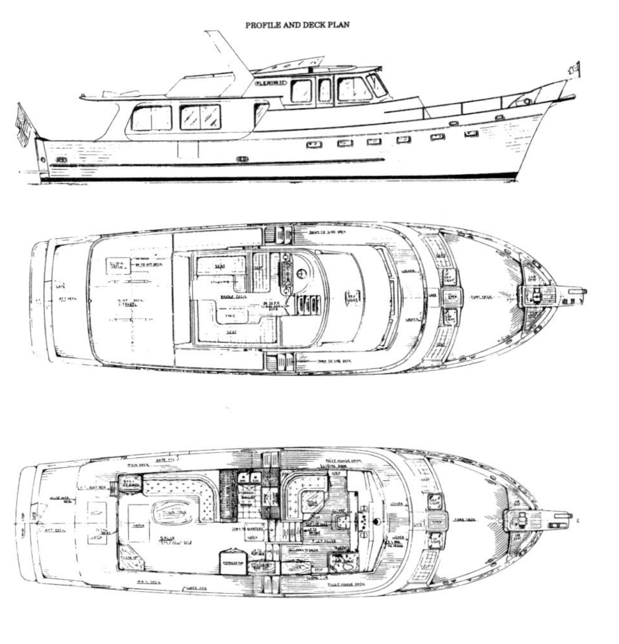

See page 238 for pricing information.

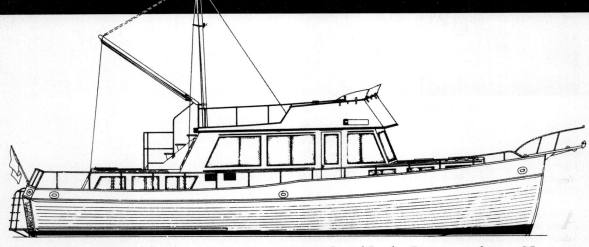

GRAND BANKS 32

SPECIFICATIONS

Length	31'11"	Water	110 gals.
LWL	30'9"	Fuel	225/250 gals.
Beam	11'6"	Hull Type	Semi-Disp.
Draft	3'9"	Designer	Ken Smith
Weight	17,000#	Production	1965–Current

A modern classic, the Grand Banks 32 is one of the most popular small trawler designs ever produced. About 900 have been built to date—fair testament to her great popularity with owners and brokers—and used models are always in demand. Her profile has remained essentially unchanged since the original wood models were introduced back in 1965. Indeed, at a distance it's difficult to distinguish between a pre-1973 wood GB 32 and one of the more recent fiberglass models. Powered with a single 6-cylinder Lehman diesel, owners report long hours of 7-knot cruising at 3 gph and a range of 500+ miles. A full-length skeg protects the underwater gear, and hard chines help to stabilize the ride. Below, the 32's straightforward and practical cabin layout will satisfy the requirements of a cruising couple. Storage is adequate, and visibility from the lower helm is very good. The high-quality hardware, furnishings, and systems are very impressive. Grand Banks 32s are dependable and seaworthy small cruisers with immense eye appeal, great popularity, and excellent resale values. ❑

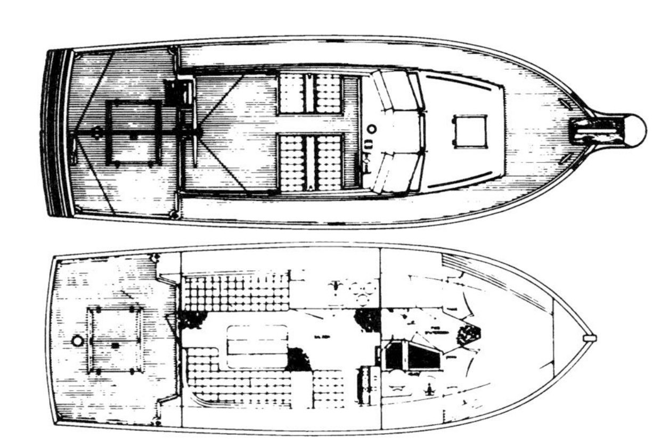

See page 238 for pricing information.

GRAND BANKS 36 CLASSIC (EARLY)

SPECIFICATIONS

Length	36'4"	Water	170 gals.
LWL	35'0"	Fuel	400 gals.
Beam	12'2"	Hull Type	Semi-Disp.
Draft	3'11"	Designer	Ken Smith
Weight	23,300#	Production	1965–87

Like all Grand Banks models, the 36 Classic has a reputation for high-quality construction, economical operation, and dependable cruising performance. Built of Philippine mahogany until mid-1973, most have been powered with a durable Lehman 6-cylinder diesel, although some 36s have been equipped with twin engines for greater speed and improved maneuverability. With the single engine, she cruises at a steady 8 knots (3 gph) and has a range of up to 1000 miles depending on conditions. A full-length keel below protects the running gear while providing a good deal of directional stability. The traditional teak interior of the GB 36 is arranged with twin berths in the aft stateroom, a small salon (thanks in part to her wide sidedecks) with built-in settees, galley and lower helm, and two heads—neither of which has a stall shower. Storage is excellent. Outside, the steadying sail mast is fully functional, and a cockpit door opens into the aft stateroom. Replaced by an all-new, slightly larger GB 36 in 1988, used models hold their values well. ❑

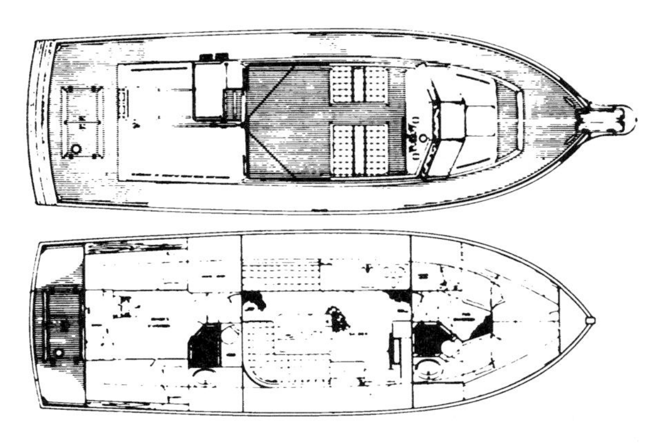

See page 238 for pricing information.

GRAND BANKS 36 CLASSIC

SPECIFICATIONS

Length	36'10"	Water	140 gals.
LWL	35'2"	Fuel	400 gals.
Beam	12'8"	Hull Type	Semi-Disp.
Draft	4'0"	Designer	Amer. Marine
Weight	26,000#	Production	1988–Current

The current Grand Banks 36 Classic has the same profile as the original GB 36 but with 6" added to the beam and length. These new dimensions allow for a slightly enlarged salon and master stateroom, and a stall shower is finally available in the aft head compartment. Note also that an alternate interior floorplan with a walkaround queen berth aft is available in the 36 for the first time. In other respects, the two boats are nearly identical, although the new GB 36 has a fiberglass bow pulpit instead of the teak platform of old. It's interesting to observe that, while other builders are constantly introducing newly redesigned models to keep up with the latest trends, Grand Banks simply improves on an already-proven design. Experienced cruisers will immediately recognize the GB 36 as a fine and seaworthy craft built to high standards. With the optional twin 210-hp Cummins diesels, the Grand Banks 36 has a cruising speed of up to 12 knots and a top speed of 14–15 knots. ❏

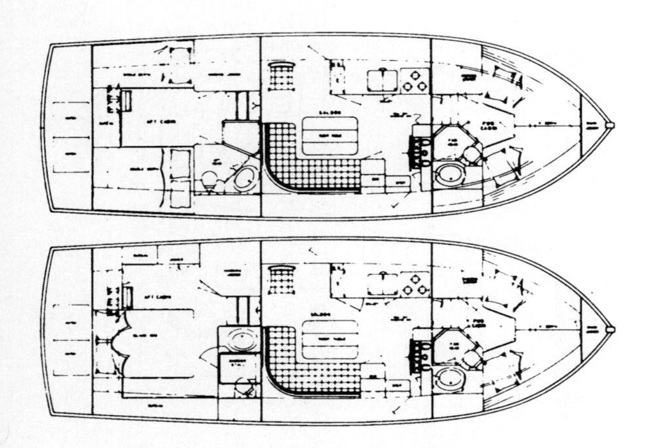

See page 239 for pricing information.

GRAND BANKS 36 EUROPA

SPECIFICATIONS

Length	36'10"	Water	170 gals.
LWL	35'2"	Fuel	400 gals.
Beam	12'8"	Hull Type	Semi-Disp.
Draft	4'0"	Designer	Amer. Marine
Weight	26,000#	Production	1988–Current

Built on the recently enlarged Grand Banks 36 hull, the Europa will appeal to those who enjoy the all-around practicality of a sedan layout. The Europa is distinguished from the less-popular GB 36 Sedan only by her attractive bridge overhangs—the sidedecks are shaded on both sides of the house and a bridge extension provides overhead cover for the cockpit; in all other respects the two boats are the same. Two floorplans are available: the standard floorplan has two staterooms, and the optional single-stateroom layout is arranged with a walkaround queen berth forward and a split head compartment with a big shower stall. Either way, the galley is to port in the salon, where it's convenient to the lower helm and cockpit, and the interior is completely finished with teak furnishings and woodwork. A single 135-hp diesel will cruise the Europa economically at 7–8 knots (10 knots top) burning about 3 gph. Twin 210-hp Cummins diesels (optional) will deliver a top speed in the 15-knot range. ❏

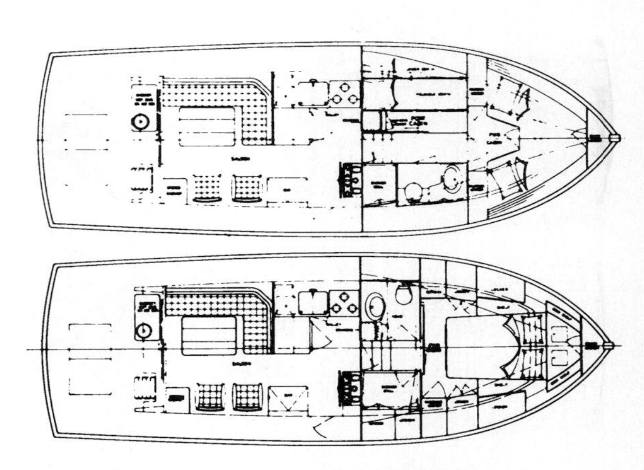

See page 239 for pricing information.

GRAND BANKS 42 CLASSIC (EARLY)

SPECIFICATIONS

Length	41'10"	Water	270 gals.
LWL	40'6"	Fuel	600 gals.
Beam	13'7"	Hull Type	Semi-Disp.
Draft	4'2"	Designer	Ken Smith
Weight	34,000#	Production	1966–91

Introduced in 1966, a total of 1,203 Grand Banks 42s were built before she was replaced with a new model in 1991. Indeed, the GB 42 is the standard by which other trawlers her size are generally measured. Her enduring popularity and strong resale values can be attributed to superb engineering and construction, timeless styling, and a proven bluewater hull design. Her fine entry is combined with a moderate beam and full-length skeg keel giving the GB 42 excellent seakeeping characteristics in a wide range of conditions. Aside from the obvious change to fiberglass construction in 1973, there were few major modifications to the basic design. Instead, continuous product refinement can best characterize her evolution from the original mid-1960s wooden version. The two-stateroom interior is finished with rich teak woodwork, and her engine room is spacious and well-designed. Most have been powered with 120/135-hp Lehman diesels and cruise economically at 8–9 knots. Optional 375-hp Cats will cruise around 17 knots and reach top speeds of 20+ knots. ❑

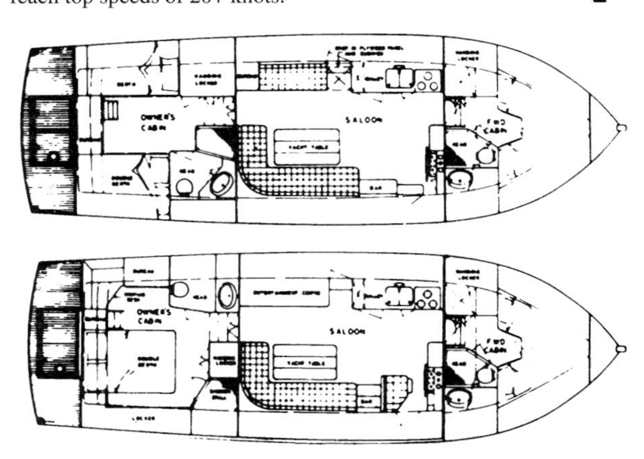

See page 239 for pricing information.

GRAND BANKS 42 CLASSIC

SPECIFICATIONS

Length	43'3"	Water	265 gals.
LWL	41'1"	Fuel	600 gals.
Beam	14'1"	Hull Type	Semi-Disp.
Draft	4'2"	Designer	Ken Smith
Weight	34,914#	Production	1991–Current

Grand Banks finally retired the hull molds for the original GB 42 Classic in 1991 after a successful 25-year production run. Beginning with hull #1204, the new 42 Classic is wider and longer than her predecessor, but few will ever notice the difference since both boats are just about identical in appearance. The floorplans are basically unchanged from earlier years, although the master stateroom's cockpit access door has been eliminated in the new model. The extra 8" of LOA and 6" of beam are seen primarily in the enlarged galley with additional counter space, a slightly roomier forward stateroom, bigger heads, more room in the engine room, and more storage. (Note that the V-berth is now 8" lower than before.) Although 135-hp Lehman diesels are standard, it's likely that most will be sold with optional Cat diesels from 210-hp to 375-hp in size. A comfortable boat, the Grand Banks 42 will cruise at 16 knots and reach a top speed of around 19 knots with big 375-hp Cats. ❑

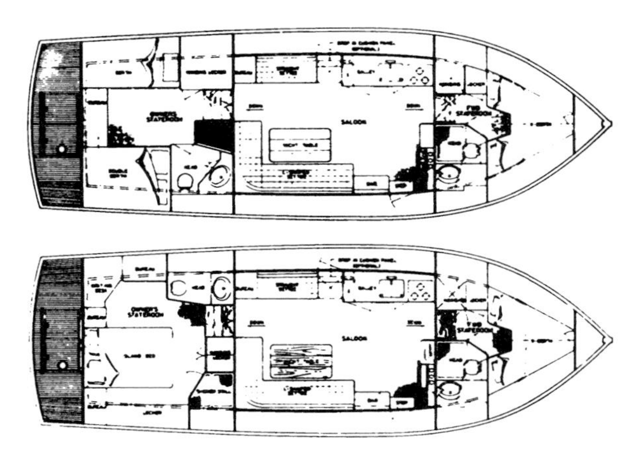

See page 239 for pricing information.

GRAND BANKS 42 EUROPA

SPECIFICATIONS

Length	42'7"	Water	270 gals.
LWL	40'10"	Fuel	600 gals.
Beam	13'7"	Hull Type	Semi-Disp.
Draft	4'2"	Designer	Ken Smith
Weight	34,000#	Production	1980–Current

The original Grand Banks 42 Europas were built of mahogany in 1970 and 1971, primarily for the European market. Several were built, but since she was never a particularly popular boat in the U.S. only a few were imported. Nothing much happened again until 1980, when Grand Banks decided to reintroduce the 42 Europa as an all-fiberglass production model. A good-looking boat with her bridge overhangs and distinctive profile, the Europa's sedan layout offers the convenience of having the salon and cockpit on the same level. Her two-stateroom floorplan is arranged with the galley forward in the salon, opposite the helm, and a built-in dinette and facing settee aft—a practical cruising layout and elegantly finished. Outside, the extended hardtop shades the cockpit, and the sidedecks are similarly protected around the house. The Europa is built on the original GB 42 Classic hull and shares similar power options and performance figures. Note that Grand Banks 42 Sports Cruiser model is essentially the same boat without the Europa's covered sidedecks and protected cockpit. ❑

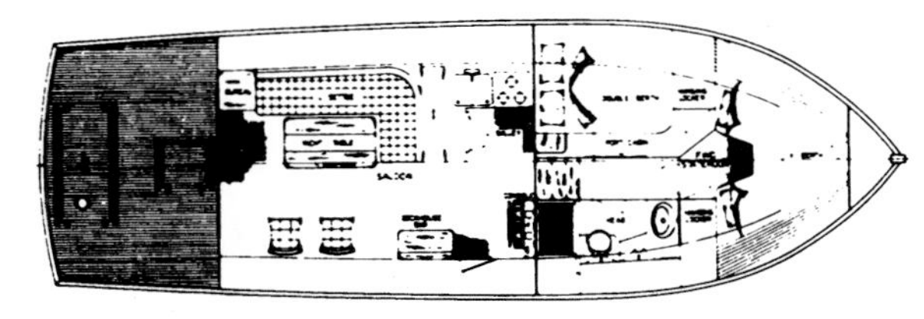

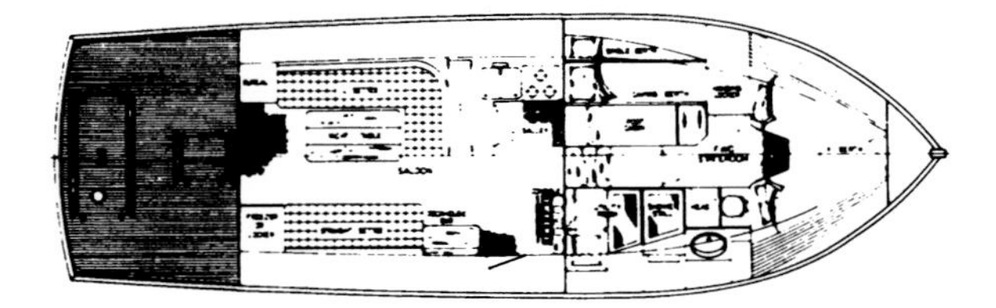

See page 239 for pricing information.

GRAND BANKS 42 MOTOR YACHT

SPECIFICATIONS

Length	42'7"	Water	260 gals.
LWL	40'9"	Fuel	600 gals.
Beam	13'7"	Hull Type	Semi-Disp.
Draft	4'2"	Designer	Ken Smith
Weight	34,000#	Production	1987–Current

Built on the same hull as the original GB 42, the features that distinguish the Motor Yacht from the 42 Classic are her three-stateroom interior layout, an enlarged master stateroom, and, of course, her raised aft deck platform (in place of the walkaround decks of the 42 Classic). Utilizing the full beam of the boat, the 42 MY's aft cabin can accommodate a queen-size berth not available in the GB 42 Classic. The portside guest stateroom is fitted with a double berth, and V-berths are located in the forward stateroom. There are port and starboard deck doors in the salon, and both heads come with separate stall showers. Note that the galley is somewhat compact and a little short of counter space. The aft deck, however, is quite roomy and able to accommodate deck furnishings without difficulty. Several engine options are offered from 135-hp Lehmans to the popular 210 to 375-hp Cats. The big Cats will cruise the Grand Banks 42 at around 16 knots and reach a top speed of 19–20 knots. ❑

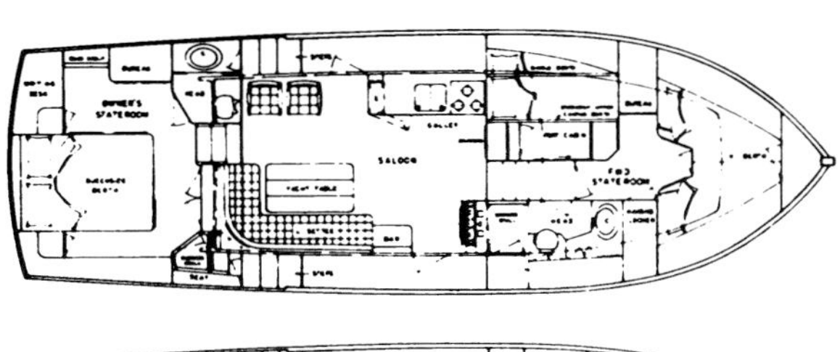

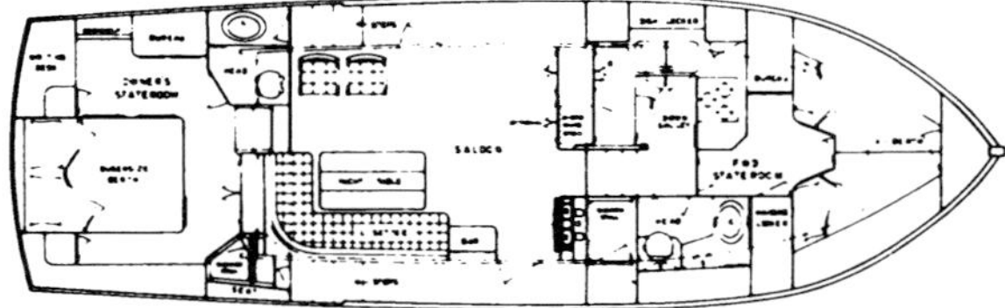

See page 239 for pricing information.

GRAND BANKS 46 CLASSIC

SPECIFICATIONS

Length	47'1"	Clearance	23'7"
LWL	44'9"	Water	278 gals.
Beam	14'9"	Fuel	600 gals.
Draft	4'2"	Hull Type	Semi-Disp.
Weight	39,000#	Production	1987–Current

Introduced in 1987 as the 46 Motor Yacht (and briefly marketed as a Cockpit MY), the Grand Banks 46 is now simply referred to as the "Classic," a designation few knowledgeable boaters would dispute. With her elegant trawler profile and proven semi-displacement hull, the GB 46 is a true offshore passagemaker. Notably, her lines are exactly those of the smaller GB 42 Classic, and it's difficult to tell the two apart from a distance. Most have been delivered with a two-stateroom, galley-down floorplan with a queen berth in the aft cabin. A three-stateroom layout with the galley up is also available. The interior is total teak; both heads have stall showers; the engine room has five feet of headroom; and the walkaround decks are wide and well-secured. (The deckhouse tooling was changed at hull #43 to provide wider sidedecks aft.) Among several engine options, 210 to 375-hp Cats have been the most popular. The 320-hp Cats, as an example, will cruise at 11 knots and reach a top speed of 14–15 knots. ❑

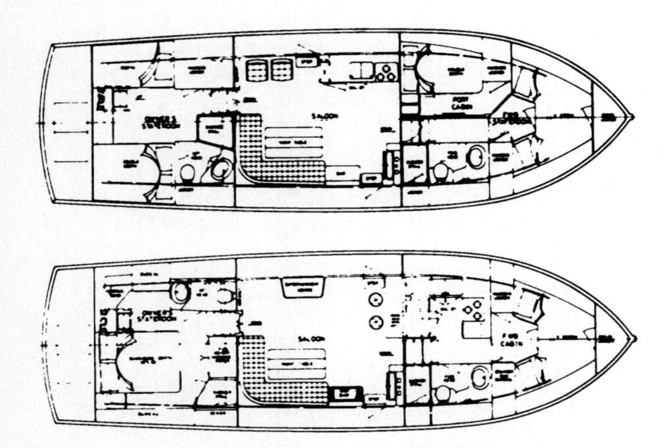

See page 239 for pricing information.

GRAND BANKS 46 MOTOR YACHT

SPECIFICATIONS

Length	47'1"	Cockpit	NA
LWL	44'9"	Water	278 gals.
Beam	14'9"	Fuel	600 gals.
Draft	4'4"	Hull Type	Semi-Disp.
Weight	39,000#	Designer	Amer. Marine
Clearance	23'7"	Production	1990–Current

The GB 46 Motor Yacht differs from the original 46 Classic only in her raised afterdeck and enlarged master stateroom. Other than that the two boats are essentially the same. Those who enjoy outdoor entertaining will find the full-width aft deck of the 46 MY a great convenience, while more traditional hands might prefer the flexibility provided by the walkaround decks and cockpit design of the 46 Classic. There are two floorplan choices for the 46 MY: a standard three-stateroom, galley-up layout, or an alternate (and more popular) two-stateroom arrangement with the galley down. Not surprisingly, the craftsmanship and attention to detail found throughout compare with the best in production yacht building. Standard power for the Grand Banks 46 is a pair of 135-hp Lehman diesels which will cruise economically at 8–9 knots. Optional 375-hp Cats will increase the cruising speed to 13–14 knots and the top speed to around 17 knots. At 9 knots, the fuel consumption is only 6–7 gph (regardless of engines), and the cruising range exceeds 800 miles. ❑

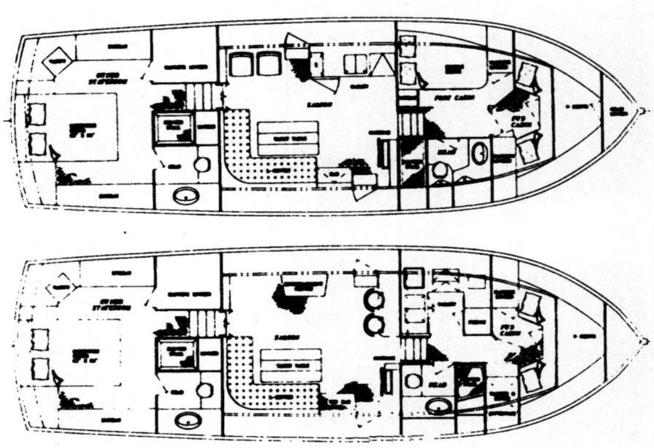

See page 239 for pricing information.

GRAND BANKS 46 EUROPA

SPECIFICATIONS

Length	47'1"	Clearance	24'0"
LWL	44'9"	Water	300 gals.
Beam	14'9"	Fuel	600 gals.
Draft	4'3"	Hull Type	Semi-Disp.
Weight	39,000#	Production	1993–Current

Introduced in 1993, the Grand Banks 46 Europa is built on the same hard-chined, deep-keel hull used for the 46 Classic and Motor Yacht models. This is the third Europa in the Grand banks fleet, and her specialty is a distinctive profile and a practical sedan floorplan. Her traditional teak interior is a showcase of workmanship and simplicity. Like all GB designs, the 46 Europa is above all a sensible and eminently conservative cruising yacht. Choose from a two- or three-stateroom floorplan. The salon is roomy enough either way, but not overly wide at the expense of her broad sidedecks. Storage is better by far than most boats her size, and the flybridge can accommodate a crowd. Generous bridge overhangs provide shade around both sides of the house (as well as the cockpit), and the stylish pilaster supports are characteristic of all Grand Banks Europa models. Standard 135-hp Lehman diesels will cruise the 46 Europa at 8–9 knots, and top speeds of 18 knots are possible with optional Cat or Cummins diesels. ❑

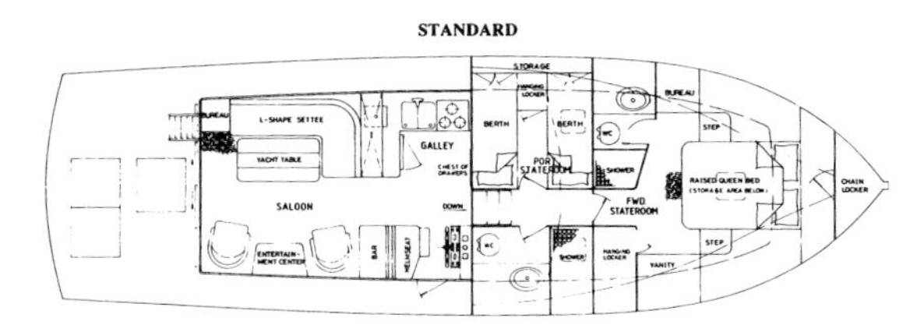

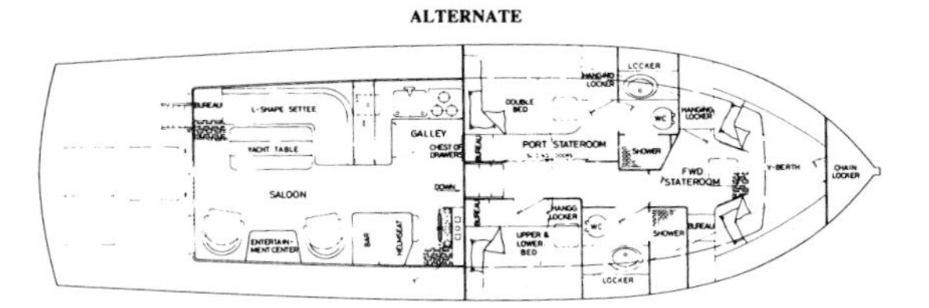

See page 239 for pricing information.

GRAND BANKS 49 CLASSIC

SPECIFICATIONS

Length	50'6"	Clearance	NA
LWL	48'9"	Water	500 gals.
Beam	15'5"	Fuel	1,000 gals.
Draft	5'2"	Hull Type	Semi-Disp.
Weight	60,000#	Production	1980–Current

A popular boat, the GB 49 Classic is a serious long-range cruiser with the characteristic trawler profile and quality workmanship typical of all Grand Banks models. Needless to say, this is not a glitzy yacht with a Euro-style interior or high-tech construction. Indeed, she comes off as a little old-fashioned when compared to most other yachts, but the 49 stands apart as a proven passagemaker with what is arguably one of the best three-stateroom, galley-up layouts found in any similar-size offshore cruiser. Highlights: the full walkaround decks are broad, secure, and easily negotiated; the stand-up engine room is spacious and carefully detailed; the teak woodwork is flawless inside and out; hardware, fixtures, and appliances are top quality; props and rudders are keel-protected; and her sea kindly hull has a long record of handling every kind of weather worldwide. Originally powered with twin 120-hp Lehman diesels, newer models have been delivered with Cat diesels from 210-hp. The popular 375-hp versions will cruise the GB 49 at 13-14 knots. ❑

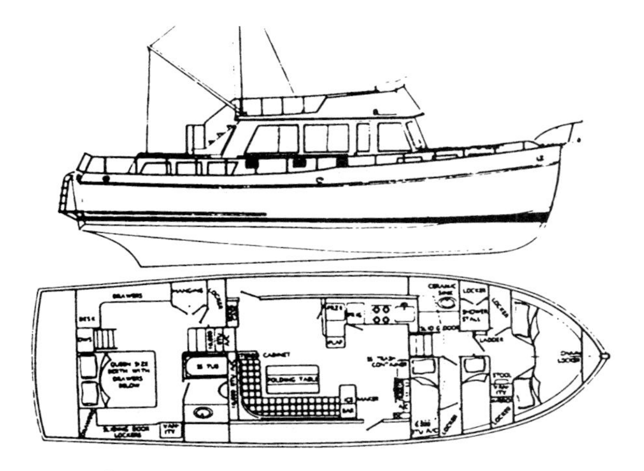

See page 239 for pricing information.

GRAND BANKS 49 MOTOR YACHT

SPECIFICATIONS

Length	50'6"	Clearance	NA
LWL	48'9"	Water	500 gals.
Beam	15'5"	Fuel	1,000 gals.
Draft	5'2"	Hull Type	Semi-Disp.
Weight	60,000#	Production	1986–Current

The Grand Banks 49 Motor Yacht can be distinguished from the original GB 49 Classic by her raised aft deck platform (the Classic has full walkaround decks and a trunk cabin profile), which gives her a notably roomy entertainment platform just a few steps down from the bridge. Below, her layout (the same as found in the 49 Classic) includes three staterooms with three heads, a stand-up engine room, a luxurious master suite, and a spacious main salon with deckhouse galley and lower helm. An optional two-stateroom arrangement has the galley down and an enlarged salon. Twin 135-hp Lehman diesels are standard, but 375-hp Cat diesels have proven popular since they raise the cruising speed to 12–13 knots and the top speed to around 15 knots. (At an 8–9-knot displacement speed the 6-7 gph fuel economy is not lost with larger engines.) The cruising range at this speed is excellent—over 1,200 nautical miles. Not inexpensive, the Grand Banks 49 MY is clearly aimed at an affluent clientele. ❑

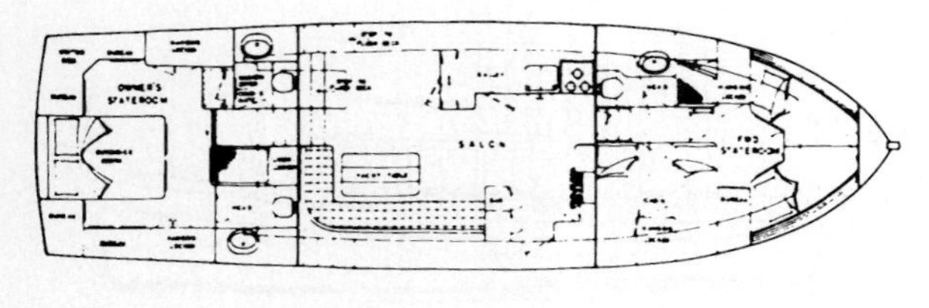

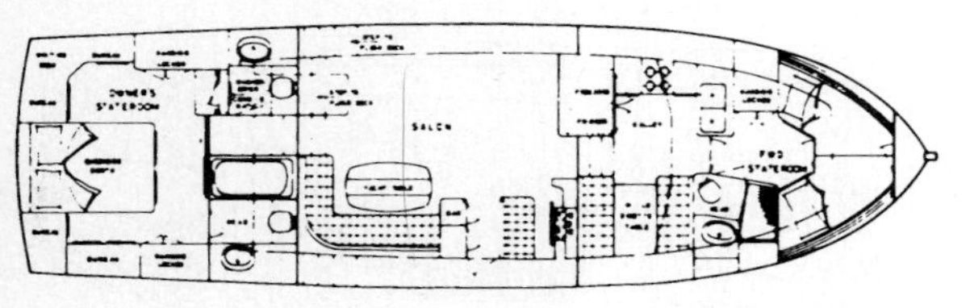

See page 240 for pricing information.

GRAND BANKS 58 MOTOR YACHT

SPECIFICATIONS

Length	58'11"	Clearance	17'6"
LWL	54'4"	Water	450 gals.
Beam	17'6"	Fuel	1,400 gals.
Draft	5'6"	Hull Type	Semi-Disp.
Weight	100,000#	Production	1990–Current

The Grand Banks 58 is a traditional flush-deck motor yacht with the distinctive styling and quality construction that Grand Banks products are known for. She's built on a solid fiberglass semi-displacement hull with a wide beam, a graceful sheer, and a deep, full-length keel below. This luxurious yacht is clearly designed with long-range passagemaking in mind, and her three-stateroom accommodations provide extravagant liveaboard comforts. The teak-paneled main salon/dining area is a full 210 sq. ft. with the galley conveniently located on the main deck level. Additional features include a pilothouse chart table and settee, a huge flybridge with extended bridge overhangs, secure walkaround sidedecks, a spacious engine room with standing headroom, and a whirlpool in the owner's head compartment. Notably, the Grand Banks 58 MY is designed to be owner-operated, and all equipment is installed for easy access. Caterpillar 375-hp diesels are standard for a 9-knot cruising speed and a top speed of 11–12 knots. The cruising range is approximately 1,400 miles. ❑

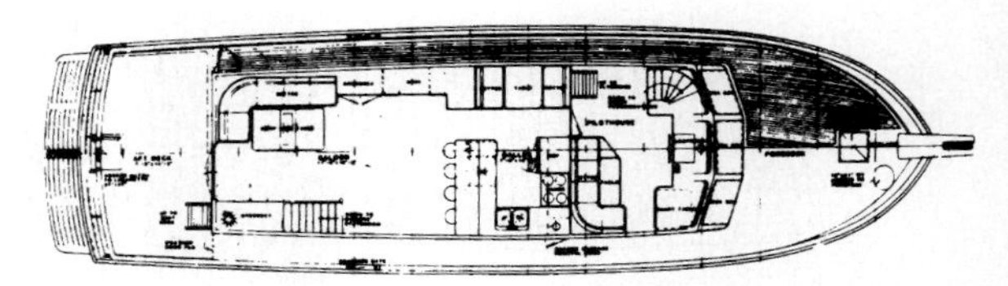

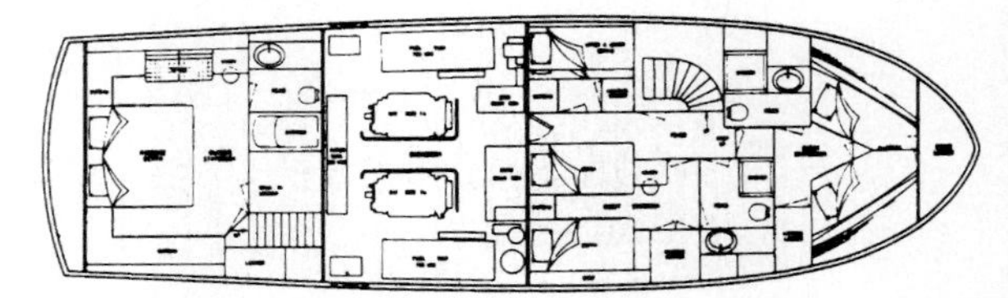

See page 240 for pricing information.

GULFSTAR 36 TRAWLER

SPECIFICATIONS

Length	36'3"	Cockpit	NA
LWL	31'0"	Water	100 gals.
Beam	12'0"	Fuel	250 gals.
Draft	3'0"	Hull Type	Semi-Disp.
Weight	16,000#	Designer	V. Lazzara
Clearance	12'0"	Production	1972–76

It's interesting to note that the Gulfstar 36 Trawler was built on the same semi-displacement hull used in the production of the Gulfstar 36 Motorsailer—a sailboat. The profiles are about the same, but the twin diesels of the Trawler provide the maneuverability and performance required of a powerboat. Not that she's fast—her 80-hp Perkins diesels will cruise at 7.5 knots and reach 9–10 knots top. The hull is solid glass and the keel is ballasted. In 1975 the Gulfstar 36 MK II version was introduced featuring a raised aft deck, a full-beam master stateroom, a slightly larger flybridge, bigger rudders, and a new teak interior to replace the previous wood-grain Formica (or mahogany) paneling. The engines were also relocated further outboard in the MK II model, which is said to improve handling. Several two-stateroom floorplans were used over the years with the latest having the dinette to port and a starboard deck access door. An easy-riding boat, the Gulfstar 36 is economical to operate and requires little maintenance. ❑

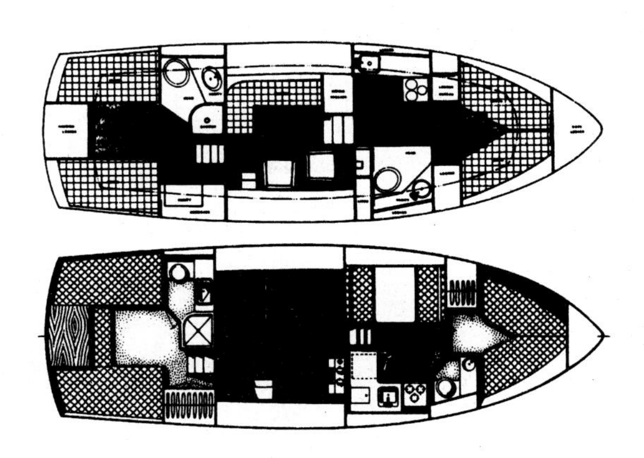

See page 240 for pricing information.

GULFSTAR 38 MOTOR CRUISER

SPECIFICATIONS

Length	38'4"	Cockpit	NA
LWL	33'6"	Water	150/294 gals.
Beam	12'5"	Fuel	250 gals.
Draft	3'3"	Hull Type	Semi-Disp.
Weight	18,000#	Designer	R. Lazzara
Clearance	12'6"	Production	1980–84

The 38 Motor Cruiser is one of Gulfstar's early applications of cored-hull construction in a powerboat—considered high-tech design and construction in 1980. A good-looking yacht with still-modern lines, the 38 Motor Cruiser weighs only 18,000 lbs. which is light compared to other yachts in this size range. With her relatively narrow 12'5" beam and high freeboard, the 38 can be a tender boat in spite of her long keel and hard aft chines. Below, the galley-down layout features a double berth in both staterooms, a tub in the aft head, full dinette, and a complete lower helm. Although the salon dimensions are compact, the overall accommodations are still impressive for a 38-foot boat. The teak woodwork found throughout the interior is notable for its quality and beauty. With twin 4-cylinder 115-hp Perkins diesels, the 38 Motor Cruiser will operate efficiently at 8.5 knots with a top speed of about 10 knots. Perkins 200-hp engines were optional and deliver a cruising speed around 14 knots (16–17 knots top). ❑

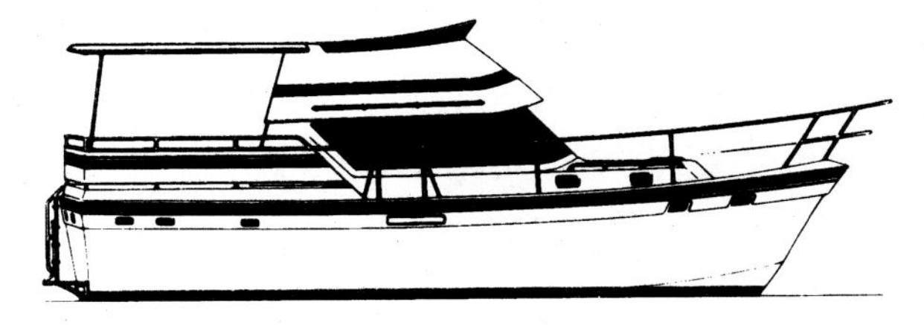

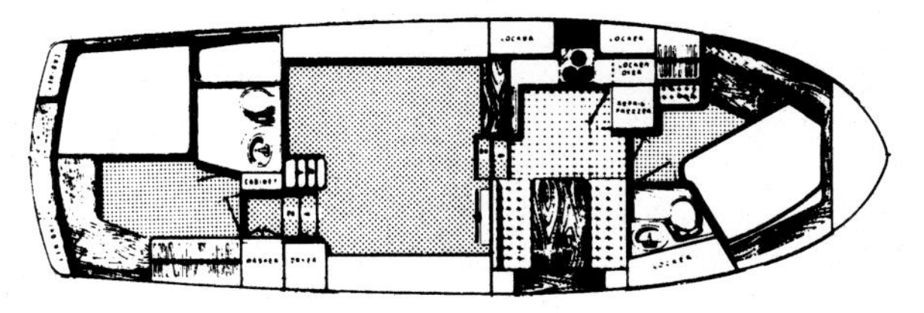

See page 240 for pricing information.

GULFSTAR 43 TRAWLER

SPECIFICATIONS

Length	43'4"	Water	130 gals.
LWL	39'2"	Cockpit	NA
Beam	13'11"	Fuel	300 gals.
Draft	3'6"	Hull Type	Semi-Disp.
Weight	21,000#	Designer	V. Lazzara
Clearance	12'0"	Production	1972–77

Built on the Gulfstar 43 Motorsailer hull, the 43 Trawler is a conservative long-range cruiser of the type popular back in the 1970s during the fuel crisis. Her solid fiberglass, semi-displacement hull features a long (but not deep) keel and a relatively modest beam. Originally a trunk-cabin design, in 1975 a raised aft deck version—the 43 MK II (pictured above)—was introduced with a redesigned flybridge and a full teak interior replacing the previous mica or mahogany decor. The MK II's full-width aft deck allows for an enlarged master stateroom below with a centerline queen berth. A comfortable cruiser, the galley and dinette in the Gulfstar 43 are down, resulting in a fairly roomy salon with complete lower helm station and deck doors port and starboard. Additional features include near-standing headroom in the engine room, wide sidedecks, and protected underwater running gear. Obviously dated by today's design standards (but a seakindly boat offshore), she'll cruise economically at 9 knots and reach a top speed of 11–12 knots with 130-hp Perkins diesels. ❑

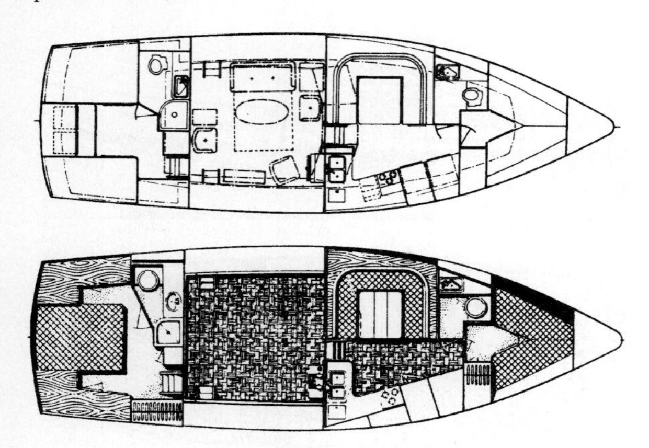

See page 240 for pricing information.

GULFSTAR 44 MOTOR CRUISER

SPECIFICATIONS

Length	44'5"	Water	250 gals.
LWL	39'0"	Fuel	500 gals.
Beam	14'6"	Hull Type	Disp.
Draft	3'6"	Designer	V. Lazzara
Weight	30,000#	Production	1978–80

In terms of styling, the 44 Motor Cruiser marked Gulfstar's entry into the modern motor yacht market when she was introduced in 1978. Looks are deceiving: while she has a genuine motor yacht look, she's actually built on a full-displacement, trawler-type hull with a deep keel and relatively flat after sections. Inside, the floorplan is arranged with the galley down and a notably spacious master stateroom aft. The salon, which is somewhat narrow due to the wide sidedecks, has a lower helm and deck access door to starboard. A ladder leads down into the 44's stand-up engine room, and a pump room forward separates the generator and air conditioning units from the motors. Twin 130-hp Perkins diesels were standard, and 160-hp Perkins turbos were optional (although they add little to performance). At her hull speed of 8.5 knots, the fuel burn is 5-6 gph. With 500 gallons of fuel (in just one tank!), the 44 has a range of approximately 700 miles. A total of 105 of these boats were built. ❑

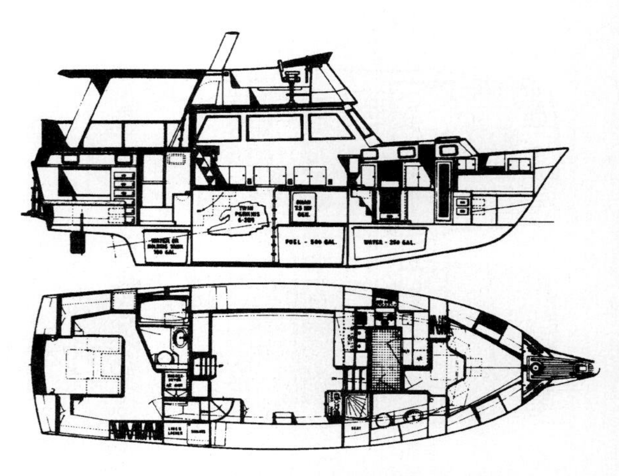

See page 240 for pricing information.

GULFSTAR 44 MOTOR YACHT

SPECIFICATIONS

Length	43'9"	Water	200 gals.
LWL	39'1"	Fuel	400 gals.
Beam	15'0"	Hull Type	Modified-V
Draft	3'6"	Deadrise Aft	NA
Weight	36,400#	Designer	R. Lazzara
Clearance	17'0"	Production	1985–86

A downsized version of Gulfstar's popular 49 Motor Yacht, the 44 Motor Yacht differs from the 44 Widebody in that she has sidedecks surrounding the fully enclosed afterdeck. Just as in the Gulfstar 49, this fully enclosed and paneled afterdeck serves as the main salon and entertainment area. In this unique layout, the (huge) galley and dinette are located on the lower level. Both staterooms have centerline double berths, and a shower stall is provided in each head. In an innovative departure from the conventional, the aft stateroom has the bed fitted against the *forward* bulkhead. Thus arranged, the Gulfstar 44 has one of the roomiest interior layouts available in a boat this size. Space for a washer/dryer is located next to the forward companionway steps, where a bulkhead door leads into the standup engine room. Note that a seawater intake chest and discharge system is installed to eliminate individual thru-hull fittings. Twin 300-hp Caterpillar diesels will cruise the Gulfstar 44 MY around 16 knots with a top speed of 19–20 knots. ❏

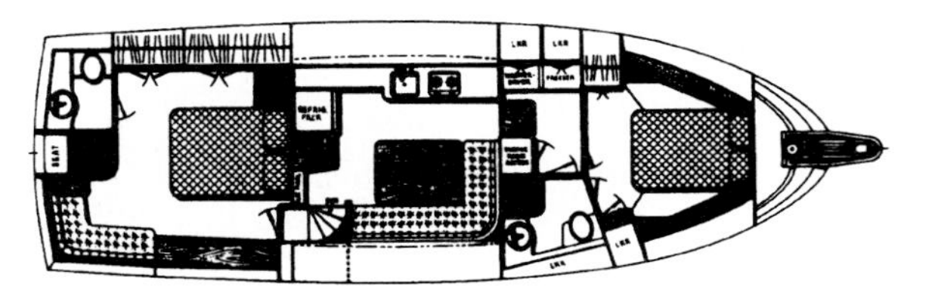

See page 240 for pricing information.

GULFSTAR 44 WIDEBODY MY

SPECIFICATIONS

Length	43'9"	Water	200 gals.
LWL	39'1"	Fuel	400 gals.
Beam	15'0"	Hull Type	Modified-V
Draft	3'6"	Deadrise Aft	NA
Weight	36,400#	Designer	R. Lazzara
Clearance	17'0"	Production	1986–88

Featuring a *full-width*, fully enclosed afterdeck with wing doors, the Gulfstar 44 Widebody (both the MK II and MK III models) replaced the original 44 Motor Yacht in 1986. A total of 25 Widebodys were built during her production run, and they remain popular today thanks to an innovative and truly practical accommodation plan. Rather than locating the salon amidships in the normal fashion, the Gulfstar 44's salon is located on the raised afterdeck—a fully enclosed, paneled, and air-conditioned cabin with a sliding glass door to the small aft deck. This is the same unique floorplan first used in the original Gulfstar 44 MY, only now— by eliminating the sidedecks—the salon dimensions are substantially increased. The mid-level galley/dinette area is huge, and centerline double berths are fitted in both staterooms. Other updates made in the 44 Widebody include the addition of a radar arch, larger standard engines, and a slightly restyled flybridge. Twin 375-hp Cat diesels will cruise at 17–18 knots with a top speed around 21 knots. ❏

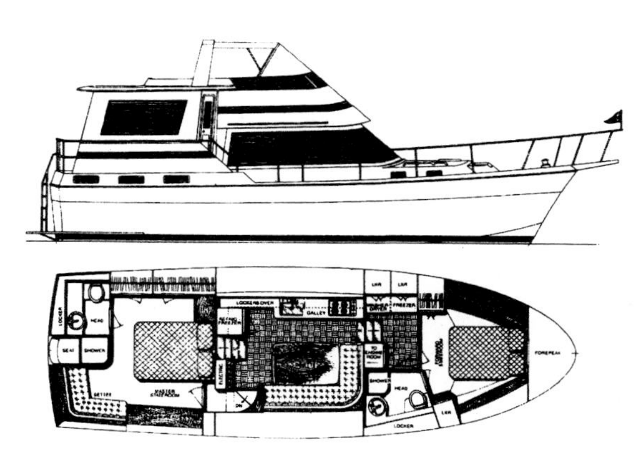

See page 240 for pricing information.

GULFSTAR 48 MOTOR YACHT

SPECIFICATIONS

Length	48'11"	Water	200 gals.
LWL	NA	Fuel	500 gals.
Beam	15'0"	Hull Type	Modified-V
Draft	3'7"	Deadrise Aft	NA
Weight	40,500#	Designer	R. Lazzara
Clearance	14'7"	Production	1981–83

Closely resembling the 44 Motor Cruiser in appearance, the 48 Motor Yacht was the first Gulfstar to be built on a planing hull and powered to attain true motor yacht speeds. Then state-of-the-art construction techniques—including balsa coring in the hullsides and unidirectional composite materials—were used in the production of the boat. Inside, her expansive three-stateroom interior is arranged with the galley and dinette down from the salon, queen berths in both aft cabins, and stacked single berths in the forward stateroom. The salon is large enough to accommodate several furniture arrangements, and there is near-standing headroom in the engine compartment. Throughout, the teak cabinetry and woodwork found in the Gulfstar 48 are exceptionally beautiful and lavishly applied. A hardtop shades the huge open aft deck, and the flybridge could be ordered with guest seating forward or aft of the helm console. Standard power for the Gulfstar 48 Motor Yacht was GM 6-71Ns (290-hp) with the 390-hp TI versions offered as an option. The latter engines cruise around 17 knots. A total of eighteen 48 MYs were built. ❑

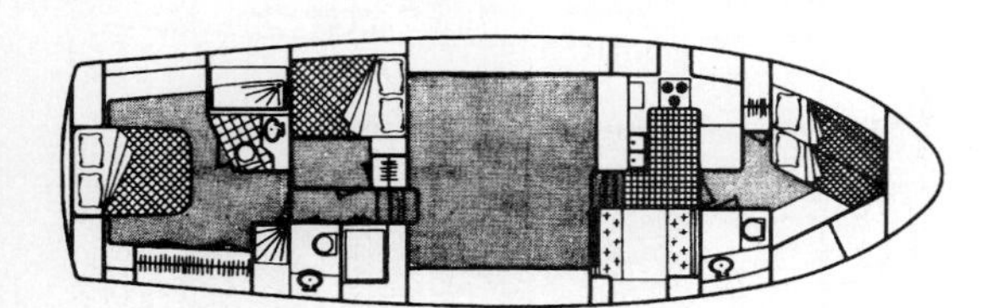

See page 240 for pricing information.

GULFSTAR 49 MOTOR YACHT

SPECIFICATIONS

Length	49'0"	Water	370 gals.
LWL	44'4"	Fuel	675 gals.
Beam	15'1"	Hull Type	Modified-V
Draft	3'10"	Deadrise Aft	NA
Weight	42,000#	Designer	R. Lazzara
Clearance	17'0"	Production	1984–87

The Gulfstar 49 MY was a further refinement of the company's popular 48 Motor Yacht. Introduced in 1984 with 350-hp Perkins diesels, the MK II version in 1985 featured optional 435-hp 6-71 GM diesels and a unique aft stateroom design with the bed fitted against the forward bulkhead. Updates in 1986 (MK III) include a standard radar arch and bigger standard engines (375-hp Cats), and the MK IV (1987) offered GM 6V92TA diesels as an option. In all models, the *main* salon and entertaining area is the afterdeck—a fully enclosed and teak-paneled indoor living area with wing doors and generous dimensions. At the deckhouse level, where the salon is normally found, the area has been converted into a huge galley and dinette. Additional features include a spacious stand-up engine room and a small aft deck for line-handling. Cruising speed with the 350-hp Perkins is 13 knots (16 knots top), and, with the 435-hp GM 6-71s, the Gulfstar 49 will cruise at 16–17 knots (about 20 knots top). ❑

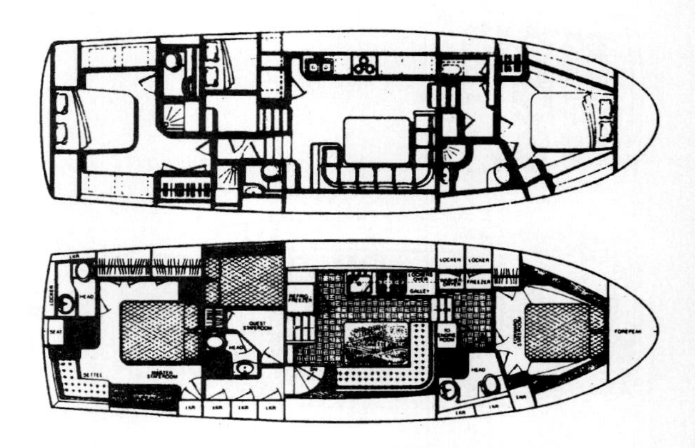

See page 240 for pricing information.

GULFSTAR 53 TRAWLER

SPECIFICATIONS

Length	53'2"	Water	300 gals.
LWL	47'3"	Fuel	1,000 gals.
Beam	15'0"	Hull Type	Disp.
Draft	4'6"	Designer	V. Lazzara
Weight	42,000#	Production	1975–76
Clearance	14'0"		

Largest of the Gulfstar trawlers, the 53 was built on the same round-bilge full-displacement hull used for the Gulfstar 53 Motorsailer—a sailboat. No award-winner when it comes to looks, she's known as a seakindly boat with an easy motion in most conditions. Her deep bilges allow the fuel tanks to be set low in the hull, reducing the center of gravity and making for a more comfortable motion. Her three-stateroom layout came with a choice of twin berths or a walkaround double in the aft cabin. The lower helm station is located on the centerline in the salon (very unusual) with deck doors port and starboard. Both heads have stall showers, and it's worth noting that while the salon in the Gulfstar 53 Trawler is somewhat compact, the lower level galley/dinette area is very spacious. Additional features include a full teak interior, stand-up engine room, broad sidedecks, and fully protected underwater running gear. Twin 160-hp Perkins diesels will cruise at 8–9 knots burning about 6 gph. Fewer than ten were built. ❑

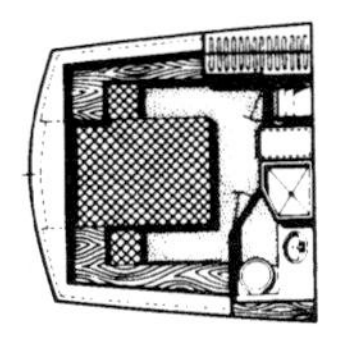

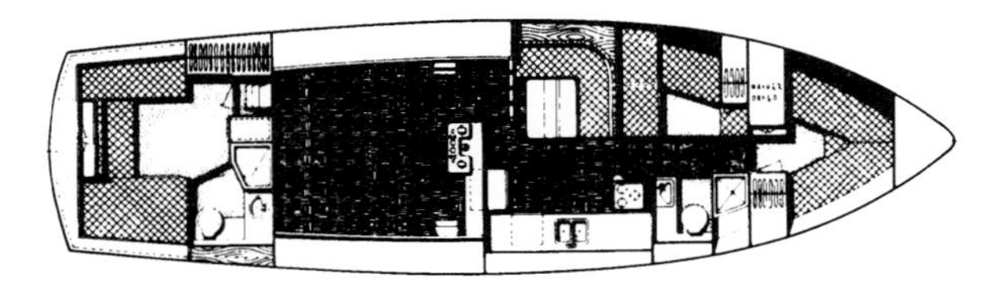

See page 240 for pricing information.

HARTMANN-PALMER 56 MY

SPECIFICATIONS

Length	56'1"	Water	600 gals.
LWL	51'0"	Fuel	1,400 gals.
Beam	17'1"	Hull Type	Modified-V
Draft	4'2"	Deadrise Aft	NA
Weight	57,350#	Designer	Seaton/Neville
Clearance	21'1"	Production	1984–87

Built by Lien Hwa in Taiwan, the Hartmann-Palmer 56 MY bears a strong resemblance (in profile, at least) to the Hatteras 53 MY. She was constructed with cored hullsides and features a wide beam, a long keel, and underwater exhausts to reduce noise. The traditional double-deck layout places the galley and dinette down from the salon level. The generous 12' x 14' salon dimensions are more than adequate, and the afterdeck is even larger. A curved staircase aft in the salon leads down to a spacious master stateroom with a tub/shower. A walk-in engine room separates the master from the guest cabins forward. Outside of the engine room, just about everything you see inside the Hartmann-Palmer 56 is either made out of teak, covered by it, or trimmed with it. No lightweight, the standard 450-hp GM 6-71s deliver an anemic 13–14 knot cruising speed when fully loaded and 17 knots flat out. The optional 600-hp 8V92TIs provide 16–17 knots cruise and 19 knots top. Note the generous fuel capacity. ❑

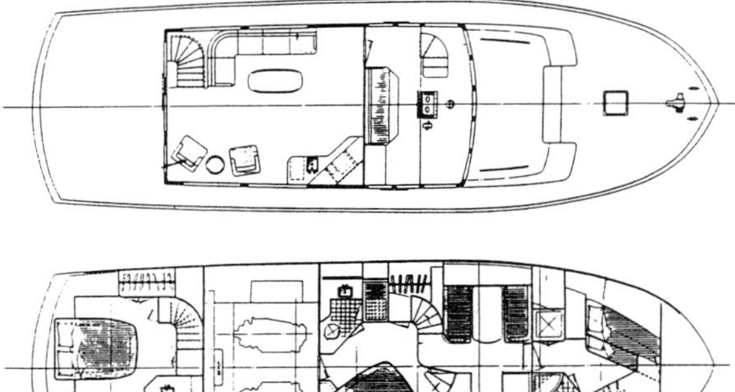

See page 240 for pricing information.

HATTERAS 38 DOUBLE CABIN

SPECIFICATIONS

Length	38'4"	Water	145 gals.
LWL	NA	Fuel	300 gals.
Beam	13'7"	Hull Type	Modified-V
Draft	3'5"	Deadrise Aft	NA
Weight	33,000#	Designer	J. Hargrave
Clearance	12'10"	Production	1973–78

The Hatteras 38 Double Cabin is a restyled version of an earlier 38 double-cabin cruiser that Hatteras introduced in 1968. This newer model has a raised aft deck (which results in a full-beam master stateroom) and places the bridge over the deckhouse—very stylish for 1973. An attractive design in spite of her stubby profile, the 38 DC is a comfortable and well-built family cruiser. Her interior accommodations are arranged around a conventional two-stateroom, galley-down layout with large wraparound salon windows and excellent storage. The owner's stateroom is fitted with twin berths (a double berth was never offered), and V-berths are in the forward stateroom. The interior is fully paneled and finished with teak woodwork, and a lower helm was optional in the salon. Her flybridge (small by today's standards) features a centerline helm and bench seating to starboard. An easy-riding boat, with standard GM 6-71N diesels, the Hatteras 38 Double Cabin will cruise at 16–17 knots and reach a top speed of around 20 knots. ❑

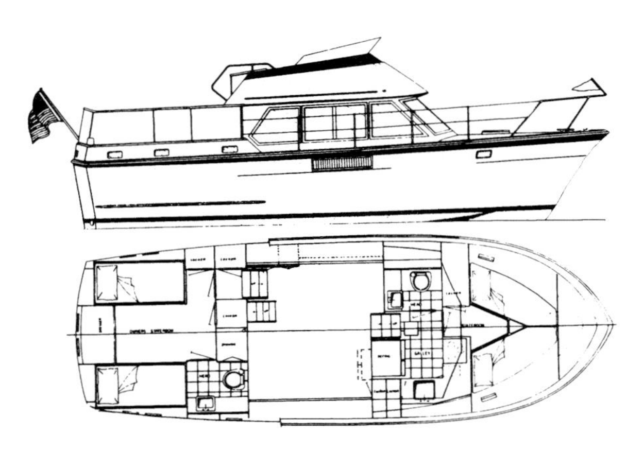

See page 240 for pricing information.

HATTERAS 40 MOTOR YACHT

SPECIFICATIONS

Length	40'10"	Water	110 gals.
LWL	NA	Fuel	359 gals.
Beam	13'7"	Hull Type	Modified-V
Draft	4'9"	Deadrise Aft	14°
Weight	38,000#	Designer	Hatteras
Clearance	15'9"	Production	1986–Current

A popular boat, the Hatteras 40 MY (originally called the Hatteras 40 Double Cabin) is a modern family cruiser with an upscale interior and excellent accommodations. She's built on conventional modified-V hull with balsa coring in the hullsides, moderate beam, and prop pockets. Her now-standard (since 1990) galley-down floorplan is arranged with a dinette in the salon and stacked single berths in the forward stateroom. The original layout (now discontinued) had a slightly smaller salon, no dinette, and V-berths forward. Notably, both heads are fitted with separate stall showers. The extensive use of teak (or ash) paneling and cabinetry is contrasted with top-quality fabrics and white Formica counters in the galley. The flybridge was redesigned in 1990 with a forward helm and swept-back windscreen, and a new full-height entry door to the salon was added at the same time. A poor performer with gas engines (13-14 knots cruise/20 knots top), 375-hp Cats are now standard. They'll deliver a cruising speed of 17 knots and a top speed of around 21 knots.❑

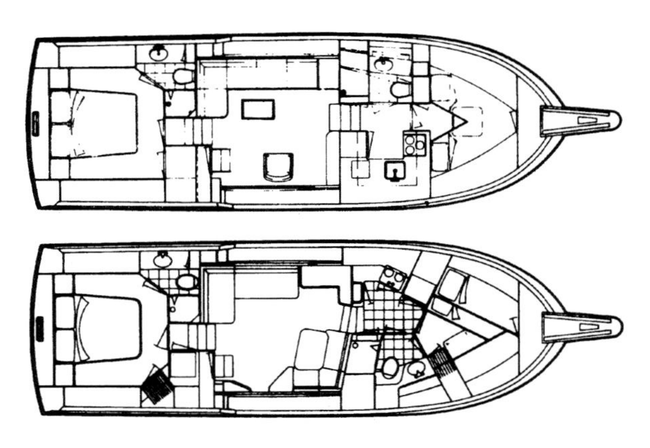

See page 240 for pricing information.

HATTERAS 42 LONG RANGE CRUISER

SPECIFICATIONS

Length	42'6"	Water	220 gals.
LWL	38'0"	Fuel	700 gals.
Beam	14'6"	Hull Type	Disp.
Draft	3'10"	Deadrise Aft	NA
Weight	36,000#	Designer	J. Hargrave
Clearance	13'6"	Production	1976–85

The Hatteras 42 LRC was introduced as one of the few alternatives to the influx of Asian trawlers imported during the era of fuel shortages. Her displacement hull form has rounded bilges forward that gradually harden to form a shallow "V" at the transom. While all trawlers will roll in a beam sea, the true displacement-hull design of the 42 LRC produces an easier, more comfortable motion than most of the more recent trawler-style boats with harder chines. Her conventional double-cabin floorplan includes fore and aft staterooms, each with a head and stall shower. The original port-side galley was replaced by a more convenient U-shaped layout in 1979. A full-width afterdeck replaced the walkaround trunk cabin in 1980, and a tapered double berth in the master stateroom became available in 1981. A superb cruising yacht, twin Lehman or GM diesels will cruise the Hatteras 42 LRC at a steady 8 knots (6 gph) with a cruising range of over 1,000 miles. At a reduced 7-knot speed the range is close to 2,000 miles. ❑

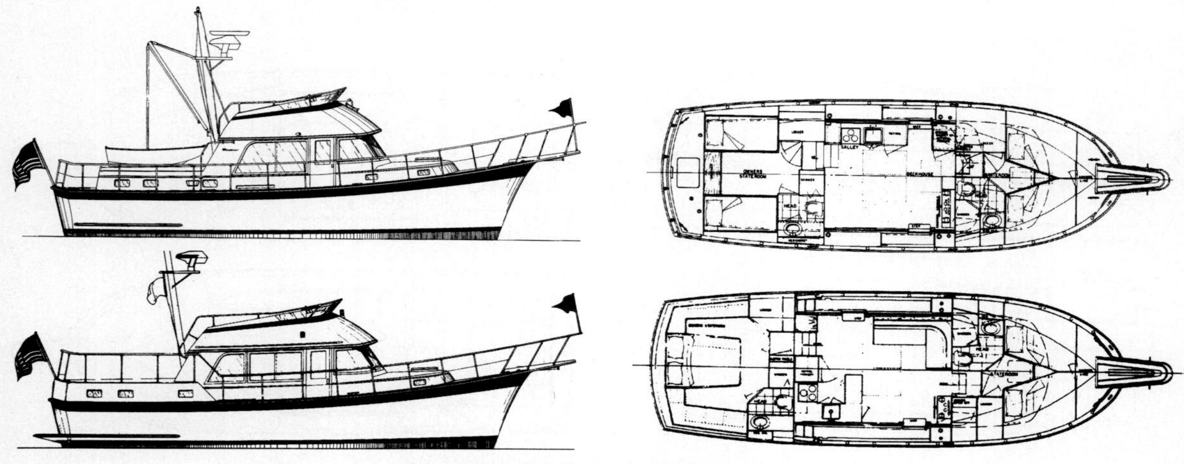

See page 240 for pricing information.

HATTERAS 42 COCKPIT MY

SPECIFICATIONS

Length	42'10"	Water	115 gals.
LWL	NA	Fuel	375 gals.
Beam	13'7"	Hull Type	Modified-V
Draft	4'9"	Deadrise Aft	NA
Weight	41,000#	Designer	Hatteras
Clearance	15'9"	Production	1993–Current

A good-looking boat with a stylish reverse transom and attractive graphics, the Hatteras 42 Cockpit MY combines the convenience and versatility of a cockpit with the cruising comforts of an aft cabin design. No lightweight for her size, the bottom is solid fiberglass, and the hullsides are cored. Her two-stateroom interior is well-arranged with the galley down and wide open to the salon. Note that the dinette is located on the deckhouse level so as not to reduce the forward head and stateroom dimensions—a practical arrangement, although it does eat into the salon space. Both heads have stall showers, and the upscale decor and woodwork is quite appealing. The engine room (below the salon) is a tight fit. Outside, the sidedecks are comfortably wide, and the full-width aft deck is large enough for an assortment of deck furnishings. The bridge is arranged with built-in seating surrounding the centerline helm position. Twin 375-hp Cat diesels (or 400-hp 6V-53s) will cruise the Hatteras 42 CMY at 18 knots and deliver 22–23 knots top. ❑

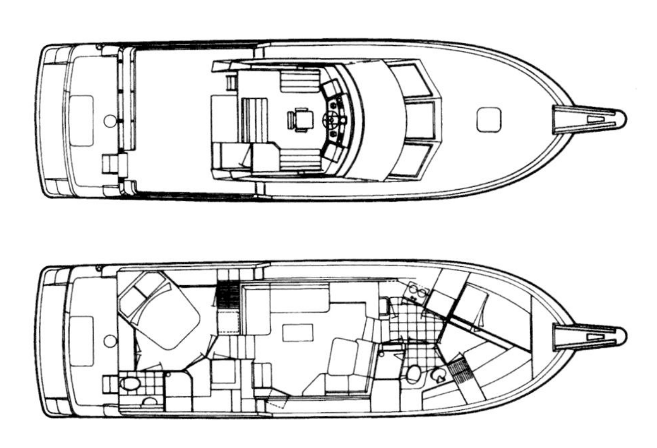

See page 240 for pricing information.

HATTERAS 43 DOUBLE CABIN

SPECIFICATIONS

Length	43'1"	Water	130 gals.
LWL	37'6"	Fuel	375 gals.
Beam	14'0"	Hull Type	Modified-V
Draft	3'5"	Deadrise Aft	8°
Weight	34,000#	Designer	J. Hargrave
Clearance	17'10"	Production	1971–84

The 43 Double Cabin was a very popular boat for Hatteras, and she had a notably successful production run. She was originally designed to replace the then-aging Hatteras 41 Twin Cabin model back in 1971, and her popularity kept her in production for the next thirteen years. Construction is solid fiberglass, and her modified-V hull features shallow deadrise at the transom and a long keel for stability. Inside, the salon is extremely open, and teak paneling and cabinetry are found in every cabin. Early models have a stall shower in the forward head; the floorplan was changed in 1979 to include a portside dinette (at the expense of that stall shower). A walkaround double berth was first offered in the master stateroom in 1978, and a three-stateroom floorplan became optional in 1980. The flybridge is small by today's standards, and visibility from the semi-enclosed lower helm is very good. With GM 6-71N diesels, the Hatteras 43 will cruise around 14 knots. The larger 390-hp TI versions cruise at 16–17 knots. ❑

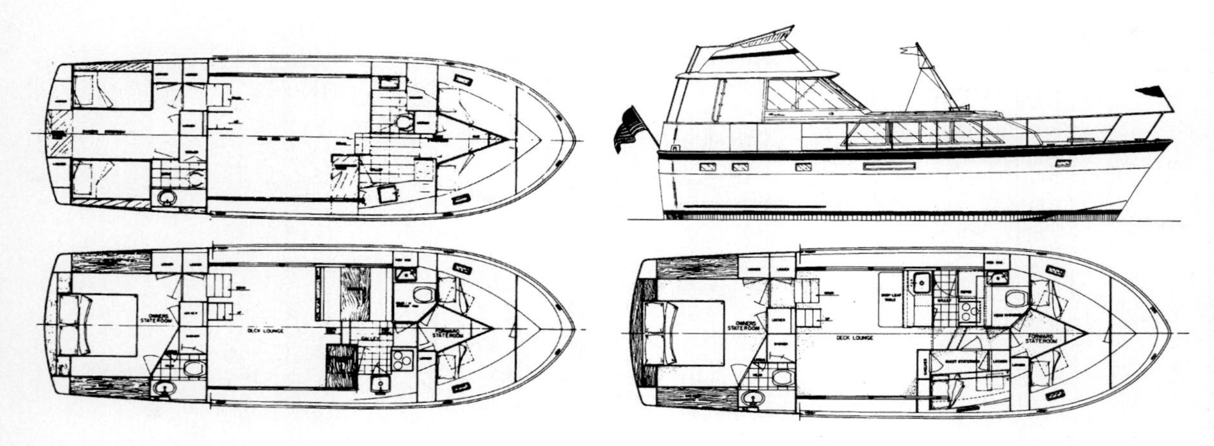

See page 240 for pricing information.

HATTERAS 43 MOTOR YACHT

SPECIFICATIONS

Length	43'1"	Water	130 gals.
LWL	37'9"	Fuel	375 gals.
Beam	14'0"	Hull Type	Modified-V
Draft	3'5"	Deadrise Aft	NA
Weight	34,500#	Designer	J. Hargrave
Clearance	16'7"	Production	1984–87

While she's built on the same hull as the earlier Hatteras 43 Double Cabin, the 43 Motor Yacht is more than a warmed-over 43 DC. The profile and interior layout are completely revised, and the exterior styling of the 43 MY is dramatically improved from that of her predecessor. The two-stateroom interior layout features a spacious main salon with a mid-level galley to port and separate stall showers in both heads. Decorator fabrics and white Formica cabinetry in the galley and heads provide a pleasant contrast to the interior's dark teak paneling and woodwork. The aft deck is very roomy for a boat of this size, and wing doors open to well-secured sidedecks. The flybridge is arranged with the helm console forward, and bench seating is provided for six. Standard 375-hp Cat diesels will cruise the Hatteras 43 MY around 16 knots with a top speed of 20–21 knots. Optional 6-71TIs cruise around 17 knots. With a small fuel capacity, the cruising range is somewhat limited. ❑

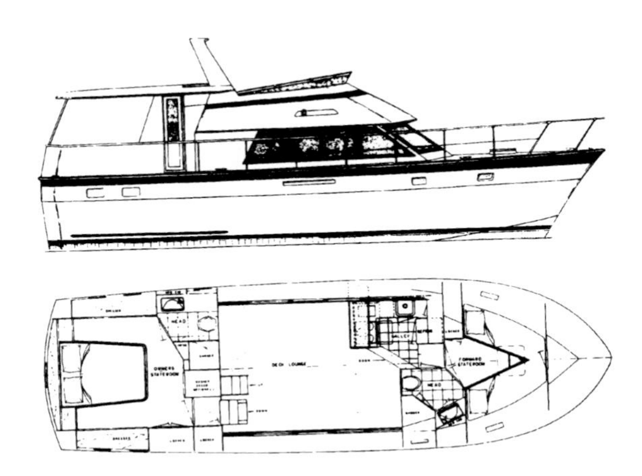

See page 240 for pricing information.

HATTERAS 48 LONG RANGE CRUISER

SPECIFICATIONS

Length	48'10"	Cockpit	NA
LWL	43'4"	Water	430 gals.
Beam	16'6"	Fuel	1,390 gals.
Draft	4'6"	Hull Type	Disp.
Weight	54,000#	Designer	J. Hargrave
Clearance	16'11"	Production	1976–81

With her raised pilothouse and sturdy profile, the Hatteras 48 LRC has the no-nonsense appearance of a bluewater cruiser. It's the raised pilothouse that sets her apart from most of the competition; from here, she can be operated in complete privacy from the rest of the boat. Deck doors provide quick outside access, and a settee/watchberth is fitted right behind the helm. The 48 LRC is heavily built on a solid-fiberglass, full-displacement hull with rounded bilges and a long, deep keel. Freeboard at the bow is quite high. Featuring a comfortable two-stateroom layout, the spacious master stateroom (with a tub in the head compartment) is located directly below the pilothouse. (A three-stateroom layout was also available.) The compact galley is to port in the salon, and the small cockpit is fitted with a boarding door. With 112-hp GM or 120-hp Lehman diesels, the Hatteras 48 LRC will cruise at her 8.8-knot hull speed burning only 9 gph. At a 7-knot cruising speed the range exceeds 2,000 miles. ❑

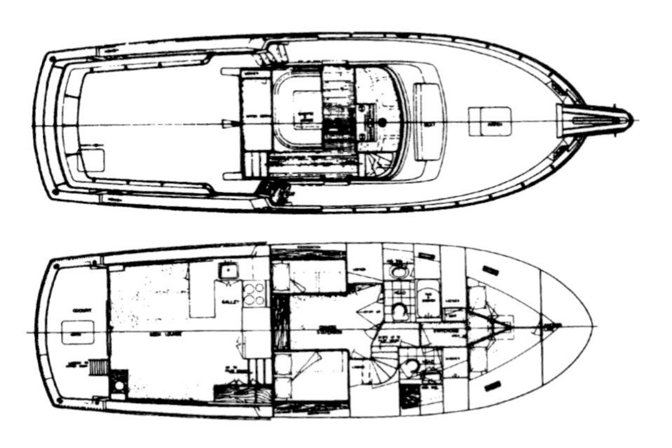

See page 240 for pricing information.

HATTERAS 48 MOTOR YACHT (EARLY)

SPECIFICATIONS

Length	48'8"	Water	190 gals.
LWL	NA	Fuel	590 gals.
Beam	15'0"	Hull Type	Semi-Disp.
Draft	3'11"	Deadrise Aft	7°
Weight	45,000#	Designer	J. Hargrave
Clearance	17'2"	Production	1981–84

It's interesting to note that the Hatteras 48 MY was built on a "dual mode" semi-displacement hull—designed for efficient low-speed operation while retaining planing-speed abilities with larger engines. (Fuel-efficiency was a serious marketing tool in the early 1980's.) Showing a slightly trawler-style profile, the 48 MY was heavily built on a solid fiberglass hull with shallow deadrise aft and a long keel for stability. Her standard three-stateroom interior is arranged with two staterooms *aft* of the salon, three heads (but only one stall shower), and a lower helm. An optional four-stateroom floorplan eliminated the dinette. Outside, the spacious aft deck is the largest found in a boat this size—the dimensions exceed those of the salon. Standard 285-hp 6-71N diesels deliver an efficient 1 mpg at a 9-knot displacement speed but lack the power to get the 48 MY on plane with a full load. The more desirable 425-hp 6V92s will cruise around 16 knots and reach 18–19 knots wide open. ❑

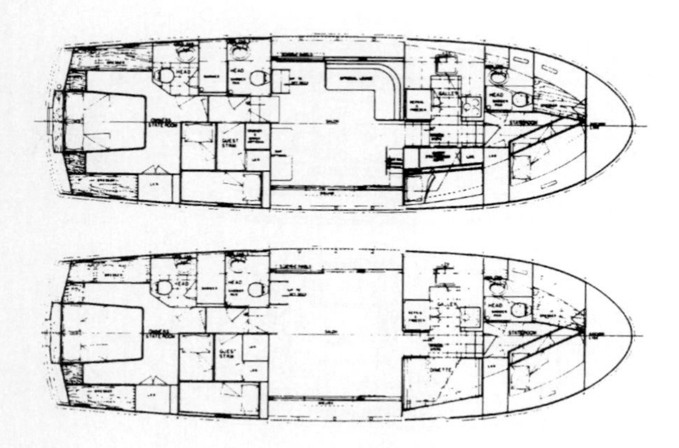

See page 241 for pricing information.

HATTERAS 48 COCKPIT MY (EARLY)

SPECIFICATIONS

Length	48'8"	Water	190 gals.
LWL	NA	Fuel	590 gals.
Beam	15'0"	Hull Type	Modified-V
Draft	3'11"	Deadrise Aft	7°
Weight	47,000#	Designer	J. Hargrave
Clearance	17'2"	Production	1981–85

A good-looking yacht with her flush-deck profile and raked bridge, the 48 Cockpit MY wasn't a particularly big seller for Hatteras. She's constructed on the same "dual mode" hull used in the Hatteras 48 MY (1981–84)—basically a semi-displacement configuration with rounded bilges forward, flat aftersections, shallow deadrise, and a long keel. Her standard layout included a centerline double berth in the master cabin (twin single berths were optional) and a second stateroom forward with twin over/under berths. An optional three-stateroom floorplan traded out the dinette for a third stateroom while leaving the galley down. Either way, the aft stateroom dimensions are impressive. Teak woodwork was used throughout the interior, and the only stall shower is aft. A lower helm station was optional. Outside, the afterdeck is very roomy with space for several pieces of deck furniture. The cockpit, however, is small. No racehorse, standard 285-hp 6-71Ns will cruise the Hatteras 48 CMY at 12–13 knots, and the optional 425-hp 6V92TAs cruise around 15–16 knots. ❑

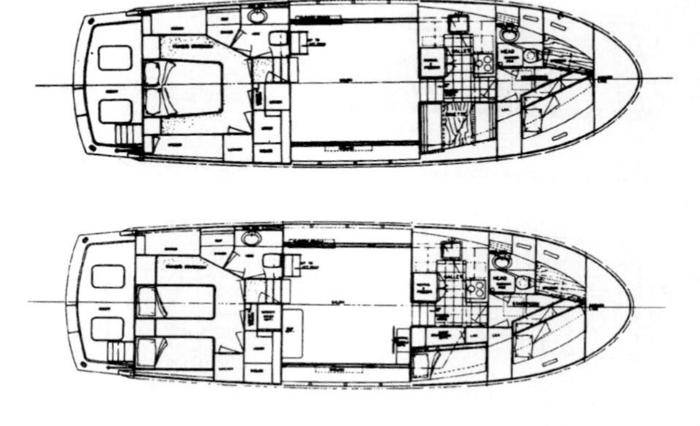

See page 241 for pricing information.

HATTERAS 48 MOTOR YACHT

SPECIFICATIONS

Length	48'9"	Water	170 gals.
LWL	NA	Fuel	750 gals.
Beam	16'0"	Hull Type	Modified-V
Draft	5'3"	Deadrise Aft	14°
Weight	63,000#	Designer	Hatteras
Clearance	16'8"	Production	1990–Current

An in-house Hatteras design, the 48 Motor Yacht is a classy offshore cruiser with a spacious interior to match her stylish profile. She's built on a modified-V hull with Divinycell coring in the hullsides although—at 63,000 lbs.—she's certainly no lightweight. The standard interior is arranged with three staterooms, two with walkaround double berths (the third is aft of the salon with stacked single berths), and three full heads. A less popular two-stateroom floorplan replaces the aft guest cabin with an enlarged master head and adds a walk-in wardrobe to port in the owner's stateroom. Originally a galley-down boat, the galley and dinette were raised to a mid level in 1993. The sidedecks are wide, and guest seating on the bridge is behind the wraparound helm. Not inexpensive, twin 535-hp 6V92 diesels were standard in very early models (16-knots cruise/19-knots top). They were soon replaced with the larger 720-hp 8V92s, which cruise the Hatteras 48 MY at a more acceptable 20 knots and reach 23 knots wide open. ❑

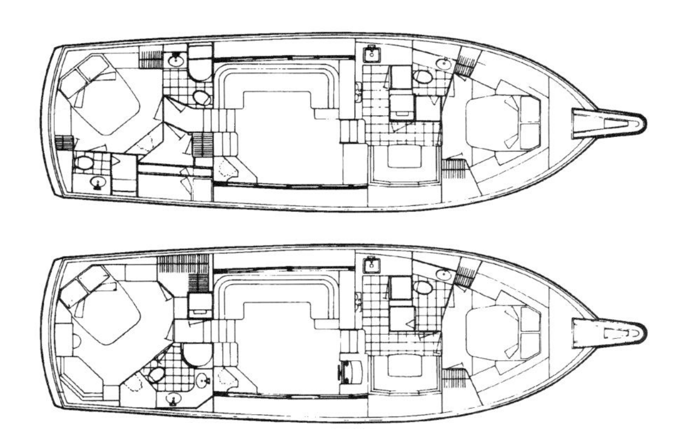

See page 241 for pricing information.

HATTERAS 48 COCKPIT MY

SPECIFICATIONS

Length	48'11"	Water	170 gals.
Beam	16'0"	Fuel	667 gals.
Draft	5'3"	Hull Type	Modified-V
Weight	59,000#	Deadrise Aft	NA
Clearance	16'9"	Production	1993–Current

A good-looking yacht with a sweptback profile and aggressive lines, the Hatteras 48 CMY is an upscale family cruiser with a wide beam and impressive accommodations. Her hull is balsa-cored from the waterline up and features a deep keel and more transom deadrise than most Hatteras designs. Note the unusually deep draft. She also has an innovative (and quiet) split exhaust system with side ports for idling and larger underwater outlets for speeds above 1,300 rpm. Below, the galley-down, two-stateroom floorplan includes walkaround berths fore and aft, a dinette, and stall showers in both heads. An optional three-stateroom layout moves the dinette up into the salon while adding the third cabin opposite the galley. Additional features include a spacious engine room, wide sidedecks, and plenty of guest seating on the flybridge. The cockpit is very small but is at least suitable for line-handling and getting on and off the boat. With standard 535-hp 6V-92s, the Hatteras 48 Cockpit MY will cruise at a respectable 18 knots and reach 21 knots wide open. ❑

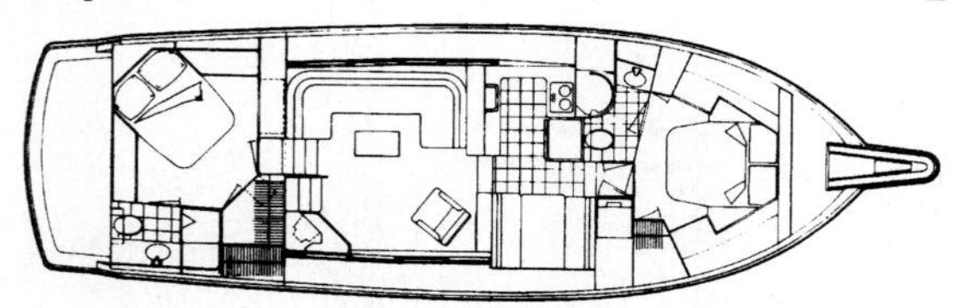

STANDARD ARRANGEMENT

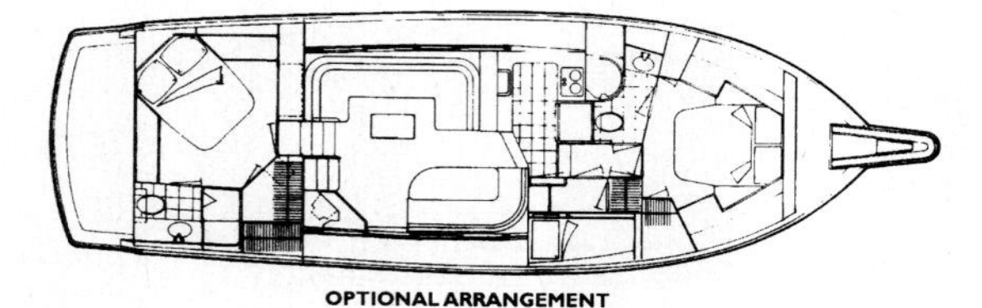

OPTIONAL ARRANGEMENT

See page 241 for pricing information.

HATTERAS 52 COCKPIT MY

SPECIFICATIONS

Length	52'9"	Water	170 gals.
LWL	NA	Fuel	994 gals.
Beam	16'0"	Hull Type	Modified-V
Draft	5'3"	Deadrise Aft	14°
Weight	66,000#	Designer	Hatteras
Clearance	16'9"	Production	1990–Current

Essentially a Hatteras 48 Motor Yacht with a cockpit, the 52 CMY is a good-looking design with a modern profile and surprisingly brisk performance. Her hull is cored from the waterline up and features a long keel and a fairly steep 14° of transom deadrise—more "V" than most Hatteras designs. She also has an innovative (and quiet) split exhaust system with side ports for idling and larger underwater outlets for speeds above 1,300 rpm. Her mid-galley, three-stateroom floorplan is the same as the Hatteras 48's with contemporary teak or light ash woodwork and three full heads. An optional two-stateroom layout trades out the aft guest stateroom and one head for a drastically enlarged master suite. Additional features include a spacious engine room, a huge aft deck platform, wide sidedecks, and lounge seating for eight on the flybridge. The cockpit is small but well-suited for boarding and line-handling duties. With standard 720-hp 8V92s, the Hatteras 52 Cockpit MY will cruise at a respectable 20–21 knots and reach 24 knots wide open. ❑

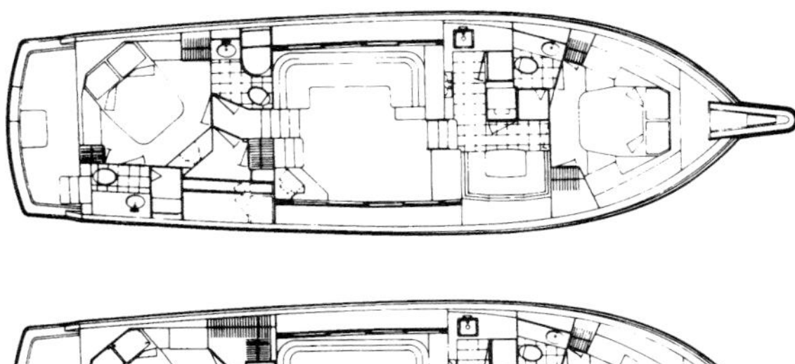

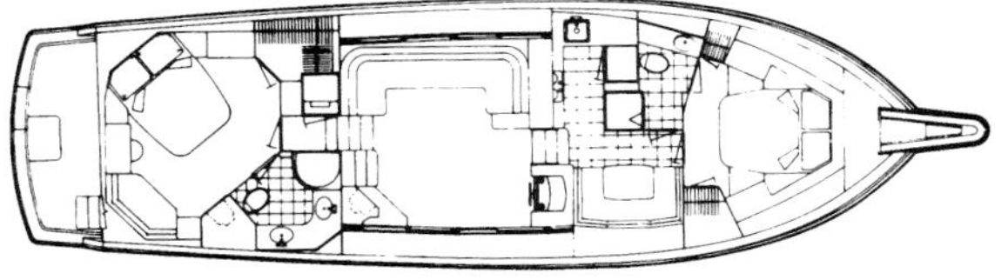

See page 241 for pricing information.

HATTERAS 52 MOTOR YACHT

SPECIFICATIONS

Length	52'9"	Water	170 gals.
Deadrise Aft	NA	Fuel	994 gals.
Beam	16'0"	Hull Type	Modified-V
Draft	5'0"	Deadrise Aft	NA
Weight	66,000#	Designer	Hatteras
Clearance	16'9"	Production	1993–Current

Aside from her sleek low-profile appearance, perhaps the most striking aspect of the Hatteras 52 Motor Yacht is her huge aft deck—a fully enclosed, paneled, and air-conditioned living area whose dimensions exceed those of the salon. (The enclosed aft deck is a $13,000 option—almost negligible in this $800,000+ boat.) At 66,000 lbs., the 52 MY is no lightweight and that, together with a low center of gravity, results in a good-riding boat. The floorplan is arranged with two of the three staterooms aft of the salon (which accounts for the huge aft deck) and a mid-level galley and dinette. Each head has a stall shower, and the interior woodwork is light ash throughout. (The stand-up engine room is accessed via the galley stairs.) Additional features include a large flybridge with a wraparound helm console and a small line-handling platform at the transom. With standard 720-hp 8V92s, the Hatteras 52 MY will cruise at a respectable 20–21 knots and reach a top speed of about 24 knots.❑

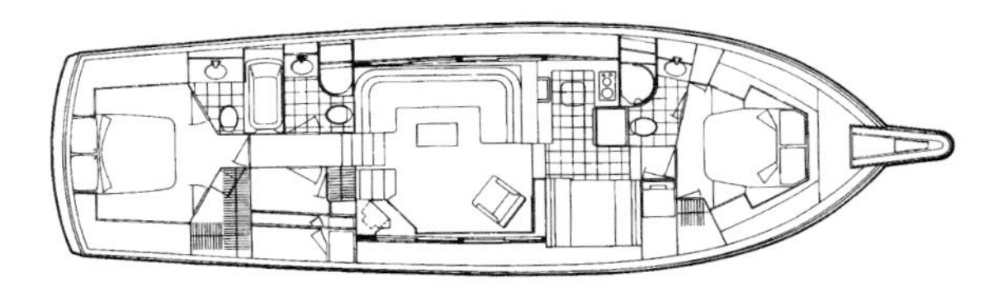

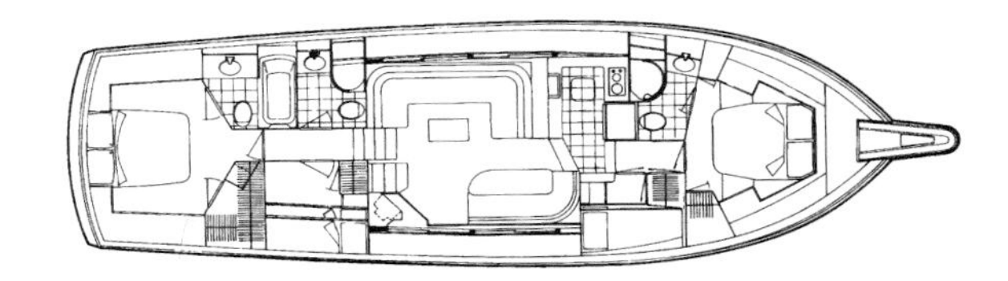

See page 241 for pricing information.

HATTERAS 53 MOTOR YACHT

SPECIFICATIONS

Length	53'1"	Water	245 gals.
LWL	47'3"	Fuel	550/600/700 gals.
Beam	15'10"	Hull Type	Modified-V
Draft	4'0"	Deadrise Aft	NA
Weight	55,000#	Designer	J. Hargrave
Clearance	18'6"	Production	1969–88

Some designs succeed so well in the marketplace that they earn the distinction of being industry classics. Such a boat is the Hatteras 53 Motor Yacht. Two decades in production, the Hatteras 53 MY became the standard by which other motor yachts in her size range were measured. Her bi-level, three-stateroom/three-head floorplan is considered ideal in a motor yacht of this size, and the split engine rooms soon became a Hatteras trademark. Notable updates include the addition of stall showers to the aft guest head and forward head in 1975, a redesigned low-profile flybridge in 1977, an optional walkaround queen bed in the master stateroom in 1978, and an increase in the fuel capacity (to 700 gals.) in 1980. Early models with the GM 8V71N diesels will cruise at 13 knots and reach 16 knots top. The larger 8V71TIs (and the later [1985] 6V92TAs) improve the cruising speed to around 16 knots with a top speed of 19 knots. Resale values for Hatteras 53 Motor Yachts remain strong in all markets. ❑

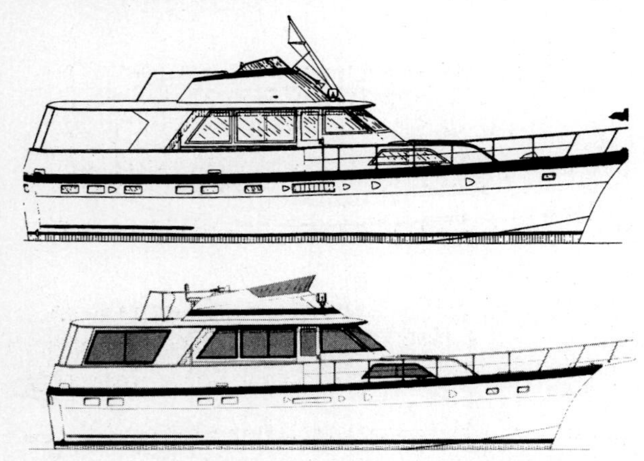

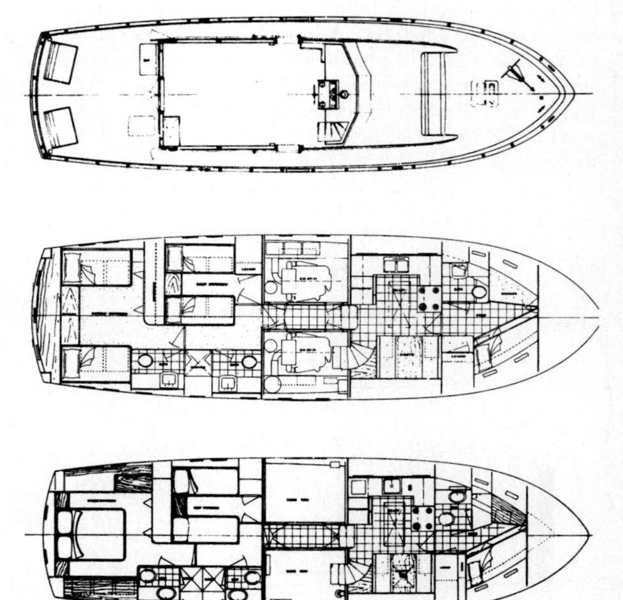

See page 241 for pricing information.

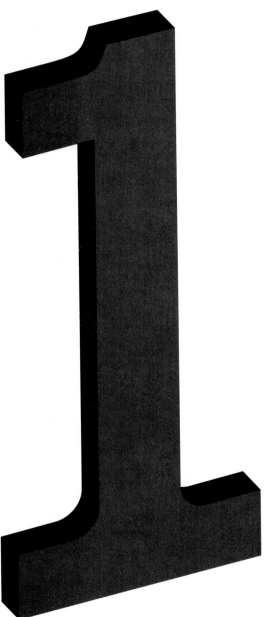

HATTERAS 53 YACHT FISHERMAN

SPECIFICATIONS

Length	52'11"	Fuel	825/1,015 gals.
Beam	15'10"	Hull Type	Modified-V
Draft	4'0"	Deadrise Aft	NA
Weight	55,000#	Designer	J. Hargrave
Clearance	18'6"	Production	1978–81
Water	235 gals.		1986–87

While never achieving the widespread popularity of the larger 58 Yacht Fisherman, the Hatteras 53 YF nevertheless combines all the comforts of a motor yacht with the added versatility of a cockpit. Interestingly, the 53 YF was withdrawn from production in 1981, only to be reintroduced in 1986 with an enlarged flybridge, updated interior decor, and the new GM 6V92TA diesels. Constructed on the same single-chine hull as the 53 Motor Yacht and Convertible models, the 53 Yacht Fisherman shares their reputation for being a wet ride offshore. She's available with a choice of two- or three-stateroom interior floorplans with the three-stateroom layout being the most popular. It's worth noting, however, that the owner's cabin in the two-stateroom version is extremely spacious. As with most Hatteras motor yachts, the engine rooms are split by the central passageway leading aft from the galley. With the early GM 8V71TI diesels or the later 6V92TAs, the cruising speed of the Hatteras 53 YF is around 16–17 knots. The fuel capacity was significantly increased in 1980. ❑

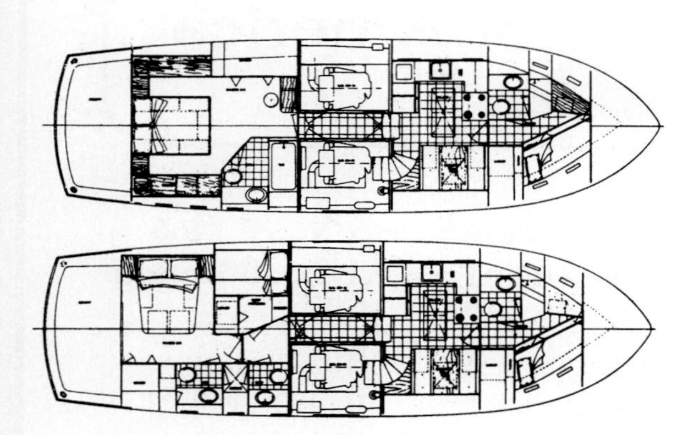

See page 241 for pricing information.

HATTERAS 53 EXTENDED DECKHOUSE MY

SPECIFICATIONS

Length	53'1"	Water	287 gals.
Beam	15'10"	Fuel	700 gals.
LWL	47'3"	Hull Type	Modified-V
Draft	4'0"	Deadrise Aft	NA
Weight	57,000#	Designer	J. Hargrave
Clearance	17'2"	Production	1983–88

Featuring a completely restyled superstructure from the Hatteras 53 MY (her sistership), the 53 Extended Deckhouse MY challenged the older model in buyer popularity soon after she was introduced in 1983. A good-looking yacht, the 53 ED's floorplan features an extended full-width main salon with an inside bridge access ladder. A partition separates the wheelhouse from the salon for privacy. The expanded salon dimensions (174 sq. ft. vs. the 53 MY's 112 sq. ft.) create a huge entertainment area for a boat of this size. The walkaround sidedecks and roomy aft deck of the 53 MY are eliminated in the ED, but a small open-air afterdeck is retained for line-handling duties. The ED's belowdeck floorplan is the same as the 53 MY with three staterooms, three heads, and separate engine rooms. Topside, the flybridge is greatly enlarged from that of her predecessor and features an improved helm console. GM 6V92TA diesels replaced the 8V71TIs as standard in 1985. Both engines will cruise the Hatteras 53 EDMY around 16 knots and reach 18 knots top. ❑

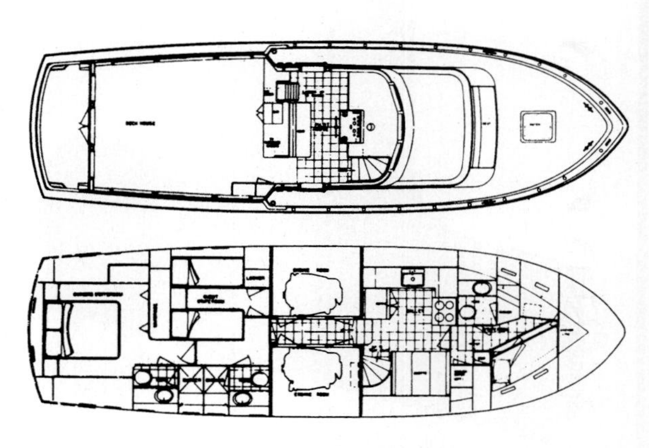

See page 241 for pricing information.

HATTERAS 54 MOTOR YACHT

SPECIFICATIONS

Length	54'9"	Water	250 gals.
LWL	NA	Fuel	800 gals.
Beam	17'6"	Hull Type	Modified-V
Draft	4'2"	Deadrise Aft	NA
Weight	62,500#	Designer	J. Hargrave
Clearance	21'3"	Production	1985–88

The Hatteras 54 MY was the first of the "new look" Hatteras motor yachts when she was introduced in 1985. Designed to replace the aging 53 Motor Yacht in the Hatteras fleet, the 54 was built on a beamy modified-V hull with cored hullsides and propeller pockets below. The 54 MY is a decidedly good-looking boat with her curved windshield and rakish, Eurostyle profile, and the interior accommodations are equally well-arranged around a practical and roomy floorplan. The lower helm is open to the main salon for an uninterrupted wheelhouse view from bow to stern. The galley is down, and the belowdeck layout includes three staterooms (each with private head and stall shower) and separate walk-in engine rooms. Outside, the semi-enclosed afterdeck is fitted with wing doors and extended windshields for weather protection, while a bridge overhang shelters the walkaround sidedecks. The early 650-hp 8V92s will cruise at a sluggish 15 knots (18 top), and the more recent 720-hp 8V92s cruise at 16–17 knots and reach a top speed of 21 knots. ❑

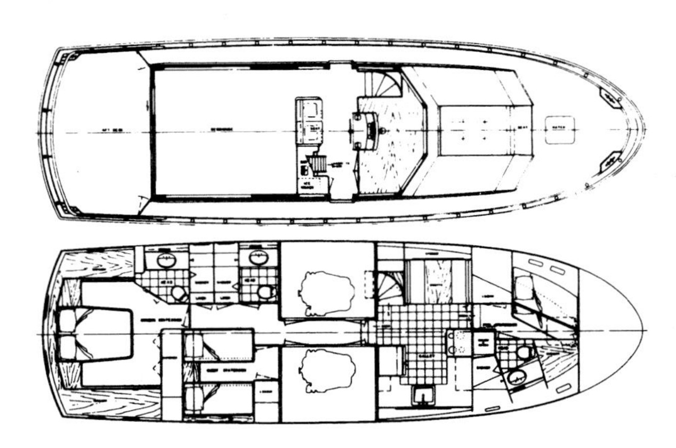

See page 241 for pricing information.

HATTERAS 54 EXTENDED DECKHOUSE MY

SPECIFICATIONS

Length	54'9"	Water	250 gals.
LWL	NA	Fuel	1,014 gals.
Beam	17'6"	Hull Type	Modified-V
Draft	4'9"	Deadrise Aft	NA
Weight	76,000#	Designer	J. Hargrave
Clearance	20'11"	Production	1989–Current

The Hatteras 54 EDMY is a maxi-volume yacht with sleek Mediterranean styling and plenty of eye appeal. By eliminating the sidedecks and moving the salon bulkhead aft, the living quarters and interior volume of the 54 ED are dramatically increased from the original 54 MY model. Hatteras has gone beyond just improving the salon size; this is an all-new interior layout. Gone is the galley-down approach so common in Hatteras motor yacht interiors. Instead, the 54 EDMY has a canted U-shaped galley in the salon with under-counter refrigeration. Unlike all Hatteras motor yachts under 60', the 54 ED has a full-width engine room rather than split compartments. Where the original 54 MY has three staterooms, the 54 EDMY has four—three with walkaround double berths—in what is surely one of the most efficient floorplans to be seen in a yacht this size. The woodwork is white ash and the cut down galley creates a very spacious deckhouse. The Hatteras 54 EDMY will cruise around 16–17 knots (about 20 knots top) with standard 720-hp 8V92s. ❑

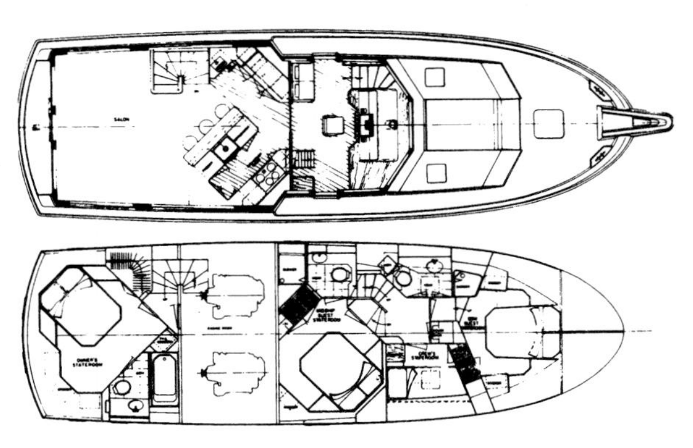

See page 241 for pricing information.

HATTERAS 56 MOTOR YACHT

SPECIFICATIONS

Length	56'3"	Water	350 gals.
LWL	NA	Fuel	1,020 gals.
Beam	18'2"	Hull Type	Modified-V
Draft	4'11"	Deadrise Aft	NA
Weight	74,000#	Designer	J. Hargrave
Clearance	18'10"	Production	1980–85

The 56 Motor Yacht is notable in that she was the first Hatteras motor yacht to use the new super-wide double-chine Hatteras hull design—the same hull later used in the production of over a dozen subsequent Hatteras models. Wide in the beam, her three-stateroom interior accommodations are truly expansive for a 56-foot boat. With the galley down, the deckhouse is arranged with the helm wide open to the salon. A corridor dividing the engine rooms leads to the spacious master stateroom with a walkaround double berth and tub/shower in the adjoining head. (Interestingly, most Hatteras motor yachts up to this time had two staterooms, not one, aft of the engine rooms.) Outside, the semi-enclosed aft deck is very spacious with wing doors opening onto broad sidedecks. The huge flybridge has space for dinghy storage and seating for a crowd. With standard 8V92TIs, she'll cruise at 15–16 knots and reach 18 knots top. Note that in 1981 a six-foot cockpit extension created the popular Hatteras 61 Cockpit MY. ❑

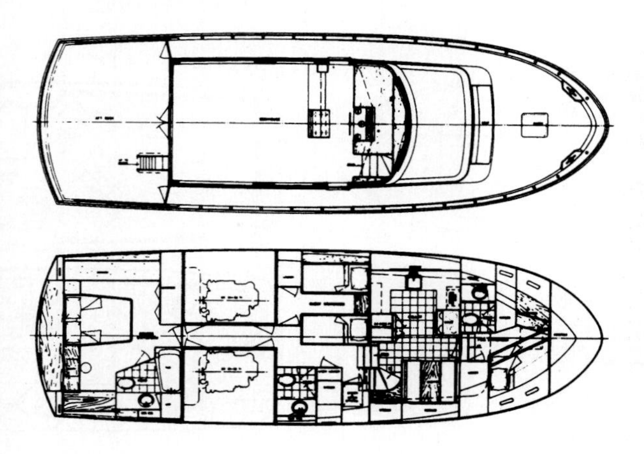

See page 241 for pricing information.

HATTERAS 58 YACHT FISHERMAN

SPECIFICATIONS

Length	58'4"	Fuel	825/1,015 gals.
Beam	15'10"	Cockpit	60 sq. ft.
Draft	4'9"	Hull Type	Modified-V
Weight	62,500#	Deadrise Aft	NA
Clearance	18'6"	Designer	J. Hargrave
Water	250/300 gals.	Production	1970–82

The Hatteras 58 Yacht Fisherman is essentially a 53 Motor Yacht with a cockpit addition. She was introduced in 1970—a year after the Hatteras 53 MY—and enjoyed a long and very successful production run during the next twelve years. The 58's cockpit makes her a more versatile (and better handling) boat than the 53 MY, and the additional fuel capacity results in a greater cruising range. Inside, her accommodation plan is the same as that found in the 53 MY. Significant improvements came in 1975 with the addition of stall showers in the aft guest and the forward head compartments. The previously optional 8V71TI diesels also became standard in 1975. In 1977 the flybridge was redesigned, and the fuel and water capacities were increased in the 1980 models. The aft deck is sheltered by a bridge overhang and fitted with wing doors for spray protection. Early Hatteras 58 YF models with 8V71Ns cruise around 13 knots, and the later versions with the 8V71TIs cruise at 16–17 knots. Still a popular yacht, resale values are strong. ❑

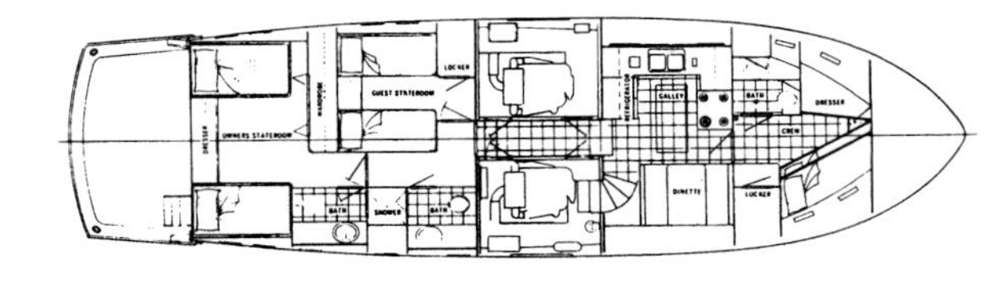

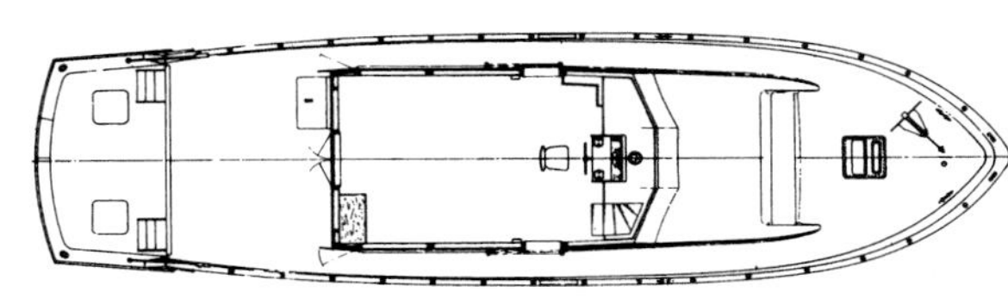

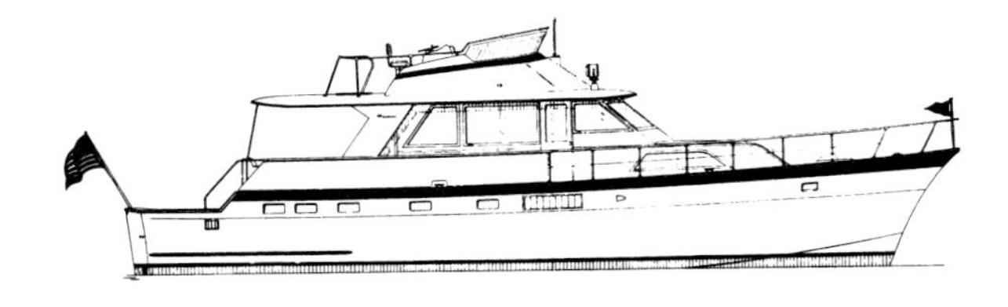

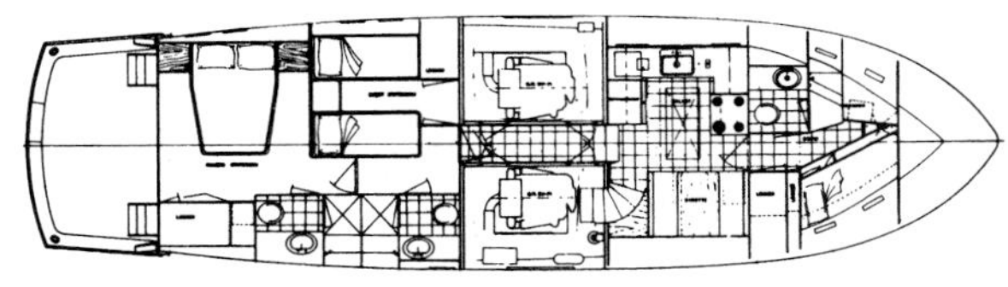

See page 241 for pricing information.

HATTERAS 58 LONG RANGE CRUISER

SPECIFICATIONS

Length	58'2"	Cockpit	NA
LWL	52'0"	Water	440/540 gals.
Beam	17'11"	Fuel	2,390 gals.
Draft	5'10"	Hull Type	Disp.
Weight	90,000#	Designer	J. Hargrave
Clearance	18'9"	Production	1975–81

Heavily built, the Hatteras 58 LRC is a true go-anywhere cruising yacht. She was the first of four Hatteras trawler-style yachts (the 42, 48 and 65 LRCs followed) built from the mid-1970s through the early 1980s. A serious and very popular cruising yacht, the 58's private wheelhouse is fully enclosed and separated from the main salon. Across from the galley is an on-deck day head—a convenient feature. On the lower level, the master stateroom is separated from the rest of the boat by an expansive engine room. Both forward staterooms are served by separate heads, each with a stall shower. Outside, wing doors protect the covered aft deck from spray, and the cockpit is fitted with a transom door. The spacious flybridge is accessed from a stairway in the pilothouse. Designed for efficient operation and possessing transatlantic range, the Hatteras 58 LRC burns only 6 gph at 8.5 knots and 8–9 gph at her 9.5-knot hull speed with standard GM 4-71 diesels. Larger 6-71Ns were optional. ❑

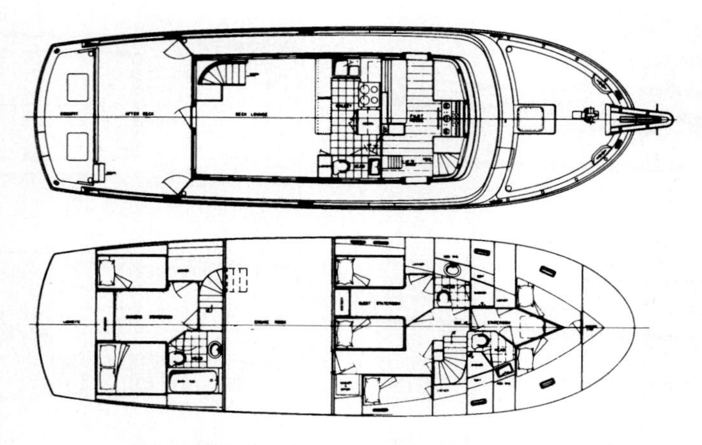

See page 241 for pricing information.

HATTERAS 58 MOTOR YACHT (EARLY)

SPECIFICATIONS

Length	58'3"	Water	300 gals.
LWL	NA	Fuel	775 gals.
Beam	15'10"	Hull Type	Modified-V
Draft	4'9"	Deadrise Aft	NA
Weight	74,000#	Designer	J. Hargrave
Clearance	16'10"	Production	1977–81

The original Hatteras 58 MY replaced the 58 Triple Cabin in the Hatteras fleet in 1977. (Note that an all-new Hatteras 58 Motor Yacht came out in 1985.) Featuring a full-width salon and offered with or without an enclosed aft deck, she was built on a stretched 53 MY hull with moderate transom deadrise and a long keel. With her narrow beam, the 58's relatively high center of gravity results in a tender ride in a beam sea when not equipped with stabilizers. The original three-stateroom, three-head floorplan was arranged with the galley and dinette down, however a more popular four-stateroom layout has the galley more conveniently located on the deckhouse level. Lacking sidedecks, the salon dimensions are very spacious. Other features of the Hatteras 58 MY include separate walk-in engine rooms, a full-size tub in the owner's stateroom, and a huge flybridge with a reverse-venturi windshield and L-shaped lounge seating. GM 8V92TI diesels replaced the original 8V71TIs in 1978 and cruise at 17–18 knots with a top speed of about 21 knots.❑

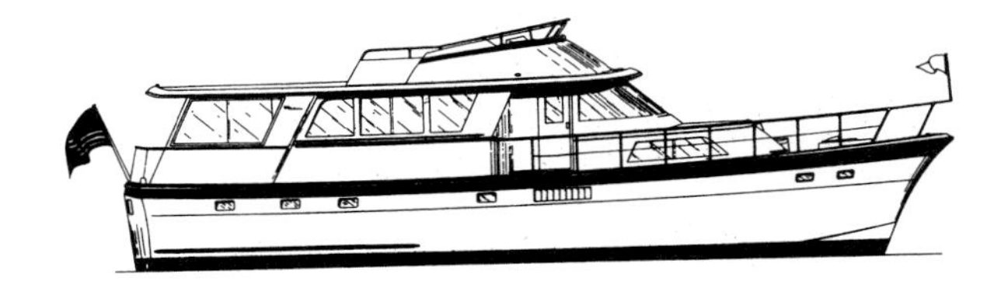

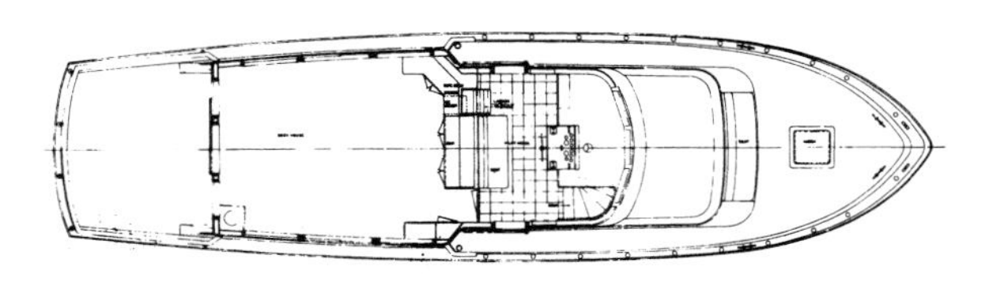

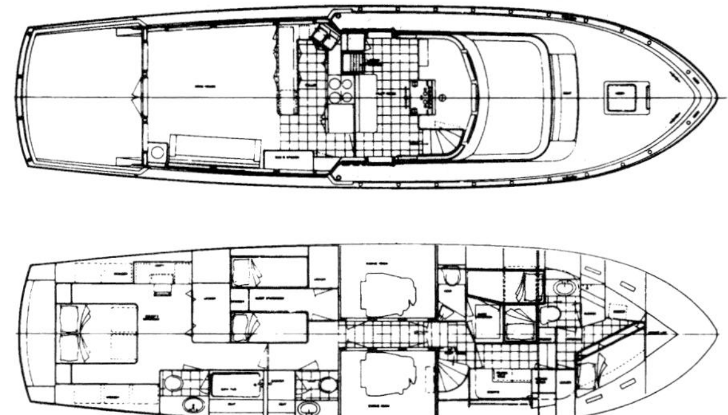

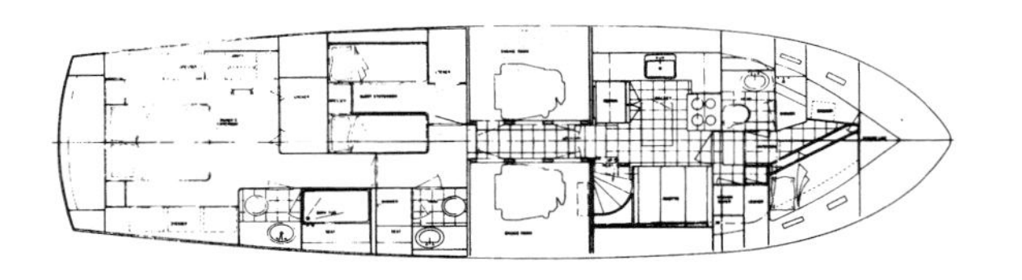

See page 241 for pricing information.

HATTERAS 58 COCKPIT MY

SPECIFICATIONS

Length	58'2"	Water	300 gals.
LWL	NA	Fuel	1,085 gals.
Beam	15'10"	Hull Type	Modified-V
Draft	4'9"	Deadrise Aft	NA
Weight	73,000#	Designer	Hargrave
Clearance	21'2"	Production	1978–81

Easily confused with the more popular 58 Yacht Fisherman, the Hatteras 58 Cockpit MY can be identified by her full-width salon and lack of walkaround sidedecks. She's built on a stretched version of the original Hatteras 53 MY hull with moderate transom deadrise and a long keel. With her enlarged salon dimensions, the 58 Cockpit MY offered a galley-up, four-stateroom floorplan (which the 58 YF never had) in addition to the standard and more popular galley-down, three-stateroom arrangement. Teak paneling and cabinetry are used throughout, and both floorplans carry one guest stateroom aft of the engine rooms. Note that a privacy door divides the wheelhouse from the salon. Bridge access is via a ladder in the wheelhouse. Additional features include a very spacious master stateroom with walk-in closets, separate engine rooms, foredeck seating, deck access doors port and starboard, and a roomy flybridge with space for a dinghy. Standard 550-hp 8V92s will cruise the Hatteras 58 Cockpit MY around 16 knots and deliver a top speed of 19 knots. ❑

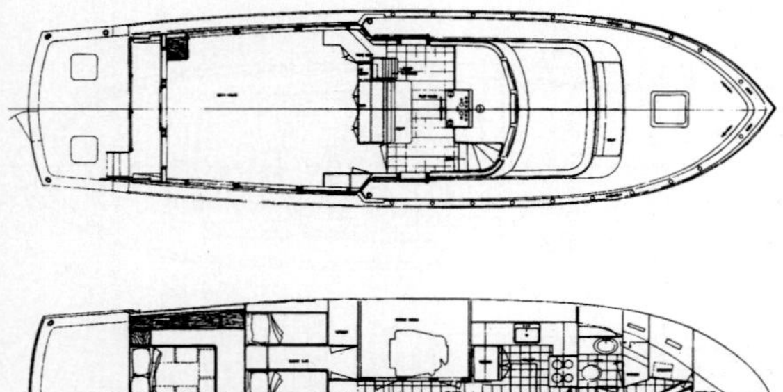

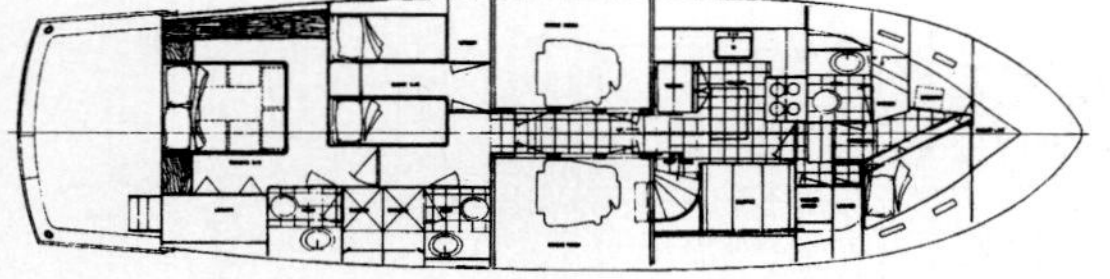

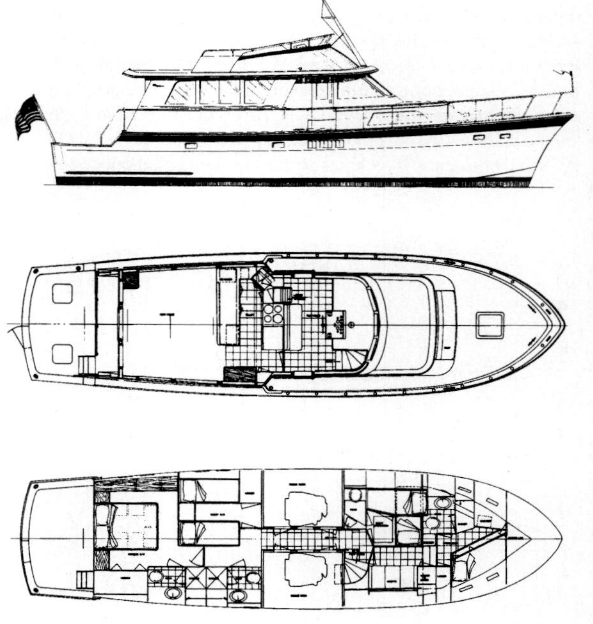

See page 241 for pricing information.

HATTERAS 58 MOTOR YACHT

SPECIFICATIONS

Length	58'8"	Water	350 gals.
LWL	NA	Fuel	1,030 gals.
Beam	18'2"	Hull Type	Modified-V
Draft	4'11"	Deadrise Aft	NA
Weight	79,000#	Designer	J. Hargrave
Clearance	18'10"	Production	1985–87

The most recent Hatteras 58 MY is basically a stretched version of the earlier Hatteras 56 MY, the boat she replaced in 1985. The three-stateroom floorplans of both boats are nearly identical, with the extra length of the 58 seen in her enlarged master stateroom and bigger aft deck dimensions. Compared with the original Hatteras 58 MY (1977–81), the newer (and more popular) model has considerably more beam (18'2" vs. 15'10"), two hull chines instead of one (for a drier ride), walkaround sidedecks, and a roomier floorplan with the galley down (not up) and a choice of twin berths or a queen-size berth in the master stateroom. The flybridge access ladder is located on the aft deck in the new 58 MY rather than forward in a wheelhouse. Throughout, the decor is lighter and more upscale than in previous Hatteras models. Standard 8V92TIs will cruise the Hatteras 58 around 17–18 knots with a top speed of about 20 knots. Note that the popular Hatteras 63 Cockpit MY is a 58 MY with a cockpit extension. ❑

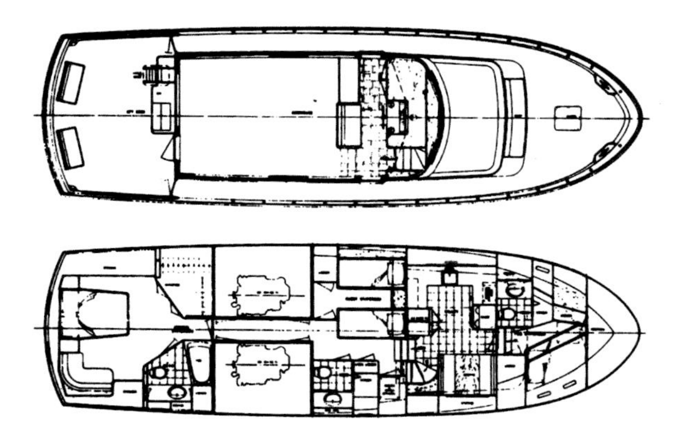

See page 241 for pricing information.

HATTERAS 60 MOTOR YACHT

SPECIFICATIONS

Length	60'9"	Water	335 gals.
LWL	NA	Fuel	1,033 gals.
Beam	18'2"	Hull Type	Modified-V
Draft	5'0"	Deadrise Aft	NA
Weight	86,000#	Designer	J. Hargrave
Clearance	20'9"	Production	1988–92

The Hatteras 60 Motor Yacht is essentially a Euro-styled version of the earlier Hatteras 58 MY (1985–87) which she replaced in 1988. The floorplans are nearly identical—three staterooms, three heads with the galley down, and the lower helm open to the salon. The master stateroom in this layout is huge. The salon is not full-width, which leaves room outside for broad sidedecks—a feature whose convenience and practicality is appreciated with experienced yachtsmen. Unlike the earlier 58 Motor Yacht with its aft deck bridge ladder, access to the flybridge in the Hatteras 60 is inside, forward in the salon near the helm. Topside, the bridge has a stylish aft-raking windshield with plenty of guest seating. Additional features include a large and easily enclosed aft deck with wing doors, separate walk-in engine rooms, dinghy storage, and foredeck seating. A respectable performer, standard 720-hp 8V92s will cruise at 18 knots and reach a top speed of 21 knots. Note that the Hatteras 67 Cockpit MY is the same boat with a cockpit extension and 12V71 engines. ❑

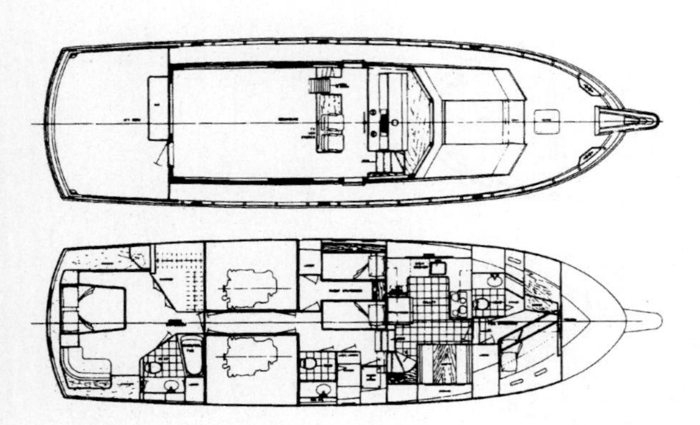

See page 241 for pricing information.

HATTERAS 60 EXTENDED DECKHOUSE MY

SPECIFICATIONS

Length	60'9"	Water	335 gals.
LWL	NA	Fuel	1,033 gals.
Beam	18'2"	Hull Type	Modified-V
Draft	5'2"	Deadrise Aft	NA
Weight	87,000#	Designer	Hargrave
Clearance	21'2"	Production	1991–Current

It isn't just the extended deckhouse that sets the Hatteras 60 EDMY apart from the 60 Motor Yacht. She's an entirely different yacht with a deckhouse galley, four staterooms, and a full-beam engine room rather than separate walk-in rooms typical of earlier Hatteras motor yachts. Although the extended deckhouse layout widens the salon dimensions, it's notable that the 60 EDMY retains at least narrow sidedecks—something most extended deckhouse models lack. By eliminating the large afterdeck found in the 60 MY, the EDMY has a spacious 200 sq. ft. deckhouse with the galley and dinette open to the salon. The wheelhouse, with direct access to the flybridge, is completely enclosed in this layout. A staircase in the salon leads down to a spacious master stateroom and engine room access door. Both the VIP and forward staterooms have double berths, and all three heads have stall showers. Standard 720-hp 8V92s will deliver a cruising speed of 16 knots (19 knots top), and optional 870-hp 12V71s will cruise around 18–19 knots and reach 21 knots wide open. ❑

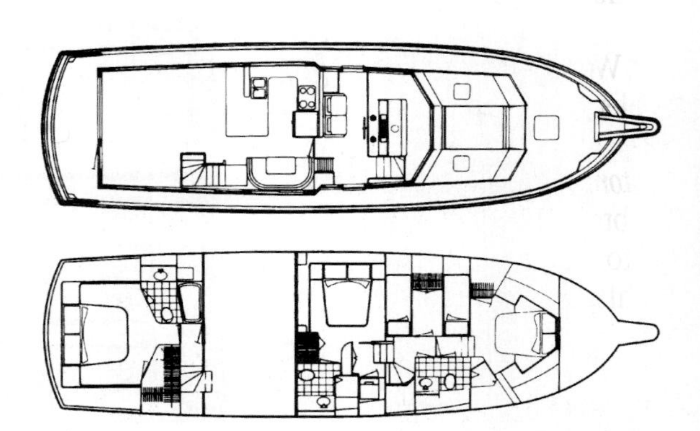

See page 242 for pricing information.

HATTERAS 61 MOTOR YACHT

SPECIFICATIONS

Length	61'3"	Water	350 gals.
LWL	55'9"	Fuel	1,150 gals.
Beam	18'2"	Hull Type	Modified-V
Draft	4'11"	Deadrise Aft	NA
Weight	82,000#	Designer	J. Hargrave
Clearance	18'10"	Production	1981–84

The Hatteras 61 MY is a popular model on the used market thanks in part to her galley-up floorplan, a feature not always found in 1980's-era Hatteras motor yachts and the preferred layout with many of today's buyers. The upper level of the Hatteras 61 is divided into four sections: an enclosed wheelhouse forward, separate galley and dinette area, a huge full-width salon, and a small aft deck platform with enough room for line-handling. A second dinette is located on the lower level along with four staterooms and four heads—a big floorplan for a 61-footer. Note that the extravagant owner's cabin with its walk-in wardrobe is the only stateroom aft of the engine rooms. Additional features include a very spacious flybridge with dinghy storage aft, a full teak interior, single berths in each of the guest cabins, foredeck lounge seating, and separate engine rooms. Powered with 650-hp 12V71 diesels, the Hatteras 61 Motor Yacht will cruise at 17 knots and reach a top speed of 21 knots. ❑

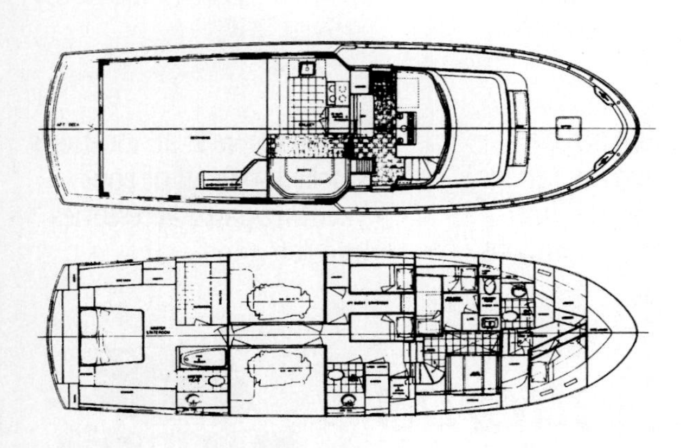

See page 242 for pricing information.

HATTERAS 61 COCKPIT MY

SPECIFICATIONS

Length	61'3"	Water	350 gals.
LWL	55'9"	Fuel	1,150 gals.
Beam	18'2"	Hull Type	Modified-V
Draft	4'11"	Deadrise Aft	NA
Weight	85,000#	Designer	J. Hargrave
Clearance	18'10"	Production	1981–84

The 61 Cockpit Motor Yacht is basically a stretched version of the Hatteras 56 Motor Yacht (1980–85) with the addition of a 5-foot cockpit. The hull is the same and so is the three-stateroom, galley-down interior layout with its split engine rooms and walkaround sidedecks. Aside from the fact that a cockpit adds great utility to any motor yacht, the additional length gives the 61 Cockpit MY a more attractive profile (and slightly better performance) than the 56 Motor Yacht. While the salon dimensions are not extravagant by today's standards, the Hatteras 61 CMY retains the wide sidedecks so valued by experienced cruisers. The helm is open to the salon, and the partially enclosed aft deck is large enough for a full set of outdoor furnishings. A transom door is fitted in the cockpit for fishing and easy stern boarding. Standard power for the Hatteras 61 Cockpit MY is a pair of GM 12V71s for a cruising speed of 17–18 knots and a top speed of about 21 knots. ❑

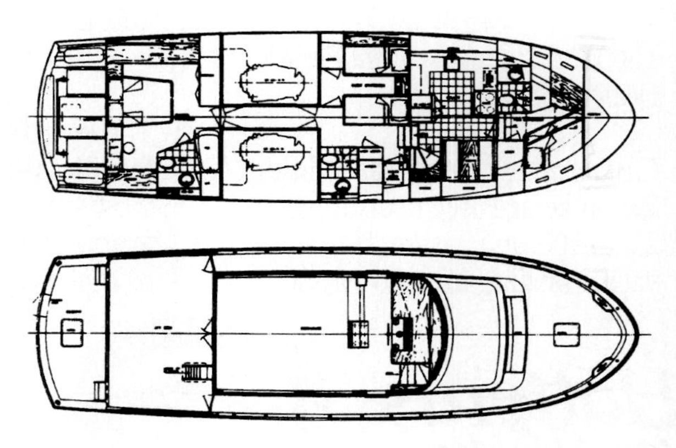

See page 242 for pricing information.

HATTERAS 63 MOTOR YACHT

SPECIFICATIONS

Length	63'10"	Fresh Water	350 gals.
LWL	NA	Fuel	1,170 gals.
Beam	18'2"	Hull Type	Modified-V
Draft	5'0"	Deadrise Aft	NA
Weight	92,000#	Designer	J. Hargrave
Clearance	21'3"	Production	1986–87

The Hatteras 63 Motor Yacht is notable because of her galley-up layout and spacious full-width salon—desirable features found in many of today's modern motor yacht floorplans. The 63 MY (and her sistership, the 68 Cockpit MY) came with a roomy four-stateroom, four-head interior with an extended salon and an enclosed galley/dinette aft of the wheelhouse. The owner's stateroom in this floorplan is exceptionally large with a separate dressing area, walk-in wardrobes, several dressers, and a full-size tub in the head. There are port and starboard deck doors in the private wheelhouse, and the interior is fully paneled and finished with teak woodwork. The maindeck salon of the 63 MY is absolutely enormous with more than 275 sq. ft. of living space—enough to create a formal dining area forward without intruding on the salon's expansive entertainment facilities. A small aft deck provides space for line-handling. With standard 650-hp 12V71s, the Hatteras 63 MY will cruise comfortably around 17 knots and reach a top speed of 20 knots. ❑

See page 242 for pricing information.

HATTERAS 63 COCKPIT MY

SPECIFICATIONS

Length	63'8"	Water	375 gals.
LWL	NA	Fuel	1,170 gals.
Beam	18'2"	Hull Type	Modified-V
Draft	4'11"	Deadrise Aft	NA
Weight	79,000#	Designer	Hargrave
Clearance	18'10"	Production	1985–87

The Hatteras 63 Cockpit MY is basically a Hatteras 58 MY with a cockpit addition. The fuel capacity was increased by 140 gallons, larger (and thirstier) 12V71 diesels maintain her cruising speed of 18 knots (about 22 knots top), and the versatility and improved appearance resulting from the cockpit extension speaks for itself. The interior accommodations are virtually the same in both boats, with the exception of the master stateroom where the 58 has a little more square footage. (Most 63 CMYs have a king-size bed in the master, but twin singles were available.) The galley and dinette are down, leaving the entire upper deck for the wheelhouse and salon. The guest staterooms are forward of the engine rooms, and all three heads are fitted with stall showers. Those who actually use their boats will appreciate the advantages of the full walkaround sidedecks and spacious, semi-enclosed aft deck. Topside, the flybridge is huge with guest seating and dinghy storage. The Hatteras 63 CMY remains a popular boat in spite of her short production run. ❑

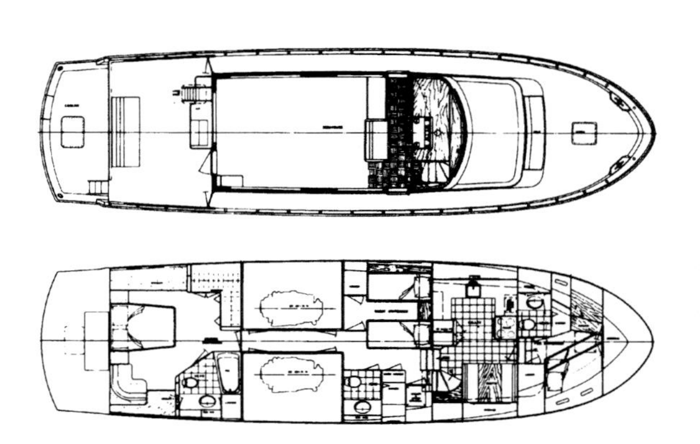

See page 242 for pricing information.

HATTERAS 64 MOTOR YACHT

SPECIFICATIONS

Length	64'1"	Water	550 gals.
LWL	NA	Fuel	1,440 gals.
Beam	18'4"	Hull Type	Modified-V
Draft	5'0"	Deadrise Aft	NA
Weight	95,000#	Designer	J. Hargrave
Clearance	22'6"	Production	1975–81

The Hatteras 64 Motor Yacht is remembered as the only Hatteras ever to use V-drives. Consequently, she hasn't been an especially popular model on the used market, although she is recognized for her good offshore handling characteristics. Designed to be operated by a crew, her unusual four-stateroom interior layout has the owner's stateroom amidships (with a tub/shower in the head), two guest cabins aft, and the crew quarters forward. With the enclosed wheelhouse separated from the main salon by the galley, the owner's party can thus enjoy complete privacy from crew. Twin doors in the salon's aft bulkhead open onto a covered, semi-enclosed afterdeck with wing doors and extended side windshields. Additional features include a concealed engine room access entrance along the portside sidedeck, an inside stairway leading to the flybridge (which was restyled in 1977), and protected sidedecks. Standard 12V71TI diesels will cruise the Hatteras 64 MY at 16–17 knots with a top speed of around 19 knots. Interestingly, the 64 is the only Hatteras ever built on this 18'4"-beam hull. ❑

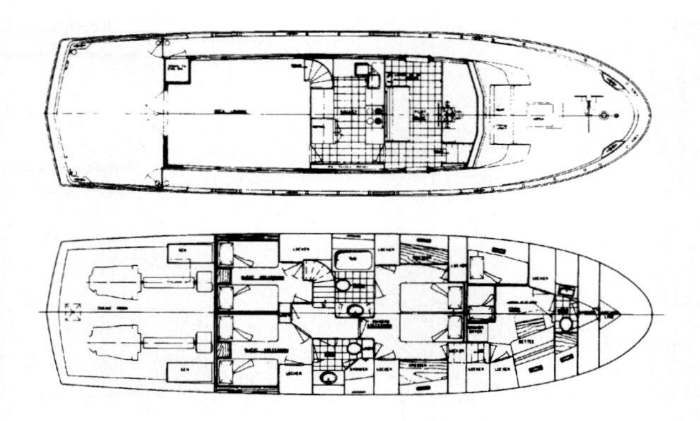

See page 242 for pricing information.

HATTERAS 65 LONG RANGE CRUISER

SPECIFICATIONS

Length	65'0"	Cockpit	NA
LWL	58'8"	Water	455 gals.
Beam	17'11"	Fuel	2,625 gals.
Draft	4'10"	Hull Type	Disp.
Weight	114,000#	Designer	J. Hargrave
Clearance	18'9"	Production	1981–85

The largest of the Hatteras Long Range Cruisers, the 65 LRC was built on a stretched 58 LRC hull—a full-displacement design with a long keel and rounded bilges that harden at the transom. Construction is solid fiberglass, and, at 114,000 lbs., the Hatteras 65 certainly qualifies as a serious heavyweight. She features a spacious four-stateroom layout with a fully enclosed pilothouse and a deckhouse galley open to the salon. A spiral teak stairway leads down to the extravagant owner's stateroom and nearby engine room access door. The 65 LRC was also offered with an optional extended deckhouse floorplan with the salon lengthened to include the aft deck, an enlarged pilothouse, and a convenient day head opposite the galley. Designed for all-weather long-range cruising, the curved protective bulwark on the foredeck provides protection against oncoming breaking seas. With a range of over 2,500 nautical miles, the Hatteras 65 LRC will cruise comfortably at her 10-knot hull speed burning only 9 gph with standard GM 6-71N diesels. ❑

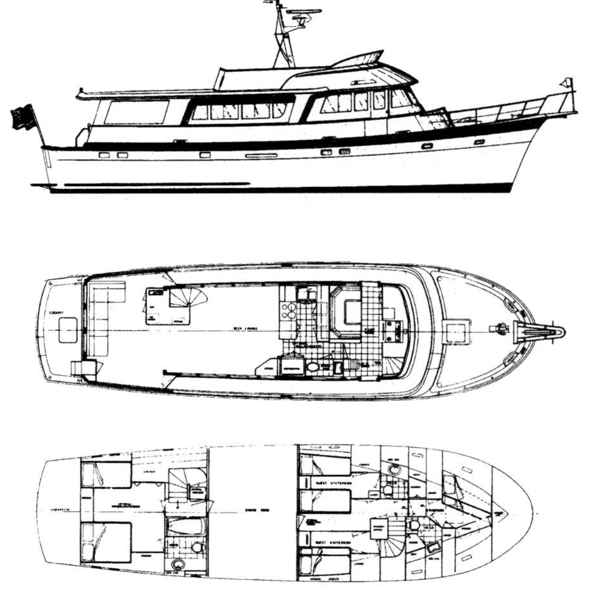

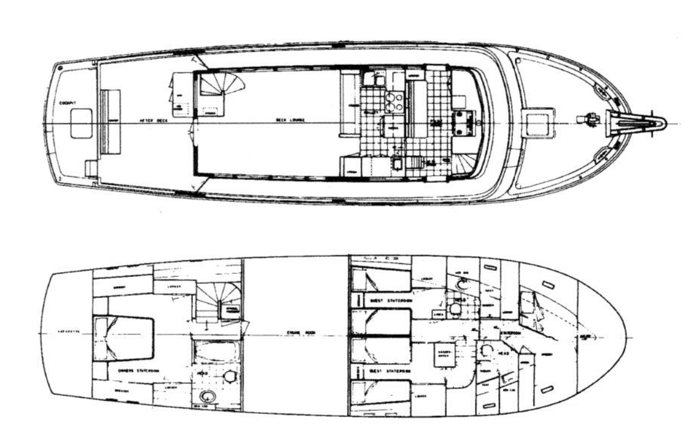

See page 242 for pricing information.

HATTERAS 65 MOTOR YACHT

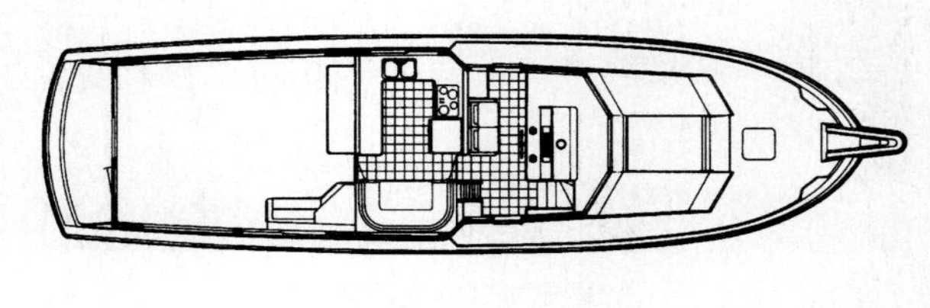

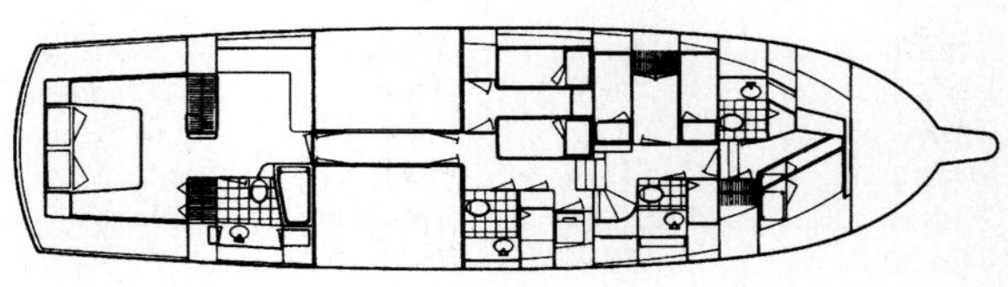

SPECIFICATIONS

Length	65'10"	Water	350 gals.
LWL	NA	Fuel	1,170 gals.
Beam	18'2"	Hull Type	Modified-V
Draft	5'5"	Deadrise Aft	NA
Weight	99,000#	Designer	Jack Hargrave
Clearance	21'5"	Production	1988–Current

The Hatteras 65 Motor Yacht was designed as an enlarged version of the earlier Hatteras 63 MY with a fresh profile, the elimination of the sheer stripe, and a revised interior. Her original floorplan has four-staterooms and heads with separate engine rooms. In 1989 an optional VIP floorplan was introduced with a more luxurious guest stateroom and a full-width engine room. Since 1992, the standard layout is organized with queen berths in both guest staterooms, a rearranged master suite with his-and-hers heads aft, and a single walk-in engine room. The galley/dinette is separated from the salon, and the small aft deck is suitable for line-handling duties. Lacking sidedecks, the deckhouse of the 65 MY is very spacious and comes with a choice of traditional teak or light ash woodwork. Above, the enormous flybridge features an aft-raking windshield and U-shaped lounge seating. Note that the Hatteras 70 Cockpit MY is the same boat with a cockpit extension. Standard 870-hp 12V71s will cruise the Hatteras 65 MY around 17 knots and reach 20 knots top. ❑

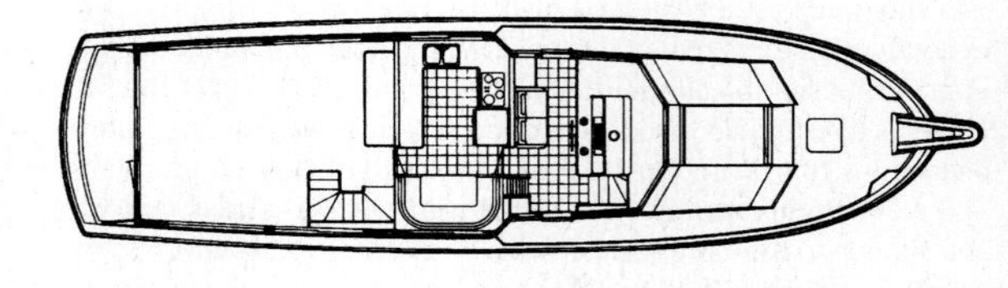

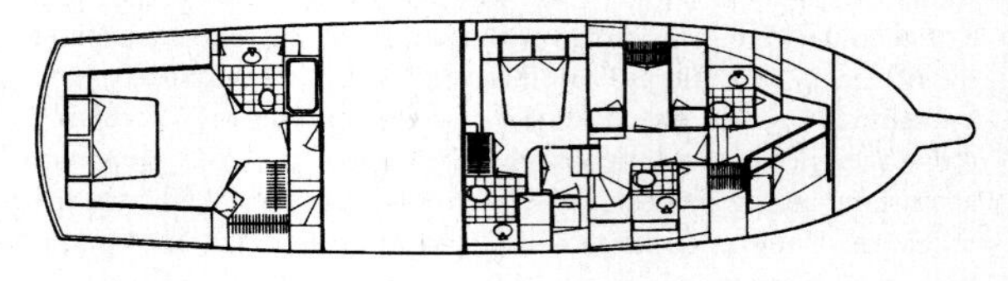

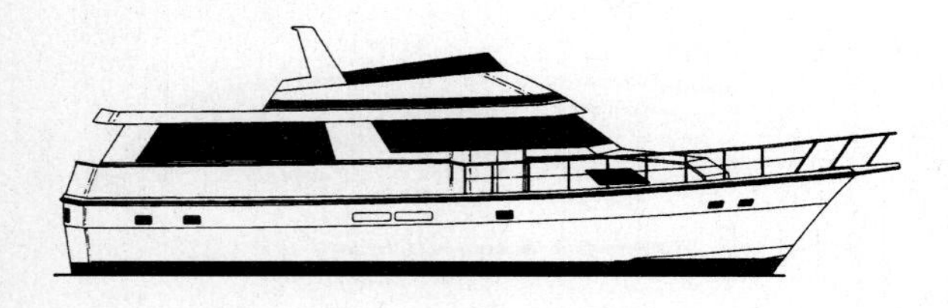

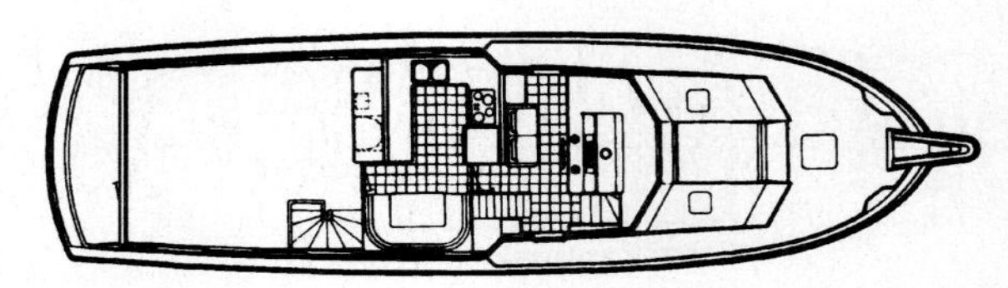

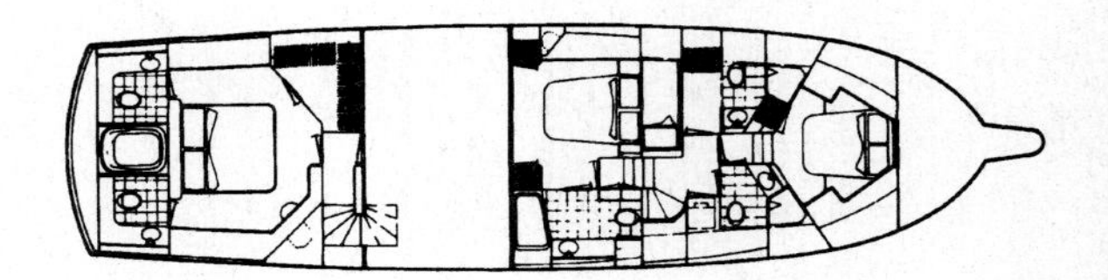

See page 242 for pricing information.

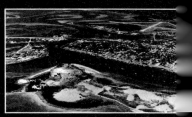

HATTERAS 67 COCKPIT MY

SPECIFICATIONS

LOA	67'8"	Water	375 gals.
LWL	NA	Fuel	1,170 gals.
Beam	18'2"	Hull Type	Modified-V
Draft	5'1"	Deadrise Aft	NA
Weight	90,000#	Designer	J. Hargrave
Clearance	20'3"	Production	1988-Current

A beautifully proportioned yacht with plenty of sex appeal, the 67 Cockpit MY is basically a Hatteras 60 MY with a cockpit extension—the largest factory-installed cockpit found in any Hatteras production model. This is, of course, the same three-stateroom floorplan used in the Hatteras 60 MY with the helm open to the salon and full walkaround sidedecks. By staying with the galley-down layout, the 67 CMY is able to retain the open-air comforts of a very large aft deck. With wing doors and side panels, this extremely popular entertainment area can accommodate a huge bar and plenty of deck furniture. With nearly 100 sq. ft. of space, the cockpit is big enough for all kinds of activities including heavy-tackle fishing. A good sea boat, standard 770-hp Detroit 12V71 diesels will cruise the Hatteras 67 CMY at 18 knots and reach around 21 knots top. Note that the newer Hatteras 67 Extended Deckhouse CMY has a galley-up floorplan with four staterooms, an enlarged salon, and a small aft deck platform. ❑

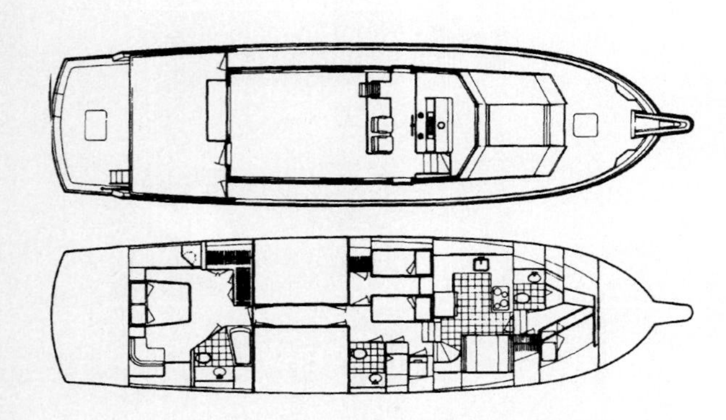

See page 242 for pricing information.

HATTERAS 67 EXTENDED DECKHOUSE CMY

SPECIFICATIONS

Length	67'2"	Water	372 gals.
LWL	NA	Fuel	1,171 gals.
Beam	18'2"	Hull Type	Modified-V
Draft	5'3"	Deadrise Aft	NA
Weight	95,000#	Designer	Hargrave
Clearance	21'5"	Production	1991–Current

The Hatteras 67 ED CMY has the distinction of being the smallest Hatteras *cockpit* model with a galley-up floorplan. Basically, this is simply a Hatteras 60 Extended Deckhouse MY with a 7-foot cockpit extension. Heavily built with cored hullsides, a long keel, and modest deadrise aft, she retains the convenience of walkaround sidedecks, although they're much narrower than in previous Hatteras models. Her extravagant galley-up, four-stateroom floorplan features light ash woodwork throughout and very generous guest accommodations. The dinette and galley are on the deckhouse level and open to the salon, while the wheelhouse—with its flybridge access—is separate. A curved salon staircase leads down to the master stateroom and engine room access door. A small afterdeck overlooks the cockpit and the flybridge dimensions are the same as the 60 MY's. Aside from the versatility provided by a cockpit, many feel that the extra LOA makes the 67 a better-looking yacht than the 60 EDMY. Standard 870-hp 12V71s will cruise at 17–18 knots, and the optional 1,040-hp 12V92s cruise around 20 knots. ❑

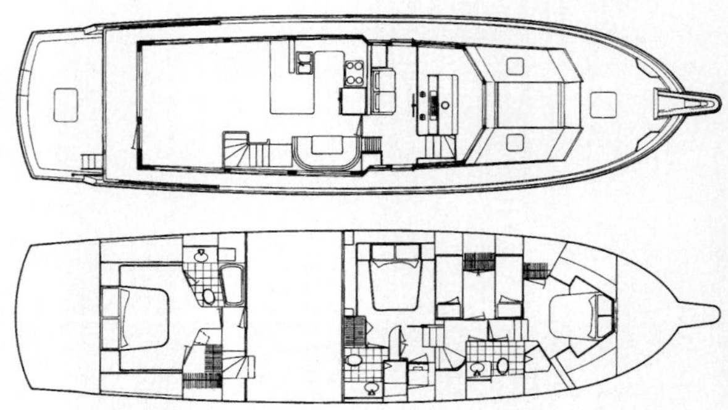

See page 242 for pricing information.

HATTERAS 68 COCKPIT MY

SPECIFICATIONS

Length	68'10"	Water	350 gals.
LWL	NA	Fuel	1,170 gals.
Beam	18'2"	Hull Type	Modified-V
Draft	5'0"	Deadrise Aft	NA
Weight	96,500#	Designer	Hargrave
Clearance	21'3"	Production	1986–87

The Hatteras 68 CMY is basically a 63 MY with a cockpit. Seven were built in her two years of production before she was replaced by the popular 70 Cockpit MY. The hull is solid fiberglass with a long keel below, and the shafts were extended to the transom when the cockpit was added. (Note that Hatteras did not increase the fuel capacity in the 68 CMY as they have in other cockpit aPlications.) Her four-stateroom floorplan follows the usual Hatteras configuration with a corridor leading aft to the master stateroom flanked by separate walk-in engine rooms. The spacious 275 sq. ft. full-width salon with its formal dining area and luxurious furnishings dominates the interior. The master suite is enormous with a king-size bed and an athwartships wardrobe dividing the living area to create a separate dressing room forward. Outside, a small covered afterdeck overlooks the cockpit. Needless to say, the flybridge is huge. Twin 770-hp 12V71 diesels will cruise the Hatteras 68 Cockpit MY at a respectable 17 knots with a top speed of 19–20 knots. ❑

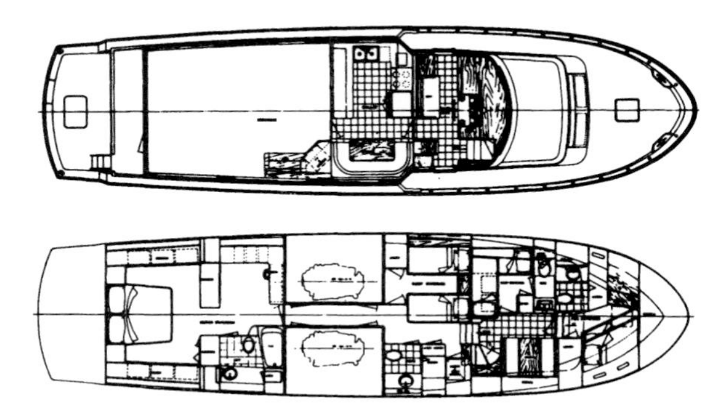

See page 242 for pricing information.

HATTERAS 70 MOTOR YACHT (EARLY)

SPECIFICATIONS

Length	70'2"	Water	400 gals.
LWL	NA	Fuel	1,440/1,650 gals.
Beam	18'7"	Hull Type	Modified-V
Draft	5'0"	Deadrise Aft	NA
Weight	109,000#	Designer	J. Hargrave
Clearance	18'10"	Production	1971–81

The Hatteras 70 Motor Yacht was the largest fiberglass production yacht of her day when she was introduced back in 1971. With her classic motor yacht profile and impressive size, the Hatteras 70 is considered a handsome yacht even by today's standards. Because of her length, Hatteras had to open a new plant in New Bern, N.C., from which to launch her (the main Hatteras plant being some 250 miles inland). Designed to be captain-operated and maintained, the Hatteras 70 will accommodate the owner and guests in three large staterooms aft of the engine room on the lower level. Crew quarters and galley areas are forward, thus affording owner and guests complete privacy. The floorplan arrangement features a spacious formal dining room detached from the galley and an enclosed wheelhouse forward of the main salon. The semi-enclosed afterdeck is fitted with wing doors and extended side windshields for weather protection. No lightweight, standard 12V71TI diesels will cruise around 14 knots and provide a top speed of 17 knots. Fuel capacity was increased in 1980. ❑

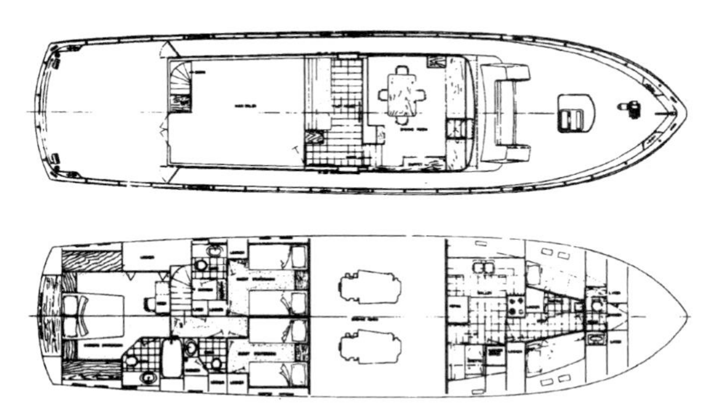

See page 242 for pricing information.

HATTERAS 70 EXTENDED DECKHOUSE MY

SPECIFICATIONS

Length	70'2"	Water	400 gals.
LWL	NA	Fuel	1,440/1,650 gals.
Beam	18'7"	Hull Type	Modified-V
Draft	5'0"	Deadrise Aft	NA
Weight	109,000#	Designer	J. Hargrave
Clearance	18'10"	Production	1976–83

Built on the same solid fiberglass modified-V hull as the earlier Hatteras 70 MY (1971–81), the Hatteras 70 Extended Deckhouse MY has a more modern flybridge and superstructure than her predecessor, although the galley-up interior floorplans of both boats are roughly similar. The pilothouse of the EDMY is set further forward on the foredeck—a feature that results in a considerably enlarged main salon. The elaborate crew quarters have been enlarged as well with separate staterooms for a captain and two mates, a small galley and dinette and direct access to the huge standup engine room. The master stateroom and both guest staterooms (located aft of the engine rooms and isolated from the crew) are the same in both boats. The semi-enclosed aft deck is very spacious, and wing doors open to broad, well-protected walkaround sidedecks. A heavy boat, 12V71TI diesels will cruise the Hatteras 70 Extended Deckhouse MY around 14 knots and provide a top speed of 17 knots. Note that fuel capacity was increased in 1981. ❑

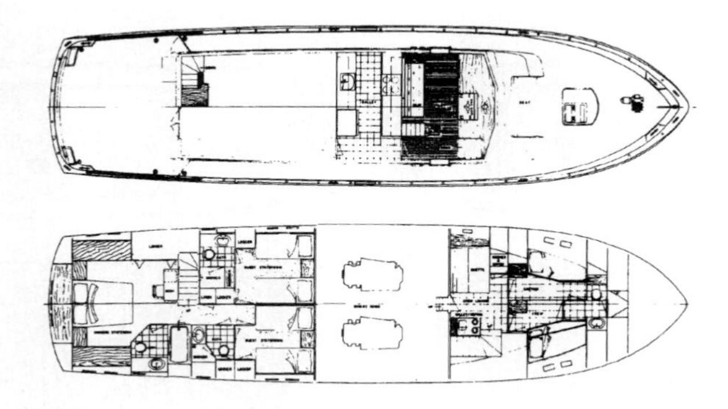

See page 242 for pricing information.

HATTERAS 70 MOTOR YACHT

SPECIFICATIONS

Length	70'11"	Water	251 gals.
LWL	NA	Fuel	1,596 gals.
Beam	18'2"	Hull Type	Modified-V
Draft	5'6"	Deadrise Aft	NA
Weight	108,000#	Designer	Hargrave
Clearance	21'2"	Production	1988–Current

A popular yacht, the graceful profile and contemporary styling of the Hatteras 70 Motor Yacht are strikingly similar to the smaller 65 MY with the additional length used to create a more spacious salon and an enlarged master stateroom. (Note that this is the largest Hatteras motor yacht built on the 18'2"-beam hull mold first used for the 56 MY.) While most Hatteras 70 Motor Yachts have been delivered with semi-custom interiors, a three-stateroom, three-head layout is standard with a four-stateroom arrangement offered as optional. Both floorplans feature an impressive VIP stateroom amidships with a tub in the head. On the upper level, the full-width salon dimensions are extravagant in spite of the fact that the galley and dinette are located on the same level. Yielding to the demands of European buyers, sidedecks became available in 1990 although few have been built. Standard 870-hp 12V71s will cruise the Hatteras 70 MY at 18 knots (21 knots top), and optional 1040-hp 12V92s will cruise at 20 knots (22 knots wide open). ❑

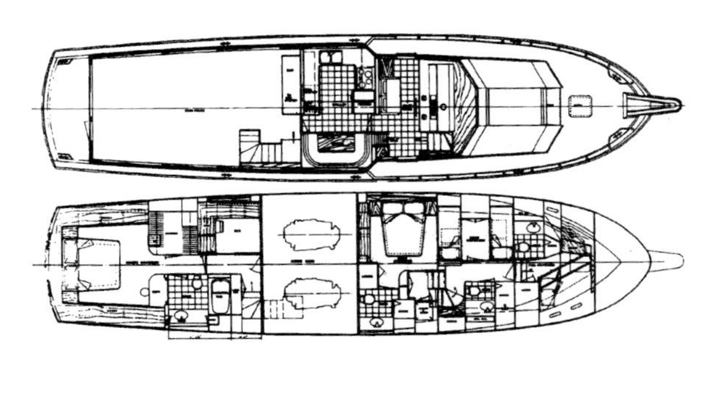

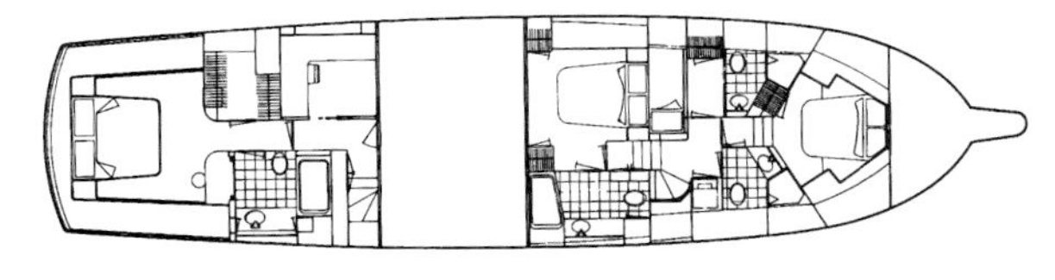

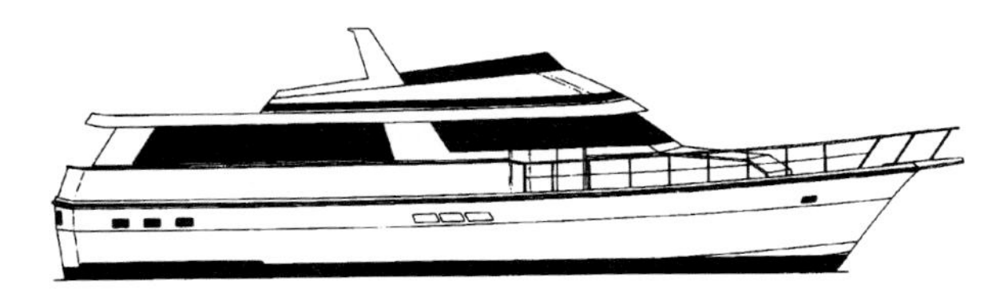

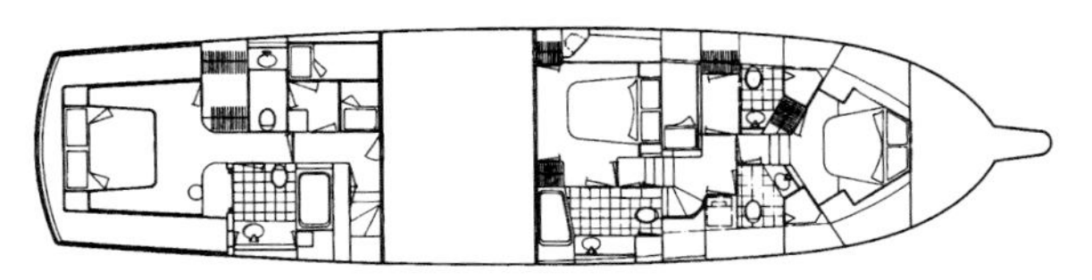

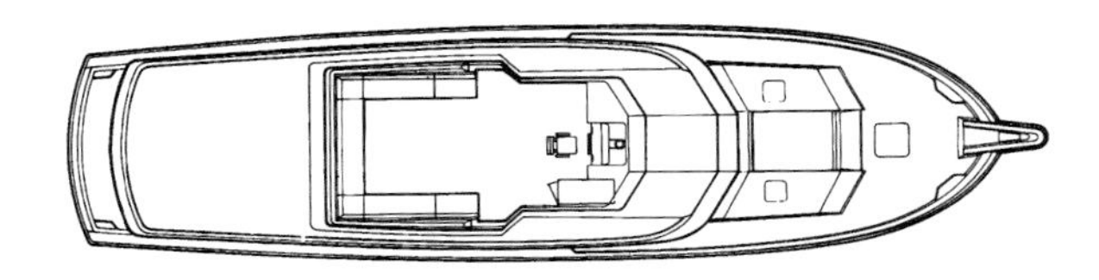

See page 242 for pricing information.

HATTERAS 70 COCKPIT MY

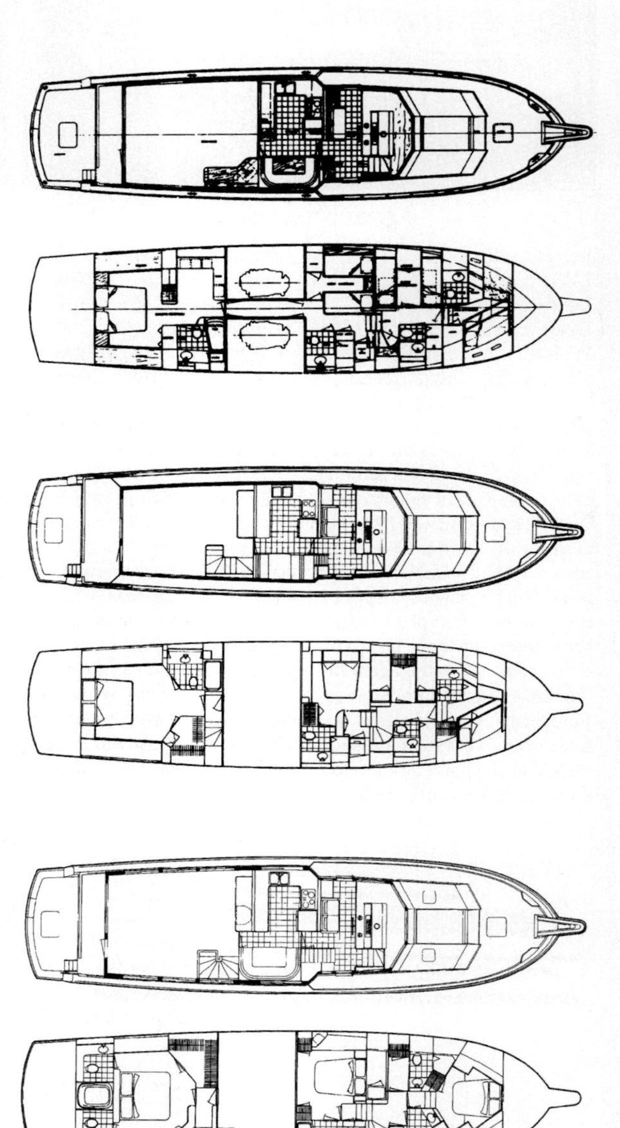

SPECIFICATIONS

Length	70'10"	Water	345 gals.
LWL	NA	Fuel	1,596 gals.
Beam	18'2"	Hull Type	Modified-V
Draft	5'4"	Deadrise Aft	NA
Weight	103,000#	Designer	Hargrave
Clearance	21'4"	Production	1988–Current

An extremely popular yacht (over 50 have been built) with a beautiful European profile and extravagant accommodations, the 70 CMY is one of the most luxurious production yachts ever offered by Hatteras. She's basically a 65 MY with a cockpit addition (the extra hull length allows for an additional 426 gallons of fuel beneath the cockpit sole). Like most of the newer Hatteras motor yachts, the floorplan is arranged with the galley and dinette on the deckhouse level. The spacious full-width salon (no sidedecks) is extended well aft leaving only a small afterdeck overlooking the cockpit. The wheelhouse and galley are both enclosed in this layout and separated from the salon. Originally designed with a standard four-stateroom floorplan with twin engine rooms, recent VIP layouts have a full-width engine room and more spacious guest accommodations. Note that many 70 CMYs have been semi-customized at the factory to meet an individual owner's requirements. Standard 870-hp 12V71s will cruise the 70 CMY at 17 knots, and optional (since 1990) 1,040-hp 12V92s cruise at 20 knots. ❏

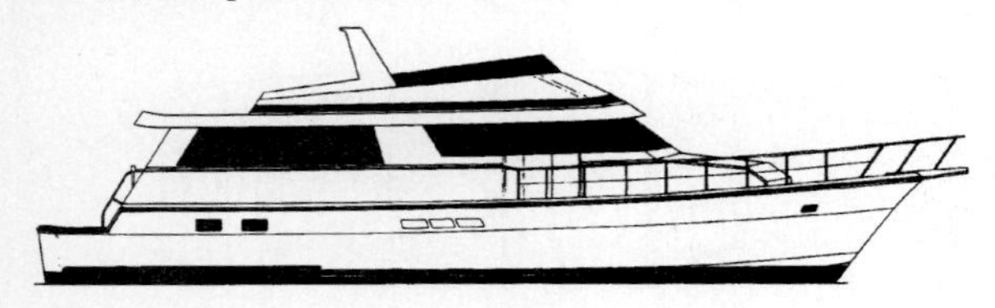

See page 242 for pricing information.

HATTERAS 72 MOTOR YACHT

SPECIFICATIONS

Length	72'8"	Water	365 gals.
LWL	NA	Fuel	1,650/1,858 gals.
Beam	18'7"	Hull Type	Modified-V
Draft	5'3"	Deadrise Aft	NA
Weight	117,000#	Designer	J. Hargrave
Clearance	26'3"	Production	1983–92

Using the slightly lengthened hull form of the original Hatteras 70 MY, the 72 Motor Yacht was introduced in 1983 as part of the Hatteras Custom Yacht Program. Each of the models in this series can be customized within the confines of hull and superstructure. With her modern flybridge and curved wheelhouse windows, the styling of the 72 is contemporary and very appealing. Although she wasn't marketed as an extended deckhouse model, the spacious salon of the Hatteras 72 is indeed full-width—the actual dimensions are an expansive 14' x 25'. With both the galley and wheelhouse enclosed and separated from the salon area, the owner and guests are allowed complete privacy from any crew interruption. Belowdecks, the roomy walk-in engine room divides the lower deck floorplan with the owner and guest staterooms aft and the crew quarters (two staterooms) forward. A small aft deck for line-handling remains at the stern. No lightweight, standard 870-hp 12V71 diesels will cruise the Hatteras 72 MY at 16–17 knots with a top speed of about 20 knots. ❑

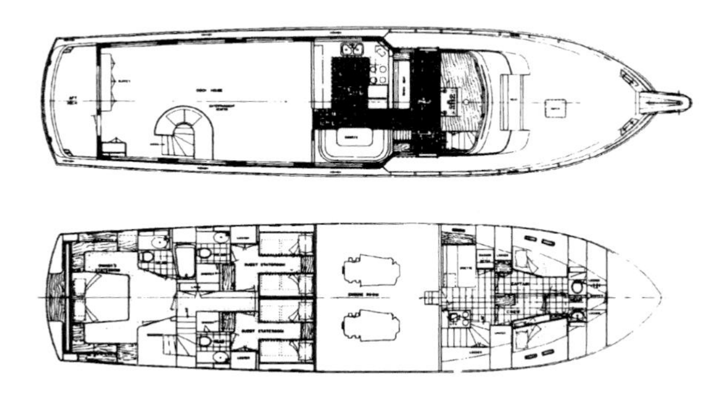

See page 242 for pricing information.

HATTERAS 77 COCKPIT MY

SPECIFICATIONS

Length	77'8"	Fuel	2,080 gals.
Beam	18'7"	Hull Type	Modified-V
Draft	5'0"	Deadrise Aft	NA
Weight	140,000#	Designer	Hargrave
Clearance	26'3"	Production	1983–86
Water	496 gals.		

The 77 CMY remains the largest production cockpit motor yacht ever offered by Hatteras. Seven of these opulent yachts were built, two with walkaround decks (pictured above) and five with a full-width salon and no sidedecks. Designed as a crewed yacht, the galley is on the deckhouse level and private from the salon. Both guest staterooms and the extravagant master suite are accessed from a salon staircase, while the crew quarters occupy the forward sections of the boat. With the engine room aft, the midships location of the master stateroom is ideal (very quiet). Needless to say, the salon dimensions of the walkaround model are extravagant; in the widebody layout—with its extended deckhouse—they're cavernous. Note that there are wing doors for the aft deck in the walkaround, while the widebody's has just a small aft deck. Additional features include a massive flybridge and cockpit access to the stand-up engine room with its separate workbench area. Standard 12V71s cruise at 16–17 knots and the optional 12V92s will cruise around 18 knots. ❑

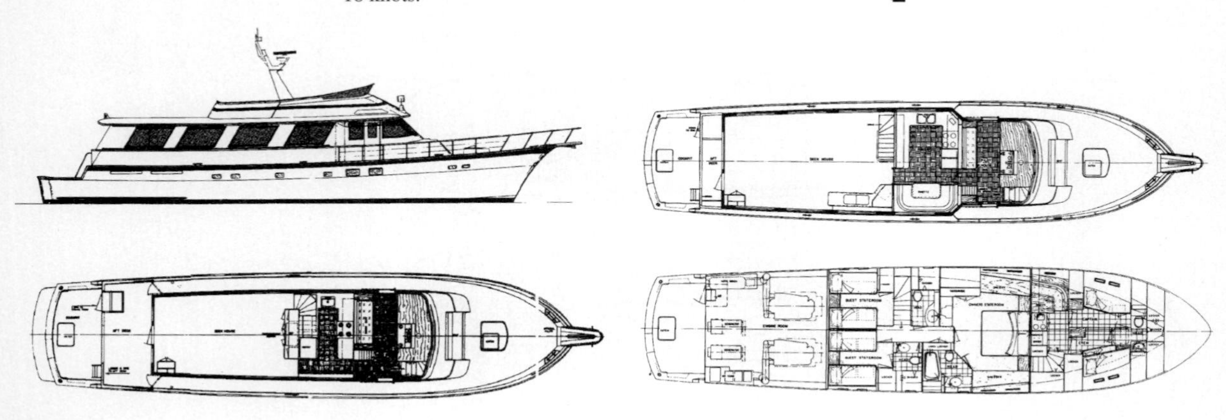

See page 242 for pricing information.

HERITAGE 40 SUNDECK MY

SPECIFICATIONS

Length	38'9"	Water	200 gals.
LWL	34'7"	Fuel	400 gals.
Beam	13'6"	Hull Type	Modified-V
Draft	3'7"	Deadrise Aft	NA
Weight	25,000#	Designer	Nova Yachts
Clearance	NA	Production	1984–90

With her upright profile and trawler-style appearance, the Heritage 40 is one of the better-looking sundeck yachts in her class. She was built in Taiwan by Nova Marine on a solid fiberglass, modified-V hull with moderate beam, simulated lapstrake hullsides, and considerable flare at the bow. (Note that her actual length is less than 39'.) Below, the efficient galley/dinette-down floorplan is practical and well-arranged. There's a deck access door in the salon (a lower helm was optional), and the forward stateroom is best described as a tight fit. The salon dimensions are about average, but the aft stateroom is quite roomy for a 39-footer. The Oriental teak carving on bulkhead doors is impressive, and all of the woodwork is well-crafted. Outside, the sidedecks are reasonably wide, and the aft deck comes with a wet bar and wing doors. Several diesel engine options were offered. The 135-hp Lehmans will cruise around 9 knots, and the popular 200-hp Volvos cruise at 13–14 knots and reach about 17 knots wide open. ❑

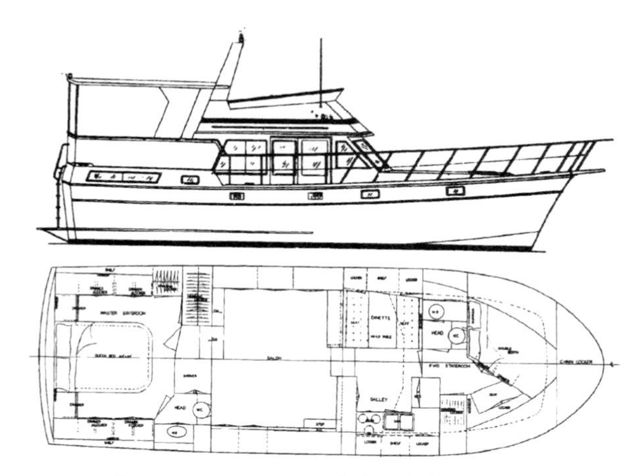

See page 243 for pricing information.

HERITAGE 44 SUNDECK MY

SPECIFICATIONS

Length	43'5"	Water	300 gals.
LWL	39'5"	Fuel	500 gals.
Beam	13'8"	Hull Type	Modified-V
Draft	3'10"	Deadrise Aft	NA
Weight	27,100#	Designer	Nova Marine
Clearance	NA	Production	1985–90

The Heritage 44 was constructed on the same solid fiberglass hull used in the construction of the Heritage 40 Sundeck with the extra length committed to an enlarged galley and dinette and enlarged salon and master stateroom dimensions. Built in Taiwan by Nova Marine, the simulated lapstrake hullsides add much to her traditional aPeal. Inside, the floorplan is arranged with the galley and dinette down from the salon level, a relatively conventional and practical layout for a cruising boat, although the lack of a stall shower in the forward head is unfortunate. Note the large master stateroom. The teak woodwork—some of which is carved—is first-rate throughout. There's a convenient sliding deck access door in the salon (the lower helm was optional), and the covered aft deck is quite spacious and fitted with wing doors. Standard engines for the Heritage 44 were 135-hp Perkins or Lehman diesels (about 8–9 knots cruise). The 200-hp Perkins diesels were a popular choice (14 knots cruise/16–17 knots wide open) among several optional diesel engines. ❑

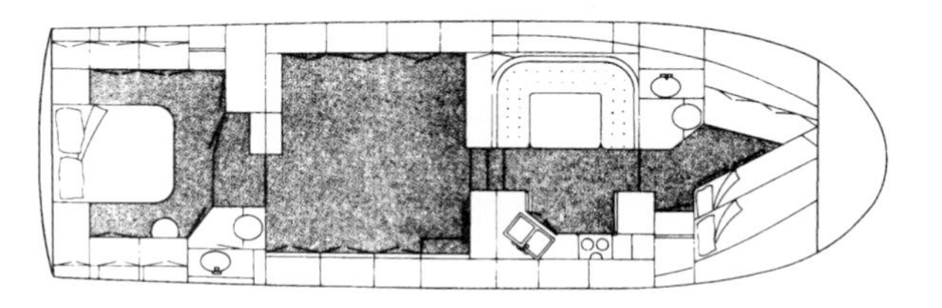

See page 243 for pricing information.

HI-STAR 48 MOTOR YACHT

SPECIFICATIONS

Length	47'9"	Water	250 gals.
LWL	41'3"	Fuel	500 gals.
Beam	15'2"	Hull Type	Modified-V
Draft	3'3"	Deadrise Aft	16°
Weight	37,000#	Designer	C. Chang
Clearance	13'6"	Production	1986–92

A popular and not inexpensive boat, the Hi-Star 48 MY is a top-of-the-line Taiwan import with a reputation for durable construction, superb engineering, and a comfortable ride. She's built on a conventional modified-V hull with moderate beam, Airex-cored hullsides, and a solid fiberglass bottom. Inside, the Hi-Star 48 has the traditional all-teak interior common to most Asian boats. Where many motor yachts this size have three staterooms, the Hi-Star has a more open two-stateroom floorplan with the galley and dinette down. This is indeed a roomy layout with big staterooms and a full-size dinette, but the forward head compartment lacks a separate stall shower—an unfortunate omission. Outside, the aft deck is large enough for plenty of deck furnishings, and the wide sidedecks make foredeck access easy. A good performer with standard 375-hp Cats, the Hi-Star 48 will cruise 18-19 knots and reach 22 knots wide open. Note that the 55 Yacht Fisherman is the same boat with the addition of a 7-foot cockpit. ❑

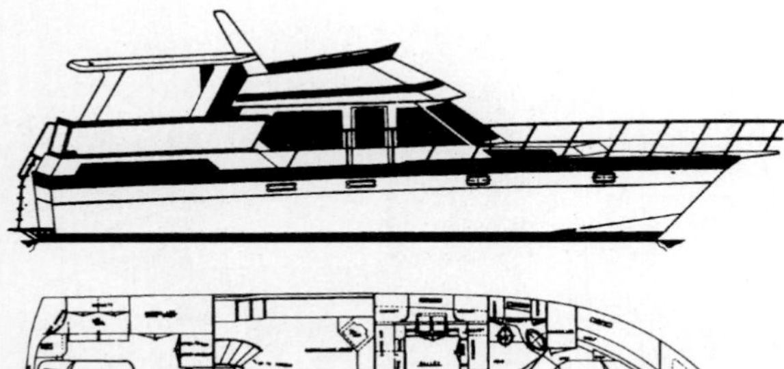

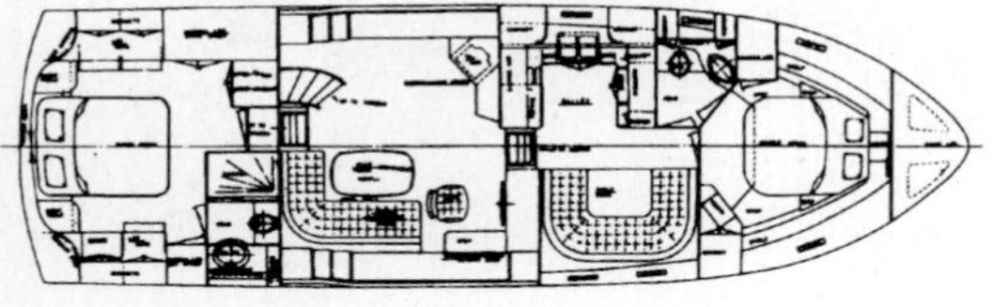

See page 243 for pricing information.

HI-STAR 55 YACHTFISHERMAN

SPECIFICATIONS

Length	54'9"	Water	250 gals.
LWL	NA	Fuel	800 gals.
Beam	48'3"	Hull Type	Modified-V
Draft	3'3"	Deadrise Aft	16°
Weight	39,000#	Designer	C. Chang
Clearance	NA	Production	1986–92

The 55 Yachtfisherman is basically a Hi-Star 48 MY with a cockpit extension, additional fuel, and a more attractive profile. A cockpit is a useful addition to just about any motor yacht, as it provides easy access to the water and easier boarding. The aft deck of the Hi-Star 55 is very spacious and comes with a wet bar and plenty of room for outdoor furnishings. Below, the two-stateroom interior is all teak (of course), and the only real complaint about the layout is the lack of a stall shower in the forward head. Both staterooms are very spacious and come with plenty of storage lockers and drawers. A lower helm station was standard along with a deck access door and wide sidedecks. Additional features include island berths in both staterooms, a spacious galley with a full-size dinette, good cabin headroom, and a well-arranged engine room. Standard 375-hp Cat diesels will cruise the Hi-Star 55 YF at a respectable 17–18 knots and reach a top speed of 22 knots. ❑

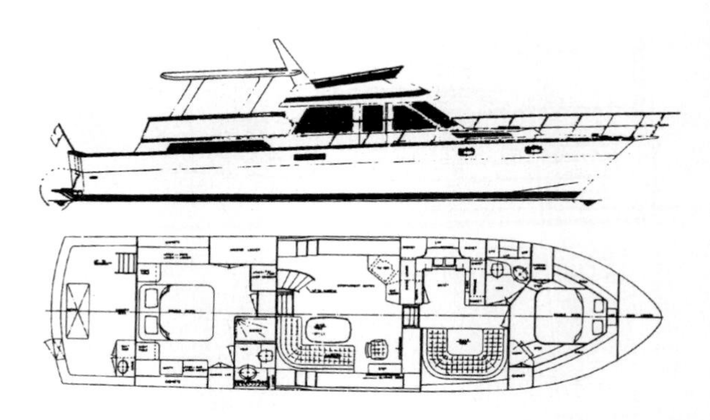

See page 243 for pricing information.

HIGH-TECH 50/55 MOTOR YACHT

SPECIFICATIONS

Length	55'0"	Fuel	550 gals.
Beam	16'0"	Hull Type	Modified-V
Draft	4'0"	Deadrise Aft	NA
Weight	40,000#	Designer	Bill Dixon
Water	200 gals.	Production	1988–Current

For those looking for something different from traditional Asian imports, the Hi-Tech 55 incorporates the European styling and impressive performance increasingly popular with U.S. buyers. Built in Taiwan by Johnson Yachts and originally marketed as a 50-footer with a conventional transom, the newer 55 introduced in 1990 has a reverse transom and integral swim platform. The hull is a modified-V design with a relatively wide beam and balsa coring in the hullsides. Inside, the Euro-style three-stateroom floorplan of the High-Tech 55 is a blend of light-colored woodwork (oak, ash, or cherry), rounded corners, and upscale hardware and fabrics. The standard layout has a queen berth in the forward stateroom, stall showers in each head, and a very classy lower helm. (An alternate arrangement has twin staterooms forward.) A canted stainless-and-glass sliding door opens to the aft deck with its teak sole, extended bridge overhang, and molded bridge steps. Standard 735-hp 8-92s will cruise either model at a fast 24 knots and reach a top speed of around 28 knots. ❑

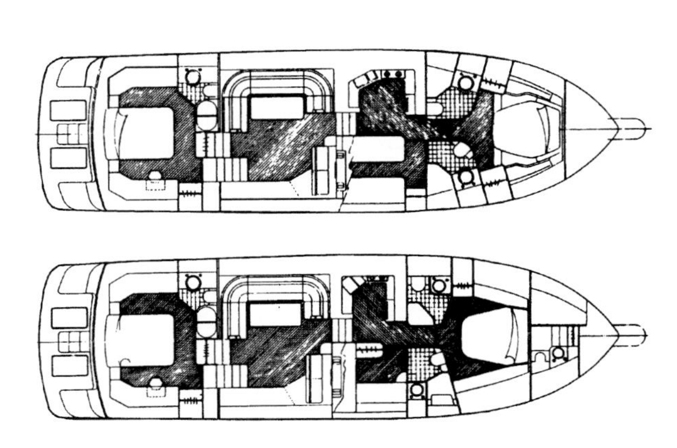

See page 243 for pricing information.

HIGH-TECH 63/65 MOTOR YACHT

SPECIFICATIONS

Length	63'0"	Fuel	1,000 gals.
LWL	55'11"	Hull Type	Modified-V
Beam	19'0"	Deadrise Aft	16°
Draft	4'5"	Designer	Bill Dixon
Weight	56,000#	Production	1990–Current
Water	400 gals.		

Imported as the High-Tech 63 MY until 1993 when the hull was stretched and she became a 65-footer, this high-style cruiser from Taiwan has the look and feel of a purebred Italian yacht. Built by Johnson Yachts (and called the Johnson 63 MY on the West Coast), she's constructed on a modified-V hull with cored hullsides, a wide beam, and a steep 16° of deadrise aft. Inside, the accommodations are as spacious as they are lush. The galley and dinette are on the pilothouse level, resulting in a wide-open and expansive salon with a wraparound settee and a full view of the cockpit. Additional features include an elegant stainless-steel-and-glass salon bulkhead, a relatively roomy aft deck with a protective bridge overhang, and a very big flybridge with plenty of guest seating. A good-selling boat with plenty of sex appeal, the Johnson/High-Tech 63 will cruise at 17–18 knots with standard 735-hp 8V92s and reach a top speed of around 22 knots. ❑

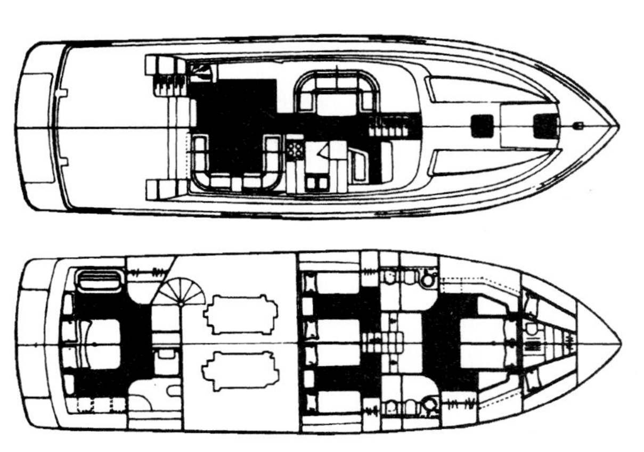

See page 243 for pricing information.

HYATT 40 SUNDECK

SPECIFICATIONS

Length	39'9"	Water	180 gals.
LWL	NA	Fuel	350 gals.
Beam	14'5"	Hull Type	Modified-V
Draft	2'10"	Deadrise Aft	NA
Weight	26,000#	Designer	Hyatt
Clearance	16'1"	Production	1988–Current

The Hyatt 40 Sundeck is a good-looking Taiwan import with a clean profile and a practical two-stateroom interior layout. Aimed at the affordable end of the family cruising market, she's constructed on a modified-V hull with balsa-cored hullsides, moderate transom deadrise, and a relatively wide beam. Two floorplans are offered: one has an offset double berth in the guest stateroom and a split forward head, and the alternate layout has stacked single berths forward and a single head. Either way, the galley and dinette are down, and a lower helm is standard. While the salon dimensions are moderate compared with other boats her size, the master stateroom is quite spacious and includes a huge hanging locker to port. The sundeck is also roomy with space for some outdoor furniture. Additional features include a full teak interior, wide sidedecks, lower helm, and a well-arranged engine room with fairly good access to the motors. Optional 300-hp Cummins diesels cruise the Hyatt 40 around 15–16 knots, and the larger 375-hp Cats cruise at 21 knots. ❑

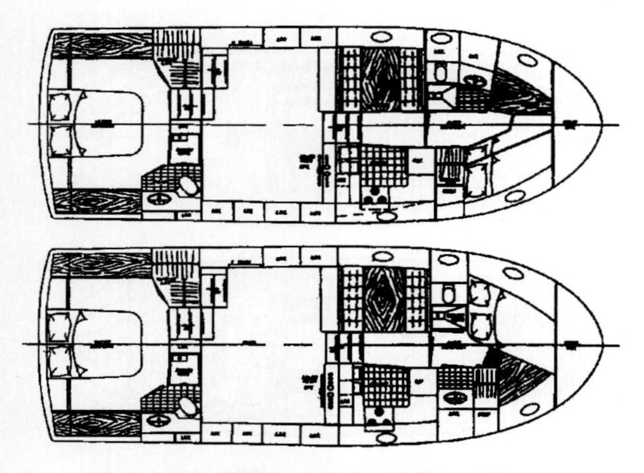

See page 243 for pricing information.

HYATT 44 MOTOR YACHT

SPECIFICATIONS

Length	44'0"	Water	165 gals.
LWL	NA	Fuel	450 gals.
Beam	14'5"	Hull Type	Modified-V
Draft	3'0"	Deadrise Aft	12°
Weight	33,500#	Designer	G. Chin
Clearance	18'5"	Production	1988–Current

A good-looking boat, the Hyatt 44 MY stands out from many of the more recent double-cabin Asian imports for her distinctive lines and affordable price tag. Built on a conventional modified-V hull with a moderate beam and transom deadrise and balsa coring in the hullsides, the Hyatt 44 comes standard with a practical two-stateroom interior floorplan with a lower-level dinette opposite the galley and walkaround island berths in both staterooms. An alternate three-stateroom layout with a deckhouse galley has also been available. The all-teak interior of the Hyatt 44 MY features wraparound cabin windows and an optional lower helm station with a sliding deck access door. The aft deck—about average in size for a 44-footer—comes with molded bridge steps, wing doors, and a built-in wet bar. Topside, the flybridge is arranged with the helm forward and bench seating aft for a crowd. Several diesel engine choices are available. The largest—375-hp Caterpillars—provide a cruising speed of 21–22 knots and a top speed of about 25 knots. ❑

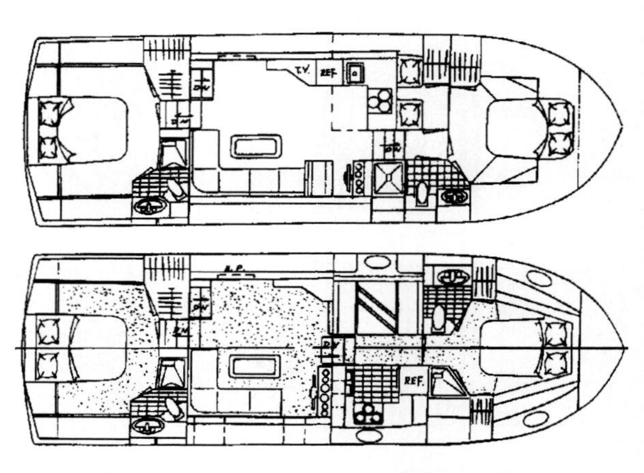

See page 243 for pricing information.

INDEPENDENCE 45 PILOTHOUSE

SPECIFICATIONS

Length	44'6"	Cockpit	NA
LWL	40'9"	Water	300 gals.
Beam	14'6"	Fuel	600 gals.
Draft	4'6"	Hull	Disp.
Weight	36,000#	Designer	J. Backus
Clearance	13'	Production	1987–Current

A good-quality import from Hans Christian with a very distinctive profile, the Independence 45 is obviously designed with long-range cruising in mind. She's a displacement yacht with a long, deep keel and a protected prop but with hard chines aft (rather than rounded bilges) to reduce roll. Built in Taiwan, she was originally called the Positive 42 Trawler. The two-stateroom teak interior of the Independence 45 is arranged with the galley in the (full-width) salon and with two heads on the lower level. The afterdeck, with its teak sole and transom door, is partially weather-protected with a bridge overhang. The raised pilothouse can be reached directly from the salon or from the outside through port and starboard deck doors. An upper deck helm is optional. There are molded steps on both sides of the house to the flybridge, where there's adequate space for dinghy storage. A single 135-hp Lehman (or Lugger) diesel is standard. She'll cruise economically at 7–8 knots with a range of over 1,000 miles. ❑

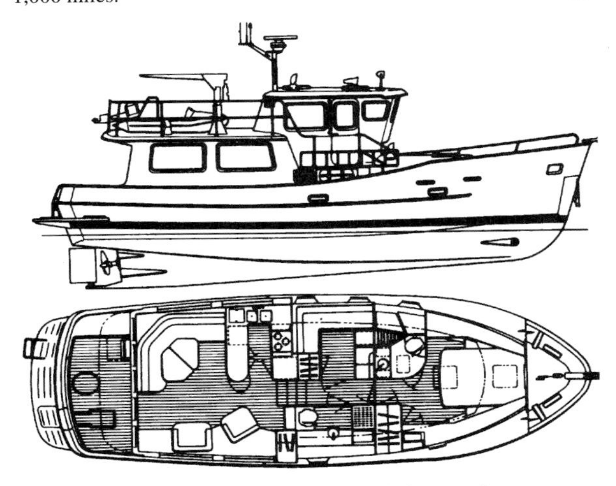

See page 243 for pricing information.

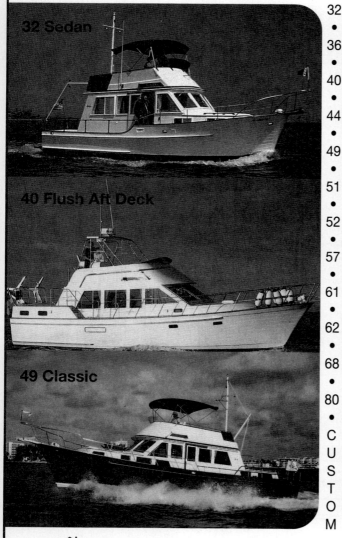

ISLAND GYPSY 30 SEDAN

SPECIFICATIONS

Length	30'0"	Water	120 gals.
LWL	27'9"	Fuel	250 gals.
Beam	11'6"	Hull Type	Semi-Disp.
Draft	3'8"	Designer	H. Halvorsen
Weight	14,400#	Production	1975–85

The Island Gypsy 30 Sedan is a salty little trawler-style design with a handsome, upright appearance. She enjoyed a good deal of popularity some years ago when high fuel prices and the strength of the U.S. dollar combined to spur the importation of efficient Asian trawlers. Built in three different configurations, the Sedan model (pictured above) began with a mahogany hull and superstructure and switched to all-fiberglass construction sometime in 1978. The hull is a hard-chined semi-displacement form with a sharp entry and a long, deep keel for directional stability and prop protection. Inside, the straightforward cabin layout includes a lower helm position opposite the galley, V-berths forward, and excellent storage space. Teak paneling and cabinetry are used liberally throughout the interior, and the joinerwork is to high standards. A flybridge overhang affords some aft deck weather protection and wide, secure teak sidedecks provide safe foredeck access. The standard 120-hp Lehman diesel will cruise the Island Gypsy 30 at her 7-knots cruising speed while burning only 2 gph. ❑

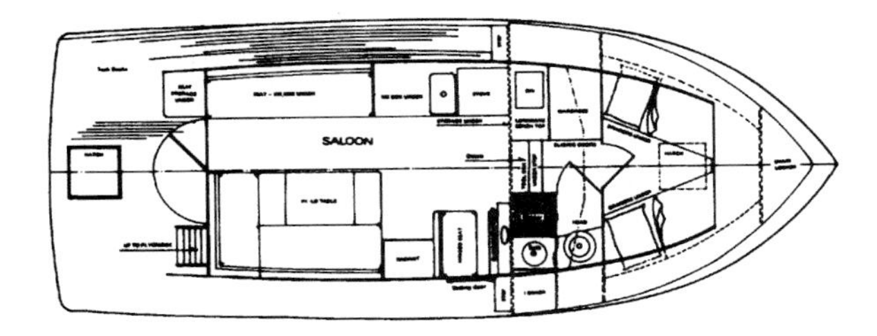

See page 243 for pricing information.

ISLAND GYPSY 32 SEDAN

SPECIFICATIONS

Length	32'1"	Cockpit	NA
LWL	29'8"	Water	120 gals.
Beam	11'6"	Fuel	250 gals.
Draft	3'8"	Hull Type	Semi-Disp.
Weight	16,500#	Designer	H. Halvorsen
Clearance	12'4"	Production	1981–Current

Like all current Island Gypsy models, the 32 Sedan is built in Mainland China. She has a distinctive trawler-style hull shape with hard-chines aft, a deep skeg, and an upright, handsome profile. Construction is solid fiberglass and her grooved hullsides require a two-piece mold. The interior of the Island Gypsy 32 is arranged in the conventional manner with a single stateroom forward (choice of V-berths or an island berth), deckhouse galley, and large wraparound cabin windows. All of the woodwork is teak, and a sliding deck access door is located next to the helm station. The aft deck is small but uncluttered and has a single seat in the shelter of the short bridge overhang. The decks are teak, the deck hardware is bronze, and the window frames are aluminum. A single Lehman 135-hp diesel will cruise the Island Gypsy 32 Sedan at 7 knots, and the twin 90-hp Lehmans cruise at 8.5 knots. Twin 130-hp or 200-hp Volvos are also available for higher speeds. ❑

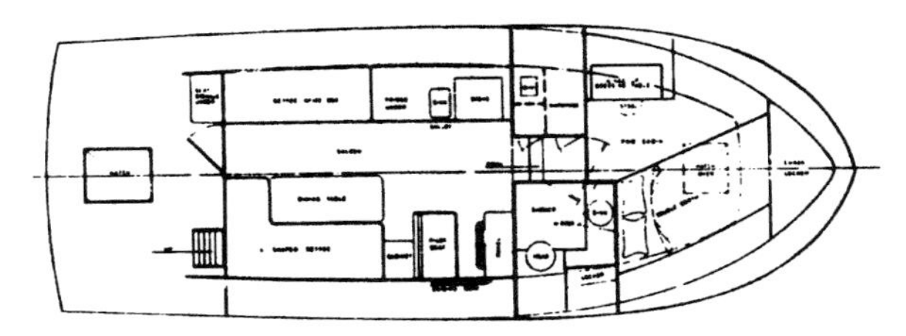

See page 243 for pricing information.

ISLAND GYPSY 36 EUROPA

SPECIFICATIONS

Length	36'0"	Cockpit	NA
LWL	32'10"	Water	200 gals.
Beam	12'6"	Fuel	450 gals.
Draft	3'11"	Hull Type	Semi-Disp.
Weight	25,520#	Designer	H. Halvorsen
Clearance	16'0"	Production	1988–Current

A salty-looking cruiser with a striking profile, the Island Gypsy 36 Europa (originally called the Island Gypsy 36 Extended Flybridge) is easily distinguished by her dramatic bridge overhangs and Euro-style supports. She's built on a solid fiberglass semi-displacement hull with moderate beam, flat aftersections, and a long keel deep enough to protect the running gear. Inside, her two-stateroom floorplan is arranged with the galley on the deckhouse level and upper/lower berths in the guest stateroom. The salon dimensions are somewhat compact (thanks to the wide sidedecks), although the cockpit is quite roomy. The decks are teak, and the interior is total teak—all very traditional and well-finished throughout. Additional features include a simulated lapstrake hull, a large flybridge with space for a small dinghy, fold-down mast, teak parquet flooring inside, and big 360° salon windows. Several engine choices are offered. The standard 135-hp Lehmans will cruise the Island Gypsy 36 Europa economically at 8–9 knots and reach a top speed of about 12 knots. ❑

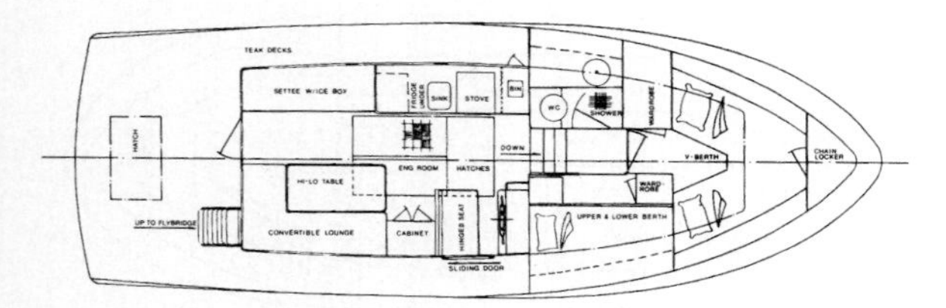

See page 244 for pricing information.

ISLAND GYPSY 36 TRI CABIN

SPECIFICATIONS

Length	36'0"	Cockpit	NA
LWL	32'10"	Water	200 gals.
Beam	12'6"	Fuel	450 gals.
Draft	3'11"	Hull Type	Semi-Disp.
Weight	25,520#	Designer	H. Halvorsen
Clearance	16'0"	Production	1988–Current

Originally called the Island Gypsy 36 Aft Cabin, the Tri Cabin designation refers to the boat's three separate cabins—two staterooms and the salon—and isn't meant to imply a three-stateroom layout. The distinguishing features of this China-built trawler are her big cockpit and abbreviated trunk cabin profile. From the outside, the aft cabin appears to have been cut short in favor of a large cockpit. In fact, the aft cabin isn't as small as one might expect. With her centerline companionway steps and flanking berths (both extending beneath the cockpit), the cabin is roomy enough to meet the needs of most couples. A choice of a double berth or V-berths is offered in the forward stateroom. The galley occupies the entire port side of the salon (plenty of counter space), and a deck access door is located at the helm. The interior is total teak, and both heads have stall showers. Powered with twin 135-hp Lehman diesels, the 36 Tri Cabin will cruise at 8–9 knots and reach a top speed of about 12 knots. ❑

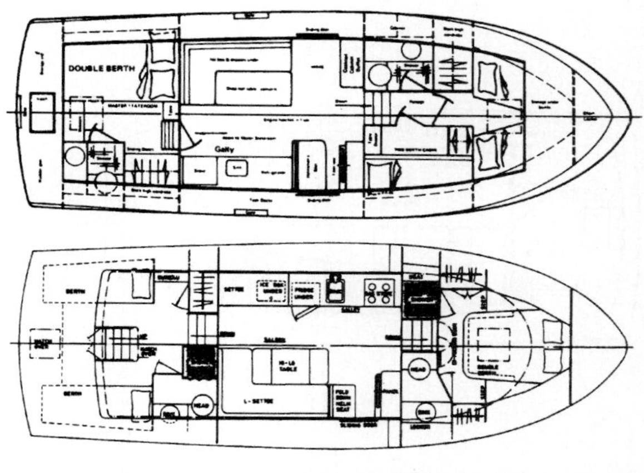

See page 244 for pricing information.

ISLAND GYPSY 36 QUAD CABIN

SPECIFICATIONS

Length	36'0"	Cockpit	NA
LWL	32'10"	Water	200 gals.
Beam	12'6"	Fuel	450 gals.
Draft	3'11"	Hull Type	Semi-Disp.
Weight	25,520#	Designer	H. Halvorsen
Clearance	16'0"	Production	1977–Current

A stylish cruiser with a distinctive profile (note the step-down sheer), the Island Gypsy 36 Quad Cabin began life as a wood boat when she was introduced in 1977, although she had become an all-fiberglass model by the following year. She's built in China on a two-piece semi-displacement hull with moderate beam and a long keel that fully protects the running gear. A popular model for Halvorsen Marine, her three-stateroom interior layout is rare in a boat this small. Needless to say, the staterooms are small, and the salon—with its galley, dinette, and lower helm—is a tight fit. There's a fair amount of storage aboard, but neither head has a stall shower (too bad). Standard features include a full teak interior, P&S deck access doors, teak decks and walkways, working mast and boom, swim platform, and 3 or 5kw generator. Powered with twin 135-hp Lehman diesels, the Island Gypsy 36 Quad Cabin will cruise easily at 8–9 knots and reach a top speed of about 12 knots. ❑

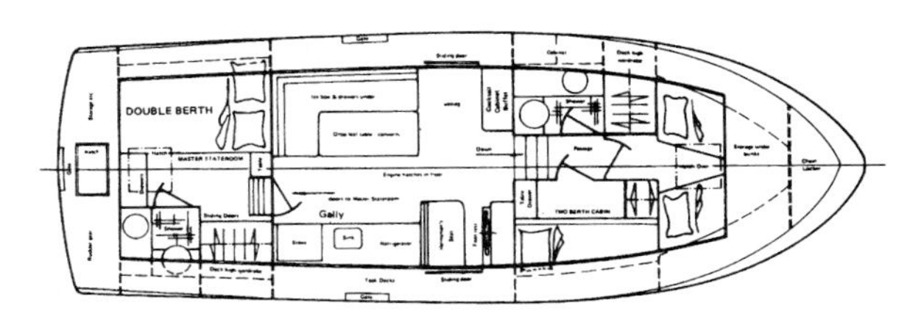

See page 244 for pricing information.

ISLAND GYPSY 40 FLUSH AFT DECK

SPECIFICATIONS

Length	40'0"	Cockpit	NA
LWL	35'3"	Water	200 gals.
Beam	14'3"	Fuel	400 gals.
Draft	3'6"	Hull Type	Semi-Disp.
Weight	33,500#	Designer	H. Halvorsen
Clearance	13'3"	Production	1986–Current

Built in Mainland China, the Island Gypsy 40 Flush Aft Deck is a traditional double-cabin cruiser with a distinctive profile (note the slightly elongated deckhouse and small aft deck) to go with her trawler-style appearance. She's built on a solid fiberglass hull with moderate beam and her full-length keel protects the running gear. The interior is arranged with two staterooms, each about the same size with centerline island berths and stall showers in the heads. The galley is aft to port in the salon (rather than forward which is usually the norm), and a deck access door is located at the lower helm. The woodwork is Burmese teak throughout. The flybridge layout is unusual: most trawlers have the helm forward on the flybridge. The Island Gypsy 40 has the helm aft with lounge seating (and a table) forward. Twin 135-hp Lehman diesels are standard (8 knots at cruise/11 knots top). Twin 210-hp Cummins (12 knots cruise/15 top), and 375-hp Cats (19 knots cruise/22 knots top) are available as options. ❑

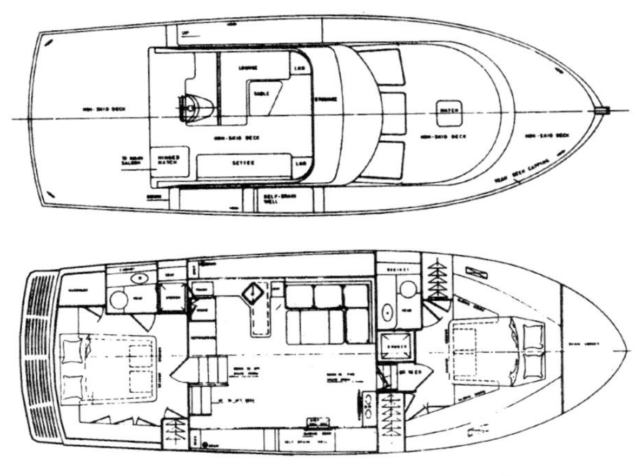

See page 244 for pricing information.

ISLAND GYPSY 40 MOTOR CRUISER

SPECIFICATIONS

Length	40'0"	Cockpit	NA
LWL	35'3"	Water	200 gals.
Beam	14'3"	Fuel	400 gals.
Draft	3'6"	Hull Type	Semi-Disp.
Weight	33,500#	Designer	H. Halvorsen
Clearance	13'3"	Production	1986–Current

A versatile boat with a conservative trawler-style appearance, the Island Gypsy 40 Motor Cruiser is built on a semi-displacement hull with a wide beam, hard chines, and a full-length prop-protecting keel. Her cockpit is on the same level as the salon resulting in a wide-open and completely practical sedan-style floorplan—the ideal cruising layout in the eyes of many. There are two staterooms forward, and the head has a shower stall. The L-shaped galley configuration is unusual in a boat this size—it *does* chop the salon up a bit, but the added counter space is a blessing. Burmese teak is liberally applied throughout the interior, and the decks and rails are teak as well. Topside, the flybridge is arranged with the helm aft and lounge seating (including a table) forward of the console. Twin 135-hp Lehman diesels are standard (8 knots at cruise/11 knots top). Twin 210-hp Cummins (12 knots cruise/15 top), and 375-hp Cats (19 knots cruise/22 knots top) are available as options. ❑

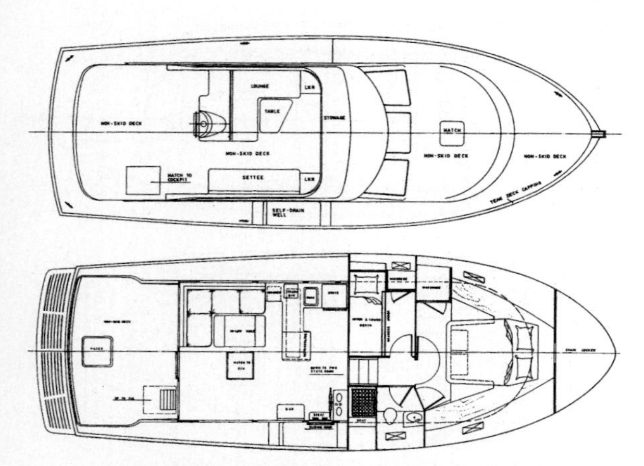

See page 244 for pricing information.

ISLAND GYPSY 40 CLASSIC

SPECIFICATIONS

Length	40'0"	Cockpit	NA
LWL	35'3"	Water	200 gals.
Beam	14'3"	Fuel	400 gals.
Draft	3'6"	Hull Type	Semi-Disp.
Weight	33,500#	Designer	M. Halvorsen
Clearance	13'3"	Production	1993–Current

The new Island Gypsy 40 Classic is actually a throwback to the traditional trunk cabin trawler made popular during the 1970s. A good-looking boat with her stepped sheerline and upright profile, she's built on a semi-displacement hull with hard chines aft and a full-length prop-protecting keel. Her teak interior is arranged with three staterooms—two forward of the salon—and a deckhouse galley. For deck access, there are port and starboard sliding doors in the salon and a cockpit passageway in the master stateroom. Both heads have stall showers, and the teak parquet flooring is impressive. The sidedecks are wide and well-secured, and the flybridge has the helm forward in the conventional manner. All-in-all, the Island Gypsy 40 is about as traditional as you're going to get short of a Grand Banks. Standard 135-hp Lehman diesels will cruise easily at 8 knots (11 knots top). Optional 210-hp Cummins will cruise at 12 knots (15 top), and 300-hp Cats cruise at about 14 knots (16–17 top). ❑

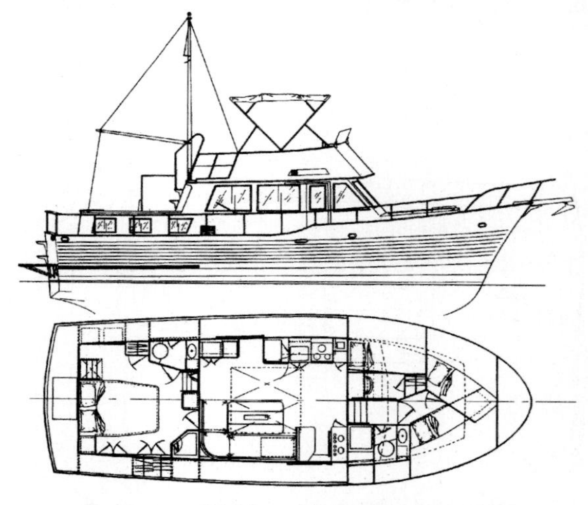

See page 244 for pricing information.

ISLAND GYPSY 44 FLUSH AFT DECK

SPECIFICATIONS

Length	44'3"	Cockpit	NA
LWL	38'9"	Water	400 gals.
Beam	15'4"	Fuel	800 gals.
Draft	4'3"	Hull Type	Semi-Disp.
Weight	38,500#	Designer	H. Halvorsen
Clearance	13'7"	Production	1979–Current

Aside from her traditional trawler-style profile and rugged construction, the Island Gypsy 44 Flush Aft Deck can claim a successful Atlantic crossing—a feat few boats of her size would ever attempt. She's built in China on a solid fiberglass semi-displacement hull with fairly high freeboard, hard chines, and a prop-protecting full-length keel. Inside, the three-stateroom floorplan is arranged with two small staterooms forward of the salon and a roomy master stateroom aft. The L-shaped galley is located aft in the salon, next to companionway stairs, where it's easily accessed from outside. Both heads have stall showers, and the interior is lavishly finished with Burmese teak woodwork. The decks (including the aft deck and flybridge) are teak as are the handrails and transom. A good-selling model, twin 135-hp Lehman diesels are standard (8–9 knots cruise/11 knots top). Optional 275-hp Lehmans (or 300-hp Cats) will cruise about 13 knots (16–17 top), and 375-hp Cats will cruise around 16–17 knots (21 top). ❑

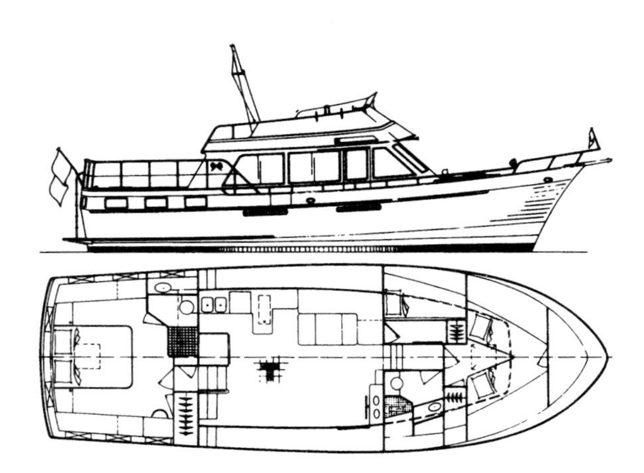

See page 244 for pricing information.

ISLAND GYPSY 44 MOTOR CRUISER

SPECIFICATIONS

Length	44'3"	Cockpit	NA
LWL	38'9"	Water	320 gals.
Beam	15'4"	Fuel	720 gals.
Draft	4'3"	Hull Type	Semi-Disp.
Weight	38,500#	Designer	H. Halvorsen
Clearance	13'7"	Production	1983-Current

With her rakish profile and roomy three-stateroom interior, the Island Gypsy 44 Motor Cruiser is a good-looking and practical sedan cruiser with just a hint of her trawler heritage. She's built in China on a wide-beam semi-displacement hull with hard aft chines and a full-length prop-protecting keel. Notably, only a few sedan-style boats in this size range have three staterooms. The owner's stateroom is amidships in this layout (where the ride is best), and the L-shaped galley is forward in the salon, opposite the lower helm. All of the interior woodwork is Burmese teak, and the outside decks and rails are teak as well. Topside, the helm console is aft on the bridge with guest seating forward. The extended flybridge shades the cockpit, and a fold-down stairwell in the bridge coaming makes foredeck access easy. A variety of engines have been offered over the years. Twin 250-hp Cummins diesels will cruise at 12 knots (15 top), and the popular 375-hp Cats will cruise at 15 knots (21 knots top). ❑

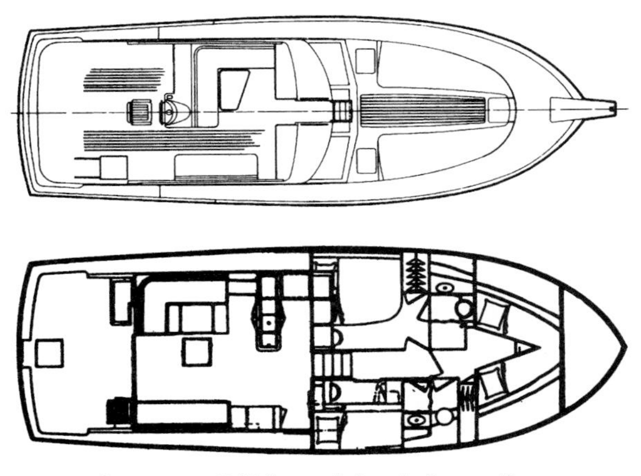

See page 245 for pricing information.

ISLAND GYPSY 49 RAISED PH

SPECIFICATIONS

Length	49'0"	Cockpit	NA
LWL	43'6"	Water	400 gals.
Beam	15'4"	Fuel	800 gals.
Draft	4'5"	Hull Type	Semi-Disp.
Weight	46,000#	Designer	Halvorsen
Clearance	13'7"	Production	1991–Current

The Island Gypsy 49 is a throwback to the old-fashioned pilothouse trawlers (mostly wood) of the 1950s and '60s. She's clearly a good-looking boat, and her strictly business appearance is more than just skin deep. Heavily built on a semi-displacement hull with hard chines and a prop-protecting keel—the same hull used for the Island Gypsy 44s, only lengthened—the raised pilothouse floorplan of the Island Gypsy 49 is greatly favored by experienced yachtsmen for its weather protection and good inside helm visibility. Two floorplans are offered (two staterooms or three), and the interior is completely finished with Burmese teak. Inside access to the flybridge is found in the pilothouse. Outside, bridge overhangs shelter the sidedecks and small cockpit from the sun and rain. Additional features include a stand-up engine room, teak decks and cabin soles, and a large flybridge with settee, table and radar mast. Several engine options are available. Among them, a pair of 375-hp Cat diesels will cruise at 13–14 knots and reach a top speed of about 17 knots. ❑

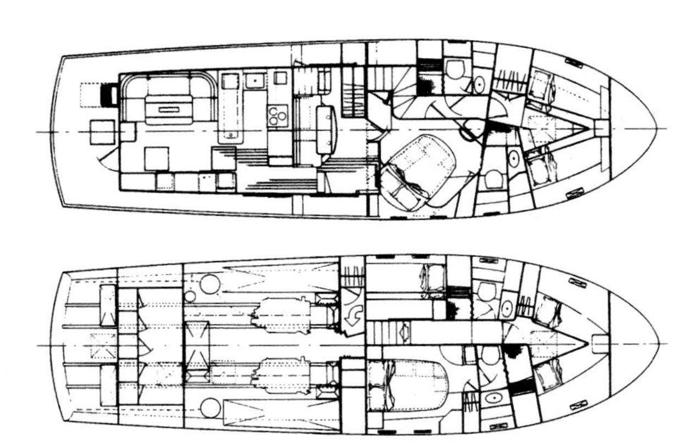

See page 245 for pricing information.

ISLAND GYPSY 51 MOTOR YACHT

SPECIFICATIONS

Length	51'0"	Cockpit	NA
LWL	44'10"	Water	600 gals.
Beam	16'6"	Fuel	1,200 gals.
Draft	4'2"	Hull Type	Semi-Disp.
Weight	74,500#	Designer	H. Halvorsen
Clearance	16'10"	Production	1983–Current

The Island Gypsy 51 MY is a traditional long range cruising yacht with a modern, somewhat rakish trawler-style appearance. She's heavily built on a solid fiberglass hull (early models used a cored hull) and features a three-stateroom, three-head floorplan with a huge stand-up engine room. The galley is on the deckhouse level—a popular choice these days—and both the helm and galley areas are open to the salon and aft deck. Access to the master stateroom is via a circular staircase in the salon. The entire interior of the Island Gypsy 51 is finished with Burmese teak. The decks (including the flybridge) are teak as well, and the sidedecks and aft deck are shaded with bridge overhangs. Topside, the extended flybridge is very spacious with room for a dinghy aft of the mast. Several diesel options are available for the Island Gypsy 51 MY including the popular 375-hp Cats (11-knots cruise/14 top). With 1,200 gallons of fuel, the cruising range at hull speed is about 1,000 miles. ❑

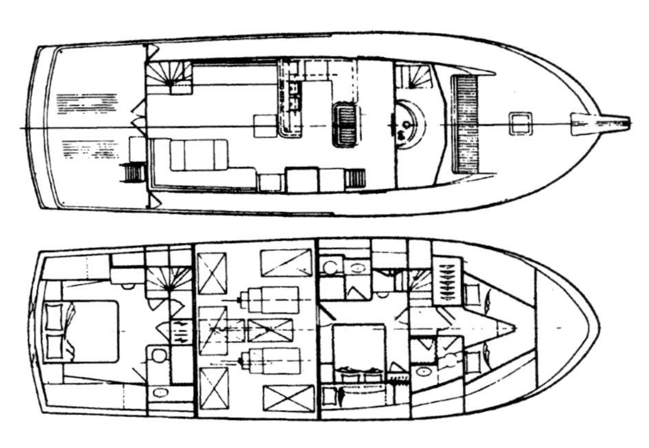

See page 245 for pricing information.

JEFFERSON 37 VISCOUNT MY

SPECIFICATIONS

Length	36'10"	Water	120 gals.
LWL	32'6"	Fuel	350 gals.
Beam	14'5"	Hull Type	Modified-V
Draft	3'0"	Deadrise Aft	NA
Weight	27,000#	Designer	Robt. Harris
Clearance	15'7"	Production	1988–Current

The Jefferson 37 Viscount is an affordable aft-cabin family cruiser with contemporary lines and an impressive list of standard equipment. As with most Jeffersons the 37 is mostly free of exterior teak. Below, however, an abundance of teak trim and cabinetry blends with off-white mica counters and surfaces for an attractive and traditional decor. The dinette floorplan is arranged with double berths in both staterooms, a comfortable salon with a built-in sofa, and lower helm. Note that the forward head is accessible from the guest stateroom only. This is a workable interior layout—it's no mean feat to design a roomy double-cabin boat on a 37-foot hull—and neither the master stateroom nor aft deck are as small as one might expect in a boat this size. Standard features include a hardtop, transom shower, swim platform, and bow pulpit. Now-standard 300-hp Cummins diesels will cruise the 37 Viscount efficiently at 19 knots (23–24 knots top). Note that the Jefferson 40 Viscount is the same boat with a cockpit extension. ❑

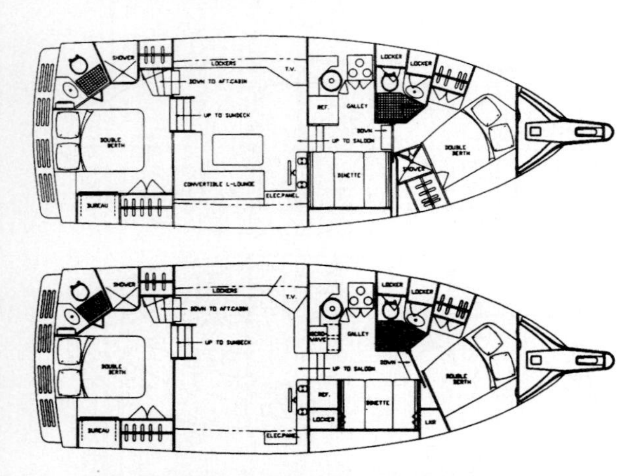

See page 245 for pricing information.

JEFFERSON 42 SUNDECK

SPECIFICATIONS

Length	41'8"	Water	200 gals.
LWL	37'8"	Fuel	350 gals.
Beam	14'3"	Hull Type	Modified-V
Draft	3'7"	Deadrise Aft	NA
Weight	30,000#	Designer	Robt. Harris
Clearance	NA	Production	1985–89

The Jefferson 42 Sundeck was built on the same hull as the Jefferson 42 Convertible. She's a fairly straightforward double-cabin design with attractive lines and a choice of interior floorplans. The "A" Plan layout features a large master stateroom aft with a walkaround queen berth and an offset double berth in the forward stateroom. A tub/shower is included in the head adjoining the owner's aft stateroom, while a separate stall shower is found in the forward head. The alternate dinette layout, however, will have a good deal of appeal to experienced cruisers who want the convenience of a built-in eating area in a boat this size. Large wraparound windows add plenty of natural lighting, and good-quality teak joinerwork is used throughout the interior. Outside, the 42 Sundeck has wide sidedecks, and the full-width aft deck is big enough for comfortable dockside entertaining. A reasonably priced and good-handling boat, 260-hp Cat diesels will cruise the Jefferson 42 Sundeck at a respectable 17–18 knots and provide a top speed of around 20 knots. ❑

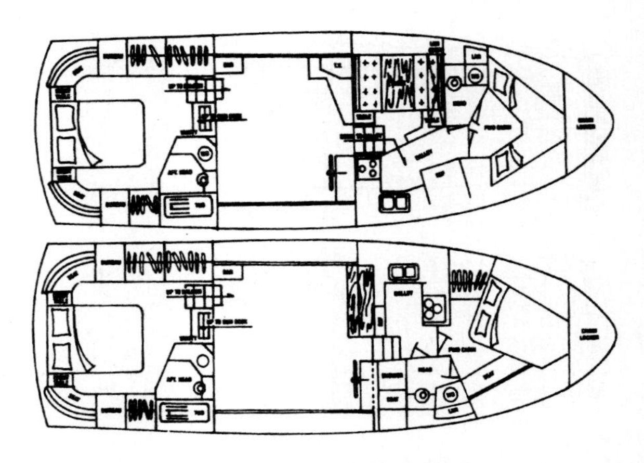

See page 245 for pricing information.

JEFFERSON 42 VISCOUNT MY

SPECIFICATIONS

Length	41'8"	Water	100 gals.
LWL	37'8"	Fuel	350 gals.
Beam	14'5"	Hull Type	Modified-V
Draft	3'0"	Deadrise Aft	NA
Weight	28,000#	Designer	Robt. Harris
Clearance	15'7"	Production	1990–Current

Built on a solid fiberglass hull with plenty of beam, the Jefferson 42 Viscount is a straightforward double-cabin family cruiser with a modern profile to go with her practical interior layout. She's a distinctive boat with her stepped sheer and dramatic hull graphics. Her two-stateroom floorplan is arranged with the galley and dinette down from the salon. A lower helm is standard, and the interior is finished with teak paneling, moldings, doors, and cabinets. (Like the 37 Viscount MY, the forward head can be entered from the stateroom only.) A tub/shower is fitted in the aft head compartment. Outside, the full-width aft deck of the 42 Viscount will accommodate several pieces of deck furniture, and the flybridge is set up with the helm forward and guest seating aft. Twin 300-hp Cummins diesels are standard in the newer models (around 18 knots cruise/22 knots top) and 375-hp Cats (21 knots cruise/24–25 knots wide open) are optional. An "SE" version with less standard equipment has been available since 1991. ❑

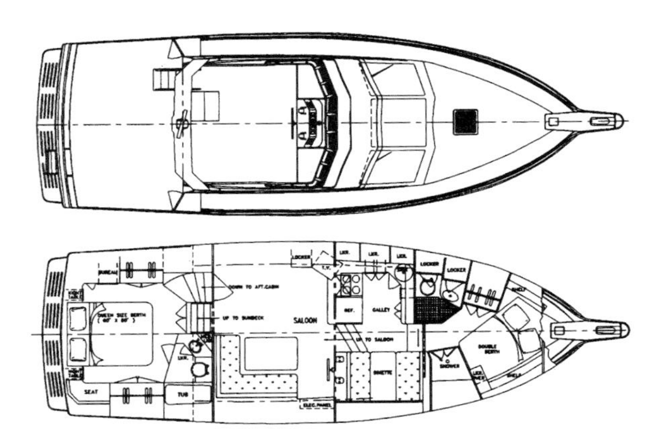

See page 245 for pricing information.

JEFFERSON 43 MARLAGO SUNDECK

SPECIFICATIONS

Length	42'10"	Water	200 gals.
LWL	NA	Fuel	420 gals.
Beam	15'0"	Hull Type	Modified-V
Draft	3'10"	Deadrise Aft	NA
Weight	31,000#	Designer	Robt. Harris
Clearance	16'7"	Production	1991–Current

A good-looking boat, the 42 Marlago is a comfortable family cruiser with a long list of standard features built into her affordable price. She's constructed on a solid fiberglass hull with modest transom deadrise and a good deal of flare at the bow. The Marlago's floorplan is arranged in the conventional manner with staterooms fore and aft and the galley and dinette down. Both staterooms have double berths (note the tub/shower in the aft head compartment), and a lower helm is standard. The 43 Marlago is a roomy boat for her size, although the forward stateroom is somewhat compact. There's space on the aft deck for a variety of deck furniture, and the flybridge is arranged with built-in guest seating behind the helm. Additional features include a teak interior, radar arch, swim platform, and a particularly attractive hardtop design. Cummins 300-hp diesels are standard (around 15–16 knots cruise) and Cat diesels to 425-hp are optional. Note that the 46 Marlago Cockpit MY is the same boat with a cockpit extension. ❑

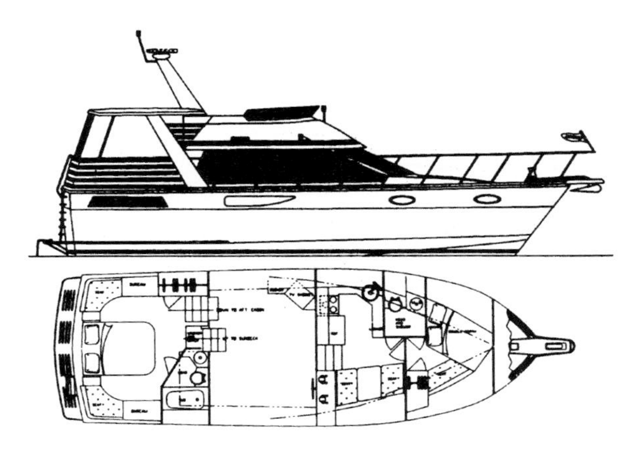

See page 245 for pricing information.

JEFFERSON 45 MOTOR YACHT

SPECIFICATIONS

Length	45'3"	Water	300 gals.
LWL	41'0"	Fuel	600 gals.
Beam	15'2"	Hull Type	Modified-V
Draft	4'5"	Deadrise Aft	NA
Weight	41,000#	Designer	Robt. Harris
Clearance	17'10"	Production	1982–89

The Jefferson 45 MY was the first boat imported and sold under the Jefferson name. A popular model (55 were sold), she's constructed on a solid fiberglass hull with moderate beam and a long keel for directional stability. Although the Jefferson 45 is a fairly conventional boat, and her dockside appearance is unlikely to stir the emotions, the secret of her success can be traced to a low price and a practical two-stateroom, galley-down interior. The owner's cabin is quite spacious and features a centerline queen berth plus good storage space. A lower helm was standard, and the interior decor is total teak throughout. Note that the original floorplan was updated in 1985 to include the tub/shower aft and a stall shower in the forward head. The full-width afterdeck has plenty of space for furniture, and wide sidedecks make for secure passage around the house. Twin 200-hp Perkins diesels were standard (11 knots at cruise/14 top) and 320-hp Cats (15–16 knots cruise/19 knots top) were a popular option. ❑

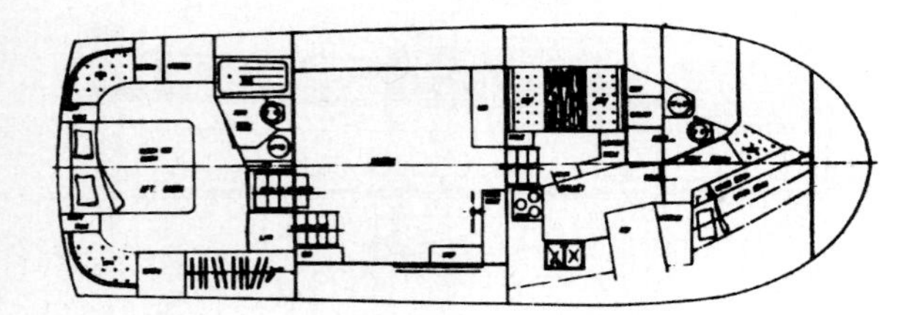

See page 245 for pricing information.

JEFFERSON 46 SUNDECK MY

SPECIFICATIONS

Length	45'8"	Water	200 gals.
LWL	41'8"	Fuel	350 gals.
Beam	14'3"	Hull Type	Modified-V
Draft	3'7"	Deadrise Aft	NA
Weight	43,000#	Designer	Robt. Harris
Clearance	NA	Production	1985–89

While the 46 Sundeck doesn't have a notably stylish profile, she was a popular model for Jefferson due to her moderate price and spacious interior accommodations. Like the rest of the Jefferson fleet, the 46 Sundeck is built on a solid glass modified-V hull form with moderate beam and a shallow keel below. A choice of several floorplans was available with two or three staterooms. (All lack a stall shower in the forward head.) Teak paneling and cabinetry are used extensively throughout the interior. The afterdeck of the Jefferson 46 is huge—one of the largest to be found in any boat this size and a superb outdoor entertainment platform. The flybridge is arranged with the helm forward and guest seating aft. Standard features included a swim platform, transom shower, deck washdown, bow pulpit, and lower helm. Small diesels were standard, but most Jefferson 46 Sundecks were sold with the 375-hp Cats which provide a cruising speed of 19–20 knots. The Jefferson 52 Cockpit Motor Yacht is the same boat with a cockpit extension. ❑

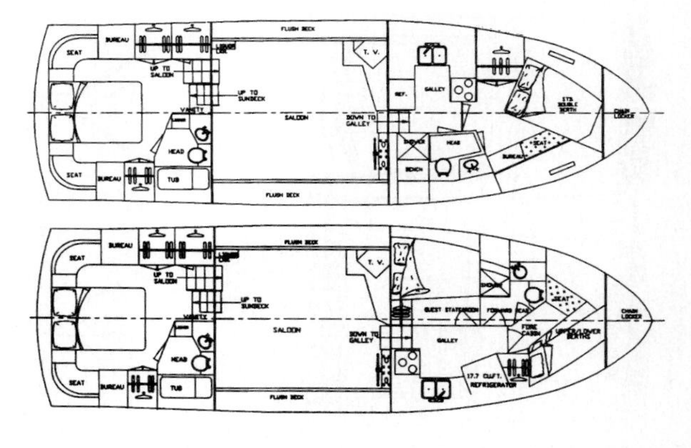

See page 245 for pricing information.

Unequalled Product Knowledge
Courteous, Professional Service

- Yacht Brokerage
- Worldwide Charters
- Construction
- Management

PALM BEACH OFFICE
Soverel Harbour
2401 PGA Blvd. Suite 160
PALM BEACH GARDENS, FL 33410
(407) 627-9500 Fax (407) 627-9503

Steve Barcsansky, Mike Brown, Alan Learch, Bill Snyder, Greg Pierce, Mark Parker, George Schilling, Jim Wallace

GULF COAST OFFICE
Orange Beach Marina
Marina Road
ORANGE BEACH, AL 36561
(205) 981-9200 Fax (205) 981-9137

Jim Greene, Dick Collier Garland Gilbert, John Haucke, Lon McCloskey, Bill Thompson

JEFFERSON 46 MARLAGO MY

SPECIFICATIONS

Length	45'10"	Water	200 gals.
LWL	NA	Fuel	420 gals.
Beam	15'0"	Hull Type	Modified-V
Draft	3'10"	Deadrise Aft	NA
Weight	34,700#	Designer	Robt. Harris
Clearance	116'7"	Production	1991–Current

A good-looking boat, the main selling features of the Jefferson 46 Marlago are a roomy interior layout and an attractive selling price. The standard floorplan consists of two staterooms with the galley and dinette down, and an alternate plan has a mid-level galley with no dinette, a larger salon, and space forward for a washer/dryer. Both staterooms are quite spacious with walkaround double berths and full heads. (Note that the forward head can only be accessed from the stateroom.) A lower helm is standard, and there's a tub/shower in the aft head compartment. There is less interior teak cabinetry and trim in the 46 Marlago than in previous Jefferson models. Additional features include wing doors, a spacious afterdeck, wide sidedecks, and a swim platform. Standard power for the 46 Marlago is 300-hp Cummins, but most buyers will likely choose the larger 375-hp or 425-hp Cats. Cruising speeds with the Cats is around 16–18 knots. The fuel capacity is a little light for a boat this size. ❑

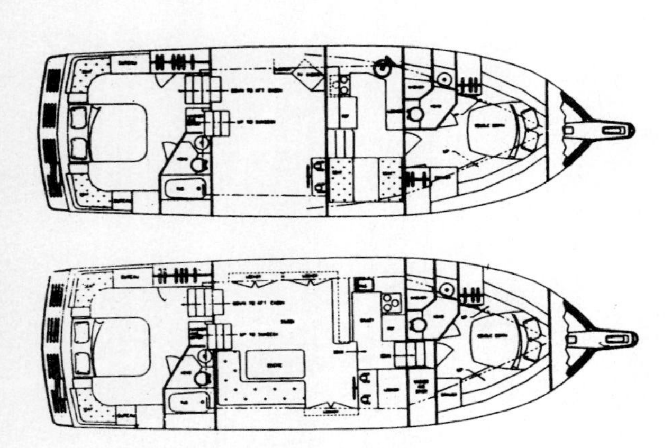

See page 245 for pricing information.

JEFFERSON 48 RIVANNA MY

SPECIFICATIONS

Length	48'4"	Water	200 gals.
LWL	NA	Fuel	600 gals.
Beam	16'0"	Hull Type	Modified-V
Draft	4'0"	Deadrise Aft	6°
Weight	42,500#	Designer	Jefferson
Clearance	13'4"	Production	1990–Current

With her modern lines and low-profile appearance, the 48 Rivanna is aimed at the growing market for affordably priced large family cruisers. She's built on a solid fiberglass modified-V hull with moderate beam, a flared bow, and a shallow keel below. A good-looking boat, the standard three-stateroom floorplan is arranged with centerline island beds in the fore and aft staterooms and a small guest cabin aft of the salon. (Note that the forward head can only be accessed from the stateroom.) The interior is completely finished with teak cabinetry and joinerwork, and a lower helm station is standard. Additional features include a hardtop and radar arch, wide sidedecks, and a huge aft deck platform with wing doors. Small Cummins 300-hp diesels are standard, but most owners will probably select the more powerful 425-hp Cats. Performance with these engines is 16 knots at cruise and 19 knots wide open. A cockpit version of this boat without the third stateroom—the Jefferson 52 Rivanna Cockpit MY—is also offered. ❑

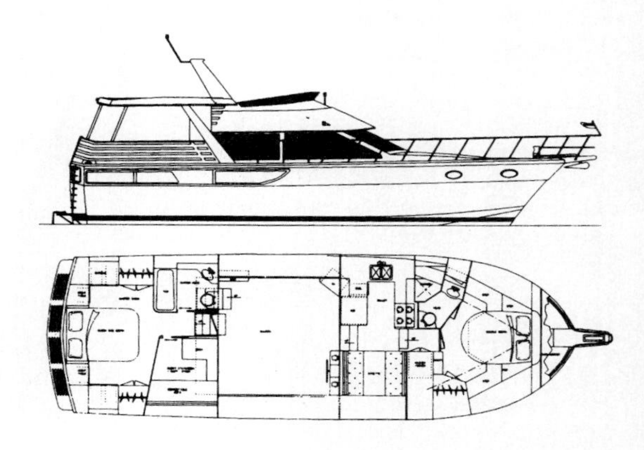

See page 246 for pricing information.

JEFFERSON 52 MONTICELLO MY

SPECIFICATIONS

Length	51'6"	Water	300 gals.
LWL	47'0"	Fuel	600 gals.
Beam	15'2"	Hull Type	Modified-V
Draft	3'7"	Deadrise Aft	NA
Weight	45,000#	Designer	Robt. Harris
Clearance	NA	Production	1986–89

A traditional design with an old-fashioned flush deck profile, the Jefferson 52 Monticello is a moderately priced cruising yacht with very generous interior accommodations considering her relatively narrow beam. Her solid fiberglass hull shows considerable flare at the bow and includes a long keel below for increased directional stability. The Monticello has a fully enclosed afterdeck/wheelhouse area—the main salon, actually—which is easily the best gathering spot on the boat. An inside ladder next to the helm leads up to the flybridge, where two large L-shaped lounges provide excellent guest seating for as many as eight. The lowerdeck salon is arranged with a built-in dinette, and a large mid-level galley is forward. Most were delivered with a three-stateroom interior which includes a spacious master stateroom and three heads, each with a stall shower enclosure. Teak paneling and cabinetry are used throughout the interior. Twin Caterpillar 375-hp diesels will cruise the Jefferson 52 Monticello at a sedate 15–16 knots with a top speed of around 19 knots. ❑

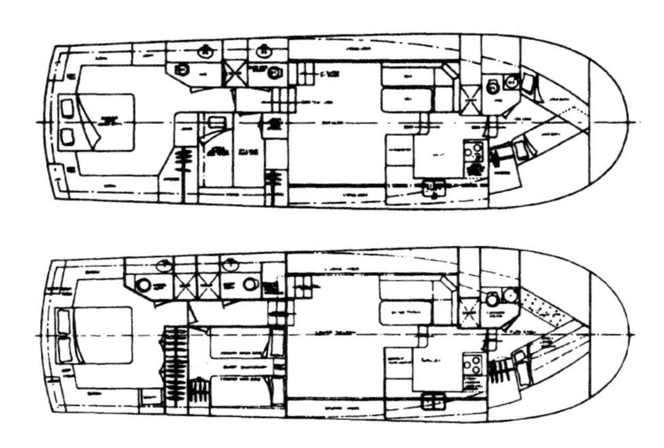

See page 246 for pricing information.

JEFFERSON 52 MARQUESSA EDMY

SPECIFICATIONS

Length	52'5"	Water	200 gals.
LWL	NA	Fuel	700 gals.
Beam	16'0"	Hull Type	Modified-V
Draft	4'0"	Deadrise Aft	6°
Weight	55,800#	Designer	Robt. Harris
Clearance	19'6"	Production	1989–Current

The Jefferson 52 Marquessa is a copy of the popular Chris Craft 501 MY, a good-looking design with a modern profile and a surprisingly spacious floorplan. Like all Jefferson yachts, the Marquessa is constructed on a modified-V hull with a shallow keel and moderate deadrise at the transom. Inside, the dominant feature is the huge full-width salon with its built-in entertainment center and wing doors. A companionway in the wheelhouse leads down to the lower level, where both the master and VIP staterooms are fitted with walkaround double berths. All three staterooms have their own heads, each with a separate stall shower. The engines are located in separate engine rooms accessed from the corridor leading to the aft stateroom. Note that there are several additional floorplans available, and a Standard Deckhouse version features full walkaround sidedecks. A good-running boat, standard 550-hp Detroit 6V92 diesels will cruise the 52 Marquessa at an impressive 19–20 knots and deliver a top speed of around 22 knots. ❑

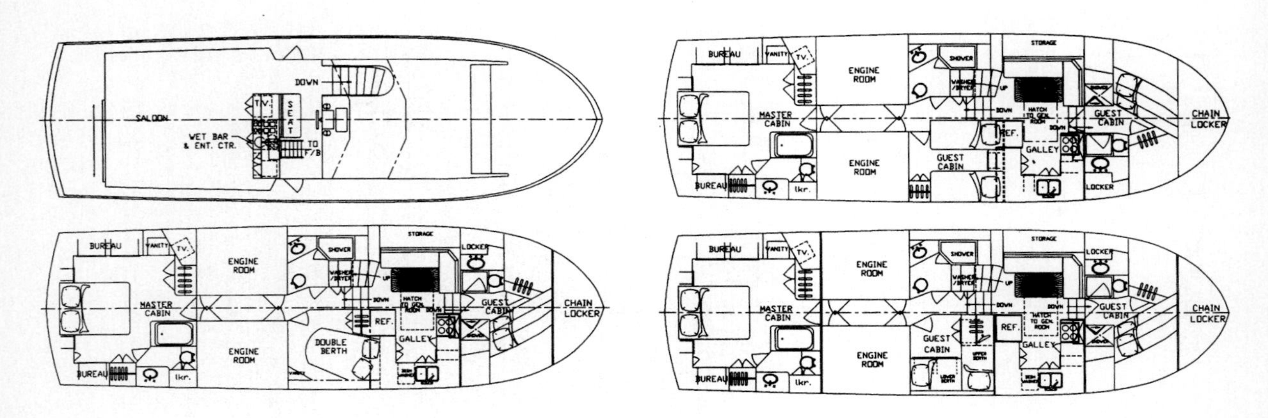

See page 246 for pricing information.

JEFFERSON 60 MOTOR YACHT

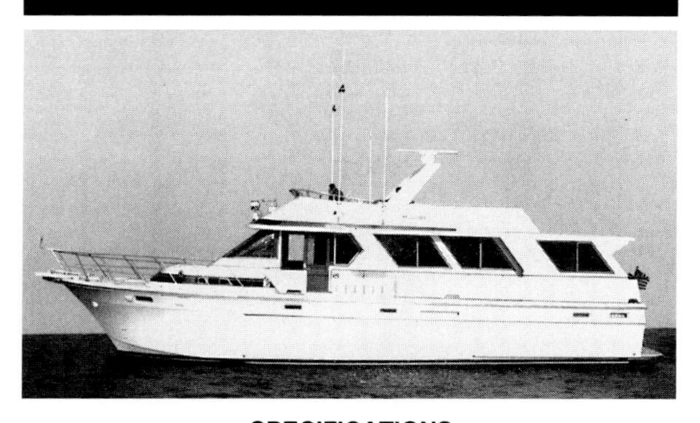

SPECIFICATIONS

Length	59'10"	Fuel	1,000 gals.
Beam	17'6"	Hull Type	Modified-V
Draft	4'7"	Deadrise Aft	NA
Weight	88,000#	Designer	Robt. Harris
Clearance	18'3"	Production	1987–Current
Water	400 gals.		

Built on a cored hull with a wide beam and moderate deadrise at the transom, the Jefferson 60 MY has been a fairly popular boat for Jefferson thanks to her roomy accommodations and affordable price tag. Her four-stateroom floorplan is arranged with the galley down resulting in a spacious and wide open salon which is open to the helm. (Note the on-deck day head.) Each stateroom has its own head compartment, and a corridor leading aft to the master stateroom provides access to the split engine rooms. The entire interior is fully paneled and finished with teak woodwork. Outside, the wide sidedecks surrounding the house are protected by a bridge overhang, and the semi-enclosed aft deck is large enough for outdoor entertaining. The flybridge comes with plenty of guest seating and space for dinghy storage aft. With standard 6V92s, the cruising speed is around 14 knots. The more popular 735-hp 8V92s provide 16–17 knots of cruising speed. Note that the Jefferson 65 Cockpit MY is the same boat with a cockpit extension. ❑

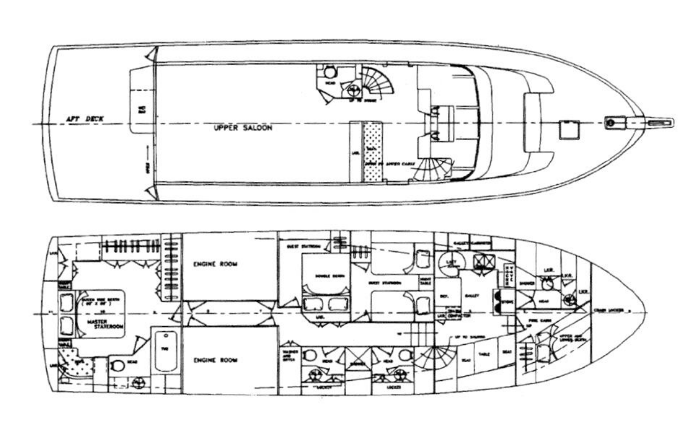

See page 246 for pricing information.

KHA SHING 40 SUNDECK

SPECIFICATIONS

Length	39'6"	Cockpit	NA
LWL	34'0"	Water	200 gals.
Beam	14'0"	Fuel	300/340 gals.
Draft	3'7"	Hull Type	Semi-Disp.
Weight	22,800#	Designer	Kha Shing
Clearance	15'1"	Production	1980–Current

Certainly one of the more popular Taiwan imports, the Kha Shing 40 is better known as the Spindrift 40 MY on the West Coast and the Vista or Southern Star 40 MY on the East Coast. She's a competent semi-displacement sundeck trawler with a solid fiberglass hull, shallow keel, and an unusually modern profile. Over 120 of these boats have been sold into the U.S. market, and most have been delivered with a two-stateroom layout with the galley down and a lower helm in the salon. The interior is teak, and the cabinetry and joinerwork are impressive. With her wraparound cabin windows and excellent headroom, the salon of the Kha Shing 40 is quite open in spite of all the teak. Notable features include deck access doors, secure sidedecks, teak handrails, and a roomy master stateroom whose dimensions are impressive for a 40-footer. Outside, the aft deck is fairly spacious with room for several chairs. Economical and capable of greater than trawler speeds, many (if not most) Kha Shing 40s are powered with 165-hp or 200-hp Volvo diesels. Cruising speeds are about 11–13 knots and top speed is around 15–16 knots. ❑

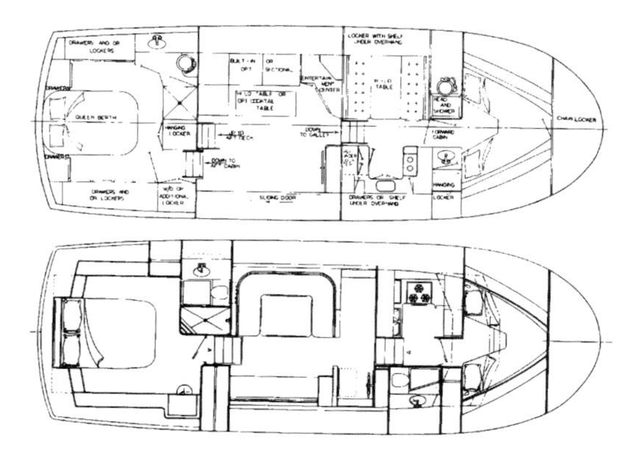

See page 246 for pricing information.

KROGEN MANATEE 36

SPECIFICATIONS

Length	36'4"	Water	300 gals.
LWL	34'0"	Fuel	280 gals.
Beam	13'8"	Hull Type	Disp.
Draft	3'2"	Designer	Jim Krogen
Weight	23,000#	Production	1984–91
Clearance	14'0"		

The Manatee 36 is a particularly well-named boat—she's as fat and slow as her namesake, and she's no award-winner when it comes to looks. Built in Taiwan, she's constructed on a cored full-displacement hull with a wide beam, deep keel, and rounded bilges aft. Perhaps the most notable feature of the Manatee 36 is the amount of usable living space packed inside this boat. The spacious salon uses the hull's full width (no sidedecks outside), and the beam is carried well forward resulting in an unusually large master stateroom. The entire layout is arranged on a single level, and the interior is finished with traditional teak cabinetry and woodwork throughout. Topside, the semi-enclosed flybridge comes with wraparound lounge seating for guests, and an extended cockpit overhang provides storage space for a dinghy. Note the rounded transom and weather-protected aft deck. A single 100-hp Volvo diesel will cruise the Manatee 36 for 1,100 miles at a leisurely 6–7 knot speed. Over 90 were built making the Manatee a very popular model indeed. ❑

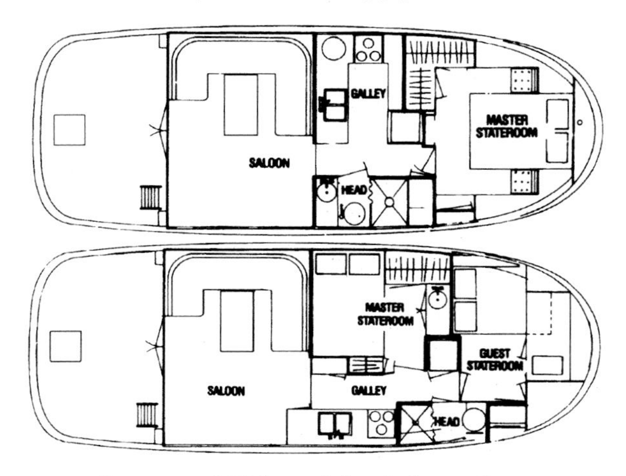

See page 246 for pricing information.

KROGEN 42 TRAWLER

SPECIFICATIONS

Length	42'4"	Water	360 gals.
LWL	39'6"	Fuel	700 gals.
Beam	15'0"	Hull Type	Disp.
Draft	4'7"	Designer	Jim Krogen
Weight	39,500#	Production	1979–Current

The Krogen 42 is one of the few trawler-style boats actually built on a full displacement hull. As such, she displays all of the characteristics normally associated with such long-range designs, including a comfortable ride, excellent seaworthiness, and the easy rolling motion typical of any soft-chined boat. The Krogen is a particularly salty-looking vessel with a distinctive profile and an upright bow. Inside, the focal point of the boat is the functional pilothouse with watch berth, located a few steps up from the salon level. Until hull #65 (1985), the Krogen 42s were built with glass-over-plywood decks with a fiberglass hull and superstructure. Beginning with hull #66, construction became all fiberglass. A single Lehman diesel is standard, and a unique "get home" feature allows the generator to be used as an emergency back-up engine. At her 8-knot hull speed, the Krogen 42 has a range of 2,000+ miles. Over 180 have been sold to date (including five with twin screws). A Wide Body model with a full-width salon was introduced in 1989. ❑

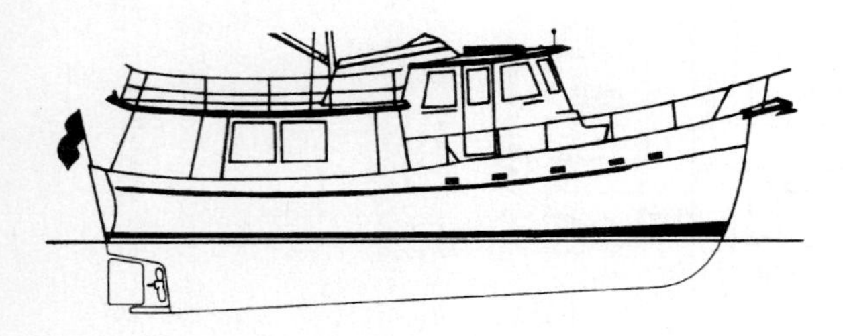

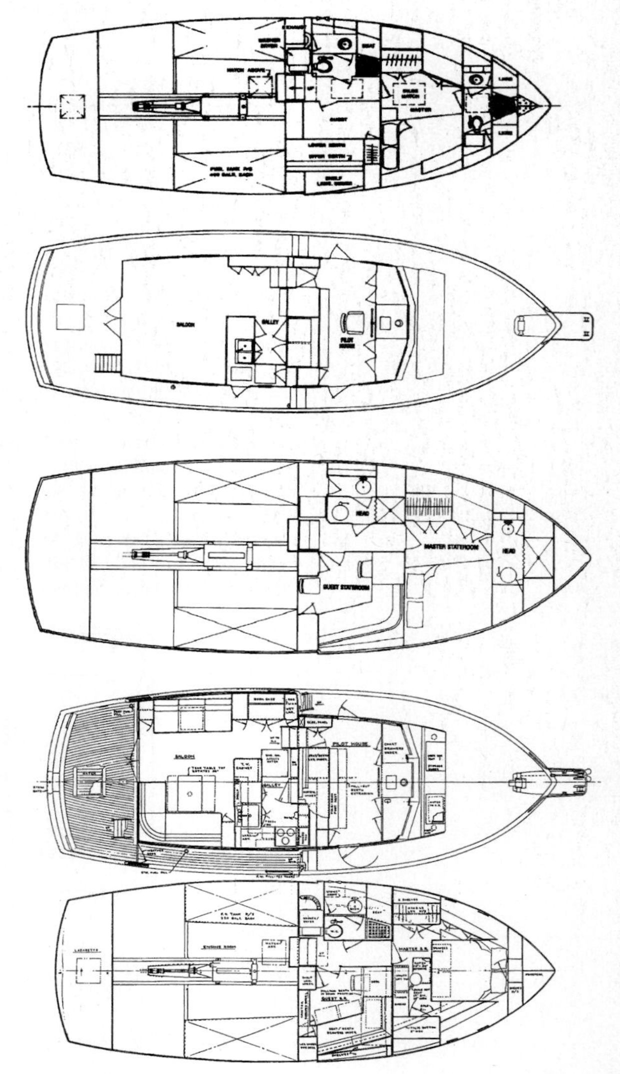

See page 246 for pricing information.

KROGEN SILHOUETTE 42

SPECIFICATIONS

Length	41'10"	Water	150 gals.
LWL	37'6"	Fuel	400 gals.
Beam	14'6"	Cockpit	NA
Draft	3'2"	Hull Type	Modified-V
Weight	28,000#	Designer	J. Krogen
Clearance	NA	Production	1987–91

The Silhouette 42 is truly an exceptional design. If her profile is unusual, the Silhouette manages to offer a unique interior layout with more living space than one might expect in a boat this size. She was built in Taiwan by Chien Hwa on a PVC-cored, modified-V hull. While she incorporates several unusual features, it's the semi-enclosed pilothouse/sundeck and the wide-open master stateroom (with its private cockpit) that attract the most attention. A real surprise is the hydraulically operated stern gate which drops to create a practically wide-open transom and swim platform. Inside, a breakfast bar separates the spacious main salon from the galley below, and both staterooms feature double berths and stall showers. The interior is finished with light ash (or teak) woodwork and mica veneers. With standard Caterpillar 375-hp diesels, the Silhouette 42 will cruise around 21 knots and reach a top speed of 25–26 knots. Never an overly popular design in spite of her unique character, only a few of these boats were sold. ❑

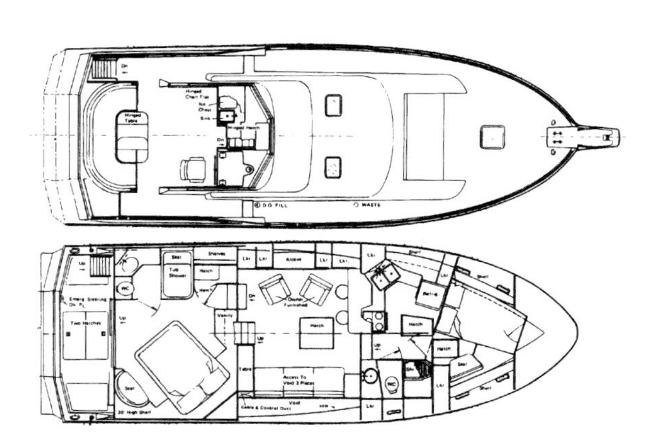

See page 246 for pricing information.

KROGEN 48 WHALEBACK

SPECIFICATIONS

Length	48'5"	Water	500 gals.
LWL	45'5"	Fuel	1,000 gals.
Beam	16'8"	Hull Type	Disp.
Draft	4'9"	Designer	J. Krogen
Weight	56,000#	Production	1993–Current

A distinctive design with a flush foredeck and Portuguese bridge, the 48 Whaleback is a full displacement cruising trawler with a deep entry, rounded bilges, and a ballasted, prop-protecting keel. With well over 16 feet of beam and no sidedecks, the interior dimensions of the Whaleback are downright extravagant for a boat this size. The floorplan—arranged on a single level from the cockpit to the forward stateroom—includes three staterooms (one of which doubles as a den) and two full heads in addition to an expansive salon and generous galley. All of the cabinetry, joinery, and paneling are fashioned of grain-matched teak. Topside, the enclosed pilothouse features full 360° visibility in addition to a spacious chart table, L-shaped settee, deck access doors, and wet bar. There's room on the extended upper deck for dinghy storage, and the mast and boom are fully functional. Note the protected walkway surrounding the pilothouse. At a 9-knot cruising speed, the single 210-hp Cat diesel delivers a range of approximately 1,400 miles. The Whaleback 48 is also available with full walkaround sidedecks. ❑

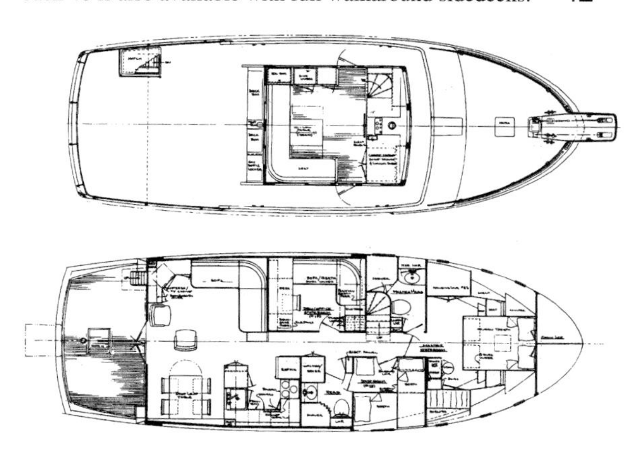

See page 246 for pricing information.

LORD NELSON VICTORY 37 TUG

SPECIFICATIONS

Length	36'11"	Cockpit	NA
LWL	33'4"	Water	185 gals.
Beam	13'2"	Fuel	250 gals.
Draft	3'6"	Hull Type	Disp.
Weight	20,500#	Designer	Jim Bachus
Clearance	12'6"	Production	1984–89

Based on a traditional New England workboat hull, the Victory 37 Tug is a long-range cruiser with a yacht-like interior to go with her salty profile. She was built in Taiwan on a solid fiberglass displacement hull with a plumb bow, full-length (ballasted) keel, and a rounded transom. High bulwarks surround the entire deck for safety. The raised pilothouse is completely enclosed and gives the helmsman a full 360° view of the horizon. The original floorplan was updated in 1988 by relocating the galley to starboard, which opened up the stateroom considerably. Companionway steps lead aft into the comfortable salon, and a pantry door in the galley opens to the spacious engine room below. Forward, the stateroom features a double berth and head with a stall shower. The original teak decks were eliminated in favor of fiberglass, and the teak handrails were replaced with stainless steel late in her production run. A popular boat with 88 delivered, the Victory 37 has an honest cruising range of 800–900 miles with a single 150-hp Cummins diesel. ❑

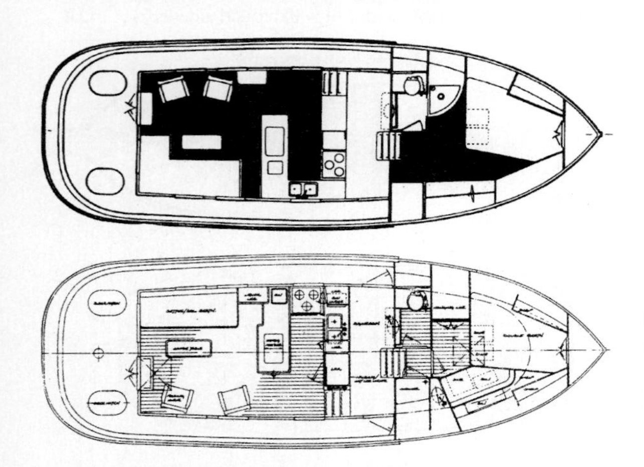

See page 246 for pricing information.

MAINSHIP 34 SEDAN

SPECIFICATIONS

Length	34'0"	Water	50 gals.
LWL	NA	Fuel	220 gals.
Beam	11'11"	Cockpit	80 sq. ft.
Draft	2'10"	Hull Type	Semi-Disp.
Weight	14,000#	Designer	Cherubini
Clearance	13'6"	Production	1978–82

The Mainship 34 Sedan is one of the most popular small cruisers ever built. She was constructed on a solid fiberglass semi-displacement hull design with a fine bow entry and a full-length keel below. First of the Mainship series, the appeal of the 34 Sedan had much to do with her trawler-style profile and affordable price tag, and her greatest attraction remains her superb economy at better than trawler speeds. With a single 160-hp Perkins 6-cylinder diesel, the Mainship's easily driven hull will cruise at 10–11 knots burning only 6 gph. At a more relaxed 7-knot speed, the fuel consumption drops to a remarkable 2 gph. The practical single-stateroom floorplan is well-suited to the needs of a cruising couple. A lower helm was standard in the salon; the galley is large enough for serious food preparation; and a stall shower is included in the head. Outside, the flybridge extends aft to provide weather protection for the cockpit. Considered a low-maintenance boat, the Mainship 34 Sedan continues to enjoy great popularity in most markets. ❑

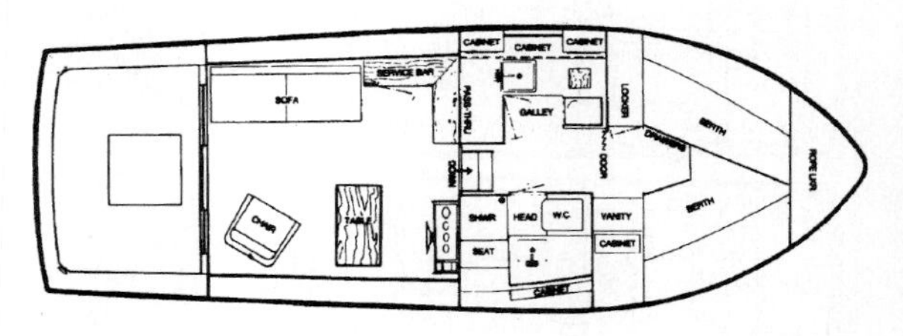

See page 246 for pricing information.

MAINSHIP 36 DOUBLE CABIN

SPECIFICATIONS

Length	36'2"	Water	100 gals.
LWL	NA	Fuel	240 gals.
Beam	13'0"	Hull Type	Modified-V
Draft	2'2"	Deadrise Aft	NA
Weight	20,000#	Designer	Mainship
Clearance	11'3"	Production	1984–89

A popular boat, the mainship 36 Double Cabin is a conservative design with a plain-Jane appearance and an efficient cabin layout. Built on a solid fiberglass hull with moderate deadrise aft, her profile displays a hint of trawler styling. The interior accommodations are arranged with the galley and dinette down a few steps from the salon level. V-berths are forward, and a walkaround double berth is fitted in the aft stateroom. Large wraparound deckhouse windows provide an abundance of natural lighting. In 1985, the original teak interior was replaced with light oak woodwork. Outside, the 36 has a full-width afterdeck with nearly 90 sq. ft. of entertainment area. The bridge is set up with the helm forward and guest seating aft. Included in a long list of standard equipment was a generator, lower helm station, bow pulpit, and swim platform. With only twin 270-hp Crusader gas engines, the Mainship 36 has a relatively good turn of speed. She'll cruise comfortably at a respectable 17 knots and reach a top speed of around 25–26 knots. ❑

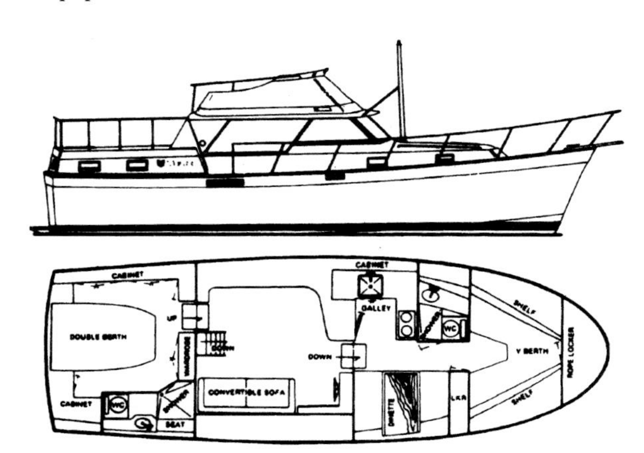

See page 246 for pricing information.

MAINSHIP 40 DOUBLE CABIN

SPECIFICATIONS

Length	40'0"	Water	140 gals.
LWL	NA	Fuel	300 gals.
Beam	14'0"	Hull Type	Modified-V
Draft	3'4"	Deadrise Aft	NA
Weight	24,000#	Designer	Cherubini
Clearance	17'6"	Production	1984–88

A popular boat with traditional styling (note the mast behind the bridge) and a straightforward aft cabin layout, the Mainship 40 DC was built on a solid glass hull with average beam and fairly high freeboard. Despite her old-fashioned Downeast profile, her performance and interior accommodations are similar to those of a modern motor yacht. The salon is particularly spacious, and the absence of a dinette in this layout is notable and perhaps even a welcome change from the norm. (Instead of a dinette, there's a convenient breakfast bar, which takes up considerably less space.) Both staterooms are arranged in the conventional manner, and a tub/shower is fitted in the aft head. An attractive light oak interior replaced the original dark teak woodwork beginning in 1985. Outside, the aft deck dimensions are somewhat confined, and guest seating is provided aft of the helm. The Mainship 40 DC will cruise at a respectable 18 knots with standard 350-hp Crusader gas engines and reach a top speed of around 27 knots. ❑

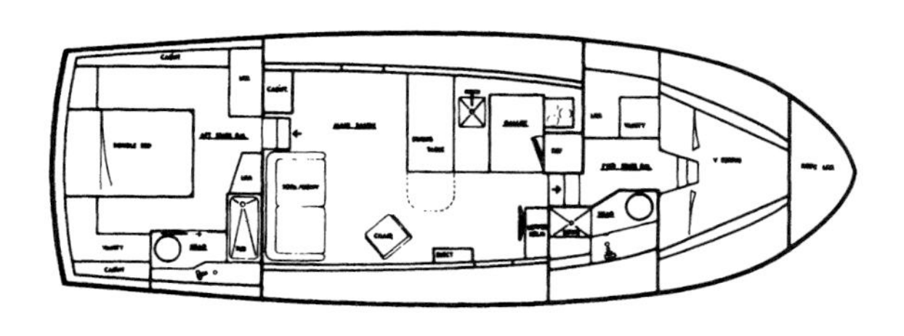

See page 247 for pricing information.

MAINSHIP 41 GRAND SALON

SPECIFICATIONS

Length........................40'11"
Beam............................14'5"
Draft...............................3'6"
Weight.....................23,000#
Clearance.....................15'0"
Water.....................130 gals.
Fuel........................375 gals.
Hull Type........Modified-V
Deadrise Aft..................12°
Designer..........Mike Peters
Production............1989–90

The Mainship 41 Grand Salon is an unorthodox Euro-style cruiser that might better be described as a floating condo. The design emphasis was obviously on the mega-volume interior. Built on a solid glass hull with a wide beam, there were two versions of the Mainship 41: the Double Cabin with a master stateroom aft, and the Grand Salon with an enormous full-length salon stretching for almost two-thirds of the boat's length. Either layout is impressive with one of the most stylish decor package ever attempted in a popular-priced production boat. Indeed, the interior of the Mainship 41 was more than just innovative—it was a giant step toward Mainship's vision of the future in mid-size U.S. yacht designs. Unfortunately, the boxy profile and townhouse accommodations failed to impress the market and production lasted only two years. Standard 454-cid gas engines will cruise the Mainship 41 at 16–17 knots (around 25 knots top), and the optional Cat 375-hp diesels cruise around 23 knots and reach 27 knots wide open. ❑

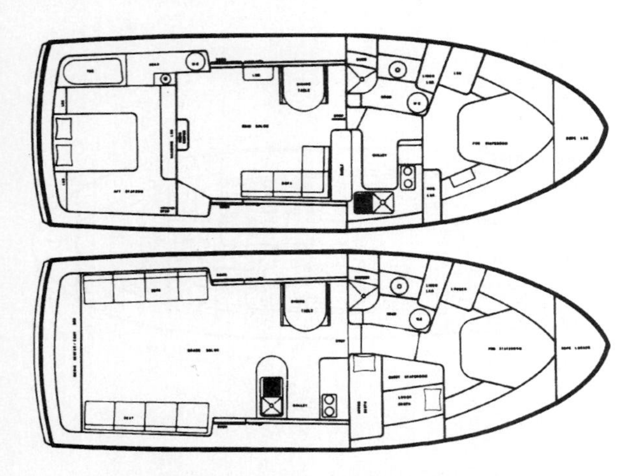

See page 247 for pricing information.

MAINSHIP 47 MOTOR YACHT

SPECIFICATIONS

Length........................46'10"
LWL...............................NA
Beam............................15'5"
Draft.............................3'10"
Weight.....................44,000#
Clearance.....................19'6"
Water.....................200 gals.
Fuel........................600 gals.
Hull Type........Modified-V
Deadrise Aft..................12°
Designer..........Mike Peters
Production...1990–Current

Introduced in 1990 and updated in 1993 with a new layout, the Mainship 47 is one of the least expensive motor yachts in her class. With her distinctive appearance (note the unusual window treatment and aggressive bridge overhang), the Mainship 47 is a modern motor yacht design with a good deal of standard equipment packed into her affordable price. Her modified-V hull is cored from the waterline up and comes with a moderate 12° of transom deadrise. Originally offered with three staterooms, the current floorplan has two staterooms with the galley and dining area *aft* in the salon and a small office/communications center forward. The salon is huge and so is the forward stateroom, but the aft cabin (and the aft deck above) is small for a boat this size. A utility room housing the washer/dryer and workbench is forward of the engine room. Additional features include an underwater exhaust system, foredeck sunpad, and light oak interior woodwork. Standard 485-hp 6-71s will cruise at 22 knots and reach a top speed of about 25 knots. ❑

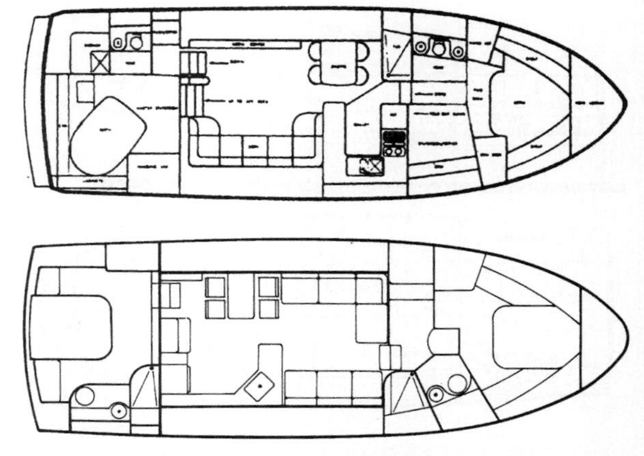

See page 247 for pricing information.

MARINE TRADER 34 DC (EARLY)

SPECIFICATIONS

Length	33'6"	Water	150 gals.
LWL	30'3"	Fuel	300 gals.
Beam	11'9"	Hull Type	Semi-Disp.
Draft	3'6"	Designer	Floyd Ayers
Weight	17,000#	Production	1972–92

The Marine Trader 34 Double Cabin is the best-selling small trawler ever imported and sold in the U.S. Built by CHB in Taiwan, there were other importers besides Marine Trader, and she may be recognized on the West Coast as the La Paz, Eagle, or CHB 34 DC. She enjoyed her best years during the fuel crisis of the 1970s when powerboats with big fuel-guzzling engines were out of favor. With only a single diesel engine and a favorable exchange rate, they were inexpensive in spite of a full teak interior and abundant exterior teak trim. They're also solidly built and fuel efficient (2–3 gph at 7 knots) with a standard 120-hp Lehman diesel. Inside, the tri-cabin floorplan includes a double berth in the aft cabin, two heads, and a small salon with a lower helm. Note the passageway from the aft cabin to the cockpit. Used 34s are easily found in most markets, and the prices are very reasonable. Those built before 1975 were constructed with glass-over-plywood decks and house. ❑

See page 247 for pricing information.

MARINE TRADER 34 DC

SPECIFICATIONS

Length	33'6"	Clearance	12'0"
LWL	30'3"	Water	150 gals.
Beam	11'9"	Fuel	300 gals.
Draft	3'6"	Hull Type	Semi-Disp.
Weight	17,200#	Production	1992–Current

The big difference between the new Marine Trader 34 Double Cabin and the previous model is immediately apparent upon boarding: the new version has been completely de-teaked on the outside. No more teak decks, window frames, rails, trim, etc. Maintenance is way down. (Note that the house and decks are now a single-piece fiberglass mold, which should eliminate the deck leakage problems common on the earlier models.) The flybridge has also been slightly redesigned, although the helm layout and seating are essentially unchanged. In other respects, the new 34 DC is very much like the original. The all-teak interior is still arranged with small head compartments fore and aft, and the passageway from the aft stateroom to the cockpit remains. Storage seems slightly improved, and, of course, the engine room is cavernous with just a single engine (twins are available). Note the wide sidedecks and functional mast and boom. Standard power in the new Marine Trader 34 DC is a single 6-cylinder Lehman diesel. She'll cruise efficiently at 7 knots burning just 2–3 gph. ❑

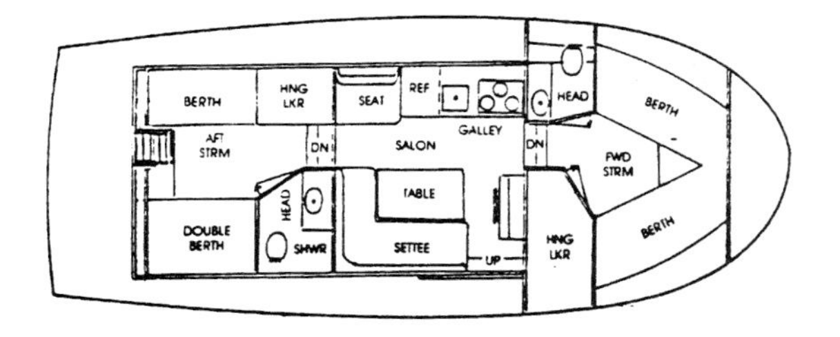

See page 247 for pricing information.

MARINE TRADER 34 SEDAN

SPECIFICATIONS

Length	33'6"	Cockpit	NA
LWL	30'3"	Water	150 gals.
Beam	11'9"	Fuel	300 gals.
Draft	3'6"	Hull Type	Semi-Disp.
Weight	19,600#	Designer	Floyd Ayers
Clearance	12'0"	Production	1973–Current

Upgraded in 1991 with a restyled flybridge and deckhouse and an updated interior, the Marine Trader 34 Sedan has long been a popular boat with economy-minded cruisers. Except for the first couple of years (when the deckhouse was glass-over-plywood), the construction is all fiberglass. The original floorplan has V-berths forward and the galley forward in the salon. The current layout has a greatly enlarged master stateroom with a walkaround bed, less hanging locker space, and the galley has been moved aft in the salon where it's more convenient to the cockpit. The entire interior is paneled and finished with teak. The sidedecks are wide and secured by raised bulwarks all around the house. Most Marine Trader 34 Sedans have been powered with a single 135-hp Lehman diesel (120-hp in older models), and she'll cruise very efficiently at 7 knots burning only 2–3 gph. The optional 210-hp Cummins can deliver up to 12 knots top. The cruising range of this boat can exceed 750 miles. ❑

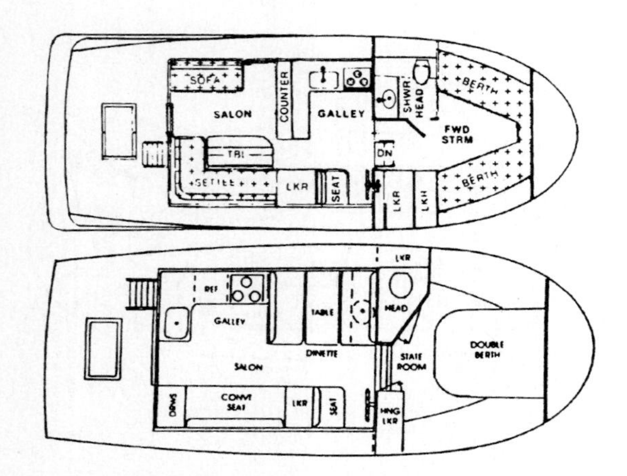

See page 247 for pricing information.

MARINE TRADER 36 DOUBLE CABIN

SPECIFICATIONS

Length	36'0"	Clearance	NA
LWL	NA	Water	150 gals.
Beam	12'2"	Fuel	400 gals.
Draft	3'6"	Hull Type	Semi-Disp.
Weight	21,000#	Production	1975–93

The Marine Trader 36 is your generic low-priced, double-cabin Taiwan trawler with a complete teak interior and full walkaround teak decks. Like most of the popular imported trawler designs, the MT 36 is built on a semi-displacement hull with hard aft chines, a flared bow, and a long prop-protecting keel. Her success can be traced to her attractive price, economical operation, and a practical two-stateroom interior. The original layout was modified in 1977 with changes in the aft cabin and in the placement of the galley. A lower helm is standard, and there's also a companionway in the aft stateroom for aft deck access. The boat has good storage for her size, and the engine room is very large. Other features of the Marine Trader 36 Double Cabin include a mast and boom, teak swim platform, wide sidedecks, and a distinctive teak foredeck hatch. Most all were fitted with a single Lehman diesel (120 or 135-hp), although twins have been available. At an 8-knot cruising speed the range is around 750 miles. ❑

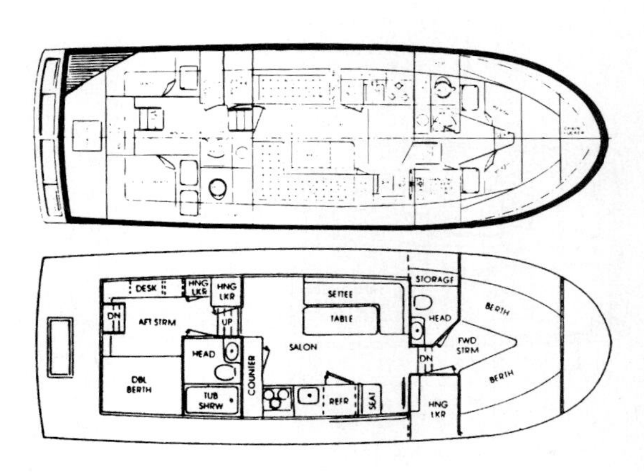

See page 247 for pricing information.

MARINE TRADER 36 SEDAN

SPECIFICATIONS

Length	36'0"	Water	150 gals.
LWL	NA	Fuel	350 gals.
Beam	12'2"	Hull Type	Semi-Disp.
Draft	3'6"	Designer	F. Ayers
Weight	21,000#	Production	1975–93

People buy boats like the Marine Trader 36 Sedan because they like the practicality of a sedan-style boat and because they're on a budget. For years Marine Trading International staked their success on importing inexpensive Taiwan trawlers, and the strategy has obviously worked since they have been successfully selling Taiwan-built boats for better than 25 years now. The Marine Trader 36 is a straightforward sedan cruiser with a classic trawler profile and sensible accommodations. The layout includes two staterooms with the owner's cabin to port rather than forward. There's a tub/shower in the head, and the galley can be aft or forward in the salon. A sliding door at the lower helm provides quick access to the decks, and everything inside is finished with teak. Note the traditional bridge overhangs that shelter the sidedecks and the aft deck. With the extended flybridge, there's space aft of the helm for a small dinghy. Most 36 Sedans have been powered with a single 135 or 120-hp Lehman diesel, although twin screw applications are common ❑

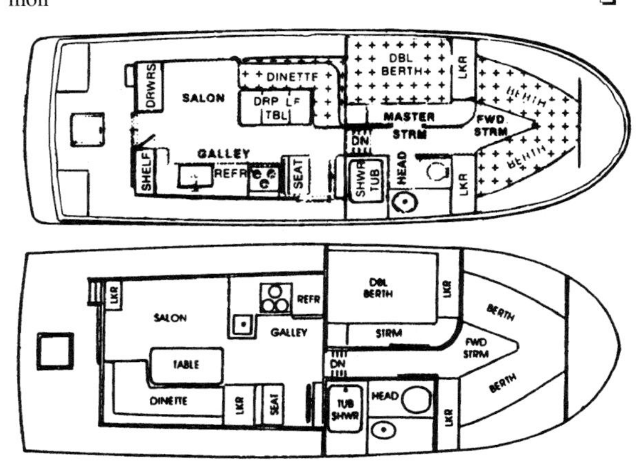

See page 247 for pricing information.

MARINE TRADER 36 SUNDECK

SPECIFICATIONS

Length	36'0"	Water	150 gals.
LWL	NA	Fuel	350 gals.
Beam	12'2"	Hull Type	Semi-Disp.
Draft	3'6"	Designer	Floyd Ayers
Weight	19,000#	Production	1985–93

A popular boat, the Marine Trader 36 Sundeck is an inexpensive trawler-style design with a balanced profile to go with her traditional appearance. She features a conventional two-stateroom layout with the galley and dinette down a few steps from the salon. The forward stateroom is fitted with a space-saving offset double berth, while the master stateroom features a walkaround double as well as a tub/shower in the head. The interior of the Marine Trader 36 Sundeck is completely finished in teak with well-crafted paneling and cabinetry applied throughout. With the galley and dinette down, the main salon is quite open and includes a sliding deck access door at the helm. Outside, there's space on the full-width sundeck for a few chairs, and there's additional seating for five on the flybridge. Engine choices (there have been several over the years) for the 36 Sundeck include twin 90-hp or 135-hp Lehman diesels or a single 210-hp Cummins diesel. All will provide an economical 7–8 knot cruising speed. ❑

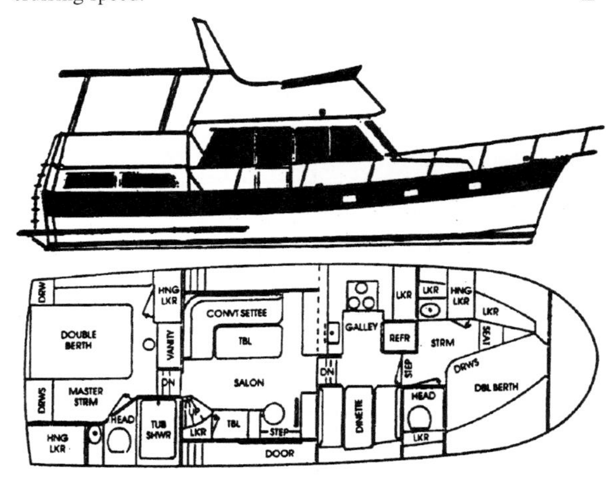

See page 248 for pricing information.

MARINE TRADER 38 DC

SPECIFICATIONS

Length	38'0"	Water	250 gals.
LWL	NA	Fuel	300 gals.
Beam	12'10"	Hull Type	Semi-Disp.
Draft	4'0"	Designer	Floyd Ayers
Weight	22,000#	Production	1980–Current

For those attracted to the traditional profile of a trunk cabin trawler, the Marine Trader 38 Double Cabin is one of only a handful of such designs still available. Although the market for trawlers has clearly declined in recent years, the Marine Trader 38 continues to sell thanks to her moderate price and economical operating costs. Construction is solid fiberglass, and the deck overlay, handrails, pulpit, and window frames are all teak. Inside, the spacious main salon features an in-line galley to starboard, a sliding deck access door at the helm, convertible settee, and a teak high-low table to port. (A galley-down floorplan is also offered.) As is the case with most Asian trawlers, the interior is completely finished in rich teak woodwork. Nearly all have been powered by a single 120/135-hp Lehman diesel and fuel consumption at a leisurely 7–8 knots is only 3 gph. With the optional twin 210-hp Cummins diesels the Marine Trader 38 DC will cruise at 12 knots and reach a top speed of about 15 knots. ❑

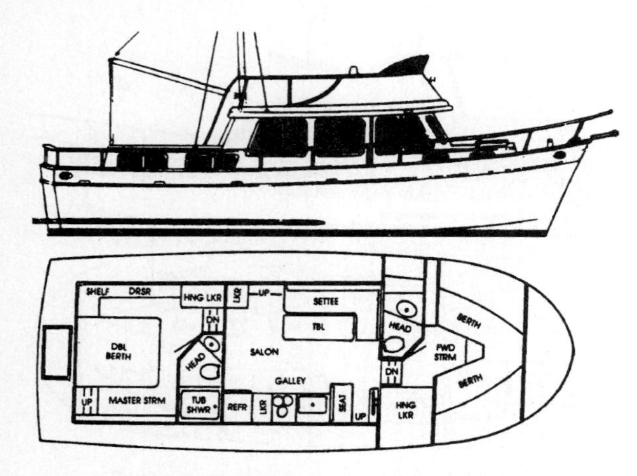

See page 248 for pricing information.

TRADEWINDS 39 SUNDECK

SPECIFICATIONS

Length	38'10"	Fuel	500 gals.
Beam	14'4"	Hull Type	Modified-V
Draft	3'1"	Deadrise Aft	NA
Weight (Approx.)	25,000#	Designer	Lien Hwa
Water	300 gals.	Production	1989–93

Because nearly all under-40' double cabin motor yachts cause a certain amount of eye strain, one can hardly fault the Tradewinds 39 Sundeck for lacking the modern profile found in larger Tradewinds models. (The problem is always the hardtop; owners love them, and rightly so, but they distort the profile of most small motor yachts.) Like all Tradewinds models, the 39 is built for Marine Trading Int'l. in Taiwan by Lien Hwa—a heavyweight among Taiwanese manufacturers. Her modified-V hull is constructed of solid fiberglass with balsa coring used in the deck and superstructure. Inside, the conventional galley-down floorplan of the Tradewinds 39 is designed for comfortable family cruising and features double berths in both staterooms. The interior gets the full Asian teak treatment, and large cabin windows admit plenty of outside lighting. With over 14 feet of beam, the accommodations are quite roomy in spite of the wide sidedecks. Twin 135-hp Lehman diesels are standard. Among several engine options, the popular 225-hp Lehmans will cruise the Tradewinds 39 around 14–15 knots. ❑

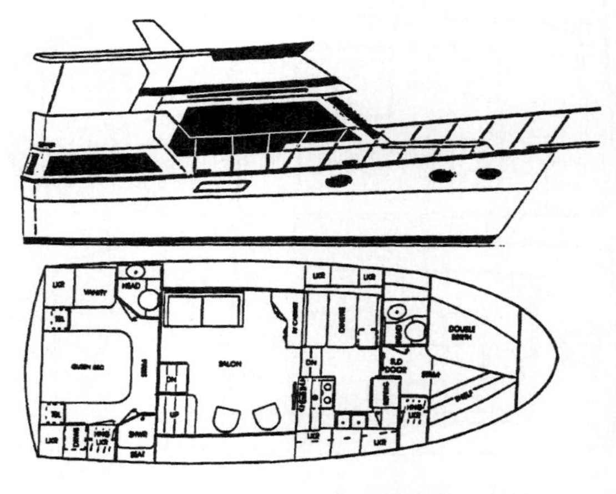

See page 248 for pricing information.

MARINE TRADER 40 DOUBLE CABIN

SPECIFICATIONS

Length	40'0"	Water	250 gals.
LWL	36'7"	Fuel	400 gals.
Beam	13'8"	Hull Type	Semi-Disp.
Draft	4'0"	Designer	Floyd Ayers
Weight	30,000#	Production	1974–86

Although she remained in production until 1986, the Marine Trader 40 Double Cabin enjoyed her greatest popularity during the fuel crisis years of the late 1970s. Trawler sales were on a roll back then, as the public turned toward the efficiency and traditional styling inherent in such a design. The Marine Trader 40 has the proper trawler "look" with enough teak trim and interior woodwork to sink a lesser boat. Plus, she was very affordable when compared to similar imported trawlers available at the time. Indeed, used models can often be found on today's market at equally attractive prices. Inside, her teak-on-teak interior is arranged in the normal double-cabin configuration with accommodations for six. Engine room access is very good, storage is excellent, the deck hardware is sturdy, but the finish work is not very impressive. A single 120-hp Lehman diesel is the usual power, and at 7–8 knots she burns about 3 gph. Twin-engine models will cruise about a knot faster and reach a top speed of 11–12 knots. ❑

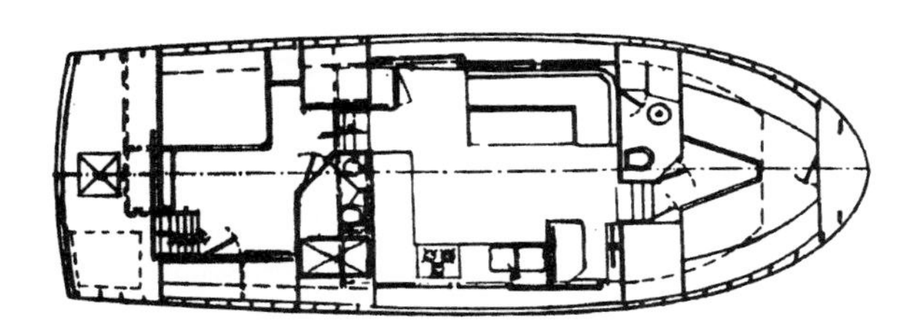

See page 248 for pricing information.

MARINE TRADER 40 SEDAN

SPECIFICATIONS

Length	39'8"	Water	200 gals.
LWL	36'5"	Fuel	400 gals.
Beam	13'8"	Hull Type	Semi-Disp.
Draft	4'0"	Designer	Floyd Ayers
Weight	24,200#	Production	1978–86

For those seeking the wide open layout of a sedan floorplan, the Marine Trader 40 can make a lot of sense. A good-looking boat with a salty profile, construction is solid glass, although some of the earliest models were built with a glass-over-plywood deckhouse. The MT 40 Sedan has a traditional two-stateroom, full-teak interior layout with a double berth in the master stateroom and a tub/shower stall in the head. All of these boats have a lower helm station with a sliding deck access door. The galley is very large and takes up about half of the salon space. Outside, bridge overhangs provide weather protection to the full walkaround decks (all of which are covered with a teak overlay). With her long deckhouse, the extended flybridge can easily accommodate a dinghy aft of the helm. An economical boat to operate, standard power for the Marine Trader 40 Sedan was a single 120-hp Lehman diesel (7–8 knots cruise), and twin engines (9–10 knot cruise) were optional. ❑

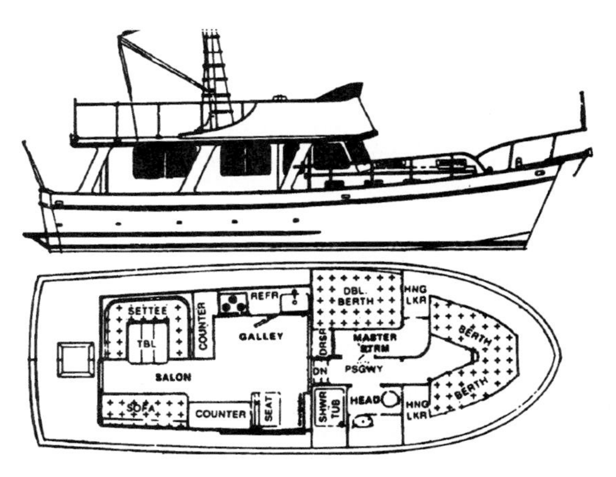

See page 248 for pricing information.

MARINE TRADER 40 SUNDECK MY

SPECIFICATIONS

Length	39'4"	Clearance	NA
LWL	NA	Water	250 gals.
Beam	12'11"	Fuel	350 gals.
Draft	4'0"	Hull Type	Semi-Disp.
Weight (Approx.)	25,000#	Production	1986–Current

A popular boat, the Marine Trader 40 Sundeck is a good-looking Taiwan import with simulated lapstrake hullsides, a rakish bridge overhang, and a solid fiberglass semi-displacement hull design. Her standard two-stateroom, galley-down floorplan is arranged with double berths in both cabins, an open salon with a serving counter overlooking the galley, and a lower helm to starboard with a sliding deck access door. The entire interior is finished with the same good-quality teak woodwork found in most Asian imports regardless of price. The master stateroom is surprisingly large and includes a tub/shower in the head and plenty of storage and locker space. Additional features include a large aft deck with teak sole, wide sidedecks, hardtop, teak handrails, and a teak swim platform. Twin 135-hp Lehman diesels (a single Lehman is standard) will cruise the Marine Trader 40 at an economical 7–8 knots burning only 5–6 gph with a top speed of about 10 knots. Note that the Marine Trader 44 Yachtfish is the same boat with a cockpit extension. ❑

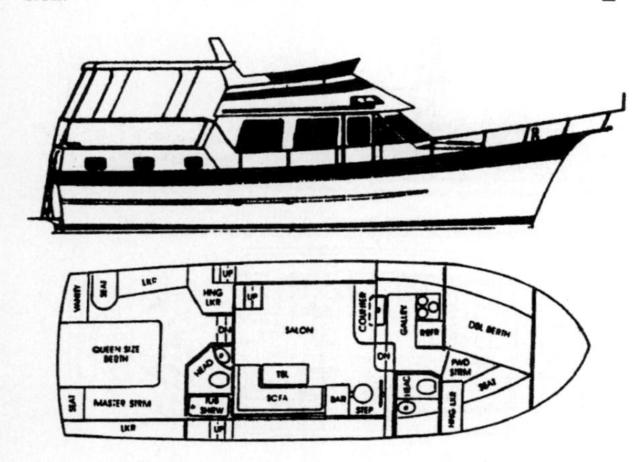

See page 248 for pricing information.

MARINE TRADER 42 SEDAN

SPECIFICATIONS

Length	42'0"	Water	250 gals.
LWL	NA	Fuel	300 gals.
Beam	12'10"	Hull Type	Semi-Disp.
Draft	4'0"	Designer	MTI
Weight	22,000#	Production	1987–Current

Sedan trawlers—with their deck-level salons and walkaround sidedecks—are practical and comfortable cruisers for those who don't require the luxury of a private aft stateroom. A handsome boat with a classic trawler profile, the 42 Sedan features an all-teak interior with two staterooms forward and a deckhouse galley. The master stateroom is quite spacious and includes a walkaround double berth, while the small guest cabin has over/under single berths. There's a deck access door at the lower helm, and the head is fitted with a tub/shower. The cockpit is extremely spacious and provides enough square footage for some light-tackle fishing pursuits. Additional features include a large flybridge with room for an inflatable, protective bridge overhangs for the decks, teak decks and cockpit sole, and a swim platform. Twin 135-hp Lehman diesels will cruise the Marine Trader 42 Sedan at an economical 8 knots and deliver a top speed of around 11 knots. Note that the Marine Trader 38 Sedan is the same boat with a smaller cockpit. ❑

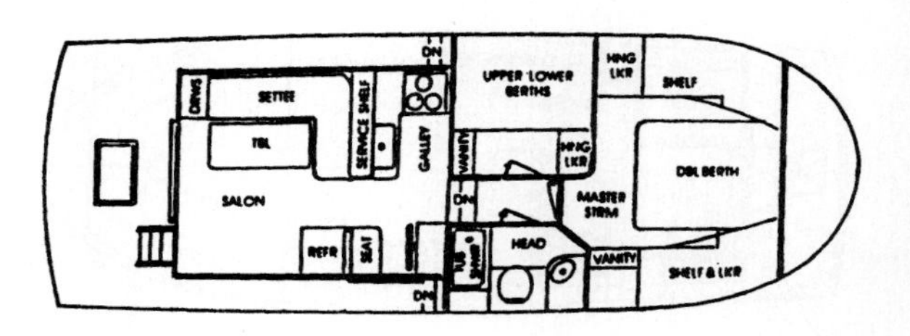

See page 248 for pricing information.

LABELLE 43 MOTOR YACHT

SPECIFICATIONS

Length	43'0"	Water	250 gals.
LWL	NA	Fuel	450 gals.
Beam	14'2"	Hull Type	Semi-Disp.
Draft	4'2"	Designer	Bestway
Weight (Approx.)	28,000#	Production	1983–1988

A distinctive design with her sweeping sheer and low-freeboard hull with simulated lapstrake grooves, the LaBelle 43 MY was built by the Bestway yard in Taiwan—an oddly named company still actively marketing boats in the U.S. under its own name. She's built on a solid fiberglass semi-displacement hull with moderate beam, a long keel, and generous flare at the bow. Most of the LaBelle 43s were sold with a wide open two-stateroom interior layout with the galley and dinette down. The interior is completely finished with teak paneling and cabinetry, and a lower helm and deck access door were standard. Note the tub/shower in the aft head and the absence of a stall shower forward. Outside, the LaBelle's raised sundeck has a teak sole and abundant entertaining space. The sidedecks are quite wide, and the flybridge is arranged with plenty of guest seating. Twin 165-hp Volvo diesels will cruise the 43 LaBelle at an economical 10 knots with a top speed of 13–14 knots. ❑

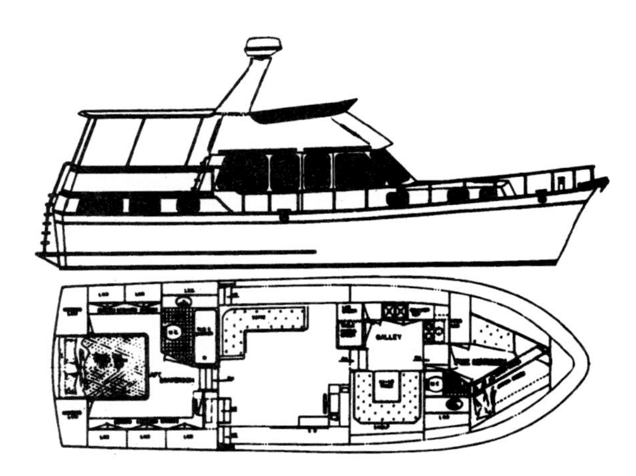

See page 249 for pricing information.

TRADEWINDS 43 MOTOR YACHT

SPECIFICATIONS

Length	42'6"	Fuel	500 gals.
Beam	14'11"	Hull Type	Modified-V
Draft	3'10"	Deadrise Aft	NA
Weight	27,500#	Designer	Lien Hwa
Water	320 gals.	Production	1986–Current

Unlike the earlier trawler-style designs that made the Marine Trading Int'l. name synonymous with low-cost Taiwan imports, the Tradewinds 43 is one of a new series of Tradewinds models with contemporary motor yacht styling and planing-speed performance. She's built by Lien Hwa on a modified-V bottom with a solid fiberglass hull and balsa coring in the deck and superstructure. Below, the all-teak interior of the Tradewinds 43 is arranged with the galley and dinette down. The result is a spacious and wide open salon with large wraparound windows and plenty of headroom. (Note that the forward head can only be accessed from the stateroom.) Standard features include a hardtop, radar arch, swim platform, and bow pulpit. With the twin 210-hp Cummins diesels, the cruising speed of the Tradewinds 43 MY is about 14 knots, and the top speed is around 17 knots. The larger 275-hp Lehmans will cruise at 16–17 knots and reach 20 knots wide open. The Tradewinds 47 YF is the same boat with a small cockpit. ❑

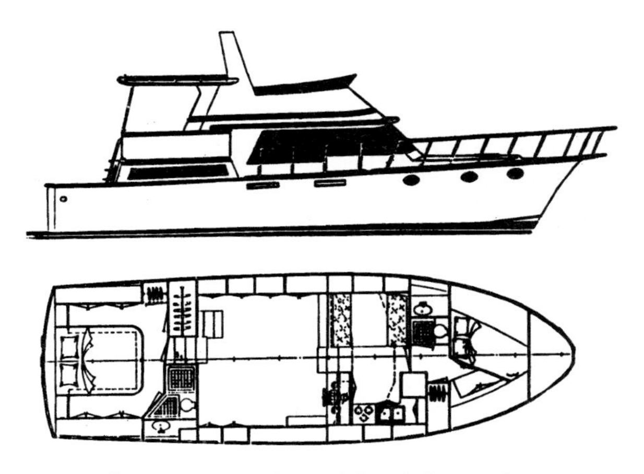

See page 249 for pricing information.

MARINE TRADER 44 TRI-CABIN

SPECIFICATIONS

Length	43'6"	Water	250 gals.
LWL	38'8"	Fuel	500 gals.
Beam	14'4"	Hull Type	Semi-Disp.
Draft	4'2"	Designer	Floyd Ayers
Weight	33,000#	Production	1977–88

The Marine Trader 44 Tri-Cabin comes about as close as you can get to describing the quintessential low-price Taiwan trawler. A popular boat, she may not have been the latest word in top-quality finish, and the wiring was sometimes a problem, but her lush all-teak interior, cheap price, and rugged fiberglass construction were enough to win over many of her detractors. Equally important, the 44 Tri-Cabin has a sensible layout. Two floorplans were offered over the years: the two-stateroom version was the most popular, but many were sold with the optional three-stateroom layout. There are deck access doors port and starboard in the salon. (Note the aft deck companionway door in the master stateroom.) Outside, the full walkaround decks are covered with a teak overlay, and the window frames, handrails, pulpit, and swim platform are all teak. Much of the hardware is bronze. The mast and boom adds much to her trawler profile. Single or twin 120-hp Lehman diesels will cruise the 44 Tri Cabin economically at 8–9 knots. ❑

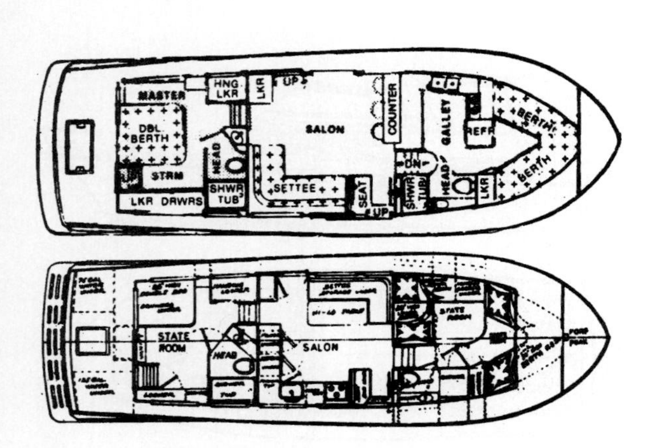

See page 249 for pricing information.

MARINE TRADER 46 DOUBLE CABIN

SPECIFICATIONS

Length	46'2"	Water	220 gals.
Beam	14'7"	Fuel	600 gals.
Draft	3'8"	Hull Type	Semi-Disp.
Weight	35,000#	Designer	MTI
Clearance	14'2"	Production	1990–Current

Marine Trading International just about owns the low-price end of the Taiwan trawler market in the U.S. Few importers continue to sell trawler-style yachts these days, but the new Marine Trader 46 DC is a reminder that some demand still exists. Actually, the 46 only looks like a trawler—given the proper horsepower, she'll achieve true planing speed thanks to her semi-displacement, hard-chined hull. Inside, the full teak interior is arranged with centerline island berths fore and aft and a wide-open salon with the galley and dinette down. A lower helm is standard. Port and starboard salon doors provide easy access to the decks, and storage is excellent. The full walkaround sidedecks, afterdeck, and flybridge are all teak planked, but the window frames are aluminum. Twin Lehman 135-hp diesels are standard but won't achieve planing speeds. The optional twin 210-hp Cummins diesels can cruise the Marine Trader 46 DC economically at 11–12 knots and reach a top speed of around 15 knots. ❑

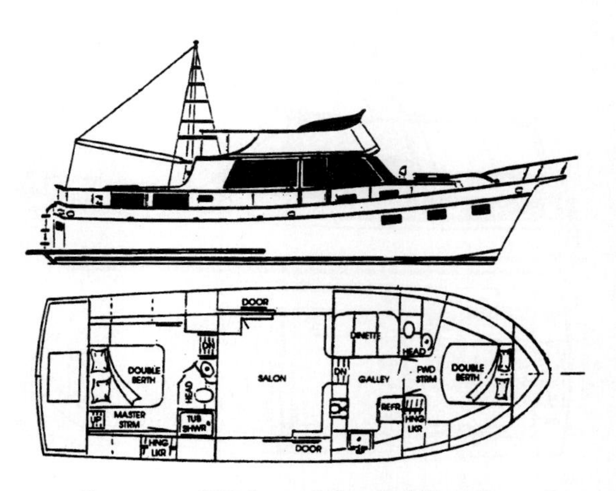

See page 249 for pricing information.

TRADEWINDS 47 MOTOR YACHT

SPECIFICATIONS

Length	46'6"	Fuel	500 gals.
Beam	14'11"	Hull Type	Modified-V
Draft	3'6"	Deadrise Aft	NA
Weight	30,500#	Designer	Lien Hwa
Clearance	NA	Production	1986–Current
Water	320 gals.		

The Tradewinds 47 MY is basically a lengthened version of the smaller Tradewinds 43. Hull construction is solid fiberglass, and a shallow keel provides increased stability underway. Until 1989, the standard floorplan had three staterooms with the galley down and the second guest cabin aft of the salon. The current two-stateroom interior is arranged with the galley and dinette down (a third stateroom can be ordered at the expense of the dinette). The lack of a stall shower in the forward head is unfortunate in a boat this size. A lower helm is standard as is a deck access door—an item seldom found on modern motor yachts. Standard features include a roomy afterdeck with wing doors, radar arch, and a full teak interior. Twin 135-hp Lehman diesels are standard. Optional 300-hp Cummins will attain a cruising speed of around 14 knots and a top speed of 18–19 knots. Note that the Tradewinds 47 was restyled in 1989 with a new superstructure and flybridge and without the original simulated lapstrake hullsides. ❑

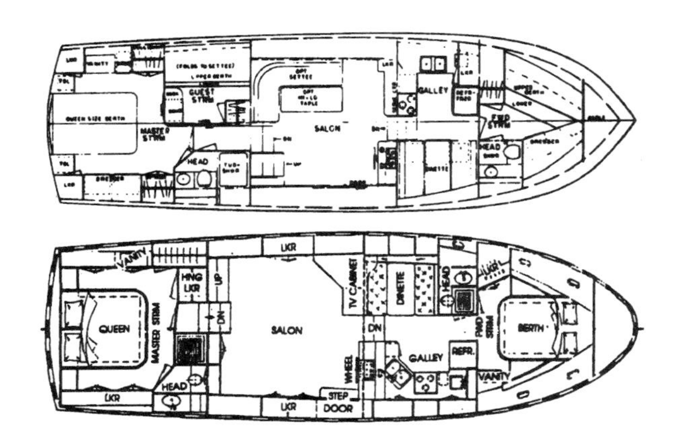

See page 249 for pricing information.

MARINE TRADER 49 PILOTHOUSE

SPECIFICATIONS

Length	48'6"	Clearance	NA
LWL	NA	Water	375 gals.
Beam	15'0"	Fuel	700 gals.
Draft	4'6"	Hull Type	Semi-Disp.
Weight	46,000#	Production	1979–93

A salty design with a prominent bow and high bulwarks around the house, the Marine Trader 49 is a classic long-range pilothouse trawler design. Construction is all fiberglass with hard chines aft, relatively flat aftersections, and simulated planking in the hullsides. Inside, the floorplan is arranged in the conventional pilothouse manner with the galley aft on the salon level and the master stateroom below the raised wheelhouse. There's a watch berth in the pilothouse, and the midships owner's cabin includes a tub in the adjoining head. The entire interior is teak and so are the decks. Additional features include a functional mast and boom and protective bridge overhangs. Price-wise, these were fairly inexpensive boats, and the detailing is not overly impressive. Note that the Marine Trader 49 Sundeck (1983–90) was built on the same hull but with a full-width salon and no sidedecks. A few 49s were powered with a single diesel, but twin 135-hp Lehmans (8-knot cruise) were standard. Optional 160-hp Perkins diesels cruise around 10 knots. ❑

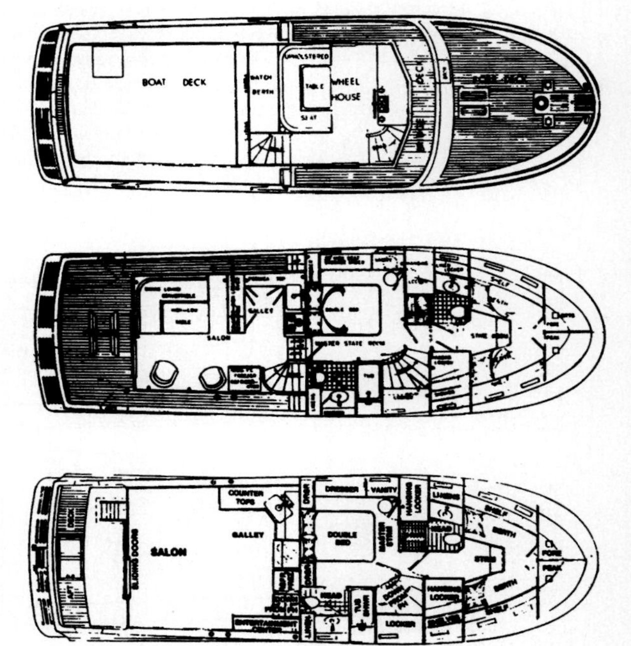

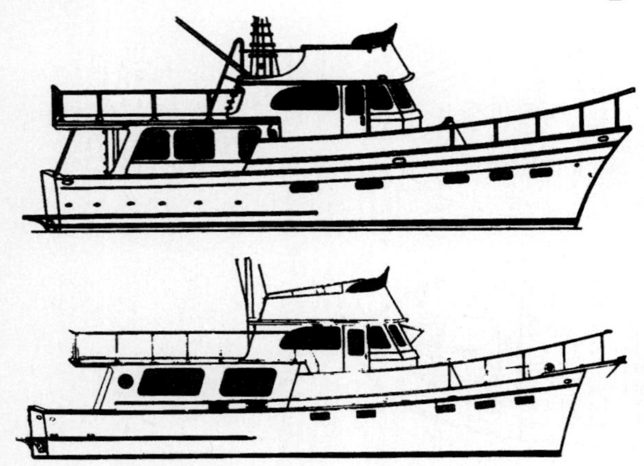

See page 249 for pricing information.

MARINE TRADER 50 MOTOR YACHT

SPECIFICATIONS

Length	50'0"	Cockpit	NA
LWL	44'0"	Water	380 gals.
Beam	15'5"	Fuel	750 gals.
Draft	4'8"	Hull	Disp.
Weight	46,000#	Designer	Floyd Ayers
Clearance	16'6"	Production	1979–93

The largest trawler-style yacht ever offered by Marine Trader, the 50 MY is a moderately priced long-range cruiser with comfortable accommodations and traditional styling. She's built at the Lien Hwa shipyard in Taiwan, which has turned out some large production motor yachts over the years. The standard Marine Trader 50 has full walkaround sidedecks and a covered aft deck. In 1985, she was joined by the 50 Wide Body model with an enlarged, full-width, and extended salon. Both versions share the same three-stateroom floorplan on the lower level, and a convenient day head is added in the salon of the Wide Body model. The engines are located in separate compartments on either side of the central corridor leading to the aft cabin. Traditional teak paneling and cabinetry are used extensively throughout the yacht's interior. A comfortable (if slow) offshore cruiser, standard 135-hp Lehman diesels will cruise around 8–9 knots. With 750 gallons of fuel capacity, the cruising range of the Marine Trader 50 Motor Yacht can exceed 1,000 miles. ❑

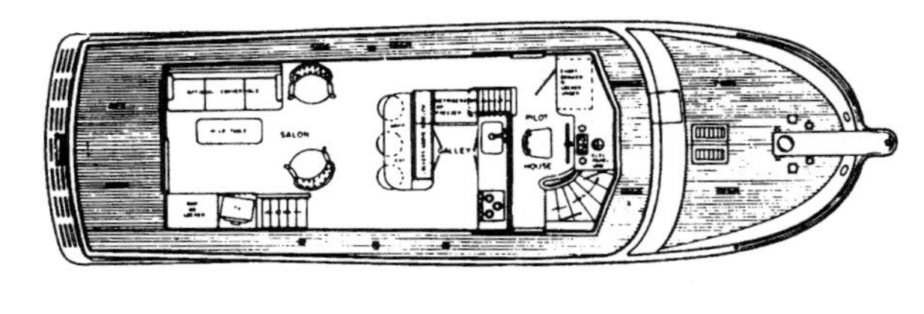

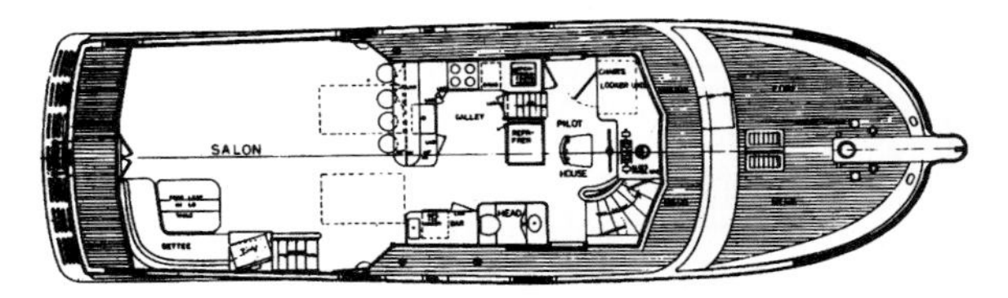

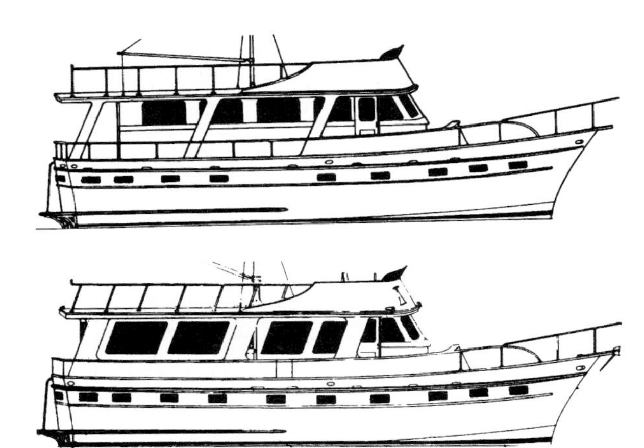

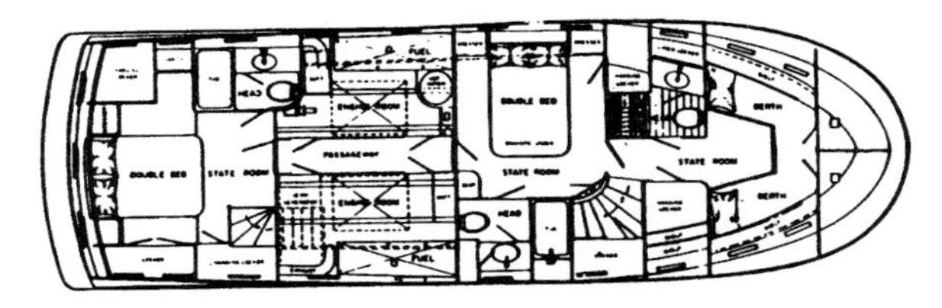

See page 249 for pricing information.

MARINE TRADER MED 14 METER

SPECIFICATIONS

Length	48'6"	Clearance	12'4"
LWL	NA	Water	160 gals.
Beam	14'5"	Fuel	475 gals.
Draft	3'11"	Hull Type	Modified-V
Weight	36,450#	Production	1991–Current

Built in Argentina, the Euro-style Med Yacht 14 Meter is a big departure from the Taiwan trawlers and motor yachts that have always characterized past imports from Marine Trading International. She definitely has the right Mediterranean *look* with her streamlined, low-profile appearance, oval ports, and very attractive reverse swim platform. The 14 Meter is built of solid fiberglass on a modified-V hull with moderate beam and a fairly steep 15° of deadrise at the transom. Her sedan-style layout is available with two staterooms and the galley down (most popular to date) or with three staterooms and the galley in the salon. Two heads are included with either floorplan, but only one has a separate stall shower—something of a disapointment in a boat this size. Additional features include a large flybridge (note the extended cockpit overhang) with wraparound venturi, huge sunpad on the foredeck, low-profile arch, and roomy cockpit. With optional Detroit 485-hp 6-71 diesels, the Med 14 Meter will cruise at 21 knots and reach a top speed of 23–24 knots. ❑

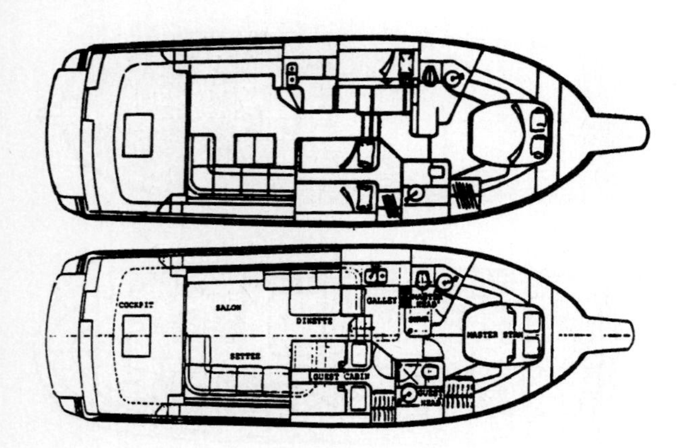

See page 249 for pricing information.

MONK 36 TRAWLER

SPECIFICATIONS

Length	36'0"	Water	120 gals.
LWL	NA	Fuel	320 gals.
Beam	13'0"	Hull Type	Semi-Disp.
Draft	4'0"	Designer	Ed Monk
Weight	18,000#	Production	1982–Current

The Monk 36 gained popularity in the 1980s as a durable and affordable Taiwan import with a classic trawler profile and a traditional all-teak interior. Construction is solid fiberglass, and the hull features a skeg-protected prop. Early models came with plenty of exterior teak—decks, window frames, hatches, etc.—but in current models only the handrails are teak. The galley-up floorplan of the Monk 36 has remained essentially unchanged over the years. The latest layout eliminates the tub in the aft head and moves the bed away from the wall in the aft stateroom—a big improvement. Outside, raised bulwarks and a high rail make movement around the house safe and secure. The original 120/135-hp Lehman diesel has recently been replaced with the 120-hp Perkins as standard power. Either way, the cruising speed is 7–8 knots (2.5 gph), and the range is about 800 miles. Note that in 1991 the molds were purchased by North Sea Yachts, and production moved to Nova Scotia. Over 175 have been built over the years. ❑

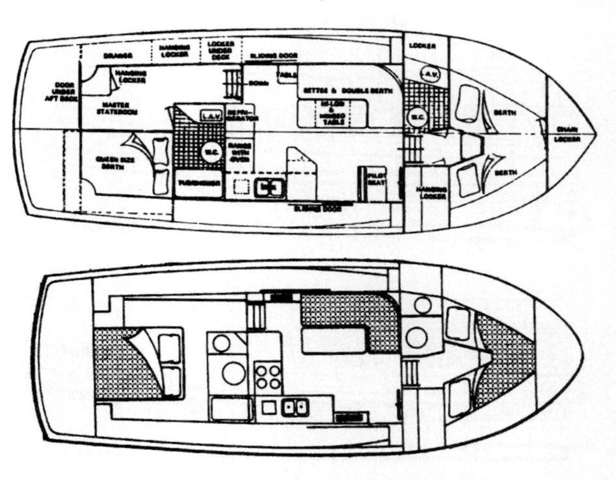

See page 249 for pricing information.

NORDHAVN 46

SPECIFICATIONS

Length	45'9"	Cockpit	NA
LWL	38'4"	Water	240 gals.
Beam	15'5"	Fuel	800 gals.
Draft	5'0"	Hull Type	Disp.
Weight	48,320#	Designer	J. Leishman
Clearance	26'0"	Production	1989–Current

The Nordhavn 46 is a top-quality Taiwan import of a type seldom seen in the production market. Indeed, she's a unique and highly practical pilothouse design with a ballasted displacement hull and the profile of a North Sea workboat. The layout of this go-anywhere offshore vessel is quite innovative. The pilothouse is amidships with full 360° helm visibility. The front windshield is canted, and a stout bridge coaming protects the house from breaking seas—a *Portugese bridge* in nautical terms. Inside, a companionway from the salon leads down to the midships owner's stateroom, while the forward guest stateroom is reached separately from the pilothouse. The glasswork, joinery, hardware, and fixtures used throughout the Nordhavn are quite exceptional. Additional features include a dry exhaust system, keel-cooler (for engine cooling), a single sidedeck (the salon extends to the portside rail), and functional mast and boom. Powered with a single 143-hp Lugger diesel, the Nordhavn 46 will cruise at 8 knots at 3 gph for a range of 2,000+ miles. Definitely a nice piece of work. ❑

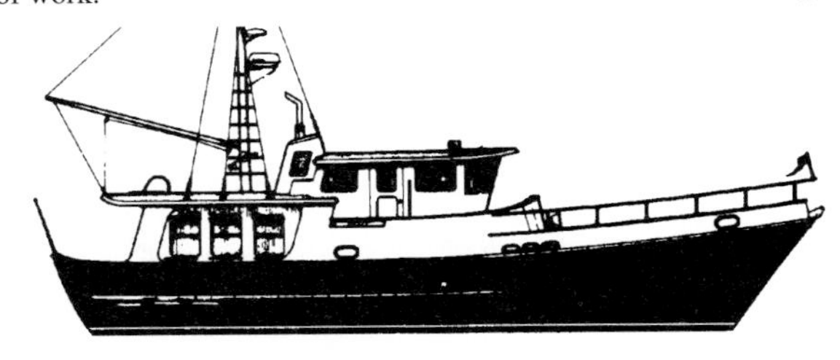

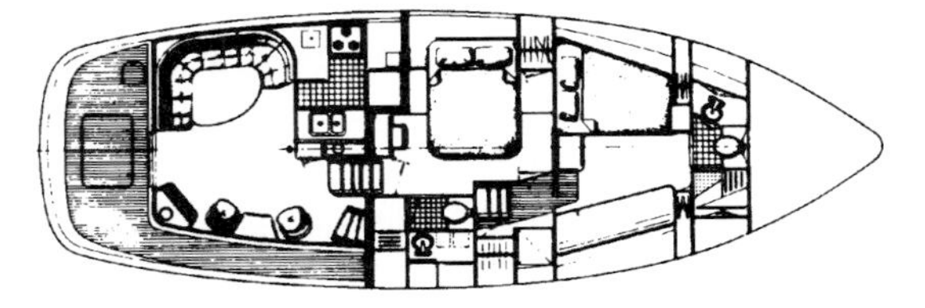

See page 249 for pricing information.

NORDIC 32 TUG

SPECIFICATIONS

Length	32'2"	Cockpit	NA
LWL	32'0"	Water	100 gals.
Beam	11'0"	Fuel	115 gals.
Draft	3'0"	Hull Type	Semi-Disp.
Weight	13,500#	Designer	L. Senour
Clearance	10'2"	Production	1985–Current

The Nordic 32 has the same tugboat profile and semi-displacement, full-keeled hull as the smaller Nordic 26. Built in America, these are true pilothouse designs with an elevated helm and excellent visibility. Both the 26 and 32 share similar interior layouts, but where the 26 is a little tight in some areas, the accommodations aboard the Nordic 32 are quite roomy. This is especially true in the salon where the large windows and traditional teak woodwork combine to produce a charming Old World ambiance. Forward (through the pilothouse), a stall shower is found in the head, and the stateroom has an offset double berth. Because the salon uses nearly the full width of the hull's beam, the sidedecks are not walkarounds. The construction of the Nordic 32 Tug is substantial, and she is fully capable of extended offshore cruising. Several diesel engines (100 to 120-hp) have been standard over the years. With the optional 210-hp Cummins diesels, she'll cruise efficiently at a brisk 13 knots and reach a 16-knot top speed. Over 60 have been built to date. ❑

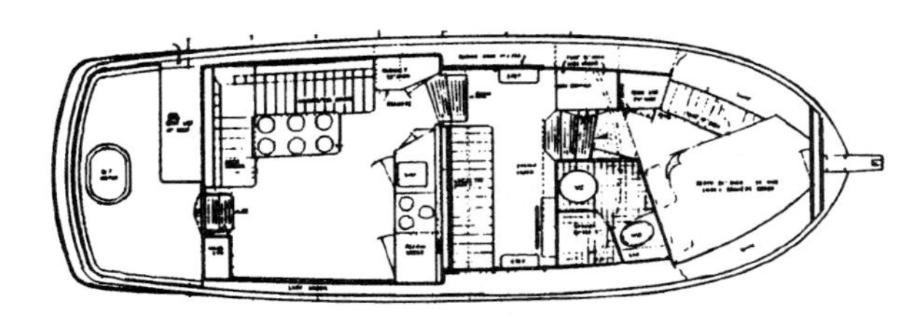

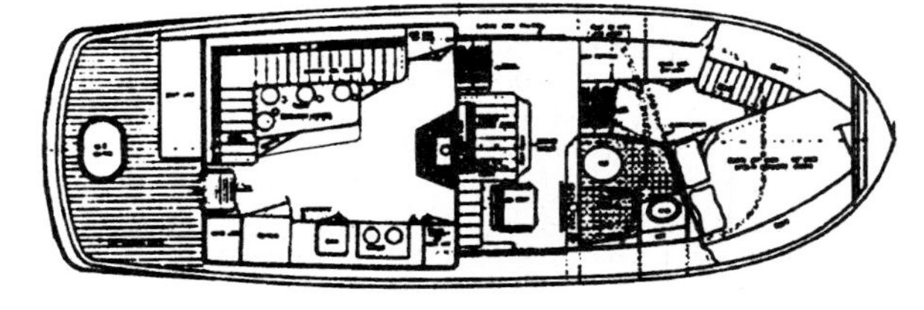

See page 249 for pricing information.

NORDIC 480 MOTOR YACHT

SPECIFICATIONS

Length	48'0"	Water	200 gals.
Beam	15'8"	Fuel	600 gals.
Draft	3'11"	Hull Type	Modified-V
Weight	39,500#	Designer	A. Nordtvedt
Clearance	13'1"	Production	1985–91

The Nordic 480 was built in Bellingham, Washington, on a semi-custom basis. Nordic Yachts wasn't a high-volume builder, but the yard (no longer in business) was known for above-average construction standards. The 480 rides on a relatively wide modified-V hull with balsa coring from the waterline up and a long keel for increased directional stability. While the interiors were often modified to suit the requirements of the buyer, most floorplans were two-stateroom arrangements with a queen berth in the master. The salon width is somewhat limited by the wide walkaround sidedecks, but the overall dimensions are certainly adequate. The grain-matched teak cabinetry and joinerwork are superb. The pilothouse features a sliding deck door and provides direct access to the bridge via a beautiful teak stairway abaft the helm. The larger Nordic 520 MY has the same interior layout as the 480 but with a larger cockpit. Most Nordic 480s were equipped with optional 375-hp Caterpillar diesels which provide a cruising speed of 17–18 knots and a top speed of around 21 knots. ❏

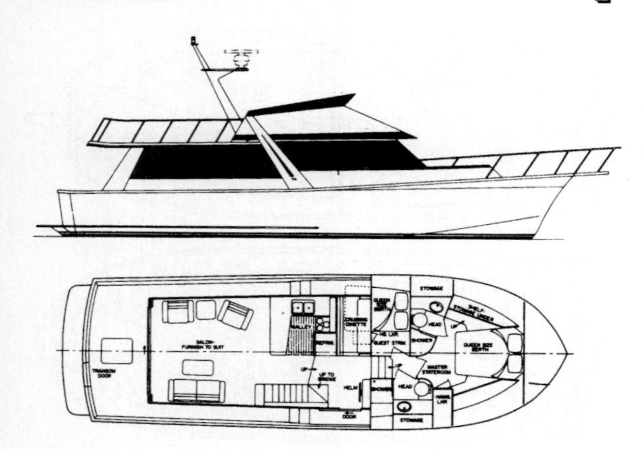

See page 250 for pricing information.

OCEAN 40+2 TRAWLER

SPECIFICATIONS

Length	42'0"	Cockpit	NA
LWL	38'0"	Water	100 gals.
Beam	14'4"	Fuel	450 gals.
Draft	3'6"	Hull Type	Semi-Disp.
Weight	30,000#	Designer	D. Martin
Clearance	12'6"	Production	1978–1980

The 40+2 was Ocean's attempt at building a trawler—a marketing decision no doubt driven by the fuel crisis of the late 1970s and then soaring trawler sales. The hull design is quite unusual: the swim platform is an integral part of the hull (the "+ 2"), and a unique reverse curve below the waterline at the stern is supposed to aid in getting the boat on plane. Sometimes called a *dual mode* hull, the 40+2's bottom is actually a semi-displacement design capable of planing speed performance. Inside, the all-teak interior is arranged with V-berths forward and twin berths aft—a practical layout but dated by today's fixation on centerline double berths. A lower helm and deck door were standard. Additional features include full walkaround decks, a small tub in the aft head, and a functional mast and boom. Most Ocean 40+2 Trawlers were powered with twin 160-hp Perkins diesels. The cruising speed is 12–13 knots, and the top speed is around 15 knots. Fewer than thirty were built during her 3-year production run. ❏

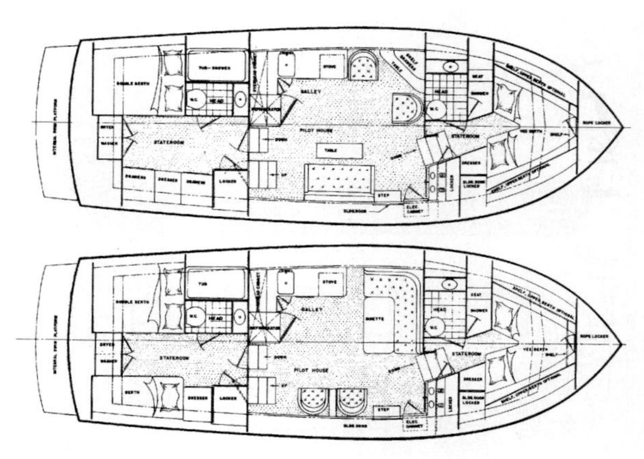

See page 250 for pricing information.

OCEAN 42 SUNLINER

SPECIFICATIONS

Length	42'0"	Water	100 gals.
LWL	38'0"	Fuel	480 gals.
Beam	14'4"	Hull Type	Modified-V
Draft	3'6"	Deadrise Aft	NA
Weight	28,000#	Designer	D. Martin
Clearance	12'0"	Production	1981–85

The Ocean 42 Sunliner was built on the same hull as the Ocean 42 Super Sport with the aft cabin replacing the Super Sport's cockpit. A good-performing family cruiser, she features a full-width aft deck above the full-beam master stateroom below with its walkaround queen berth. V-berths are fitted in the forward cabin, and both head compartments are equipped with separate stall showers. The 42 Sunliner came with few options. Standard equipment included a generator, garbage disposal, washer/dryer, vacuum system, instant hot water, microwave oven, and much more. Like the 42 Super Sport from which she was designed, the engine room in the Sunliner is compact with limited outboard engine access. Interior accommodations are comfortable and attractively finished with an abundance of teak cabinetry and woodwork throughout. Several diesel options were offered during her production run. Popular GM 6-71TIs will cruise around 26 knots and reach a top speed of 29–30 knots. No award-winner in the looks department, the 42 Sunliner is probably the best-looking of the three Sunliner designs. ❑

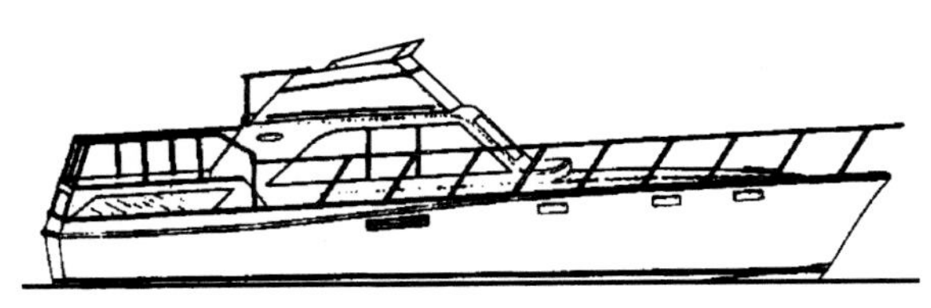

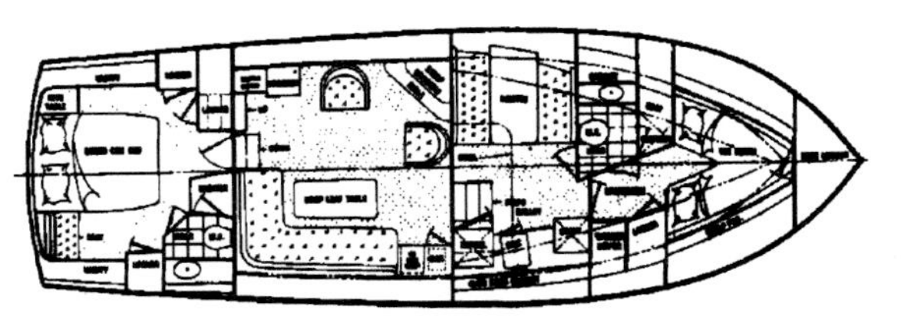

See page 250 for pricing information.

OCEAN 44 MOTOR YACHT

SPECIFICATIONS

Length	44'0"	Water	100 gals.
LWL	38'9"	Fuel	466 gals.
Beam	15'0"	Hull Type	Modified-V
Draft	3'7"	Deadrise Aft	5°
Weight	40,000#	Designer	D. Martin
Clearance	12'0"	Production	1992–Current

Based on the hull of the 42 Super Sport, the Ocean 44 MY is a high-performance motor yacht with a stylish profile and a very innovative floorplan. She's built on a low-deadrise hull with a relatively wide beam, tapered aftersections, and a well-flared bow. Most motor yachts in this size range have two-stateroom floorplans with the galley down, however the Ocean 44 boasts three staterooms—each with a double berth—with the galley and dinette on the upper level. Note that the starboard guest stateroom extends aft below the salon dinette base (at some expense to the engine room dimensions). Both heads are fitted with stall showers, and clever use of the doors provides either of the two forward staterooms with private head access. Additional features include a built-in radar arch, comfortable bridge seating, and a very upscale interior decor. A good performer with optional 485-hp 6-71s (425-hp Cats are standard), the Ocean 44 MY will cruise around 25 knots and reach a top speed of 27–28 knots. ❑

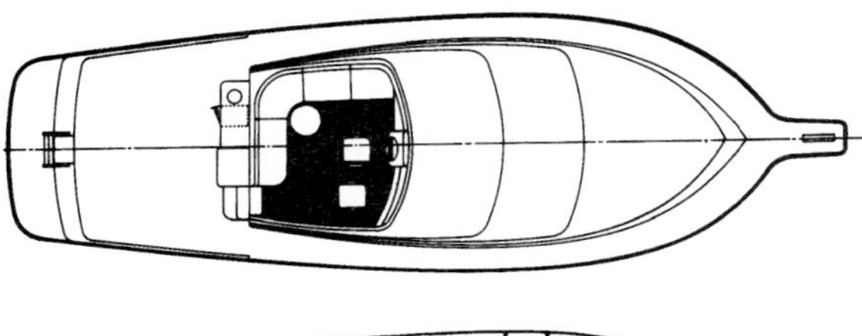

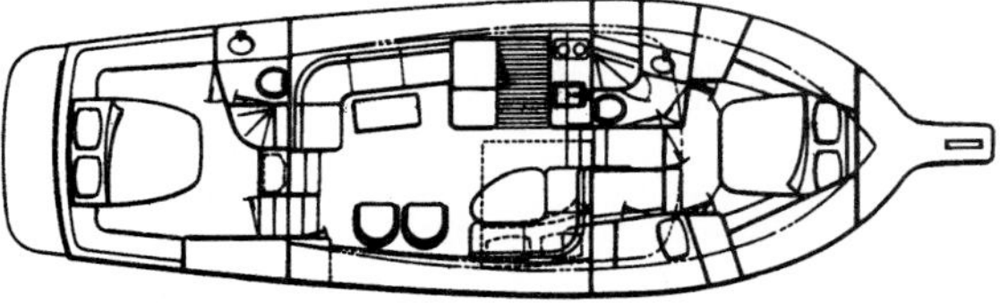

See page 250 for pricing information.

OCEAN 46 SUNLINER

SPECIFICATIONS

Length	46'0"	Water	150 gals.
Beam	15'2"	Fuel	480 gals.
Draft	3'6"	Hull Type	Modified-V
Weight	40,000#	Designer	D. Martin
Clearance	13'3"	Production	1983–86

The Ocean 46 Sunliner is a high-performance aft-cabin motor yacht built on the relatively flat-bottom hull used in the production of the popular Ocean 46 Super Sport. Whether one likes or dislikes her exterior lines, the Sunliner's all-teak interior and colorful fabrics were considered very attractive back in the 1980s. A roomy boat, the two-stateroom, galley-down floorplan is arranged with a dinette opposite the galley and double berths in both staterooms. An optional three-stateroom interior trades the dinette for a third stateroom. With either floorplan, the built-in salon settee makes it impossible to upgrade the furnishings. Note that both heads are fitted with separate shower stalls. Standard features included a central vacuum system, wet bar, entertainment center, washer/dryer, and garbage disposal. The flybridge is three steps up from the aft deck with guest seating for eight. A fast and economical boat for her size, the Ocean 46 Sunliner will cruise at around 24 knots with standard GM 6-71TI diesels and reach a top speed of 27 knots. ❑

See page 250 for pricing information.

OCEAN 48 MOTOR YACHT

SPECIFICATIONS

Length	48'6"	Water	150 gals.
LWL	NA	Fuel	500 gals.
Beam	16'4"	Hull Type	Modified-V
Draft	4'0"	Deadrise Aft	5°
Weight	51,000#	Designer	D. Martin
Clearance	14'10"	Production	1989–92

The 48 MY is a scaled-down version of the popular Ocean 53 MY with one less stateroom, a smaller salon, and a boxier profile. She's a true double-deck motor yacht (rare in such a small boat) with a full-width salon and three staterooms and heads. Indeed, the accommodations are so expansive and well-designed that it's hard to believe this is just a 48-footer. The master stateroom is accessed via a spiral staircase in the salon and another staircase forward leads up to the bridge. The galley and helm are open to the salon, and the teak woodwork and decorator furnishings are lush. Both guest staterooms have double berths and are quite large. A door in the midships guest cabin provides access to the stand-up engine room. Topside, there's plenty of guest seating on the large flybridge. Standard 485-hp 6-71 diesels will cruise the Ocean 48 at 23–24 knots and reach 27 knots top. Note that the Ocean 56 Cockpit MY is basically the same boat with a cockpit. ❑

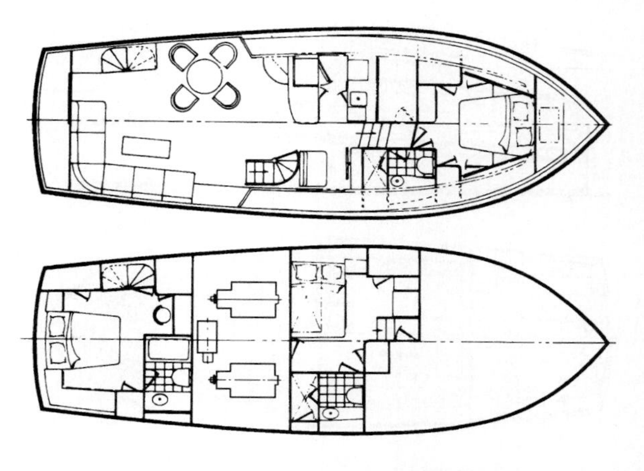

See page 250 for pricing information.

OCEAN 53 MOTOR YACHT

SPECIFICATIONS

Length	53′0″	Water	300 gals.
Beam	17′2″	Fuel	750 gals.
Draft	4′6″	Hull Type	Modified-V
Weight	64,000#	Designer	D. Martin
Clearance	16′0″	Production	1988–91

Introduced in 1988, the Ocean 53 Motor Yacht represented an obvious departure from Ocean's tradition of building Jersey-style sportfishing boats. The first of a new series of Ocean motor yachts, the 53 was built on a relatively flat-bottom hull with Divinycell coring in the hullsides and a wide beam. Aside from her luxurious and innovative interior accommodations, the Ocean 53 MY has the distinction of being one of the fastest production motor yachts available. With a pair of 750-hp 8V92s, she'll cruise around 26 knots and reach a top speed of 29–30 knots. Two floorplans were offered, each with four staterooms. Plan A has a fully extended, full-width main salon of impressive size, and Plan B shortens the salon area while leaving a semi-enclosed afterdeck. Lacking sidedecks or a workable aft deck area in Plan A, line-handling space is severely restricted. The moderate price and surprising performance of the Ocean 53 Motor Yacht do much to offset her boxy appearance. A total of 29 were built during her four-year production run. ❑

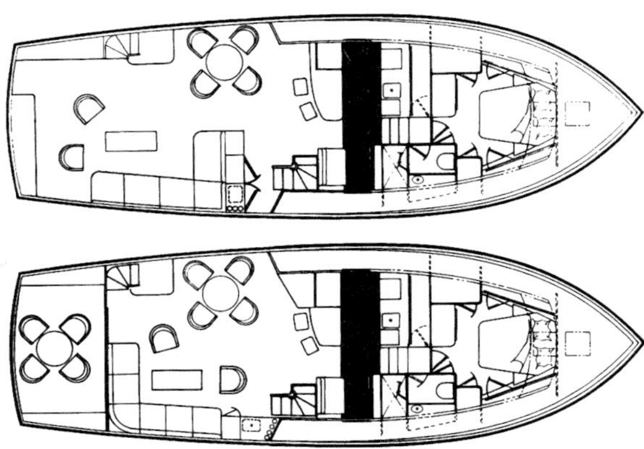

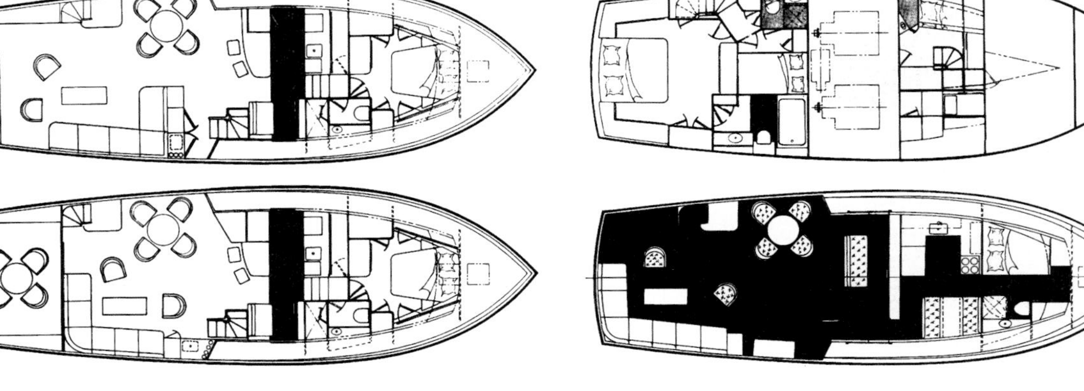

See page 250 for pricing information.

OCEAN 55 SUNLINER

SPECIFICATIONS

Length	55'8"	Water	200 gals.
LWL	50'0"	Fuel	750 gals.
Beam	16'4"	Hull Type	Modified-V
Draft	4'4"	Deadrise Aft	4°
Weight	60,000#	Designer	D. Martin
Clearance	14'6"	Production	1983–86

The Ocean 55 Sunliner was built using the hull and deckhouse of the Ocean 55 Super Sport. While she's a good performer for a motor yacht, her profile is certainly less than inspiring. Indeed, she *looks* like she was originally designed as a convertible, and it's unusual to see this conversion attempted in such a large boat. Below, the Sunliner's three-stateroom layout is arranged with the galley and dinette down. Notably, the salon dimensions are modest for a 55-footer, and the built-in settee makes it impossible to rearrange the furnishings. The master stateroom, however, is very spacious with plenty of storage space and a tub/shower in the adjoining head. Topside, the full-width aft deck is fitted with wing doors and enclosure panels for weather protection. There's plenty of guest seating on the bridge forward of the helm console. GM 8V92s were standard, and the 600-hp versions will cruise the Ocean 55 Sunliner at 22–23 knots. The more powerful 675-hp versions cruise around 25 knots and turn 28+ wide open. ❑

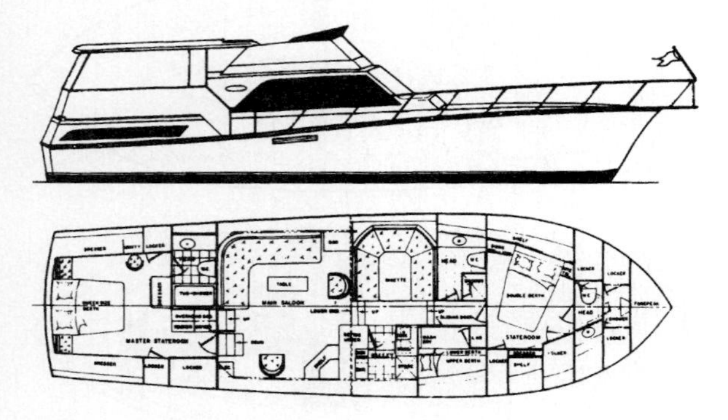

See page 250 for pricing information.

OCEAN 56 COCKPIT MY

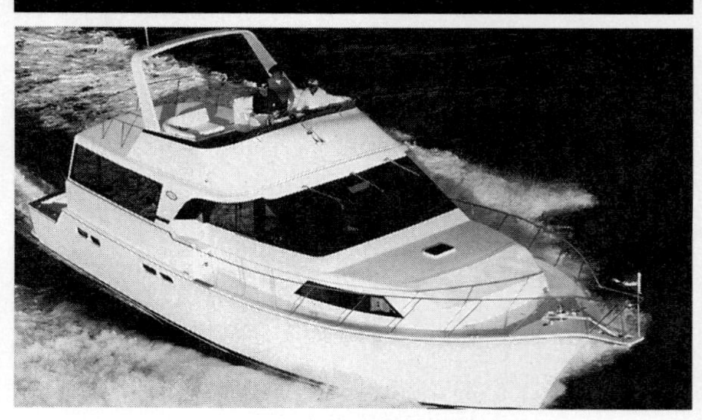

SPECIFICATIONS

Length	56'0"	Fuel	525 gals.
Beam	16'4"	Cockpit	80 sq. ft.
Draft	4'0"	Hull Type	Modified-V
Weight	54,500#	Deadrise Aft	5°
Clearance	14'10"	Designer	D. Martin
Water	150 gals.	Production	1991–Current

The Ocean 56 Cockpit MY is basically an Ocean 48 MY with a cockpit. To create this boat, the original 48's hull was lengthened, and a small skeg was added to the bottom to improve steering. Constructed on a relatively flat-bottom hull with Divinycell coring above the waterline, the tapered shape of the 56 CMY carries the maximum beam well forward of the salon. Inside, the efficient galley-up, three-stateroom floorplan is arranged with double berths in each stateroom and three full heads. Note that the helm is open to the deckhouse. There's a sliding deck door at the lower helm, and a salon staircase leads up to the flybridge. A small afterdeck overlooks the cockpit and is shaded by a bridge overhang. With some 80 sq. ft. of space, the cockpit is quite large with plenty of room for fishing. A molded tackle center and transom door are standard. A good performer, standard 485-hp 6-71 diesels will cruise the Ocean 56 Cockpit MY around 22 knots and reach a top speed of 25–26 knots. ❑

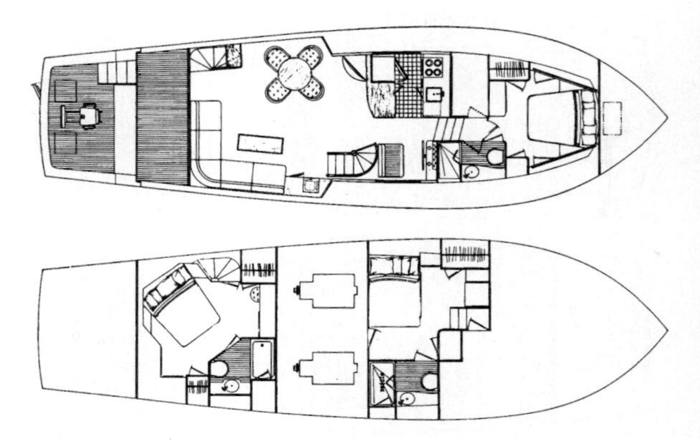

See page 250 for pricing information.

OCEAN ALEXANDER 38 DC

SPECIFICATIONS

Length	38'4"	Water	200 gals.
LWL	NA	Fuel	300 gals.
Beam	13'4"	Hull Type	Semi-Disp.
Draft	3'2"	Designer	Ed Monk
Weight	21,500#	Production	1984–87

The Ocean Alexander 38 is a good quality trawler-style cruiser with a clean-cut profile and a traditional layout. She was built on a solid fiberglass semi-displacement hull with hard aft chines and a moderate keel deep enough to provide protection to the running gear. Note that her exterior is relatively free of teak. Inside, the accommodations are typical of those found in most double-cabin trawlers with the exception of the unusual mid-level galley configuration to port in the salon. A lower helm is standard, and there are two deck access doors. The master stateroom comes with a centerline queen berth, separate shower and head compartments, and a convenient access door to the cockpit. Teak woodwork is used throughout the interior, and the joinerwork is excellent. Outside, raised bulwarks provide protection for the walkaround decks, and the flybridge has adequate guest seating. There's space on the cabintop for a dinghy, and the mast was standard. Powered with twin 135-hp Lehman diesels, the Ocean 38 DC will cruise economically at 7–8 knots. ❑

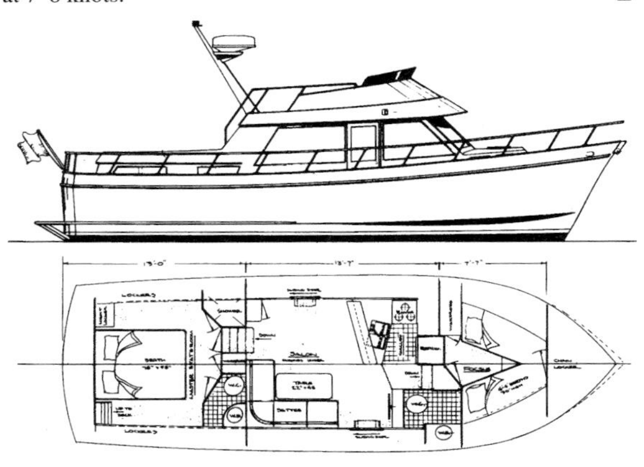

See page 250 for pricing information.

OCEAN ALEXANDER 390 SUNDECK

SPECIFICATIONS

Length	39'3"	Water	150 gals.
LWL	NA	Fuel	300 gals.
Beam	13'11"	Hull Type	Modified-V
Draft	3'2"	Deadrise Aft	12°
Weight	24,800#	Designer	Ed Monk
Clearance	14'0"	Production	1986–Current

The introduction of the 390 Sundeck in late 1986 marked the beginning of what has become a successful series of mid-size Alexander sundeck models. Built on a modified-V hull with cored hullsides and moderate transom deadrise, the low-profile design of the Alexander 390 makes her a more attractive boat than many of her competitors. Inside, the galley-down floorplan is arranged with double berths in both staterooms (V-berths forward are optional), and there are stall showers in both heads. The salon dimensions are more than adequate for a boat of this size, and the interior can be finished with teak or light oak woodwork. Topside, the helm console is located well forward on the bridge, and the full-width aft deck is large enough for a few pieces of deck furniture. Several engine options have been offered over the years including 275-hp Lehmans, 250-hp GMs, 306-hp Volvos, and 300-hp Cummins. Depending upon the engines, cruising speeds range from 15–17 knots, and top speeds are around 19–21 knots. ❑

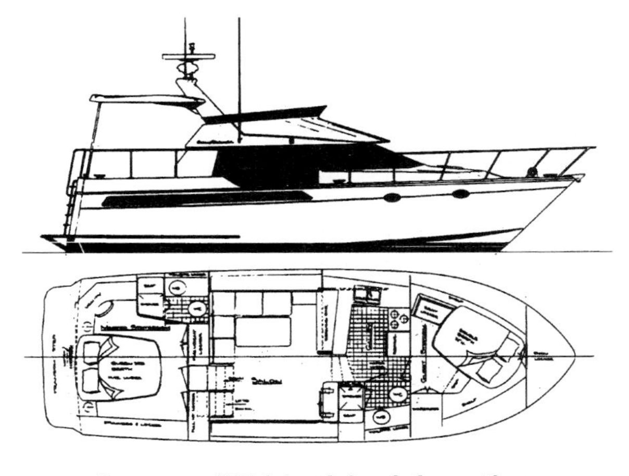

See page 250 for pricing information.

OCEAN ALEXANDER 40 DOUBLE CABIN

SPECIFICATIONS

Length	40'10"	Water	240 gals.
LWL	36'0"	Fuel	400 gals.
Beam	13'4"	Hull Type	Semi-Disp.
Draft	3'6"	Designer	Ed Monk
Weight	22,500#	Production	1980–85

A good-looking cruiser, the Ocean Alexander 40 DC was introduced in 1980 (during the era of fuel shortages) to meet the public demand for fuel-efficient trawlers. She's built on a solid fiberglass semi-displacement hull with hard chines, moderate beam, and a long keel of sufficient depth to offer good protection to the running gear. The Ocean 40 has an abundance of exterior teak—decks, doors, window frames, and rails—and those who enjoy the nautical appeal of a well-crafted teak interior will find much to admire in the Alexander 40's grain-matched cabinetry and paneling. The floorplan is fairly traditional with a walkaround queen berth in the master stateroom and V-berths forward. Port and starboard deck doors were standard in the salon and visibility from the lower helm is good. Note that the engine room is quite spacious. A well-built boat, twin 120-hp Lehman diesels will cruise the Ocean Alexander 40 DC at 7–8 knots at only 6 gph for a cruising range of approximately 500 miles. ❑

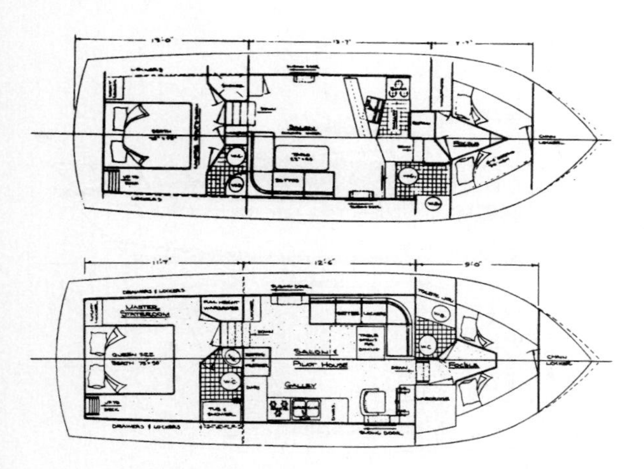

See page 250 for pricing information.

OCEAN ALEXANDER 40 SEDAN

SPECIFICATIONS

Length	40'10"	Water	240 gals.
LWL	36'0"	Fuel	400 gals.
Beam	13'4"	Hull Type	Semi-Disp.
Draft	3'6"	Designer	Ed Monk
Weight	22,500#	Production	1980–85

The Ocean Alexander 40 Sedan is a great-looking sedan style family cruiser with distinctive lines and a handsome, slightly European profile. Her open sedan layout will appeal to those looking for the convenience and improved entertaining capabilities of a larger salon with direct aft deck access. In the 40 Sedan, the galley is located forward, a step up from the salon level, to better accommodate the engines below. A raised double berth is located in the large master stateroom forward, and over/under bunks are fitted into the small guest cabin. Both staterooms are served by a double-entry head with a separate stall shower. The interior is finished with solid teak woodwork in every cabin, and the joinerwork is impressive throughout. Outside, the modest aft deck has room for a couple of folding chairs, but that's about it. The sidedecks are wide and well-protected with bridge overhangs and raised bulwarks. The flybridge is very spacious for a 40-footer. With the standard twin 120-hp Lehman diesels, the Alexander 40 Sedan will cruise economically at 7–8 knots.

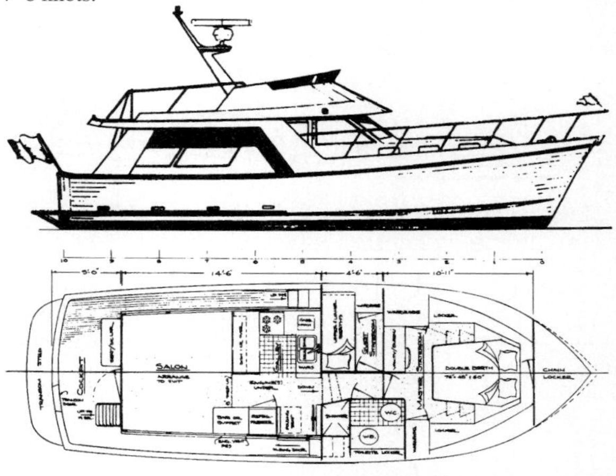

See page 250 for pricing information.

OCEAN ALEXANDER 420/440 COCKPIT

SPECIFICATIONS

Length	42'3"/43'9"	Water	150 gals.
Beam	13'11"	Fuel	300 gals.
Draft	3'2"	Hull Type	Modified-V
Weight	27,000#	Designer	Ed Monk
Clearance	NA	Production	1987–Current

The Alexander 420 (pictured above) and 440 are basically the same boat with the cockpit size accounting for the different lengths. Both share the same modified-V hull with cored hullsides, a shallow keel, and moderate transom deadrise. Inside, the two-stateroom, galley-down layout is offered with either an offset double berth or V-berths in the forward stateroom. The interior is finished in teak (or light oak) woodwork, and the joinerwork is excellent throughout. Both heads are fitted with stall showers, and a lower helm is standard. The afterdeck has enough space for a table and a few chairs, and the flybridge will seat six. Additional features include a spacious engine room, hardtop and radar arch, wide sidedecks, and good storage. A popular boat, several diesel options have been offered in the Alexander 420/440 over the years. Twin 250-hp GM and Cummins engines will cruise at 14–15 knots (about 19 knots top), and optional 375-hp Cats will cruise around 20 knots and reach a top speed of 24 knots. ❑

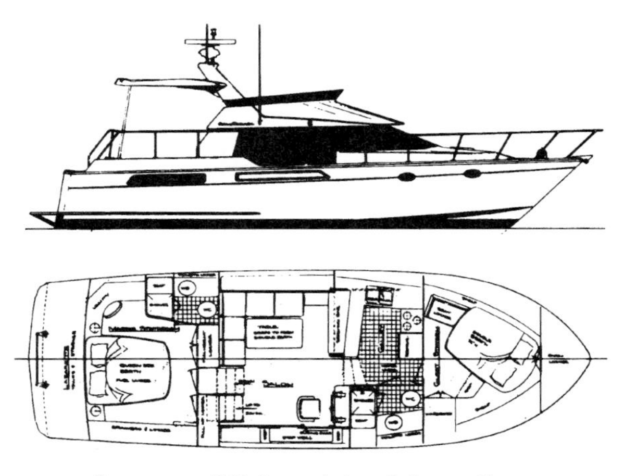

See page 250 for pricing information.

OCEAN ALEXANDER 42/46 SEDAN

Ocean Alexander 42 Sedan

Ocean Alexander 46 Sedan

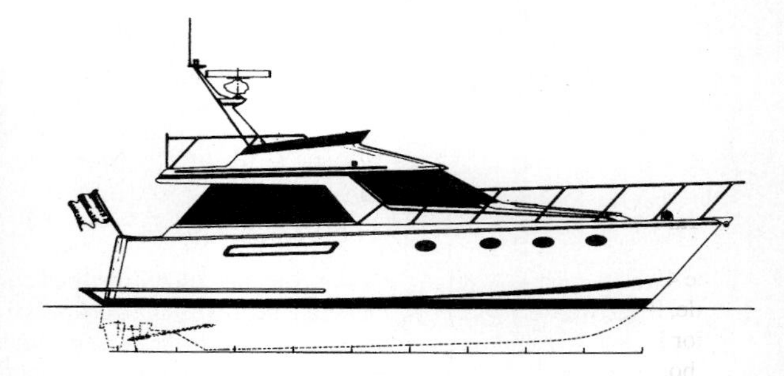

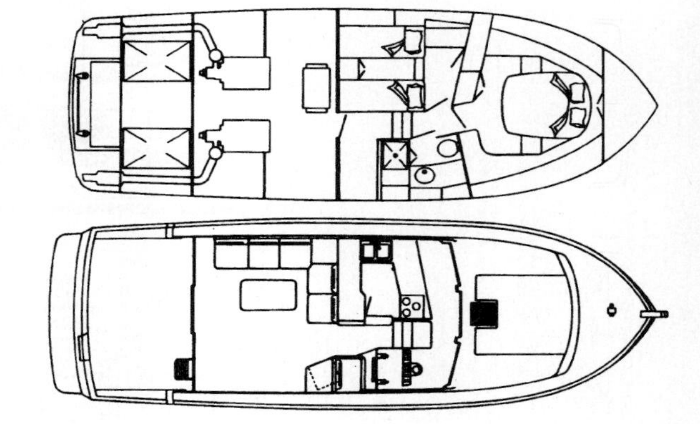

SPECIFICATIONS

Length	46'0"	Fuel	500 gals.
Beam	14'4"	Cockpit	NA
Draft	3'2"	Hull Type	Modified V
Weight	23/26,000#	Designer	Ed Monk
Clearance	11'6"	Production	1987–Current
Water	150 gals.		

Beginning as the 390 Sedan in 1985, Alexander stretched the hull a couple of years later to create the 42 Sedan (1987–93) as well as the 46 Sedan model. A popular boat (especially the 42), the hull is cored from the waterline up. She's a relatively light boat for her size. While neither the 42 or 46 is particularly beamy, the salon and living areas are well-proportioned and beautifully finished throughout. Two staterooms and one head are forward along with a unique storage room where the engine room access door is found. Originally designed with the galley to port opposite the lower helm, an alternate floorplan eliminates the helm, adds a dinette, and moves the galley into the salon. Updates in 1990 included a restyled superstructure and arch, wider sidedecks, and a revised interior decor. Note that production of the 46 ended in 1993. Standard 250-hp Cummins will cruise either boat at 13–14 knots, and the optional 425-hp Cats will cruise at 22–23 knots (about 26 top). ❑

See page 250 for pricing information.

OCEAN ALEXANDER 423 CLASSICO

SPECIFICATIONS

Length	42'3"	Water	160 gals.
LWL	NA	Fuel	550 gals.
Beam	14'8"	Hull Type	Semi-Disp.
Draft	3'10"	Designer	Ed Monk
Weight	34,200#	Production	1993–Current
Clearance	NA		

The 423 Classico is a well-crafted cruiser with a trunk cabin profile, full walkaround decks, and a straightforward two-stateroom interior layout. Alexander's reintroduction of the classic trawler-style boat will appeal to those enthusiasts who enjoy the look and feel of a traditional trawler but want something better than trawler-speed (7–8 knot) performance. The relatively soft chines and wide, wineglass keel of the 423 are unusual in a semi-displacement design. A distinctive boat with her rakish bridge overhang and radar mast, the Classico's house is set well forward on the foredeck in order to expand the salon and galley dimensions. Inside, a deck access door is adjacent to the lower helm, and both heads are fitted with stall showers. Not surprisingly, the interior is all teak. Topside, the flybridge is huge for a 42-footer with room for a crowd. Standard 210-hp Cat diesels will cruise the 423 Classico at an efficient 12 knots. Note that the keel is deep enough to provide some protection to the running gear in the event of a grounding. ❑

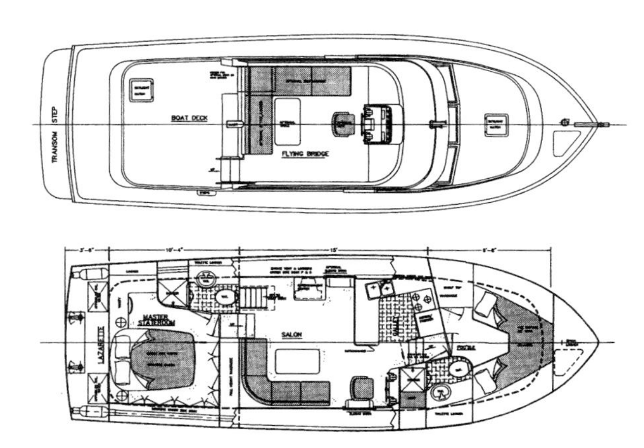

See page 251 for pricing information.

OCEAN ALEXANDER 43 DC

SPECIFICATIONS

Length	42'6"	Water	200 gals.
LWL	38'2"	Fuel	500 gals.
Beam	14'6"	Hull Type	Semi-Disp.
Draft	3'6"	Designer	Ed Monk
Weight	29,000#	Production	1980–85

The Ocean Alexander 43 DC earned a reputation early on as one of a handful of better-quality Asian trawler-style imports. Along with the earlier 50 Pilothouse, she did much to establish the Ocean Alexander name in the minds of brokers and dealers. Unlike most Taiwan trawlers of the early '80s, the 43 is relatively free of exterior teak—a big plus in the minds of most owners. Inside, the all-teak interior is arranged with a centerline queen berth in the full-width aft stateroom and stall showers in both heads. The galley is down, and the salon dimensions are very adequate. Note the port and starboard deck doors. Visibility from the lower helm is excellent. Additional features include simulated lapstrake hullsides, a big flybridge with the helm forward, and keel-protected running gear. A popular model, standard 120-hp Lehman diesels will cruise the Ocean 43 DC efficiently at 7-8 knots with a range of about 600 miles. Optional 255-hp Lehmans cruise at 13 knots and reach a top speed of around 17 knots. ❑

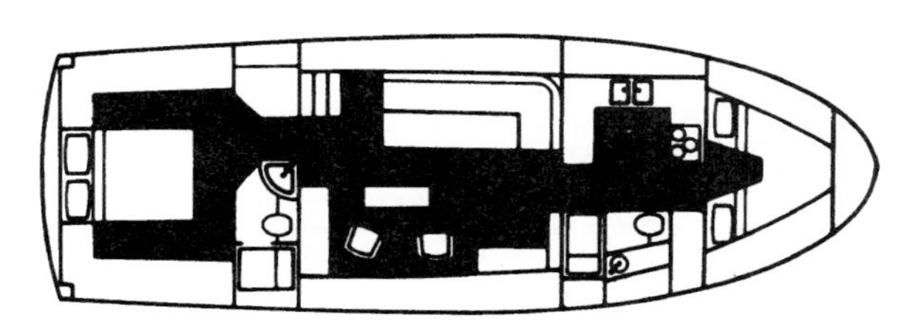

See page 251 for pricing information.

OCEAN ALEXANDER 456 CLASSICO

SPECIFICATIONS

Length	45'6"	Water	250 gals.
Beam	15'8"	Fuel	550 gals.
Draft	4'0"	Hull Type	Semi-Disp.
Weight	40,000#	Designer	Ed Monk
Clearance	12'9"	Production	1992–Current

First in the new Classico series of traditional trawler-style boats from Ocean Alexander, the 456 is a well-built family cruiser with a semi-displacement hull capable of planing speed performance. She's built on a wide-beam hull with cored hullsides, double chines (unusual in a trawler-style design), and a cutaway keel that protects the running gear. With the house set well forward on the hull, the Classico is a very roomy boat inside. The floorplan is arranged with an aft-facing queen bed in the master stateroom—a configuration almost never seen in a production design—separate stall showers in both heads, and a mid-level galley. Note the port and starboard deck doors in the salon. Additional features include a full teak interior, a spacious flybridge with excellent seating, radar mast, teak handrails (the only exterior teak on the boat), and a deep step-down cockpit with transom door. With standard 210-hp Cat (or Cummins) diesels, the Alexander 456 Classico will cruise economically at 10–11 knots. No lightweight, the cruising range exceeds 600 miles at hull speed. ❑

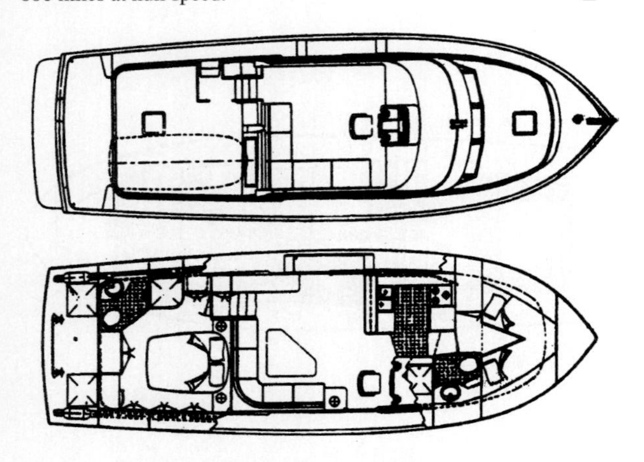

See page 251 for pricing information.

OCEAN ALEXANDER 48/50 SEDAN

SPECIFICATIONS

Length	48'0"	Fuel	410/500 gals.
Beam	15'6"	Cockpit	NA
Draft	4'6"	Hull Type	Modified-V
Weight	38,000#	Designer	Ed Monk
Water	210 gals.	Production	1988–91

A good-looking sedan, the Alexander 48 is a well-designed family cruiser with an upscale interior to go with her sleek, Euro-style profile. Her modified-V hull is constructed with cored hullsides, and the 15'6" beam is about average for a boat this size. Inside, the two-stateroom floorplan features a spacious salon open to the helm and a big master stateroom forward. Both heads are fitted with stall showers, and a deck door is adjacent to the lower helm. The light oak (or teak) woodwork is impressive. The cockpit—partially covered with a bridge overhang—comes with a transom door, and the sidedecks of the 48 are notably wide. Additional features include oversize salon windows, an extended flybridge with space for a small dinghy, and good access to the motors in the low-overhead engine room. A good performer with 375-hp Cat diesels (or 400-hp 6V53s), the cruising speed is 18 knots, and the top speed is around 21 knots. Note that the Alexander 50 Sedan is the same boat with a slightly enlarged cockpit. ❑

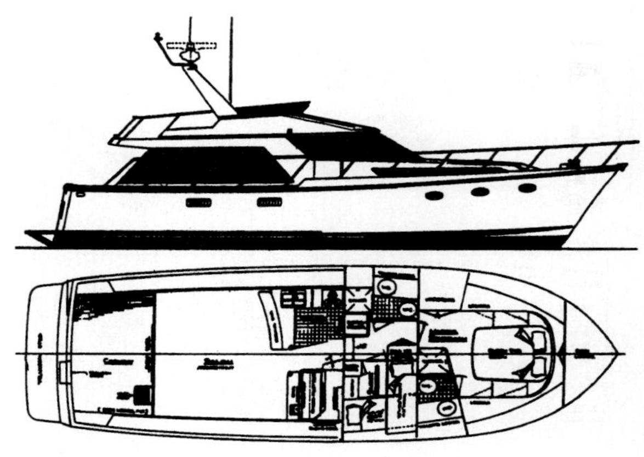

See page 251 for pricing information.

OCEAN ALEXANDER 486 CLASSICO

SPECIFICATIONS

Length	48'0"	Water	300 gals.
Beam	15'8"	Fuel	700 gals.
Draft	4'0"	Hull Type	Semi-Disp.
Weight	48,000#	Designer	Ed Monk
Clearance	NA	Production	1993–Current

Largest in Alexander's Classico series of traditional trawlers, the 486 is a sturdy raised-pilothouse design with the strictly-business appearance of a bluewater passagemaker. She's heavily built on a modern semi-displacement hull with cored hullsides and a keel deep enough to protect the props—basically a stretched version of the smaller 456 Classico hull. Many pilothouse boats this size have three stateroom layouts, but the 486 has only two, both of which are very spacious and about equal in size with walkaround double beds, excellent storage, and private heads with stall showers. Aft of the pilothouse are the galley and salon with large cabin windows, teak parquet flooring, and sliding glass doors opening to the cockpit. The sidedecks are wide and well-protected, and there's room on the extended flybridge for a dinghy. Additional features include plenty of interior teak woodwork, pilothouse deck access door, and a cockpit transom door. A good-looking yacht with a Pacific Northwest flavor, optional 375-hp Cat diesels will cruise the Alexander 486 Classico at an easy 13–14 knots. ❑

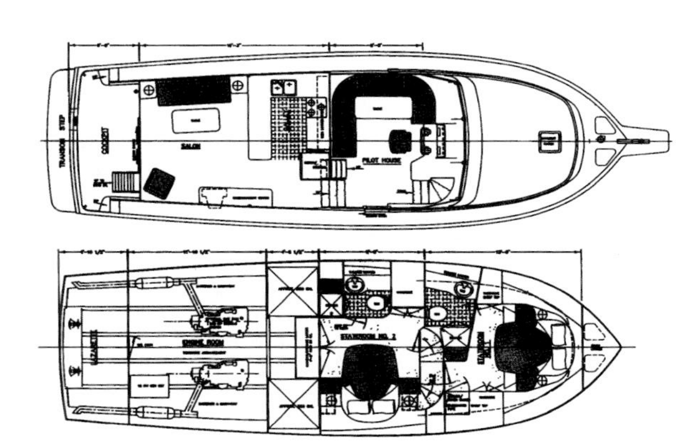

See page 251 for pricing information.

OCEAN ALEXANDER 50 PILOTHOUSE

SPECIFICATIONS

Length	50'3"	Cockpit	NA
LWL	45'8"	Water	420 gals.
Beam	15'6"	Fuel	1,000 gals.
Draft	4'6"	Hull Type	Semi-Disp.
Weight	46,500#	Designer	Ed Monk
Clearance	19'0"	Production	1977–83

The first Ocean Alexander design to be imported and sold in the U.S., the 50 Pilothouse was a very successful design (about 90 were sold during her production run) and is now something of a classic on the West Coast. Her appeal had as much to do with an affordable price as it did with her rugged, go-anywhere appearance. Construction is solid fiberglass, and the Portugese bridge surrounding the pilothouse is quite distinctive. Inside, the two-stateroom floorplan (fully paneled with grain-matched teak woodwork and cabinetry throughout) is arranged with a huge amidships owner's cabin on the lower level and a wide-open salon and galley on the deckhouse level. The raised pilothouse is private and comes with sliding deck doors port and starboard as well as a built-in dinette. Additional features include a covered cockpit area, well-secured sidedecks, an extended flybridge with space for a dinghy, and a simulated lapstrake hull. A good performer, the Ocean Alexander 50 will cruise around 12 knots with the optional 270-hp Cummins diesels. ❑

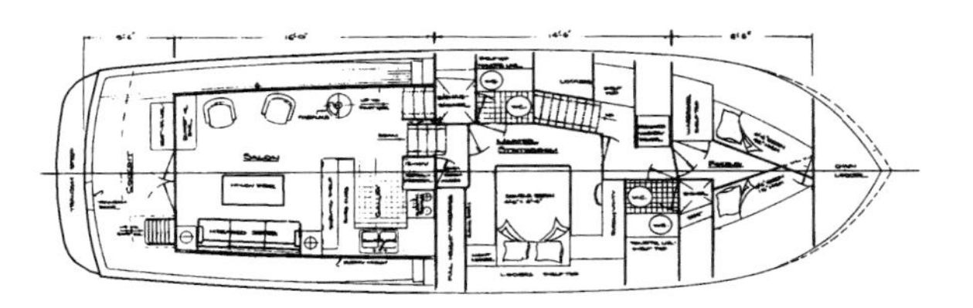

See page 251 for pricing information.

OCEAN ALEXANDER 50 MK II PH

SPECIFICATIONS

Length	50'0"	Fuel	600 gals.
Beam	15'6"	Hull Type	Modified-V
Draft	4'6"	Deadrise Aft	NA
Weight	42,000#	Designer	Ed Monk
Water	220 gals.	Production	1984–90

A good-looking yacht, the 50 MK II PH replaced the original 50 PH model in the Ocean Alexander lineup in 1984. The 50 MK II displays a modern Euro-style appearance in contrast to the very traditional profile of the original model. She was built in Taiwan on a modified-V hull with balsa-cored hullsides, moderate beam, and a shallow keel deep enough to offer some protection to the props. Inside, the MK II features a spacious three-stateroom layout with double berths in two of the cabins (the original 50 PH had only two staterooms), two heads, and a fully enclosed raised pilothouse. The salon is aft of the galley on the deckhouse level and wide-open to the cockpit. Ocean Alexanders have always had a reputation for quality interior woodwork and the MK II is no exception. The teak paneling, countertops, and cabinets are of custom furniture quality. A good performer with optional 375-hp Cat (or 400-hp 6V53) diesels, the 50 MK II will cruise at 16–17 knots and reach about 20 knots top. ❏

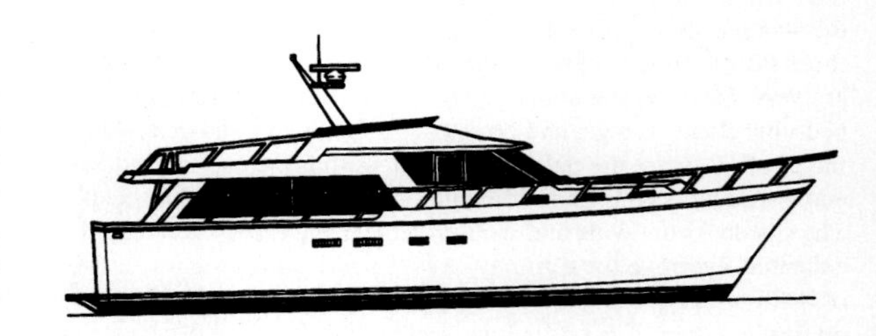

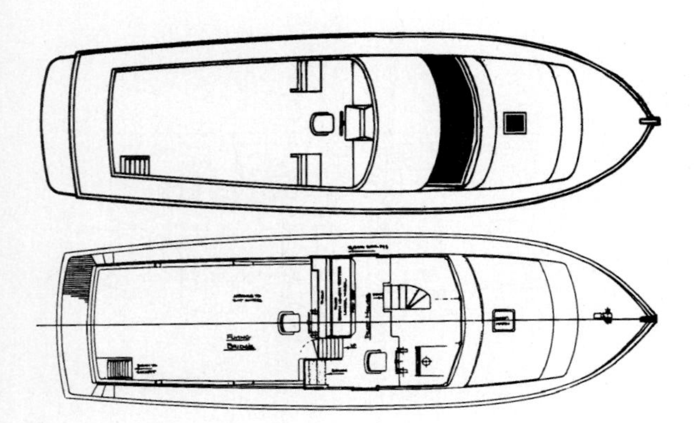

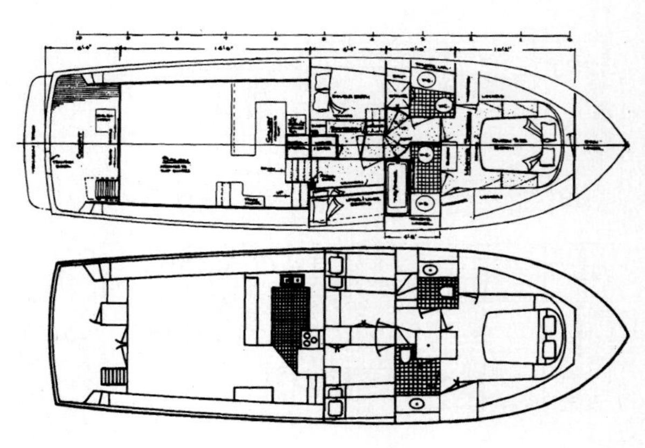

See page 251 for pricing information.

OCEAN ALEXANDER 51/53 SEDAN

SPECIFICATIONS

Length	51'1"/53'0"	Water	250 gals.
Beam	16'4"	Fuel	500 gals.
Draft	3'2"	Hull Type	Modified-V
Weight	45,500#	Designer	Ed Monk
Clearance	12'6"	Production	1989–Current

This good-looking sport cruiser from Ocean Alexander features an aggressive Euro-style profile and a raised pilothouse floorplan—a potent combination in today's upscale yachting market where pilothouse designs are becoming increasingly popular. Offered in two versions, the 51 Sedan pictured above has a conventional squared-off transom, while the 53 model (introduced in 1991) has a reverse transom that adds an extra two feet to her LOA. Instead of locating the galley on the salon level—the traditional layout in pilothouse floorplans—the 51/53 has the galley to port on the pilothouse level in the European fashion. The result is a huge salon with extravagant entertaining capabilities. The master stateroom is forward in this layout, and both guest cabins are located beneath the pilothouse sole. Additional features include a large cockpit with a transom door, wide sidedecks, a foredeck sunpad, and a spacious flybridge with plenty of guest seating. Standard 400-hp 6V53s will cruise the Alexander 51 Sedan at 16 knots, and optional 735-hp 8V92s will cruise at a fast 24–25 knots. ❑

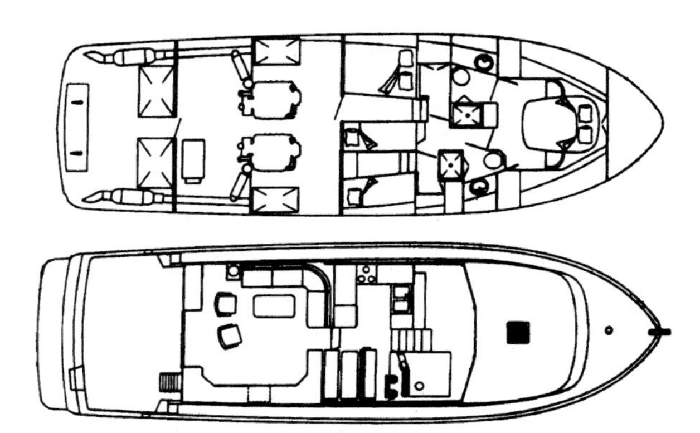

See page 251 for pricing information.

OCEAN ALEXANDER 520/540 PH

SPECIFICATIONS

Length	52'5"/54'0"	Water	300 gals.
Beam	15'6"	Fuel	600 gals.
Draft	4'0"	Hull Type	Modified-V
Weight	42,500#	Designer	Ed Monk
Clearance	14'0"	Production	1991–Current

Offered in two versions, the Alexander 540 Sedan (pictured above) comes with a Euro-style reverse transom while the original 520 model retains the squared-off transom seen in more traditional sedan motor yachts. She's built on a moderate-deadrise modified-V hull with cored hullsides and a relatively modest beam. There are two floorplans offered, both with the same three-stateroom layout on the lower level. The spacious master stateroom is forward, and both guest cabins are located beneath the pilothouse sole. The original interior has a dinette opposite the helm in the raised pilothouse, and the other (less popular) layout has no dinette and port and starboard pilothouse deck doors. Additional features include a huge flybridge with plenty of seating and room for a Whaler, wide sidedecks, well-crafted interior woodwork, and excellent storage. A good-running boat with standard 400-hp 8V53s (or 425-hp Cats), she'll cruise at 17–18 knots and reach a top speed of about 20 knots. The optional 485-hp 6-71s are reported to cruise at 21–22 knots. ❑

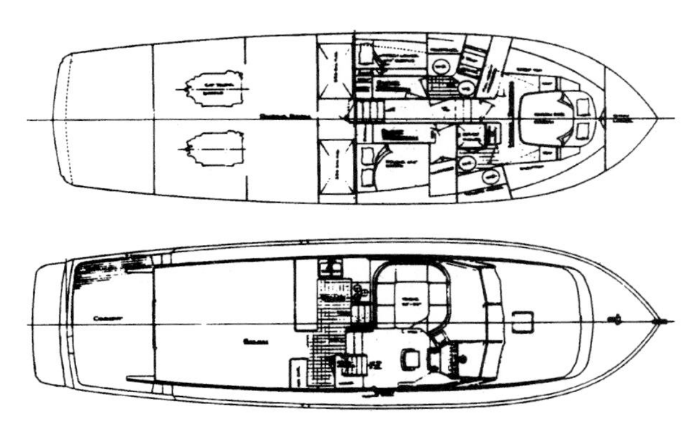

See page 251 for pricing information.

OCEAN ALEXANDER 54 COCKPIT MY

SPECIFICATIONS

Length	54'0"	Water	230/260 gals.
LWL	NA	Fuel	540/700 gals.
Beam	15'6"	Hull Type	Modified-V
Draft	4'6"	Designer	Ed Monk
Weight	48,000#	Production	1985–89

Whatever the Alexander 54 CMY lacks in styling (her appearance is slightly boxy), she more than recovers in living space and interior accommodations. Indeed, the 54's tri-level floorplan is unusual in a motor yacht and will appeal to those looking for something other than the standard motor yacht layout. The salon is aft and up from the intermediate level, where the galley and lower helm are located. The result is a spacious galley and dining area and a completely separate full-width salon. There are three staterooms on the lower level, and both aft staterooms share a common head—a rather unattractive compromise in a boat of this size. A staircase provides easy bridge access from the salon, and a small afterdeck is provided for line handling. The engine room is unusually spacious with excellent outboard access. Among several engine options, 450-hp 6-71s will deliver a 15-knot cruising speed and 17–18 knots top. Note that the Ocean Alexander 50 MY is the same boat without the 54's cockpit. ❑

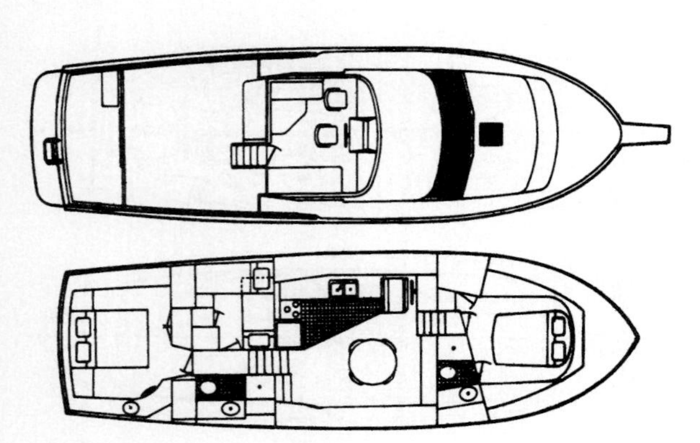

See page 251 for pricing information.

OCEAN ALEXANDER 60 MY (EARLY)

SPECIFICATIONS

Length	60'0"	Water	365 gals.
LWL	NA	Std. Fuel	1,200 gals.
Beam	18'0"	Hull Type	Modified-V
Draft	4'10"	Deadrise Aft	12°
Weight	65,000#	Designer	Ed Monk
Clearance	17'0"	Production	1984–87

Designed by Ed Monk, Jr., on an efficient modified-V hull with simulated lapstrake hullsides, the Alexander 60 is a handsome bluewater yacht with an excellent cruising layout to go with her sturdy offshore character. The Portugese bridge and rakish bridge overhangs give her a decidedly serious persona. Her roomy galley-up pilothouse floorplan is arranged with four staterooms on the lower level including a VIP queen stateroom forward. The galley is in the salon, across from a convenient on-deck day head. A staircase aft in the salon leads down to the master stateroom with its queen-size berth, tub/shower, and engine room access door. Additional features include full teak decks, port and starboard deck doors in the pilothouse, foredeck seating, wide sidedecks, and a unique transom that opens to provide access to the swim platform via molded steps. While 550-hp 6V92 diesels were standard, most of the ten Alexander 60 MYs built were delivered with the larger 650-hp 8V92s. The cruising speed is around 15 knots, and the top speed is 18–19 knots. ❑

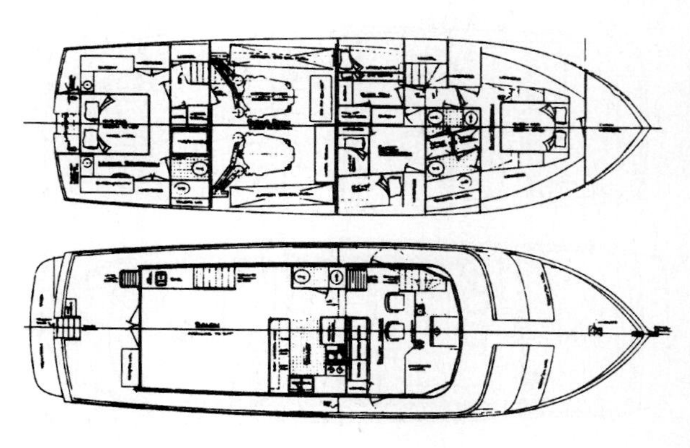

See page 251 for pricing information.

OCEAN ALEXANDER 60 MY

SPECIFICATIONS

Length	60'0"	Water	365 gals.
LWL	NA	Std. Fuel	1,000 gals.
Beam	18'0"	Hull Type	Modified-V
Draft	4'10"	Deadrise Aft	12°
Weight	65,000#	Designer	Ed Monk
Clearance	17'0"	Production	1987–Current

Unlike the earlier Alexander 60 MY model with its trawler-like profile and four-stateroom layout, the Alexander 60 pictured above is a modern flush-deck design with a contemporary Euro-style appearance and a pilothouse floorplan. She's built in Taiwan on a modified-V hull form with a notably wide beam, full-length keel, and moderate transom deadrise. The Alexander 60's spacious galley-up layout and her generous salon and master stateroom dimensions are impressive. The wheelhouse is separated from the salon, and there's an on-deck day head opposite the galley. A unique office alcove located beneath the pilothouse comes at the expense of a third head on the lower level—a sacrifice that some might argue over. The engine room is reached from the huge (opulent) master stateroom with its king-size bed, excellent storage, and apartment-size head. The dinette is ideally located next to the lower helm, and traditional teak woodwork and paneling are used extensively throughout the interior. Optional 735-hp 8V92s will cruise the Alexander 60 at a steady 17–18 knot cruising speed. ❑

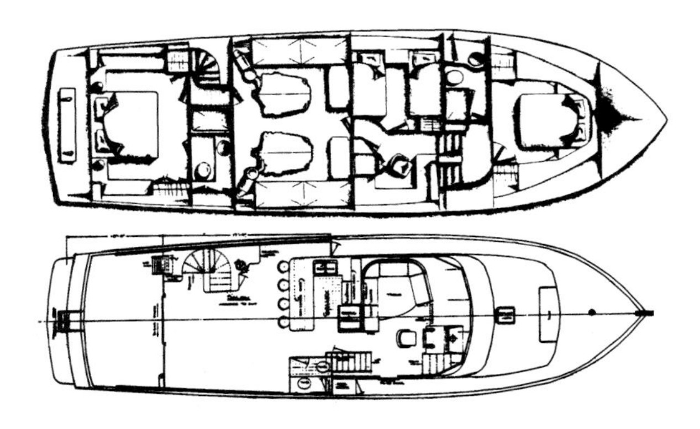

See page 251 for pricing information.

OCEAN ALEXANDER 63 MY

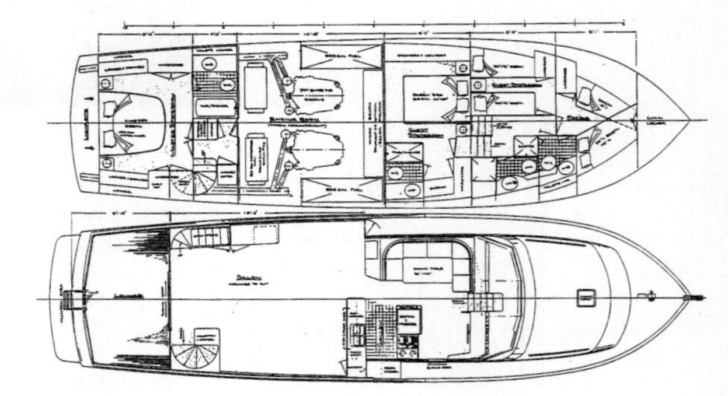

SPECIFICATIONS

Length	63'0"	Water	400 gals.
LWL	NA	Fuel	1,200 gals.
Beam	18'0"	Hull Type	Modified-V
Draft	4'10"	Deadrise Aft	12°
Weight	65,200#	Designer	Ed Monk
Clearance	23'3"	Production	1986–92

A popular boat, the 63 MY became one of Ocean Alexander's longest-running production yachts. Aside from her sleek lines, a key to the 63's success is her enormous full-width salon and enclosed afterdeck. When the sliding glass salon doors are opened, the entire deckhouse level is wide open from the aft deck forward past the galley to the lower helm. The standard four-stateroom layout is arranged with the master stateroom aft with private salon access. A full-size VIP guest cabin is amidships, and there is a total of four heads. (Many of these boats have been delivered with semi-customized interiors.) Additional features include direct salon access to the flybridge, an engine room workbench, and a huge flybridge with room for a crowd. Standard power is 550-hp 6V92s, but most 63's have been powered with the 735-hp 8V92s which will cruise around 15 knots. Note: the Ocean Alexander 63 YF is the same boat with a cockpit and smaller salon, and the Ocean Alexander 70 CMY is a 63 MY with a cockpit. ❑

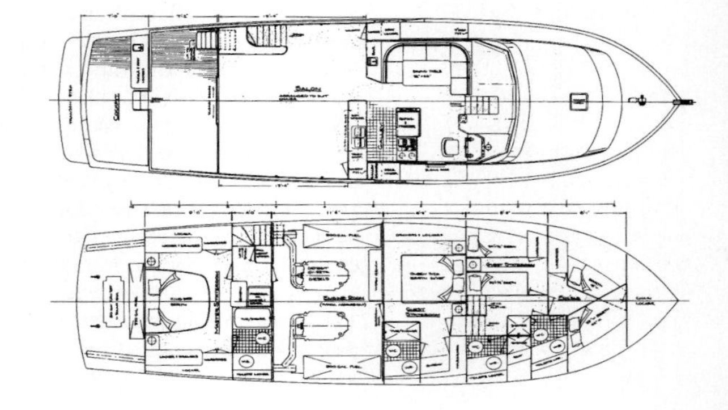

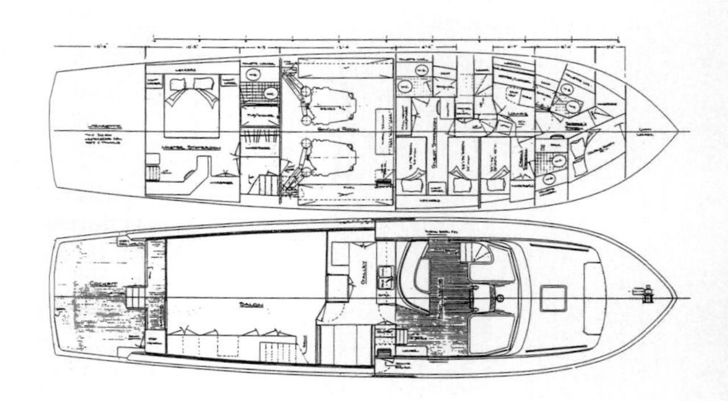

See page 251 for pricing information.

OCEAN ALEXANDER 630 MY

SPECIFICATIONS

Length	63'0"	Water	300 gals.
LWL	NA	Fuel	1,100 gals.
Beam	17'6"	Hull Type	Modified-V
Draft	4'4"	Deadrise Aft	NA
Weight	68,000#	Designer	Ed Monk
Clearance	NA	Production	1992–Current

Introduced as the Ocean Alexander 600 MY in 1992, the 630 MY is certainly one of the more dramatic and beautifully styled production motor yachts available on the market. She's built on a moderate deadrise hull with a shallow keel and a relatively wide beam. Inside, the 630's three-stateroom floorplan is arranged Euro-style with the galley and dinette on the pilothouse level, two steps up from the salon. The opulent master stateroom (there's a Jacuzzi in the head) is amidships where the ride is best, and the bow stateroom has an island bed and private head access. Note the crew quarters aft beneath the cockpit sole with twin berths, an enclosed head, and an entryway to the engine room. Additional features include pilothouse wing doors, a roomy cockpit, weather-protected walkaround sidedecks, a well-arranged but small flybridge (for a 63-footer) with space for a dinghy, and twin curved stairways molded into the transom. A good performer, standard 735-hp 8V92 Detroits cruise the 630 MY at a brisk 21–22 knots and reach a top speed of about 24 knots. ❑

See page 251 for pricing information.

OCEAN ALEXANDER 66 MY

SPECIFICATIONS

Length	66'0"	Water	400 gals.
LWL	NA	Fuel	1,300 gals.
Beam	18'0"	Hull Type	Modified-V
Draft	4'10"	Designer	Ed Monk
Weight	71,000#	Production	1986–Current
Clearance	NA		

With her striking European profile and aggressive megayacht appearance, the Ocean Alexander 66 is one of the best-looking production motor yachts in her size range. If the published weight of 71,000 lbs. is even close, she's also one of the lightest—a fact which helps account for her good performance with relatively small (for a 66-footer) 8V92 Detroit diesels. Inside, the layout is arranged with the full-width salon and glass-enclosed afterdeck on the same level with the added convenience of an on-deck day head forward, opposite the galley. The huge master stateroom is opulent, and the woodwork used throughout the interior of Alexander 66 is very impressive. Additional features include four full heads (two with tubs), a well-arranged engine room, port and starboard pilothouse wing doors, and enough space on the extended flybridge for a crowd. Beautifully proportioned and built to high standards, the Ocean Alexander 66 MY will cruise at a brisk 17 knots with the standard 735-hp diesels and reach a top speed of about 21 knots.❑

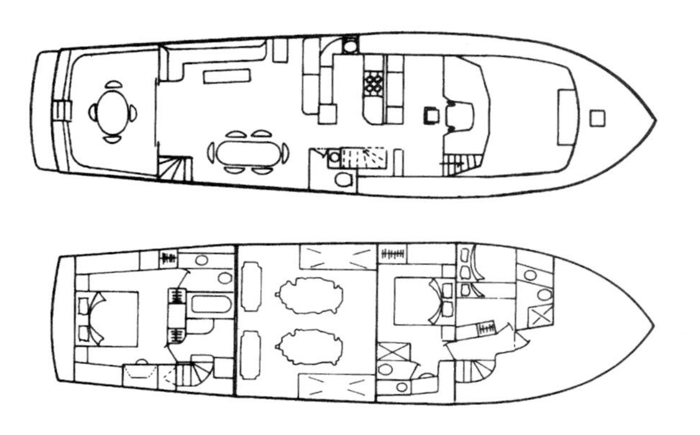

See page 251 for pricing information.

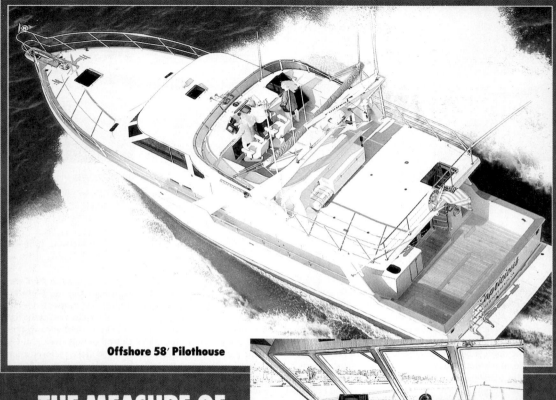

OFFSHORE 48 SEDAN

SPECIFICATIONS

Length	48'6"	Water	325 gals.
LWL	43'0"	Fuel	600 gals.
Beam	15'6"	Hull Type	Modified-V
Draft	3'6"	Deadrise Aft	NA
Weight	38,000#	Designer	Wm. Crealock
Clearance	15'2"	Production	1987–Current

Built by Tung Hwa, the Offshore 48 is an attractive and practical sedan with several features of interest to experienced cruisers. While she's not designed with the mega-volume interior found in some other yachts her size, the Offshore does place a priority on outdoor recreation as her generous deck areas suggest. Not only are the walkaround sidedecks wide and the cockpit spacious, but the flybridge dimensions are extremely generous for a 48-footer. Constructed on a modified-V hull with balsa coring from the waterline up, the interior is arranged with the salon and galley on the same level. A stairway abaft the lower helm leads up to the bridge and two double staterooms and two heads are forward. Varnished teak woodwork and cabinetry are used throughout, and the wraparound cabin windows make the salon seem more expansive than it really is. Note that bridge overhangs provide weather protection for the cockpit and walkways. Among several diesel options, 375-hp Cats will cruise the Offshore 48 at 16–17 knots and reach a top speed of about 20 knots top. ❑

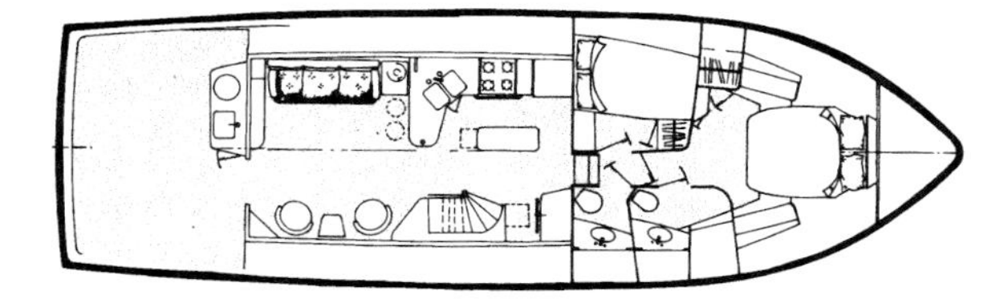

See page 251 for pricing information.

OFFSHORE 48 YACHTFISHER

SPECIFICATIONS

Length	48'3"	Water	325 gals.
LWL	43'0"	Fuel	600 gals.
Beam	15'6"	Hull Type	Modified-V
Draft	3'6"	Deadrise Aft	NA
Weight	38,000#	Designer	Wm. Crealock
Clearance	15'2"	Production	1985–Current

The Offshore 48 Yachtfish (also marketed as the Tung Hwa or Hartmann-Palmer 48 YF) is a traditional double-cabin design with a trawler-style appearance and a practical cruising layout. Her low-freeboard modified-V hull is cored from the chine up and carries a relatively wide 15'6" of beam. Inside, the 48's two-stateroom, galley-down floorplan is available with or without a salon dinette. A serving counter overlooks the galley, and the wide-open salon seems big for a boat this size. Additional features include a full teak interior, deck doors port and starboard in the salon, and an entryway door from the cockpit to the aft cabin—a convenient feature. The walkaround sidedecks are quite wide and protected with raised bulwarks. Although she's called a *Yachtfisher*, the cockpit is far too small for any serious fishing and is better suited for boarding and swimming. The full-width aft deck is quite roomy, and the flybridge is arranged with the helm forward. Cat 375-hp diesels will cruise the Offshore 48 at 16–17 knots and reach about 20 knots top. ❑

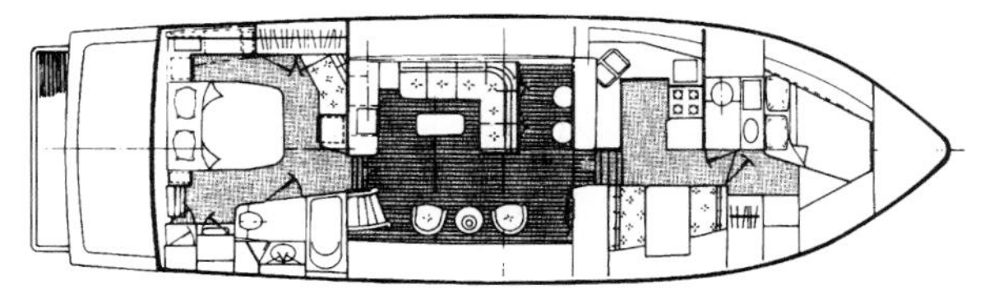

See page 251 for pricing information.

OFFSHORE 52 SEDAN MY

SPECIFICATIONS

Length	52'0"	Fuel	700 gals.
Beam	16'10"	Hull Type	Modified-V
Draft	3'10"	Deadrise Aft	12°
Weight	52,000#	Designer	Wm. Crealock
Water	400 gals.	Production	1991–Current

Sedan-style yachts may lack the mega-volume interiors of their motor yacht counterparts, but they're nearly always better-looking boats. The low-profile deckhouse and modern styling of the Offshore 52 give her a distinctive and altogether pleasing appearance. Built by Camargue in Taiwan, her modified-V hull incorporates a very wide beam with a cutaway keel and generous flare at the bow. Inside, her three-stateroom floorplan is arranged sedan-style with the galley forward in the salon opposite the lower helm. (Note the portside sliding deck door in the galley—a very unusual arrangement indeed.) A salon staircase provides easy access to the flybridge, and the salon dimensions are quite generous thanks to the wide beam. Two of the staterooms are fitted with double berths, and both heads have separate stall showers. Additional features include a good-size cockpit with transom door and storage bins, wide and well-secured sidedecks, and a spacious flybridge with room for a dinghy. Standard 485-hp 6-71s will cruise the 52 Sedan MY at 17 knots and reach about 20 knots top. ❑

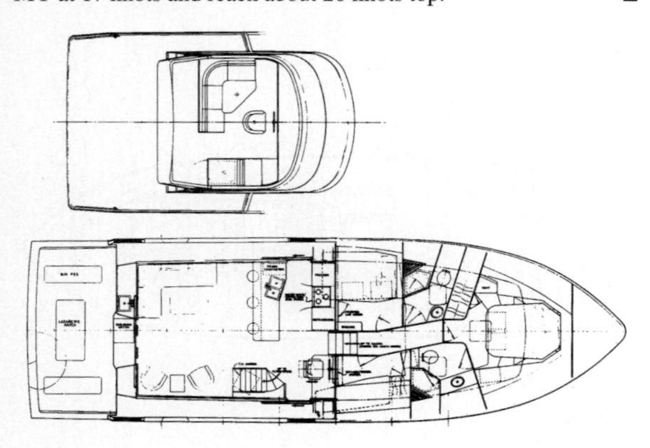

See page 252 for pricing information.

OFFSHORE 55/58 PILOTHOUSE

SPECIFICATIONS

Length	55'0'	Fuel	700 gals.
Beam	16'10"	Hull Type	Modified-V
Draft	4'4"	Deadrise Aft	12°
Weight	52,000#	Designer	Wm. Crealock
Water	400 gals.	Production	1990–Current

Pilothouse yachts have a strong following in the Northwest where the advantages of an indoor helm can be appreciated year-round. Built on a modified-V hull with moderate deadrise aft and cored hullsides, the Offshore 55 PH is a particularly good-looking yacht with plenty of beam and interior volume. Her three-stateroom layout is conventionally arranged with the galley in the salon and the master stateroom beneath the pilothouse. Both heads have shower stalls, and the forward VIP stateroom has a centerline island berth. The pilothouse has wing doors port and starboard and a convenient pass-thru to the galley. Having the salon on the cockpit level makes the entire deckhouse area into a vast entertainment platform at water's edge. Topside, the extended flybridge has plenty of guest seating and space for a dinghy. Standard 485-hp Detroit 6-71s will cruise the 55 PH at 16–17 knots and optional 550-hp 6V92s cruise around 18 knots. Note that the Offshore 58 PH is basically the same boat with an enlarged salon. ❑

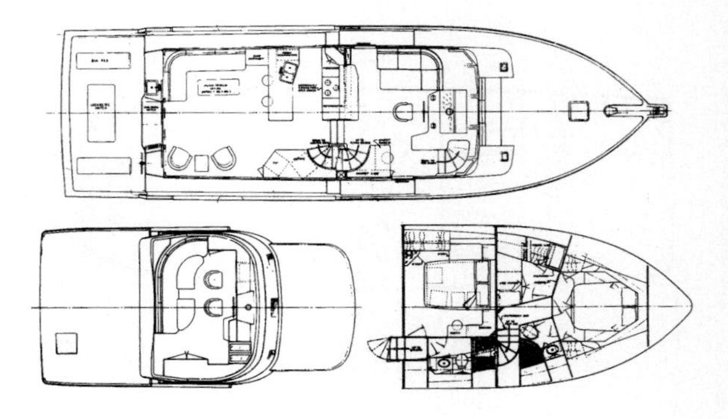

See page 252 for pricing information.

PT 35 SUNDECK

SPECIFICATIONS

Length	35'4"	Water	100 gals.
LWL	31'10"	Fuel	300 gals.
Beam	12'6"	Hull Type	Semi-Disp.
Draft	3'0"	Designer	J. Norek
Weight	20,000#	Production	1984–90

The PT 35 will appeal to those seeking a compact double-cabin trawler without the exterior teak woodwork found on most Asian imports of her vintage. She's heavily built on a solid fiberglass hull with a relatively wide beam and a cutaway keel. Two floorplans were offered: the galley-up layout features a huge forward stateroom at the expense of a very confining salon, while the galley-down arrangement opens up the salon and results in a more practical layout. Either way, a lower helm was standard, and the full-beam master stateroom is comfortable and well-arranged. Note the sliding deck doors in the salon. Just about everything below is built of teak, trimmed with it, or covered by it, and the quality of the joinerwork is quite impressive. There were several engine options available including both single- and twin-diesel applications. She may look like a trawler, but, with a single 200-hp Perkins, the PT 35 Sundeck can achieve a top speed of 12–13 knots. Twin-diesel models can reach about 20 knots. ❑

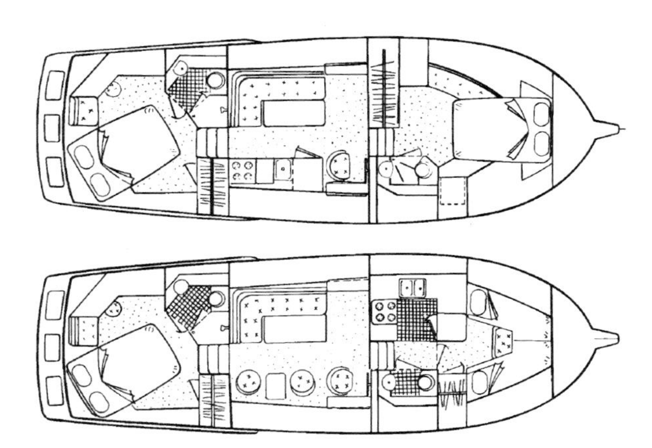

See page 252 for pricing information.

PT 42 COCKPIT MOTOR YACHT

SPECIFICATIONS

Length	42'0"	Water	150 gals.
LWL	38'3"	Fuel	300 gals.
Beam	13'6"	Hull Type	Semi-Disp.
Draft	3'6"	Designer	John Norek
Weight	25,000#	Production	1984–90

The PT 42 Cockpit MY was a fairly popular model a few years back with a good deal of value built into her affordable price tag. Imported from Taiwan, she's basically a PT 38 Sundeck with a cockpit. She was constructed on a solid fiberglass modified-V hull with moderate beam and an extended keel below for stability. The overall quality of the PT 42 is about average, and she has a modern, clean-cut deckhouse profile. Below, the all-teak interior is arranged in the conventional manner with a double berth aft and V-berths forward. The galley is down and almost hidden from the salon, so the layout isn't as open as some. Deck access doors are port and starboard in the salon, and a lower helm was standard. The small cockpit makes boarding easy, and the full-width aft deck is large enough for several pieces of outdoor furniture. With twin 225-hp Lehman diesels, the PT 42 will cruise at 14 knots and reach a top speed of about 17. Larger 250-hp Cummins diesels will add one to two knots to those speeds. ❑

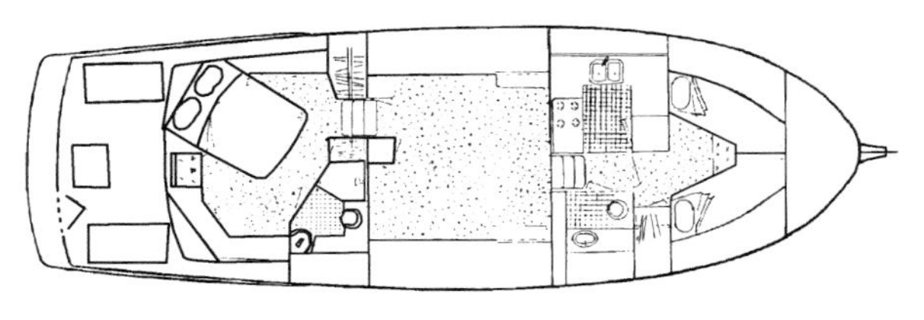

See page 252 for pricing information.

PT 52 COCKPIT MOTOR YACHT

SPECIFICATIONS

Length	52'4"	Water	200 gals.
LWL	46'4"	Fuel	600 gals.
Beam	15'9"	Cockpit	NA
Draft	3'0"	Hull Type	Modified-V
Weight	32,500#	Designer	John Norek
Clearance	NA	Production	1986–90

The PT 52 is a straightforward Taiwan import with a clean-cut profile and surprisingly expansive interior accommodations. Built on a solid fiberglass hull with moderate beam, the 52 CMY is actually a PT 46 Sundeck model with a cockpit extension. Two floorplans were offered: the standard two-stateroom layout has a full dinette opposite the galley on the lower level, and the three-stateroom version eliminates the dinette in favor of a third stateroom with a double berth. The main salon is especially open and expansive with wraparound cabin windows and teak paneling. Stall showers are fitted in both heads, and a door in the master stateroom opens directly into the cockpit—a particularly useful feature. Outside, the full-width aft deck and flybridge are about average in size with plenty of entertaining space for guests. With a pair of relatively small 270-hp Cummins VT-555 diesels (one of several engine options), the PT 52 will cruise at a very respectable 16 knots and reach a top speed of around 20 knots. ❑

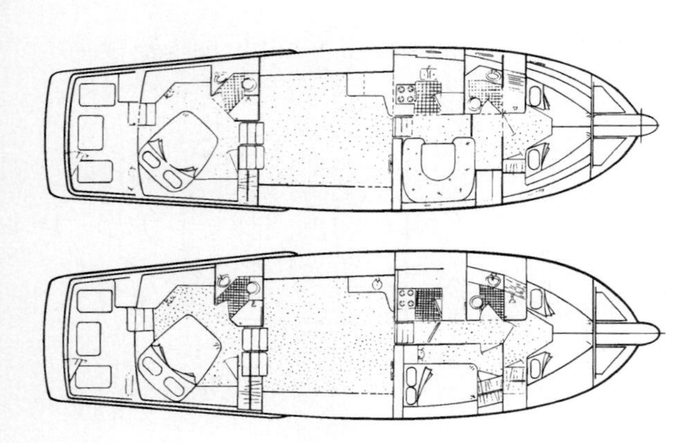

See page 252 for pricing information.

PACEMAKER 40 MOTOR YACHT

SPECIFICATIONS

Length	39'11"	Water	82 gals.
LWL	NA	Fuel	260/300 gals.
Beam	13'11"	Hull	Modified-V
Draft	2'11"	Deadrise Aft	8°
Weight	24,000#	Designer	Pacemaker
Clearance	12'10"	Production	1972–80

Built on the same hull as the Pacemaker 40 SF, the Pacemaker 40 MY proved to be a popular boat with mid-1980s motor yacht enthusiasts. Her traditional flush deck profile is obviously dated today, but she's a comfortable boat inside with enough room for cruisers and liveaboards alike. Early models had the galley aft in the salon with twin single berths in the owner's stateroom. In 1976, teak replaced the mahogany interior, and the layout was revised with the galley moved forward from the salon for a notably larger and more open entertainment area. A walkaround queen berth in the master stateroom also became available in 1976. Seating for five is provided on the flybridge, and visibility from both helm positions is very good. Standard 454-cid gas engines will cruise the Pacemaker 40 Motor Yacht at 15–16 knots and reach a top speed of around 23 knots. The GM 6-71N diesels (17–18 knots cruise) were a popular option. Note the fuel capacity was increased to 300 gallons in 1977. ❑

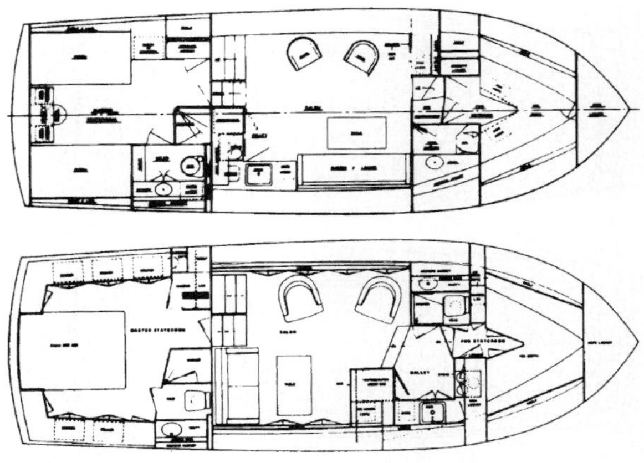

See page 252 for pricing information.

PACEMAKER 46 MOTOR YACHT

SPECIFICATIONS

Length	46'3"	Fuel, Std.	300 gals.
Beam	15'3"	Fuel, Opt.	500 gals.
Draft	4'0"	Hull Type	Modified-V
Weight	42,000#	Deadrise Aft	4°
Clearance	17'1"	Designer	D. Martin
Water	160 gals.	Production	1977–80

There are those who believe the Pacemaker 46 to be one of the best-looking flush-deck motor yachts ever designed. (Indeed, when Pacemaker went out of business in 1980, the molds to the 46 were picked up by Uniflite and later by Chris Craft.) She's built on a solid fiberglass hull with moderate beam and an almost-flat bottom aft. Her standard three-stateroom floorplan is uniquely arranged: most were built with *two* salons with the aft deck fully enclosed and paneled. The step-down galley is open to the lower salon, and the forward head is located in the forepeak—very unusual. A guest cabin adjacent to the master stateroom doubles as a den/office with a desk, bookshelves, and cabinets. The interior is finished with traditional teak woodwork, and large cabin windows provide good natural lighting. Additional features include wide sidedecks, wing doors, foredeck seating, and a roomy bridge. A good performer, 435-hp 8V71 diesels will cruise the Pacemaker 46 MY at a respectable 19 knots and deliver a top speed of about 22 knots. ❑

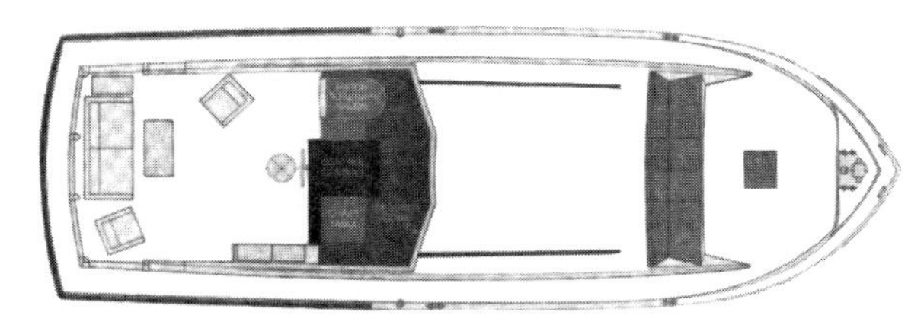

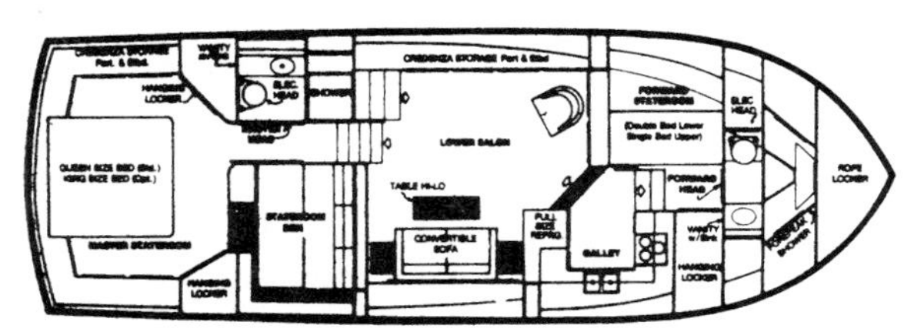

See page 252 for pricing information.

PACEMAKER 57 MOTOR YACHT

SPECIFICATIONS

Length	57'2"	Fuel	980 gals.
LWL	52'0"	Hull Type	Modified-V
Beam	17'2"	Deadrise Aft	6°
Draft	4'0"	Designer	D. Martin
Weight	66,300#	Production	1977–80
Clearance	NA		

The Pacemaker 57 MY is best described as a Pacemaker 62 CMY minus the cockpit. Like the earlier Pacemaker 62 with her combination glass-over-wood and fiberglass superstructure, the Pacemaker 57 is built with a fiberglass hull and flybridge with full teak decks. She features the same interior layout as the 62 CMY with the owner and guest staterooms aft and a single stateroom forward. The mid-level salon and separate galley area are three steps down from the wheelhouse. This lower salon is generally used as a formal dining room. On the deckhouse level, the spacious full-width main salon is open to the helm and extends aft nearly to the transom. A small afterdeck leaves room for line-handling duties. The crew quarters forward are fitted with stacked single berths with an adjoining head and stall shower. A washer/dryer unit and engine room access door are also forward. A handsome yacht, standard 8V71TI diesels provide a comfortable cruising speed of 16–17 knots and a top speed of about 20 knots. ❑

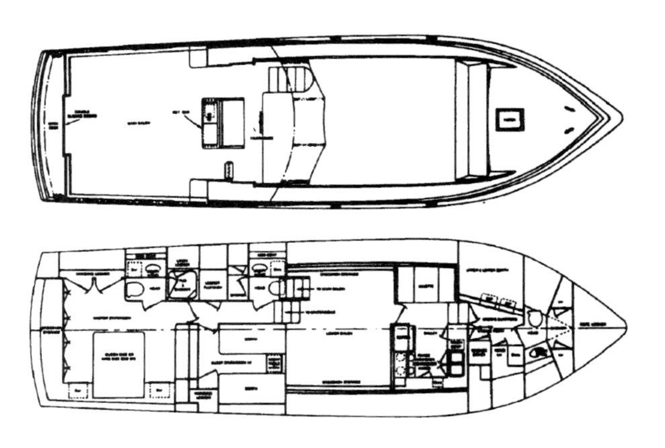

See page 252 for pricing information.

PACEMAKER 62 COCKPIT MY

SPECIFICATIONS

Length	62'0"	Fuel	980 gals.
Beam	17'2"	Hull Type	Modified-V
Draft	4'0"	Deadrise Aft	6°
Weight	73,000#	Designer	D. Martin
Clearance	NA	Production	1973–80
Water	300 gals.		

The Pacemaker 62 Cockpit MY is basically a Pacemaker 57 MY with a cockpit extension. (Interestingly, the 62 CMY preceded the Pacemaker 57 MY by several years—it's usually the other way around.) The hull and flybridge are solid fiberglass, and the superstructure is a combination of fiberglass and glass-over-mahogany. Unlike most flush-deck motor yachts with their open aft decks, the 62 CMY features a fully enclosed and paneled aft deck which is essentially a full-width main salon. The second salon on the lower level is used as a formal dining area. The galley is forward of the lower-level salon, and a guest stateroom is forward. (Note that the engine room is accessed from the crew quarters.) There are two big staterooms aft, and the master stateroom has a full-height door leading directly to the cockpit. Outside, the decks are teak, and a small observation platform overlooks the cockpit. Standard 8V71TIs will cruise the Pacemaker 62 CMY at 15–16 knots, and optional 8V92TIs will cruise at 19–20 knots. ❑

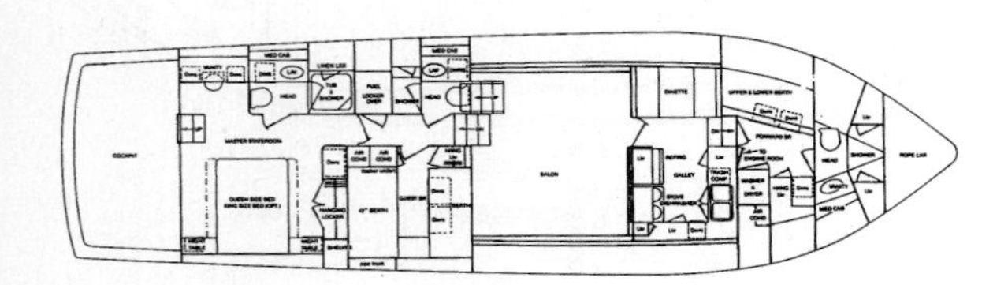

See page 252 for pricing information.

PACEMAKER 62 MOTOR YACHT

SPECIFICATIONS

Length	62'0"	Fuel	980 gals.
Beam	17'2"	Hull Type	Modified-V
Draft	4'0"	Deadrise Aft	6°
Weight	73,000#	Designer	D. Martin
Water	300 gals.	Production	1973–80

With her classic flush-deck profile and opulent interior, the Pacemaker 62 MY is a traditional cruising yacht with genuine liveaboard accommodations. The hull and flybridge are solid fiberglass, and the superstructure is a combination of fiberglass and glass-over-mahogany. Unlike most flush-deck motor yachts with their open-air aft decks, the 62 MY features a fully enclosed and paneled aft deck which serves as the main salon. This expansive entertainment area uses the full width of the hull and is open to the helm. The second salon on the lower level is used as a formal dining area. Designed as a crewed yacht, the galley is forward of the lower-level salon, and the crew quarters are at the bow. (The engine room is accessed from the crew quarters.) There are two guest staterooms aft together with the master stateroom with its *en suite* head. Topside, the extended flybridge is huge with plenty of room for a dinghy. Note the teak decks. Standard 8V71TIs will cruise the Pacemaker 62 MY at 15–16 knots. ❑

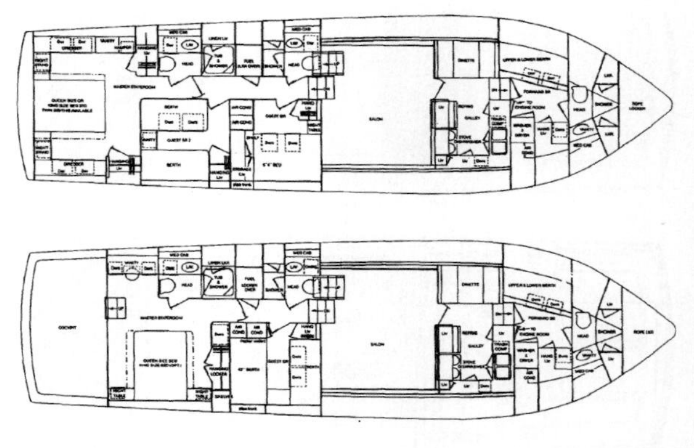

See page 252 for pricing information.

PACEMAKER 66 MOTOR YACHT

SPECIFICATIONS

Length	66'0"	Water	300 gals.
LWL	60'10"	Fuel	1,100/1,400 gals.
Beam	17'2"	Hull Type	Modified-V
Draft	4'0"	Deadrise Aft	6°
Weight	73,000#	Designer	D. Martin
Clearance	NA	Production	1977–1980

The Pacemaker 66 Motor Yacht has the upper-class elegance that few other limited-production motor yachts can match. Sharing the same construction characteristics as the Pacemaker 62s, the hull is rounded at the chines and employs only modest deadrise at the transom. The 66 MY remained popular right up until Pacemaker went out of business in 1980. Uniflite then acquired the molds, but before they delivered the first boat (three years later), Uniflite itself had gone under. Chris Craft then took over, eventually building two boats. With her extended full-beam main salon, separate formal dining area, palatial master stateroom, and sumptuous guest accommodations, the Pacemaker 66 MY offers extravagant entertaining and cruising possibilities. While the aft deck is somewhat small, the flybridge dimensions are expansive. Standard power was a pair of GM 12V71N diesels for a cruising speed of 15–16 knots and a top speed of about 18 knots. The 650-hp TI versions of the same motors will cruise at 18–19 knots. The fuel capacity was increased in 1978. ❏

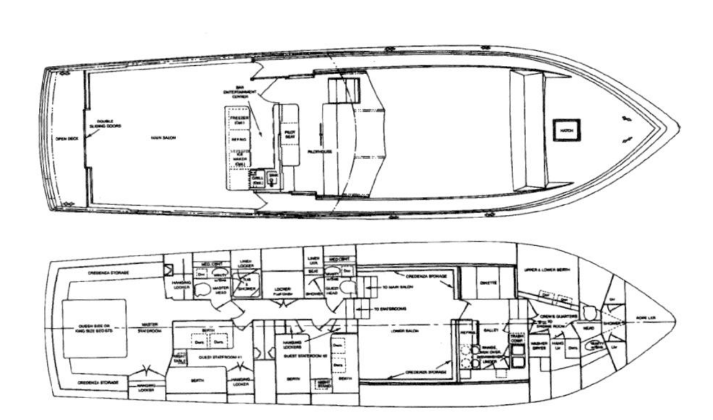

See page 252 for pricing information.

PEARSON 38 DOUBLE CABIN

SPECIFICATIONS

Length	37'9"	Water	100 gals.
LWL	31'11"	Fuel	300 gals.
Beam	13'10"	Hull Type	Deep-V
Draft	3'9"	Deadrise Aft	19°
Weight	25,000#	Designer	Hunt Assoc.
Clearance	13'0"	Production	1988–91

The Pearson 38 DC is a straightforward aft-cabin design with a modern profile to go with her conventional cabin layout. She was built on the same deep-V (19° deadrise) hull used in the production of Pearson's 38 Convertible, with a solid fiberglass bottom and balsa coring in the hullsides and decks. Deep-V hulls are rarely seen in family boats since they tend to roll more than a conventional modified-V hull. The Pearson 38 will excel, however, in poor sea conditions when other cruising boats are running for cover. Inside, the floorplan is arranged with the galley down and walkaround queen berths in both staterooms—a neat trick in such a small boat. Note that both heads are fitted with stall showers (another big plus). While the interior dimensions are quite generous for a boat this size, the 38's aft deck is too small to be put to any good use. Twin 320-hp gas engines were standard (16 knots cruise/24–25 top), and 320-hp Cat diesels were optional (20-knots cruise/24 knots top). ❏

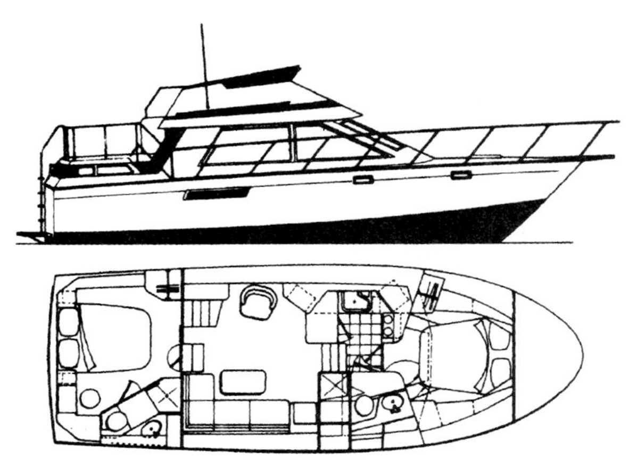

See page 252 for pricing information.

PEARSON 43 MOTOR YACHT

SPECIFICATIONS

Length	43'0"	Fuel	300 gals.
LWL	37'6"	Hull Type	Modified-V
Beam	14'10"	Deadrise Aft	5°
Draft	3'0"	Designer	Bill Shaw
Weight	22,000#	Production	1971–76
Water	200 gals.		1984–87

Pearson originally introduced this houseboat-style cruiser back in 1973, calling her the Portsmouth 43. Designed for inland and coastal waters, she was built on a flat-bottom modified-V hull with hard chines, a flared bow, and very low freeboard at the transom. The Portsmouth 43 remained in production through 1976 when she was retired from the fleet only to reappear eight years later as the Pearson 43 MY. The interior—arranged on a single level from the helm aft and featuring a hidden mini-cabin below the forward salon sole— was updated in the MY to include double berths in the staterooms and a brighter decor. Bridge overhangs shade the walkaround decks, and the extended flybridge is huge with room for a wet bar, sunbathers, and a dinghy. (Note that all but the latest models have a false stack on the bridge.) A hard-riding boat in any kind of chop, twin 454-cid gas engines (with V-drives) will cruise at 15–16 knots and reach a top speed of around 25 knots. Several diesel options were available. ❑

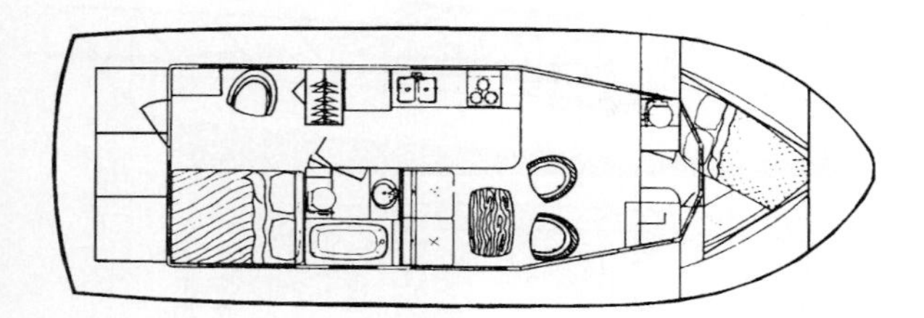

See page 252 for pricing information.

PILGRIM 40

SPECIFICATIONS

Length	40'0"	Water	240 gals.
Beam	14'0"	Fuel	142 gals.
Draft	3'6"	Hull Type	Disp.
Weight	25,000#	Designer	Ted Gozzard
Clearance	22'0"	Production	1984-91

With her tugboat profile, fake smokestack, and trolley-car windows the Pilgrim 40 is a genuine eye-catcher in a sea of look-alike designs. Over fifty were built at Pilgrim Marine in Ontario during her production. She's built on a full keel, full displacement hull with a plumb bow and rounded bilges. A bow thruster was standard, and the prop is fully protected by the deep rudder and skeg. Inside, the Pilgrim's cozy pilothouse—three steps above the salon and galley level—is particularly well-arranged with excellent visibility ahead and sliding deck access doors port and starboard. The galley is huge for a 40-footer (there's a big storage area beneath the galley sole), and the head compartment is just forward of the salon. The cockpit is shaded by the extended flybridge, and 30" raised bulwarks provide excellent sidedeck security around the house. An economical and altogether practical liveaboard cruiser, most Pilgrim 40s were fitted with a 100-hp Westerbeke diesel, which provides a steady 8-knot cruising speed at 3 gph. Note that a flybridge was optional. ❑

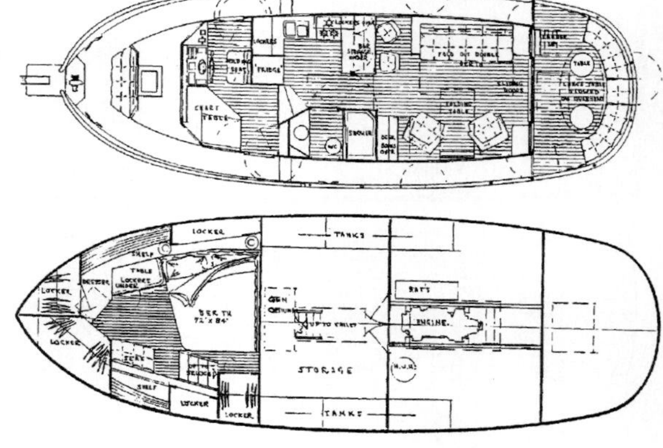

See page 252 for pricing information.

PRESENT 42 SUNDECK MY

SPECIFICATIONS

Length	41'10"	Clearance	NA
LWL	38'0"	Water	200 gals.
Beam	13'8"	Fuel	450 gals.
Draft	3'6"	Hull Type	Semi-Disp.
Weight	20,000#	Production	1983–87

Chung Hwa is generally regarded as one of the more experienced Taiwan builders with a string of successful models sold into the U.S. market. The Present 42 (also known as the CHB 42 MY) is an affordable double-cabin sundeck design with a trawler-style profile, conventional semi-displacement hull, full teak interior, and simulated lapstrake hullsides. Aside from the traditional interior woodwork and practical two-stateroom dinette layout, the Present features a fairly large salon separated from the galley with a convenient breakfast bar. A lower station was standard along with a convenient deck access door, and the aft head contains the only stall shower. Present offered a range of diesel options for the 42 Sundeck, but most came with twin 135-hp or 225-hp Lehmans or twin 200-hp Perkins. Economical to operate, the larger engine options will provide cruising speeds in the 12–14 knot range while the less powerful Lehmans reduce the Present 42 Sundeck to trawler speeds. At hull speeds, the cruising range exceeds 600 miles. ❑

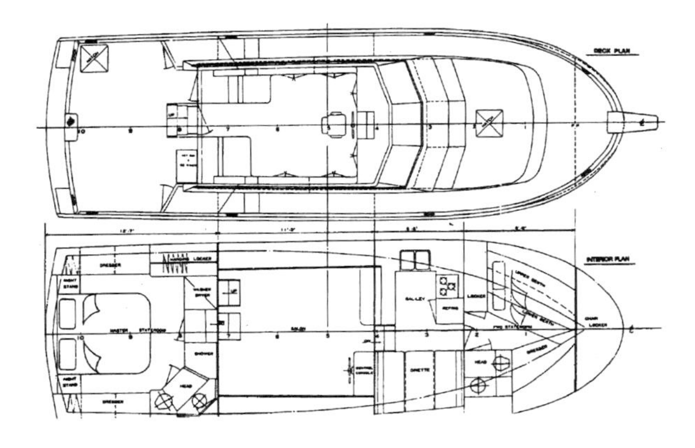

See page 252 for pricing information.

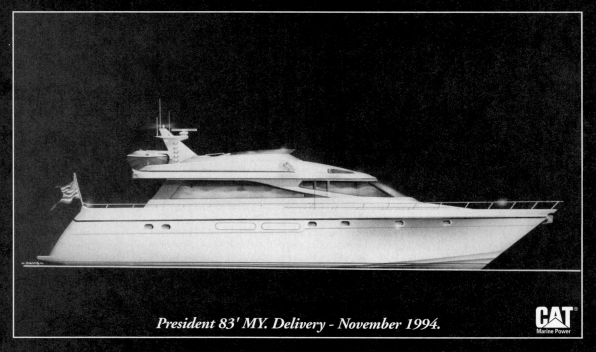

PRESIDENT 395 DOUBLE CABIN

SPECIFICATIONS

Length	38'5"	Fuel	300 gals.
Beam	13'0"	Hull Type	Modified-V
Draft	3'2"	Deadrise Aft	NA
Weight	24,600#	Designer	President
Water	100 gals.	Production	1992–Current

The President 395's stylish reverse transom and integral swim platform combine to make her one of the better-looking under-40' double-cabin designs on the market. Smallest model in the current President fleet, the 395 is built on a beamy modified-V hull with a shallow, full-length keel and a solid fiberglass bottom. Two galley-down floorplans are currently available: a dinette layout with an offset double berth in the forward stateroom, and an alternate arrangement with an inside helm, no dinette, and separate stall showers in both heads. The light oak woodwork applied throughout the interior is impressively finished, and the big wraparound cabin windows create a bright and cheerful salon. Outside, the aft deck is large enough for entertaining, and there's adequate guest seating on the bridge. A good performer with optional 300-hp Cummins diesels, the President 395 will cruise at 21 knots and reach 24–25 knots top. Note that the President 37 Double Cabin (1986–92) is basically the same boat without the Euro-style reverse transom. ❑

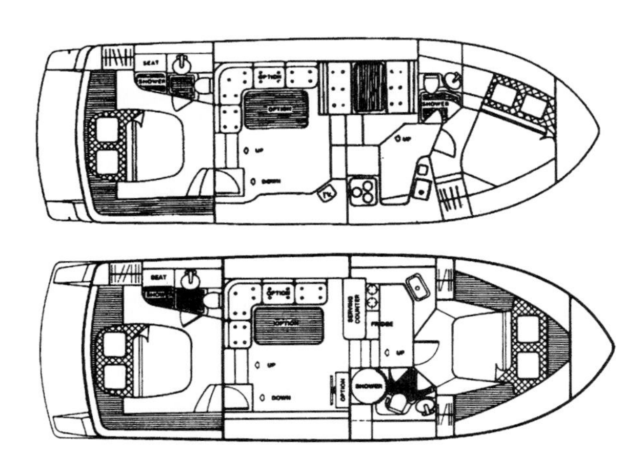

See page 252 for pricing information.

PRESIDENT 41 DOUBLE CABIN

SPECIFICATIONS

Length	40'6"	Water	120 gals.
LWL	35'10"	Fuel, Std.	420 gals.
Beam	13'5"	Hull Type	Modified-V
Draft	2'10"	Deadrise Aft	14°
Weight	22,500#	Designer	G. Stadel
Clearance	NA	Production	1982–88

With 116 delivered in the U.S., the President 41 was one of the most popular Asian imports during the 1980s due to her moderate price and roomy accommodations. While she may look a lot like a trawler, she's actually built on a conventional modified-V hull with a deep forefoot and a prop-protecting 18" keel. Inside, the two-stateroom interior of the President 41 is unusually spacious and completely finished with plenty of teak paneling and cabinetry. A lower station was standard, and the galley and dinette are down from the salon level. Notably, a double berth is fitted in the forward stateroom and twin singles were originally standard in the aft cabin—definitely an unusual layout. Topside, the helm console is set well forward on the bridge with guest seating for five. Most President 41s were sold with the 120/135-hp Lehman diesels which provide an economical 8–9 knot cruising speed. The larger 225-hp Lehmans will cruise at around 13 knots and reach a top speed of 15–16 knots. ❑

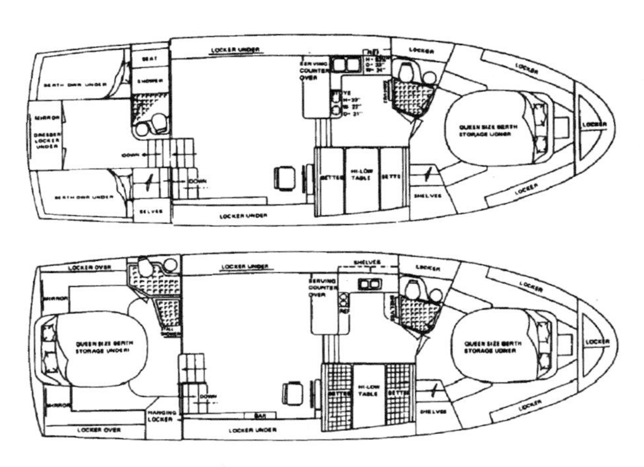

See page 252 for pricing information.

PRESIDENT 43 DOUBLE CABIN

SPECIFICATIONS

Length	42'6"	Water	120 gals.
LWL	37'10"	Fuel	420 gals.
Beam	13'10"	Hull Type	Modified-V
Draft	3'2"	Deadrise Aft	NA
Weight	28,000#	Designer	G. Stadel
Clearance	12'4"	Production	1984–90

A good-looking boat with a trawler-style profile and set-back house, the President 43 MY is built on a solid fiberglass hull with moderate beam and a shallow full-length keel below. At just 28,000 lbs., the President is a relatively lightweight boat for her size. Her two-stateroom, galley-down floorplan was available with queen berths in both cabins (an alternate layout has stacked single berths forward and the convenience of a second stall shower in the guest head), and the interior is fully finished with teak woodwork and cabinetry. The staterooms in both floorplans are quite large, while the salon dimensions are about average. Note that a lower helm station was optional. Additional features include a spacious aft deck platform, secure sidedecks, a fairly small flybridge, and a standard factory hardtop. Several diesel options were offered over the years including Lehmans, 8.2 GMs, and Cummins. The popular 275-hp Lehmans will cruise the President 43 MY at 14 knots and reach about 17 knots top. A total of 42 were brought into the U.S. ❑

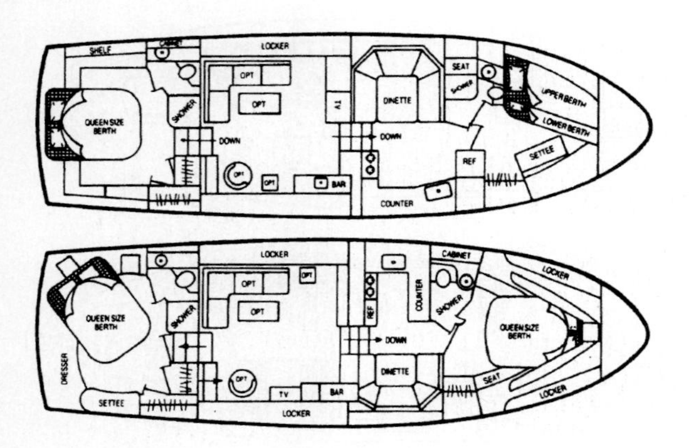

See page 253 for pricing information.

PRESIDENT 46/485 MOTOR YACHT

SPECIFICATIONS

Length	48'9"	Water	200 gals.
Beam	15'9"	Fuel	500 gals.
Draft	4'2"	Hull Type	Modified-V
Weight	37,840#	Deadrise Aft	16°
Clearance	15'6"	Production	1987–Current

The President 485 MY is actually an updated version of the popular and long-running President 46 Sundeck (1987–92), an affordably priced family cruiser noted for her wide beam, sleek profile, and generous interior accommodations. The only difference between the two models is the new Euro-style integral swim platform of the 485—a modest design change that adds a couple of feet to the hull length and much to her already stylish appearance. Inside, the two-stateroom floorplan of the 485 is essentially unchanged from her predecessor. The galley and dinette are down, and a lower helm is standard. (Several alternate layouts are available including a four-stateroom layout for charter-boat applications.) Regardless of the floorplan, the wide-open interior is very roomy thanks to the reasonably wide beam. The woodwork (teak or light oak) is well-crafted and very appealing. Note that the spacious engine room is accessed via a door in the aft cabin wardrobe. Powered with 375-hp diesels, the President 485 will cruise at 16–17 knots and reach 19 knots wide open. ❑

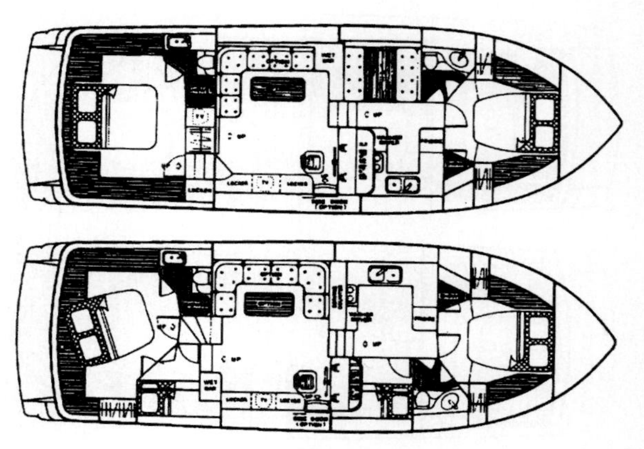

See page 253 for pricing information.

PRESIDENT 52/545 COCKPIT MY

SPECIFICATIONS

Length..........................54'9"	Water.....................200 gals.
Beam ....15'9"	Fuel........................500 gals.
Draft...............................4'4"	Hull Type........Modified-V
Weight.....................46,110#	Deadrise Aft ....16°
Clearance.....................15'6"	Production...1987–Current

A popular boat for President, the good-looking 545 Cockpit MY pictured above is basically a President 485 MY with a six-foot cockpit extension. She's built on a beamy modified-V hull with cored hullsides, an integral swim platform, and a relatively steep 16° of deadrise at the transom. Note here that the earlier President 52 CMY (1987–91) is the same boat without the 545's Euro-style transom. The standard two-stateroom, galley/dinette-down floorplan is arranged with walkaround double berths in both staterooms and a surprisingly roomy salon with a lower helm. (Several alternate floorplans are available.) The teak (or oak) interior woodwork is well-finished, and the oversize cabin windows give the interior a wide open and spacious character. Outside, the aft deck and flybridge dimensions are about average for a boat this size. The cockpit, however, is big enough for some serious fishing and includes a transom door. Standard 550-hp 6V92 Detroits will cruise the President 545 CMY at 17 knots and reach a top speed of 22–23 knots. ❏

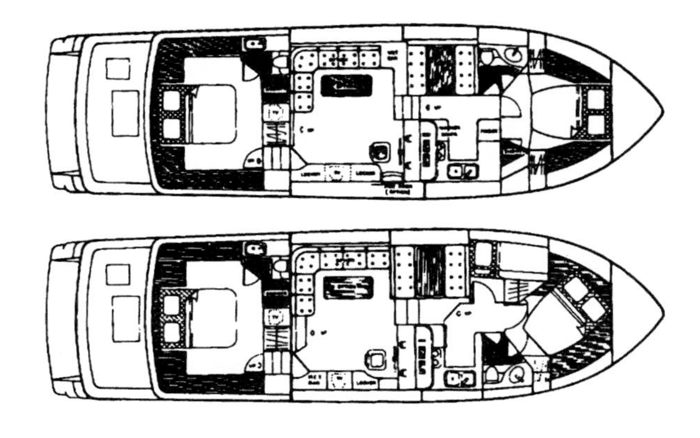

See page 253 for pricing information.

SABRELINE 34

SPECIFICATIONS

Length	34'0"	Fuel	250 gals.
Beam	12'6"	Cockpit	70 sq. ft.
Draft	3'3"	Hull Type	Modified-V
Weight	17,800#	Deadrise Aft	14°
Clearance	12'8"	Designer	Sabre
Water	160 gals.	Production	1991–Current

An upright, slightly Downeast profile and an attractive interior characterize the Sabreline 34 Sedan, a modern trawler-style design from Sabre Yachts. She's built on the same modified-V hull used in the production of the original Sabreline 36 with moderate beam and a shallow keel. Like all Sabre boats (sail or power), there's no shortage of craftsmanship and attention to detail in the way she's put together. Her traditional teak interior is arranged with the galley forward in the salon opposite the lower helm and a walkaround island berth in the stateroom. Everything is carefully arranged with an eye toward practicality and comfort. The large salon windows all open (so does the center windshield forward) and ventilation in this boat is exceptional. The 34 isn't a wide-beam boat, so the interior dimensions are somewhat compact. Features include good storage, excellent engine access, stall shower with wet locker (nice touch), a roomy cockpit, and a quiet underwater exhaust system. Economical to operate and a good performer, the Sabreline 34 will cruise at 18 knots with twin 210-hp Cummins diesels. ❑

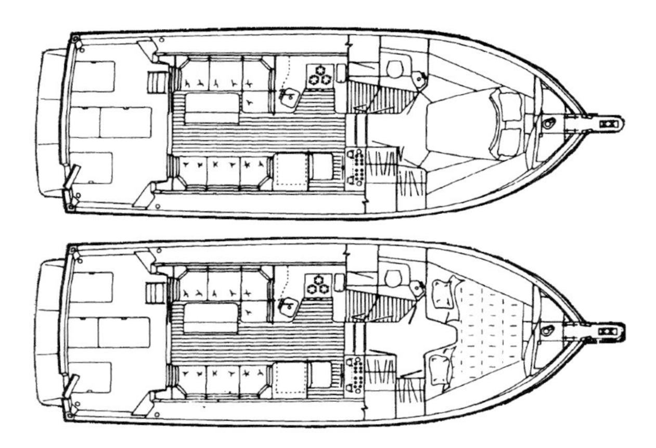

See page 253 for pricing information.

SABRELINE 36

SPECIFICATIONS

Length	36'0"	Water	225 gals.
LWL	32'4"	Fuel	250 gals.
Beam	12'6"	Cockpit	NA
Draft	4'3"	Hull Type	Modified-V
Weight	20,000#	Designer	Sabre
Clearance	12'6"	Production	1989–Current

Marketed as a "Fast Trawler," the Sabreline 36 is a well-crafted family cruiser with a trunk cabin profile and a traditional all-teak interior. She's built on a fully cored modified-V hull with hard chines, a relatively fine entry, and moderate transom deadrise. Unlike many of today's powerboat designs, the Sabreline 36 is not a notably roomy boat inside (the 12'6" beam is about average for a 36-footer), although her white Formica counters and large cabin windows give the interior a wide open and very inviting character. The standard layout has twin berths in the aft cabin (a double berth is optional), and the galley is aft in the salon. Note that the master stateroom has a companionway exit to the aft deck. Additional features include wide sidedecks, a deck access door at the lower helm, underwater exhausts, and a very large flybridge. A good-running boat, standard 250-hp 8.2 GM diesels will cruise the Sabreline 36 a very respectable 17–18 knots and reach 22 knots wide open. ❑

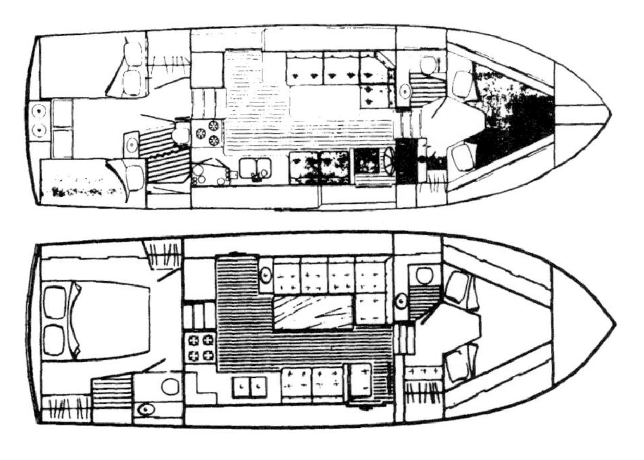

See page 253 for pricing information.

SEA RANGER 36 SUNDECK

SPECIFICATIONS

Length	35'8"	Water	150 gals.
LWL	30'10"	Fuel	300 gals.
Beam	13'4"	Hull Type	Semi-Disp.
Draft	3'3"	Designer	J. Griffin
Weight	19,000#	Production	1980–86
Clearance	NA		

This plain-Jane Taiwan trawler offers up a decidedly unattractive profile (note the slab-sided hull and the midships wood planking—very tacky) in exchange for an exceedingly roomy tri-cabin layout. Also known as the C&L 36 Trawler, the Sea Ranger 36 was Taiwan-built on a solid fiberglass hull with a generous beam, hard chines, and a full-length prop-protecting keel. Her spacious all-teak interior is quite impressive for a 36-footer, especially the wide-open salon and oversize master stateroom with its walkaround island bed and excellent storage. Indeed, the 36 Sundeck has more interior square footage than most double cabins her size. Outside, the greatly elevated sundeck (for headroom below) is only three steps down from the flybridge with its centerline helm and guest seating. (Here again, the sundeck dimensions are very expansive for a boat this size.) A single or twin 120-hp diesels will cruise the Sea Ranger 36 efficiently at 7.5 knots and speeds to 13 knots can be attained with twin engines. ❑

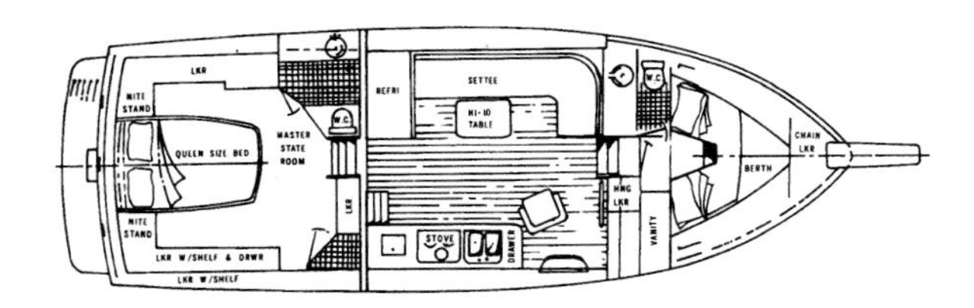

See page 253 for pricing information.

SEA RANGER 39 SEDAN

SPECIFICATIONS

Length	38'3"	Water	150 gals.
LWL	33'4"	Fuel	600 gals.
Beam	13'8"	Hull Type	Semi-Disp.
Draft	3'3"	Designer	J. Griffin
Weight	23,000#	Production	1980–89
Clearance	NA		

Sedan-style trawlers are popular with those who are willing to trade the luxuries of a private aft stateroom for the versatility of an enlarged single-level deckhouse layout. A good-looking boat, the Sea Ranger 39 has a distinctive profile (note the rakish flybridge with its Euro-style overhang supports) to go with her practical floorplan. Her all-teak interior is arranged in the conventional manner with the portside galley forward and a built-in dinette and settee aft. While most 39 Sedans were sold with a two-stateroom layout, an alternate single-stateroom floorplan with a centerline double bed and tons of storage was also available. Outside, the wide sidedecks (partially shaded by a bridge overhang) are well-secured with raised bulwarks all around the house. The extended flybridge is enormous. With 600 gallons of fuel, the Sea Ranger 39 Sedan (also known as the C&L 39 Sedan) is capable of a thousand-mile cruising range (at her 7.5-knot hull speed) with twin 124-hp Volvo or 120-hp Lehman diesels. ❑

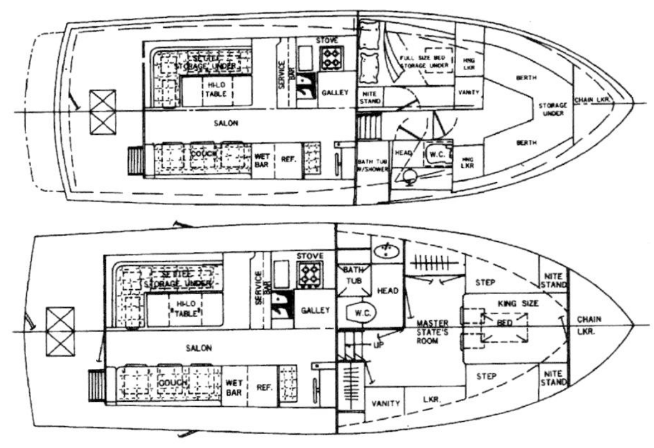

See page 253 for pricing information.

SEA RANGER 39 SUNDECK

SPECIFICATIONS

Length	38'3"	Water	150 gals.
LWL	33'4"	Fuel	600 gals.
Beam	13'8"	Hull Type	Semi-Disp.
Draft	3'3"	Designer	J. Griffin
Weight	23,000#	Production	1983–89
Clearance	NA		

There isn't a lot to distinguish the Sea Ranger 39 Sundeck from the raft of other inexpensive 39-to-40' Asian imports of the mid-1980s. Like most trawlers of her era, she was built on a solid fiberglass semi-displacement hull with flat aftersections and a full-length keel. Sundeck designs like the Sea Ranger 39 are popular with those who want the added open-air deck space of a sundeck together with the enlarged stateroom dimensions. (The trade-off for the full-width sundeck is the convenience of having full walkaround decks.) Note that there is no companionway from the sundeck to the salon; the entryways are the port and starboard salon doors. Inside, the all-teak interior is typical of Asian boats. Additional features include a functional mast and boom, teak decks, pulpit and swim platform, and a big 600-gallon fuel capacity (for an honest 1,000-mile cruising range). Twin 120-hp Lehman diesels (or 124-hp Volvos) will cruise at 7–8 knots and reach a top speed of about 10 knots. ❑

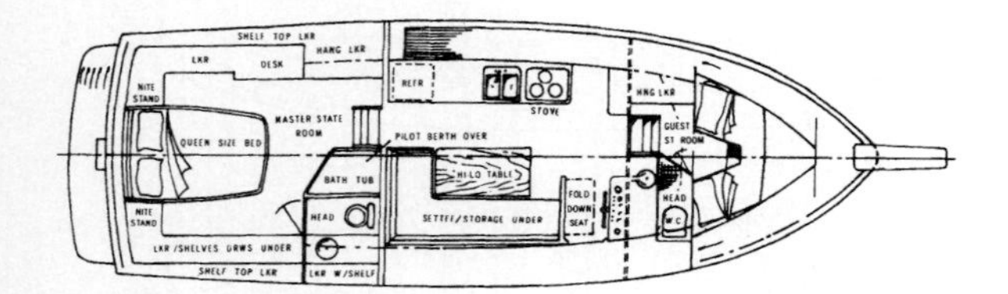

See page 253 for pricing information.

SEA RANGER 45 SUNDECK

SPECIFICATIONS

Length	45'0"	Water	350 gals.
LWL	40'0"	Fuel	850 gals.
Beam	15'3"	Hull Type	Semi-Disp.
Draft	4'0"	Designer	J. Griffin
Weight	37,400#	Production	1983–89
Clearance	NA		

A popular boat, the Sea Ranger 45 is a conventional aft-cabin Taiwan trawler with a big interior, long-range cruising capabilities, and an inexpensive price. She was built on a solid fiberglass semi-displacement hull and the running gear is well-protected by her full-length keel. The 45 has one of the larger aft decks to be found in a boat this size (over 140 sq. ft.), and most were sold with the optional hardtop and waist-high enclosure panels. Two floorplans were offered: a galley-down two-stateroom layout with a wide-open salon, and a more popular three-stateroom arrangement with the U-shaped galley aft in the salon. Either way, the master stateroom is fitted with an island berth, and both heads have stall showers. Additional features include teak decks, a large flybridge, and wide sidedecks. Optional 235/255-hp Volvo diesels will cruise at 14 knots (about 17 knots top), and 375-hp Cats will cruise around 19 knots. Note that the Sea Ranger 50 Cockpit MY is the same boat with a cockpit extension. ❑

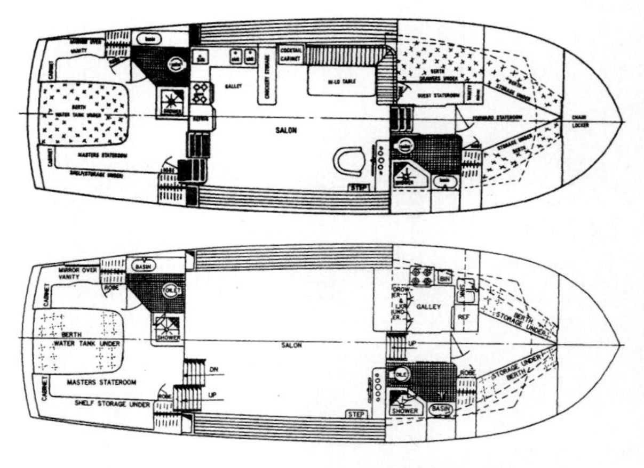

See page 253 for pricing information.

SEA RANGER 47 PILOTHOUSE

SPECIFICATIONS

Length	47'3"	Water	280 gals.
LWL	43'0"	Fuel	710 gals.
Beam	14'8"	Hull Type	Semi-Disp.
Draft	4'4"	Designer	C&L
Weight	27,000#	Production	1980–1989
Clearance	NA		

Pilothouse yachts have long been popular with northern boaters for their weather-protected inside helm. Recent years, however, have seen the popularity of these boats on the increase in warmer climates where many have come to admire the strictly business appearance of a raised pilothouse design as well as her all-weather cruising capabilities. The Sea Ranger 47 was built on a solid fiberglass semi-displacement hull with a relatively narrow beam, flat aftersections, and a full-length prop-protecting keel. Her three-stateroom interior features a roomy full-width midships master stateroom with direct access to the engine room. Both heads have stall showers, and a staircase in the pilothouse provides inside access to the bridge. Additional features include an on-deck day head (a great cruising convenience), a full teak interior, dinghy storage on the extended flybridge, radar arch, and teak decks. Twin 135-hp Lehman diesels (among several engine options) will cruise the Sea Ranger 47 PH efficiently at 7–8 knots and reach a top speed of about 10 knots. ❑

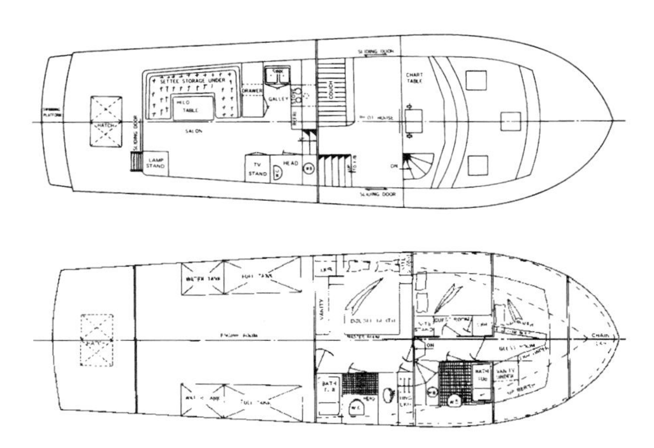

See page 253 for pricing information.

SEA RANGER 51 MOTOR YACHT

SPECIFICATIONS

Length	51'0"	Water	420 gals.
LWL	45'0"	Fuel	1,000 gals.
Beam	16'8"	Hull Type	Semi-Disp.
Draft	4'3"	Designer	J. Griffin
Weight	55,000#	Production	1983–89

The Sea Ranger 51 (also known as the C&L 51 Motor Yacht) enjoyed a good deal of popularity in the mid-1980s due to her attractive price and sensible layout. This is a straightforward flush-deck motor yacht with a wide beam, full walkaround decks, and an extended flybridge—a good cruising design with a choice of several floorplans. Construction is solid fiberglass, and a long keel protects the underwater gear in the event of a grounding. The galley is on the upper level in the four-stateroom layout, and it's down in the three-stateroom floorplan. Either way, the salon is open to the helm, and the interior woodwork is teak throughout. Note that there is no inside access to the bridge. The aft deck and sidedecks are sheltered with bridge overhangs, and all of the decks are teak. Twin 260-hp Cat (and 265-hp 6-71) diesels will cruise the Sea Ranger 51 MY at 11 knots (14 top), and the 650-hp 8V92s will cruise at 15–16 knots (18+ knots top). ❑

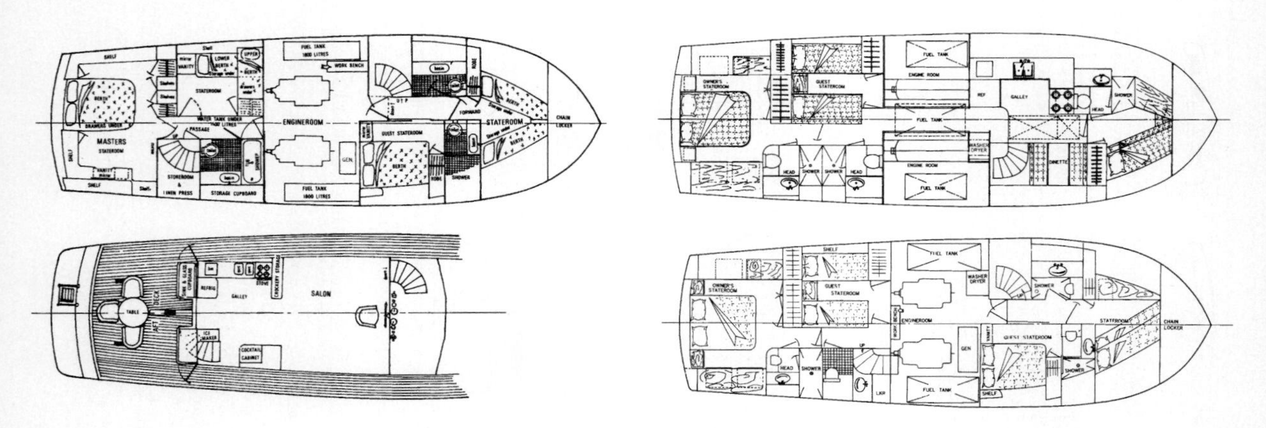

See page 253 for pricing information.

SEA RANGER 55 PH MOTOR YACHT

SPECIFICATIONS

Length	55'0"	Water	300 gals.
LWL	48'6"	Fuel	1,800 gals.
Beam	18'7"	Hull Type	Semi-Disp.
Draft	4'9"	Designer	Herb David
Weight	64,900#	Production	1976–89
Clearance	NA		

Built by C&L Marine in Taiwan and aimed at the West Coast cruising market, the Sea Ranger 55 is a classic low-profile pilothouse design with a wide beam and genuine long-range cruising capabilities. She's constructed on a solid fiberglass hull with hard chines, flat aftersections, and a full-length keel. Notably, with her 18'7" beam, the Sea Ranger 55 is wider than any other production boat under 60 feet. Inside, the all-teak interior is arranged with a full-width master stateroom below the raised pilothouse, two guest staterooms forward, and the galley in the salon—the standard floorplan of most pilothouse yachts. Needless to say, the Sea Ranger 55 is a roomy boat inside thanks to her super-wide beam, and the cabin dimensions are very generous. Additional features include teak decks, a huge extended flybridge with dinghy storage aft, shaded sidedecks, and radar arch. About average in quality, 310-hp 6-71N diesels will power the Sea Ranger 55 to speeds of 10–12 knots. At her 9-knot hull speed, the cruising range is over 2,000 miles. ❑

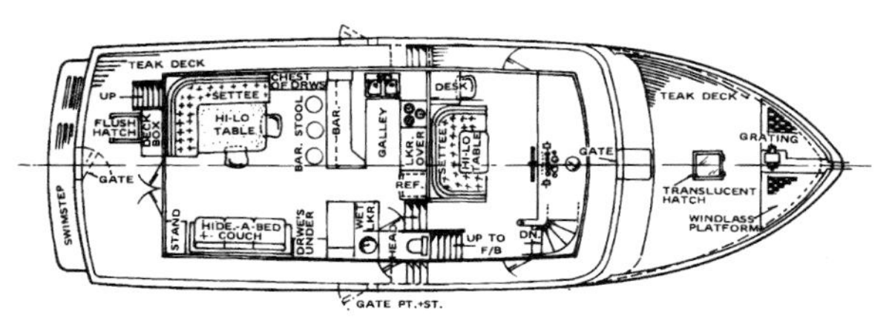

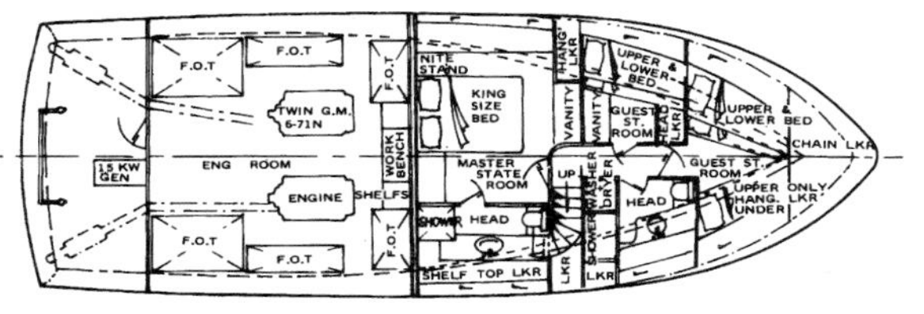

See page 254 for pricing information.

SEA RANGER 65 MOTOR YACHT

SPECIFICATIONS

Length	65'3"	Water	350 gals.
LWL	59'0"	Fuel	1,800 gals.
Beam	18'7"	Hull Type	Semi-Disp.
Draft	4'9"	Designer	Herb David
Weight	72,000#	Production	1976–89
Clearance	NA		

An inexpensive import from Taiwan, the Sea Ranger 65 MY is a capable long-range cruiser with a distinctive trawler-style profile. She's built on a solid fiberglass semi-displacement hull with hard chines, flat aftersections, and a full-length keel. Note the simulated lapstrake hull grooves and the rakish bridge overhangs. Inside, her all-teak interior is arranged with an on-deck galley and fully enclosed wheelhouse on the upper level (plus a convenient day head/powder room) and four staterooms below. The huge aft stateroom (accessed from a teak staircase in the salon) is lavish, and two of the three guest staterooms forward are fitted with double berths. The salon dimensions are about average for a yacht this size, with enough room for a separate dining area as well as a fairly roomy aft deck platform. The walkaround decks are teak, and the extended flybridge is simply huge. Standard 350-hp 8V71Ns will cruise the Sea Ranger 65 MY at an easy 10–11 knots and reach a top speed of about 13 knots. ❑

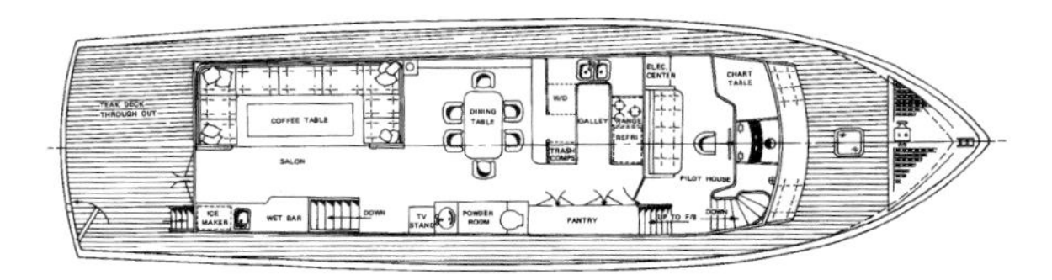

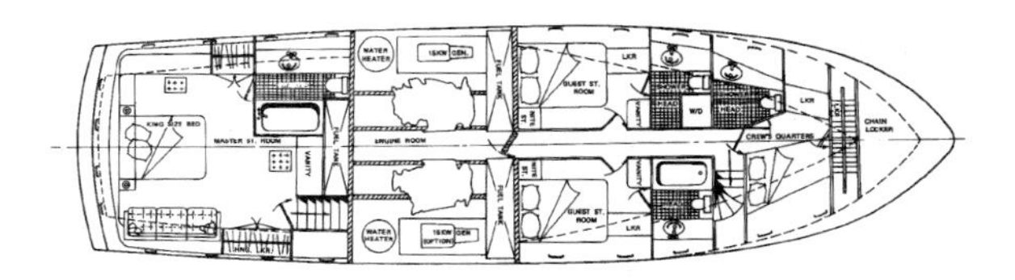

See page 254 for pricing information.

SEA RAY 360 AFT CABIN

SPECIFICATIONS

Length	36'3"	Fuel	270 gals.
Beam	12'6"	Hull Type	Modified-V
Draft	2'11"	Deadrise Aft	9°
Weight	15,100#	Designer	Sea Ray
Water	120 gals.	Production	1983–87

The 360 Aft Cabin was built on a "dual mode" modified-V hull capable of trawler-style economy at 7–8 knot displacement speeds while still able to achieve planing-speed performance. The low-deadrise hull is solid glass and features propeller pockets, moderate beam, and a shallow full-length keel. She was a fairly popular model for Sea Ray thanks to her roomy twin-cabin interior and reasonable price. The layout is somewhat unusual in that a walkaround double bed is located in the forward cabin, while twin berths are fitted into the smallish aft stateroom (a double berth aft was optional). With the galley down and no dinette, the salon is quite open and roomy. Outside, the small afterdeck has enough room for a only a couple of deck chairs. (Most 360s were sold with the optional hardtop.) Standard 260/270-hp gas engines cruise the Sea Ray 360 Aft Cabin at 15 knots and reach a top speed of 23–24 knots. Optional 200-hp Perkins diesels provide an efficient cruising speed of 15–16 knots. ❑

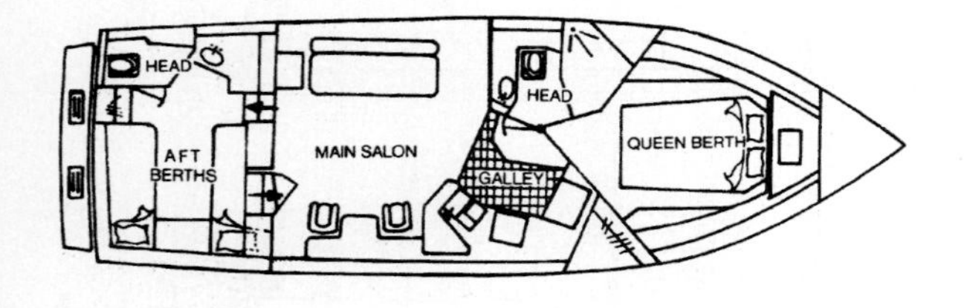

See page 254 for pricing information.

SEA RAY 380 AFT CABIN

SPECIFICATIONS

LOA	42'7"	Water	100 gals.
Hull Length	37'9"	Fuel	300 gals.
Beam	13'11"	Hull Type	Deep-V
Draft	2'7"	Deadrise Aft	19°
Weight	20,000#	Designer	Sea Ray
Clearance	NA	Production	1989–91

The Sea Ray 380 Aft Cabin is a modern maxi-cube family cruiser with a slightly topheavy profile and a very spacious interior layout. Built on a fully cored hull with a fairly wide beam, prop pockets, and an integral bow pulpit, she has as much living space below as most 40-footers. Indeed, the accommodations are impressive with large staterooms (each with a double berth), a full dinette, and a wide-open salon capable of seating a small crowd in comfort. A blend of pastel fabrics, Formica countertops and cabinetry, and teak trim presents an appealing and upscale interior decor. Features include a vanity in the forward stateroom, a stall shower enclosure in the aft head, and a serving counter overlooking the large U-shaped galley. Outside, the flybridge will seat six, and the aft deck is large enough for several deck chairs. A popular boat in spite of her sluggish performance, twin 340-hp MerCruisers will cruise the Sea Ray 380 Aft Cabin around 16 knots and reach a top speed of 25+ knots. ❑

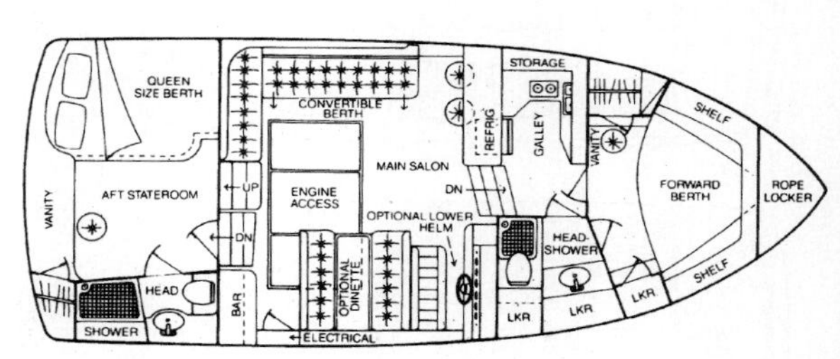

See page 254 for pricing information.

SEA RAY 440 AFT CABIN

SPECIFICATIONS

LOA	45'11"	Water	130 gals.
Hull Length	43'6"	Fuel	400 gals.
Beam	13'11"	Hull Type	Deep-V
Draft	3'2"	Deadrise Aft	17°
Weight	23,000#	Designer	Sea Ray
Clearance	15'3"	Production	1986–91

The Sea Ray 440 Aft Cabin pictured above began life in 1986 as the 410 Aft Cabin. She became the 415 the next year, when Sea Ray added a molded bow pulpit, and in 1988 the marketing people re-measured the boat and settled on the 440 designation which lasted for the duration. She's constructed on a deep-V hull with prop pockets—the same hull used in the 440 Convertible. Inside, her floorplan is arranged with an offset double berth in the forward stateroom instead of the over/under bunks found in the earlier 410/415 models. In 1990, new off-white laminates replaced the teak paneling and cabinetry used in previous years. The dinette was also eliminated then in favor of a wraparound deckhouse lounge. Outside, both the flybridge seating and aft deck dimensions are limited. A popular model, standard 454-cid gas engines will cruise the 440 Aft Cabin at a sedate 15 knots with a top speed of 24 knots. Optional 375-hp Cats cruise at 20 knots and reach 24 knots wide open. ❑

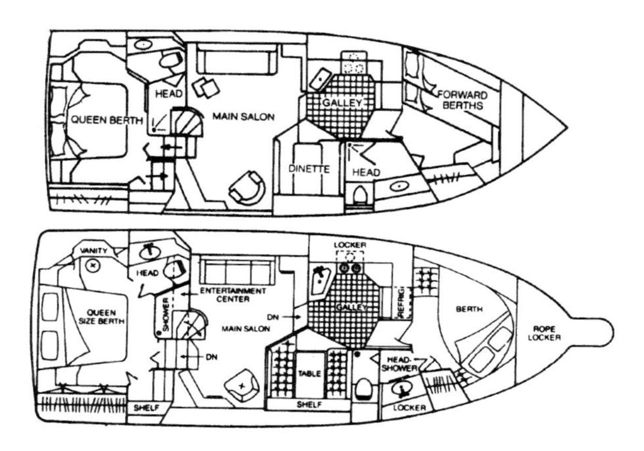

See page 254 for pricing information.

SEA RAY 500/550 SEDAN BRIDGE

Sea Ray 500 Sedan Bridge

Sea Ray 550 Sedan Bridge

SPECIFICATIONS

LOA	49'11"/57'10"	Water	200 gals.
Hull Length	55'4"/54'10"	Fuel	600 gals.
Beam	15'0"	Hull Type	Deep-V
Draft	4'2"	Deadrise Aft	17°
Weight	45,000#	Designer	J. Michalak
Clearance	17'6"	Production	1989–Current

The original Sea Ray 500 Sedan Bridge and the newer 550 model (introduced in 1992) differ only at the transom: the 500 has a bolt-on swim platform, and the 550 has a more stylish integral platform. Built on a fully cored deep-V hull with moderate beam and prop pockets, the key to the popularity of this boat lies with her expansive indoor and outdoor accommodations. The three-stateroom floorplan is arranged with the galley forward in the salon and two full heads below. The salon is impressive—a wide-open living area with a cut-down galley, a long leather sofa, and big wraparound cabin windows. A portside lower helm is optional. Topside, the party-style 20-foot-long flybridge is set up for serious entertaining with the helm forward and innovative guest seating aft. A good performer with standard 550-hp 6V92 diesels, the 500/550 Sedan Bridge will cruise around 23 knots with a top speed of 27 knots. Optional 650-hp 6V92s will cruise at 25 knots and reach about 29 top. ❑

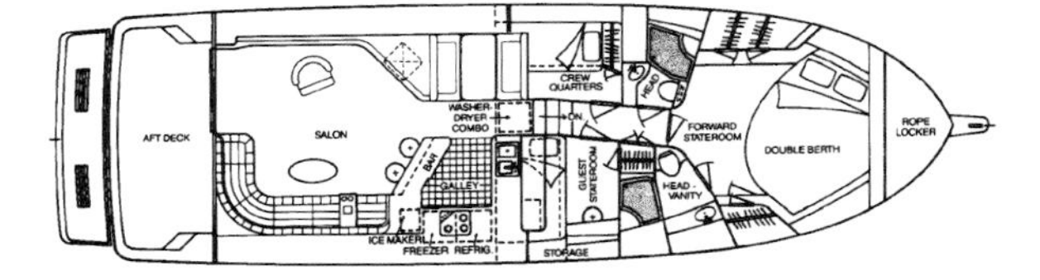

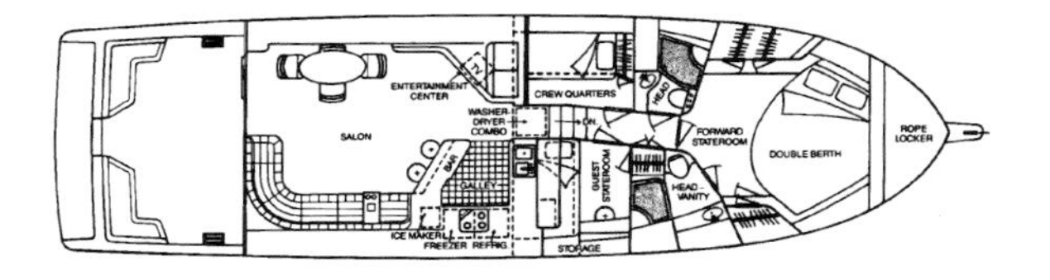

See page 254 for pricing information.

SEA RAY 650 MOTOR YACHT

SPECIFICATIONS

Length ....64'6"	Fuel.....................1,000 gals.
Beam ....18'1"	Cockpit ....95 sq. ft.
Draft.............................4'10"	Hull Type..............Deep-V
Weight ....67,500#	Deadrise Aft ....18°
Clearance.....................25'3"	Designer................Sea Ray
Water.....................275 gals.	Production...1992–Current

The 650 MY is an impressive creation from a company that little over a decade ago ranked a 36-footer as their largest model. Built on a fully cored deep-V hull, she's an elegant yacht with a Mediterranean profile and an innovative four-stateroom, galley-up floorplan. What makes this layout unusual is the aft engine room—only a few builders have ever built a motor yacht with V-drives. The stylish interior of the 650 MY is very contemporary and features a Jacuzzi-style tub in the owner's head, a huge master stateroom, and a spacious full-width salon with a formal dining area to port and a bridge stairway forward. The cockpit is quite roomy and provides easy access to the engine room below. The flybridge, on the other hand, is small compared to other 65-foot motor yachts. A good-running boat, 870-hp (or 900-hp) 12V71s will cruise at 21 knots (25 knots wide open), and the now standard 12-cylinder 1,000-hp MTUs will cruise at a fast 24 knots (28 top). ❑

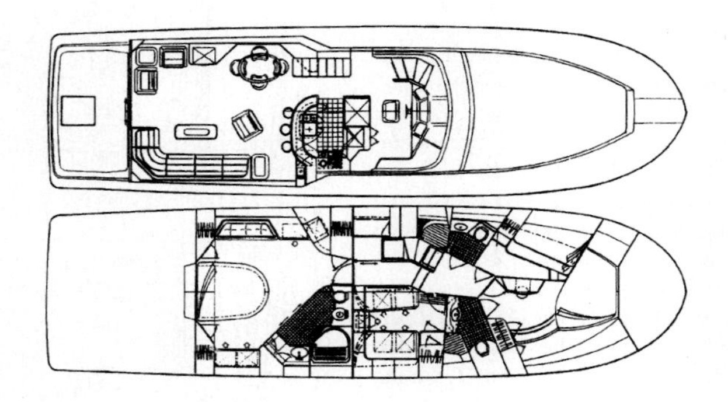

See page 254 for pricing information.

SEAMASTER 48 MOTOR YACHT

SPECIFICATIONS

Length	47'8"	Clearance	16'10"
LWL	43'6"	Water	200 gals.
Beam	15'4"	Fuel	520 gals.
Draft	3'10"	Hull Type	Modified-V
Weight	39,000#	Production	1983–89

Also known as the CHB 48 MY, the Seamaster 48 is basically an inexpensive Taiwan copy of the original Hatteras 48 Motor Yacht (1981–84). She's built on a solid fiberglass modified-V hull with moderate beam, hard chines, and a shallow, full-length keel below. The spacious three-stateroom, three-head floorplan—with only a *single* stall shower—is virtually identical to the Hatteras 48. The interior is finished with an abundance of teak cabinetry and woodwork throughout. A lower helm was optional, and a work bench is found in the stand-up engine room. The Seamaster features a truly spacious afterdeck with wet bar and wing doors. (Note that the sidedecks on this boat are quite wide.) Topside, there's seating for eight on the large flybridge. At 39,000 lbs., the Seamaster is slightly lighter than the Hatteras 48 MY she so blatantly imitates, and most were imported with too-small 375-hp Cat diesels to keep the price down. The cruising speed is a sluggish 13–14 knots, and the top speed is around 17 knots. ❑

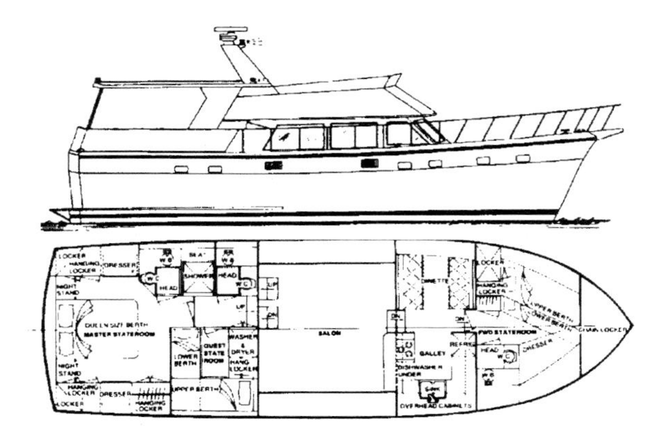

See page 254 for pricing information.

SHANNON VOYAGER 36

SPECIFICATIONS

Length	35'7"	Fuel	325 gals.
Beam	13'3"	Cockpit	75 sq. ft.
Draft	3'0"	Hull Type	Modified-V
Weight	17,500#	Deadrise Aft	NA
Clearance	12'6"	Designer	W. Schultz
Water	150 gals.	Production	1991–Current

The Shannon 36 is one of the few trawler-style cruisers capable of 20-plus knot performance. Built on a semi-production basis by Shannon Yachts (a firm whose reputation was made in the upper end of the sailboat market), her sedan profile and upright appearance make the Shannon 36 an easy boat to spot in a crowd. The hull is cored from the waterline up and features a sharp entry with a graceful sheerline, plenty of beam, and a shallow keel. Inside, the two-stateroom, galley-up floorplan is surprisingly roomy (considering the wide sidedecks) and includes an island berth in the forward stateroom, over/under bunks in the guest cabin, a single head compartment (with a tub/shower), and an inside helm with a sliding deck access door. Other features include a fairly large cockpit (partially shaded by the bridge overhang), radar mast, and a well-arranged flybridge with L-shaped guest seating. A good performer with the now-standard 300-hp Cat 3116 diesels, she'll cruise economically at 20 knots and reach a top speed of about 23–24 knots. ❑

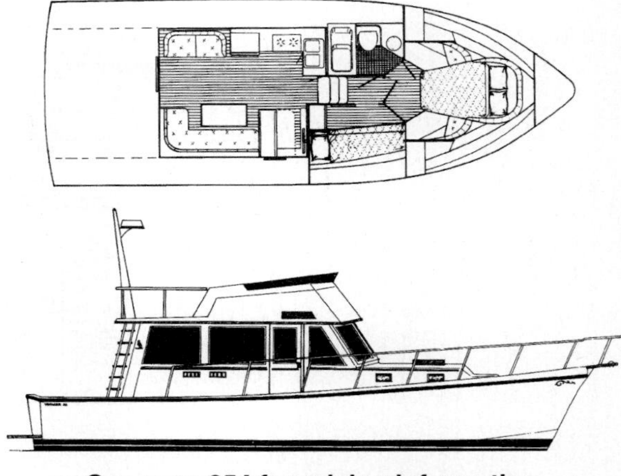

See page 254 for pricing information.

SILVERTON 34 MOTOR YACHT

SPECIFICATIONS

Length	34'6"	Fuel	260 gals.
Beam	12'11"	Cockpit	NA
Draft	2'11"	Hull Type	Modified-V
Weight	19,000#	Deadrise Aft	17°
Clearance	15'1"	Designer	Silverton
Water	74 gals.	Production	1993–Current

The Silverton 34 MY is one of the smallest aft-cabin models on the market. A stylish but boxy-looking boat with a fairly wide beam, she somehow manages to include two reasonably sized staterooms inside along with a full-size dinette and stall showers in both heads—no small achievement in a 34' hull. The salon dimensions are small, but the forward stateroom is surprisingly roomy and very much the equal of the aft cabin when it comes to floorspace. On the downside, storage space is slim. Outside, the full-width aft deck is too small for anything but a couple of chairs, but the sidedecks are wide enough for secure passage forward. The well-arranged flybridge is arranged with guest seating forward of the helm. Additional features include a hardtop and arch, oak interior woodwork, molded steps from the aft deck to the swim platform, and a steep 17° of transom deadrise. A good-running boat with standard 454-cid Crusader gas engines, the Silverton 34 MY will cruise at 17–18 knots and reach a top speed of around 28. ❏

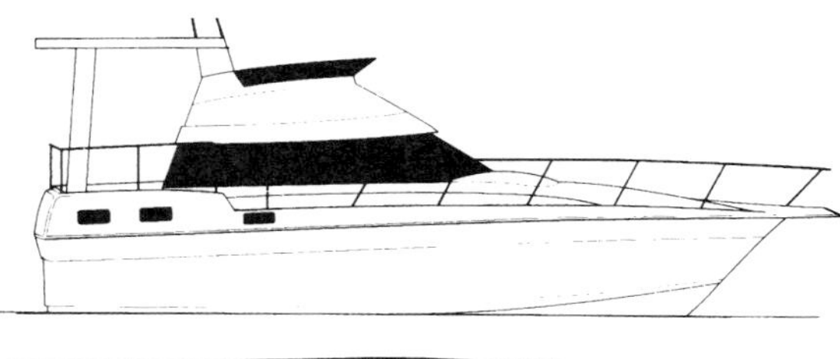

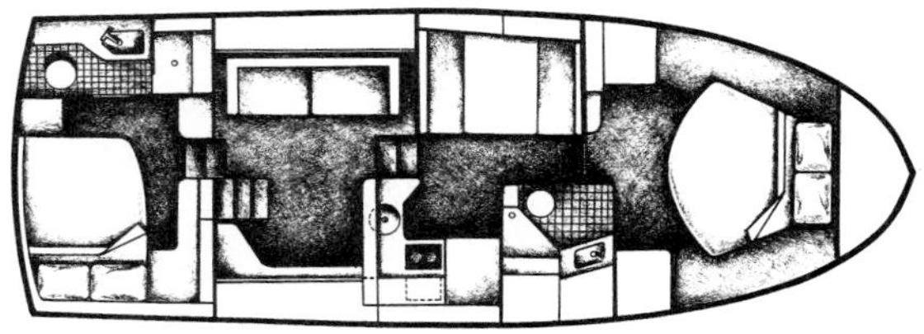

See page 254 for pricing information.

SILVERTON 37 MOTOR YACHT

SPECIFICATIONS

Length	37'6"	Water	100 gals.
LWL	NA	Fuel	300 gals.
Beam	13'9"	Hull Type	Modified-V
Draft	3'8"	Deadrise Aft	NA
Weight	22,000#	Designer	M. Peters
Clearance	16'0"	Production	1988–89

Motor yachts under 40 feet have been popular in recent years, and the Silverton 37 was one of the more affordable (if less attractive) models available. Built on a conventional modified-V hull, the interior dimensions of the 37 are quite impressive. Both staterooms have queen berths, and the forward stateroom is nearly as spacious as the master stateroom. Notably, stall showers are found in each head—an unexpected convenience in a 37' boat. The galley is only a step down from the salon/dinette level, thus opening up the interior considerably. The light oak paneling and woodwork applied in each cabin are especially attractive. Topside, the theme of comfortable family cruising is continued in the huge bridge with lounge seating for up to ten. Boarding is somewhat difficult, however, due to the unusually high freeboard. (The cabin entryway is located on the flybridge.) The cruising speed of the Silverton 37 MY with standard 350-hp Crusaders is around 15 knots, and the top speed is 23–24 knots. Note that this boat remained in production for only two years. ❏

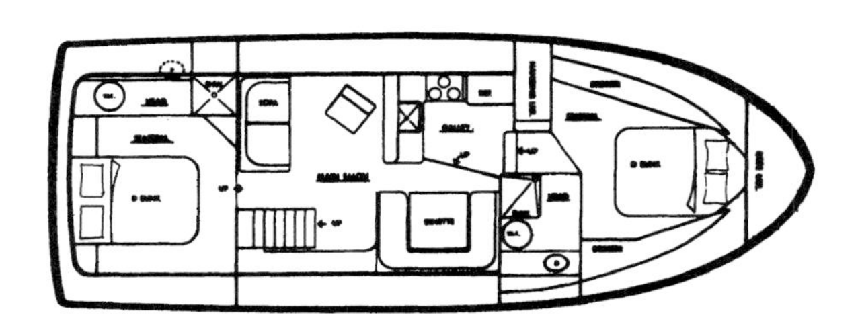

See page 254 for pricing information.

SILVERTON 40 AFT CABIN

SPECIFICATIONS

Length	40'0"	Water	100 gals.
LWL	NA	Fuel	300 gals.
Beam	14'0"	Hull Type	Modified-V
Draft	3'0"	Deadrise Aft	14°
Weight	24,000#	Designer	Bob Rioux
Clearance	13'6"	Production	1982–90

Introduced in 1982, the Silverton 40 Aft Cabin was Silverton's first double-cabin design. She was built on a solid fiberglass, modified-V hull with relatively high freeboard, considerable bow flare, and 14° of deadrise at the transom. Below, her two-stateroom interior is arranged with the galley and dinette down, which results in a fairly spacious salon. The full-width aft cabin has a walkaround double berth, and V-berths are installed in the small forward stateroom. Both heads in the Silverton 40 are fitted with shower stalls, and light oak interior woodwork replaced the original teak interior beginning with the 1985 models. The aft deck dimensions are about average for a boat this size, and the flybridge (with seating for only three) is on the small side. Note that the two-piece entryway door from the aft deck is a tight fit for most people and somewhat awkward to negotiate. An inexpensive boat, standard 454-cid gas engines will cruise the Silverton 40 at a respectable 17 knots and deliver a top speed of 25–26 knots. ❑

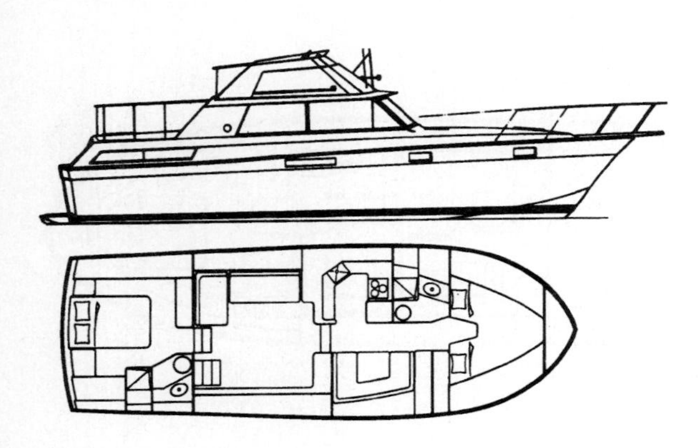

See page 254 for pricing information.

SILVERTON 41 AFT CABIN

SPECIFICATIONS

Length	41'3"	Water	200 gals.
LWL	NA	Fuel	408 gals.
Beam	14'10"	Hull Type	Modified-V
Draft	3'9"	Deadrise Aft	17°
Weight	28,000#	Designer	Silverton
Clearance	16'3"	Production	1991–Current

A modern and notably affordable flush-deck cruiser, the Silverton 41 MY is basically a scaled-down version of the Silverton 46 MY. She's built on a solid fiberglass, modified-V hull with a wide beam, shallow keel, and a steep 17° of transom deadrise. Silverton has packed a lot of floorplan into the 41 MY, including queen berths in both staterooms, a fair-size salon, a full dinette, and stall showers in both heads. (Note that the forward head is larger than the aft head—a useful feature when cruising with guests.) A tempered glass partition divides the galley from the salon, and light oak woodwork and cabinetry are applied throughout. Additional features include side exhausts, a tubular radar arch, and adequate sidedecks. The standard 502-cid gas engines are barely up to the job in a boat this large, and it takes a thirsty 3,500 rpm to get a respectable 15–16 knots of cruising speed. Optional 375-hp Cats will cruise in the neighborhood of 21–22 knots and reach a top speed of about 26 knots. ❑

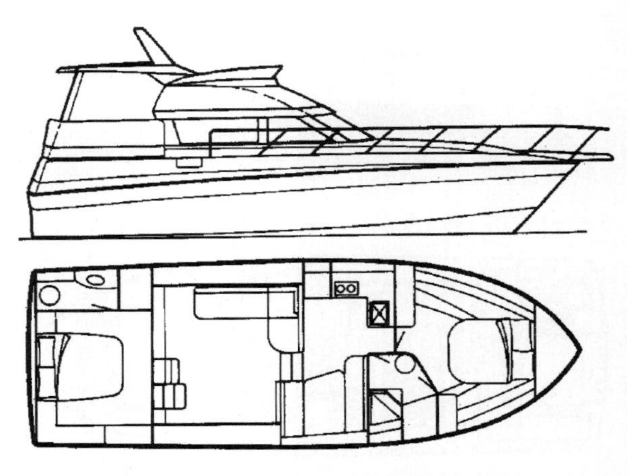

See page 255 for pricing information.

SILVERTON 46 MOTOR YACHT

SPECIFICATIONS

Length	46'8"	Water	200 gals.
LWL	NA	Fuel	580 gals.
Beam	16'2"	Hull Type	Modified-V
Draft	3'9"	Deadrise Aft	17°
Weight	40,500#	Designer	M. Peters
Clearance	17'8"	Production	1989–Current

Introduced in mid-1989, the 46 Motor Yacht is the largest production hull Silverton has built to date. She's a good-looking boat with modern lines and rakish profile, and her affordable price makes the 46 MY a lot of boat for the money. The hull is a modified deep-V design with a wide beam, a solid fiberglass bottom, and foam-cored hullsides. Inside, the light oak interior is arranged with the relatively small salon wide open to the galley and dinette forward. Both staterooms are fitted with double berths, and both heads have separate stall showers. The aft deck is rather small for a 46-footer, but there's plenty of guest seating on the bridge. Additional features include foredeck seating, washer/dryer, hardtop and arch, and a bathtub in the owner's head. Standard 485-hp 6-71s will cruise the Silverton 46 MY around 22 knots with a top speed of 25 knots. True to Silverton's long-standing reputation for building affordable boats, the 46 comes standard with items normally considered options. ❑

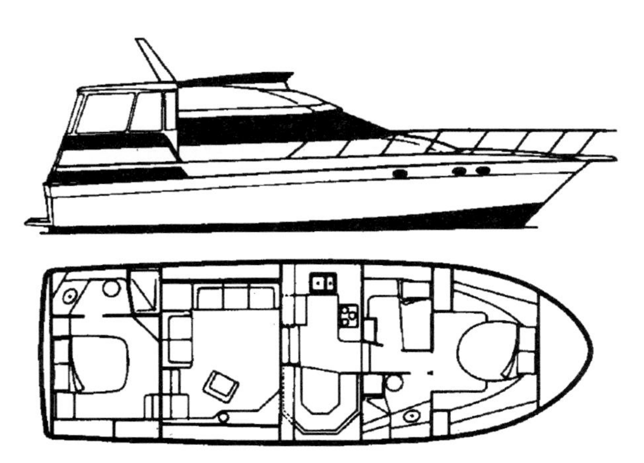

See page 255 for pricing information.

STEVENS 59 MOTOR YACHT

SPECIFICATIONS

Length	58'10"	Water	300 gals.
LWL	NA	Fuel	1,000/1,200 gals.
Beam	17'10"	Hull Type	Modified-V
Draft	4'9"	Deadrise Aft	10°
Weight	70/75,000#	Designer	S&S
Clearance	17'1"	Production	1989–92

The Stevens 59 MY incorporates more innovative design concepts than most other motor yachts this size. Granted, her big-yacht profile is a little less than graceful, but the unique raised pilothouse layout, expansive full-width deckhouse, and four-stateroom floorplan provide the comforts one might expect on a larger boat. Built in Taiwan, the Stevens rides on a modified-V hull with a deep keel and a wide beam. A look at the four-stateroom floorplan reveals an aft engine room—an unusual feature made possible by reversing the engines and using V-drives to apply the power. The central pilothouse configuration of the Stevens 59 allows for the addition of a formal dining area forward on the deckhouse level. The master stateroom is huge, and both guest cabins have double berths and private heads. The crew quarters are accessed from the foredeck. Optional 735-hp 8V92s will cruise at a respectable 19 knots and reach 22 knots top. Note that the Stevens 67 Cockpit MY has a completely different floorplan arrangement and conventional straight-drive motors (no V-drives). ❑

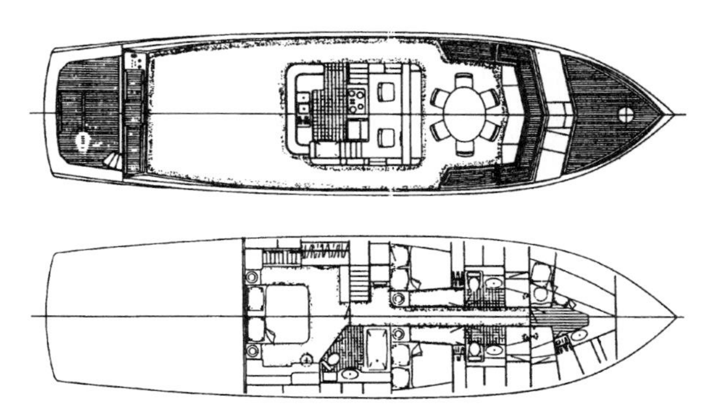

See page 255 for pricing information.

NEW

FOR

1995

40' SPORT SEDAN

45' COCKPIT MOTORYACHT

48' MOTORYACHT

50' COCKPIT MOTORYACHT

57' COCKPIT MOTORYACHT

57' WALK - A - ROUND MOTORYACHT

65' WALK - A - ROUND MOTORYACHT

NEW COMPANY, NEW DECOR'S, NEW MODELS

56 YEAR OLD YACHT BUILDING TRADITION

SEE YOUR LOCAL TOLLYCRAFT YACHTS DEALER

OR CONTACT TOLLYCRAFT YACHTS , 2200 CLINTON AVE

KELSO, WA 98626-5526. PHONE 1-800 -536-5160

TOLLYCRAFT 34 TRI CABIN

SPECIFICATIONS

Length	34'0"	Water	77 gals.
LWL	30'6"	Fuel	200 gals.
Beam	12'6"	Hull Type	Modified-V
Draft	2'10"	Deadrise Aft	13°
Weight	17,000#	Designer	Ed Monk
Clearance	12'0"	Production	1975–85

Although she's been out of production for several years, the Tollycraft 34 Tri Cabin is one of the more desirable small aft-cabin cruisers to be found on the used market. Her traditional lines are somewhat trawler-like, and, at 17,000 lbs., she's a heavy boat for her size. Tollycraft designers managed to pack a lot of living space inside the 34 including two heads, a complete lower helm station with a deck access door, and even some useful storage space. The large salon windows provide an abundance of natural lighting, and visibility from the lower helm is good. Early models have twin single berths in the aft cabin, and a double bed became standard in 1982. An all-teak interior was introduced in 1984—a big improvement over the teak-grain mica veneers of earlier models. The nonskid is an imitation-teak fiberglass surface, and the wide sidedecks make passage around the house very secure. With twin 270-hp gas engines, the 34 Tri Cabin will cruise at 17–18 knots with a top speed of around 25 knots. ❑

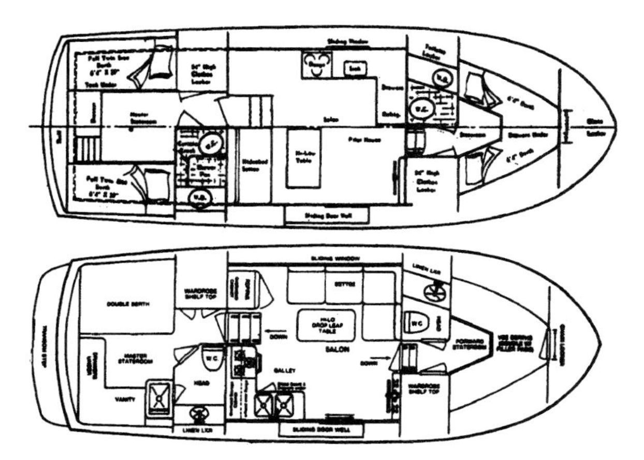

See page 255 for pricing information.

TOLLYCRAFT 34 SUNDECK

SPECIFICATIONS

Length	34'0"	Water	77 gals.
LWL	30'6"	Fuel	200/296 gals.
Beam	12'6"	Hull Type	Modified-V
Draft	2'10"	Deadrise Aft	NA
Weight	17,000#	Designer	Ed Monk
Clearance	12'0"	Production	1986–88

The 34 Sundeck Cruiser was introduced in 1986 as a replacement for Tollycraft's aging 34 Tri Cabin. This all-new model received a fresh deckhouse and flybridge design resulting in a rather stylish profile, especially when compared to most other 34-foot double-cabin designs. Below, the interior remains similar to the older Tri Cabin layout, but with an enlarged aft cabin featuring a walkaround double berth and a rearranged head compartment. The teak woodwork and cabinetry are well-crafted throughout, and the fabrics, hardware, and appliances are all good quality. The efficient use of interior space in the 34 Sundeck is quite remarkable, and, when combined with the rich furnishings, it would be easy to confuse this interior with that of a larger boat. Built on the then-new Quadra-Lift hull, the Tollycraft 34 Sundeck weighs in at a hefty 17,000 lbs. and provides a dry, stable ride in most sea conditions. She'll cruise around 20 knots with 454-cid gas engines and reach a top speed of 28 knots. Note the significant fuel increase to 296 gallons in 1988. ❑

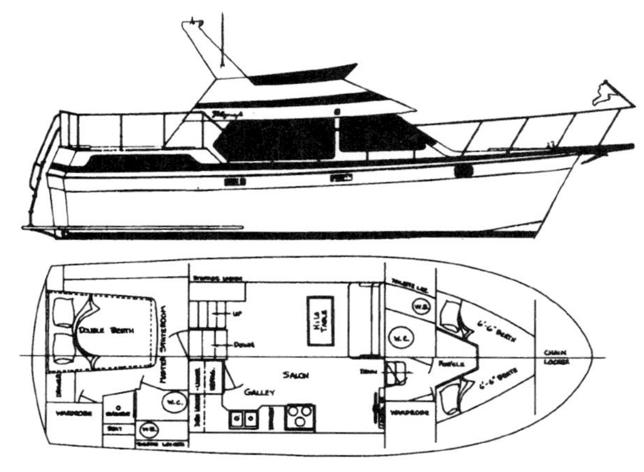

See page 255 for pricing information.

TOLLYCRAFT 40 TRI CABIN MY

SPECIFICATIONS

Length	40'2"	Fuel	440 gals.
Beam	13'4"	Cockpit	40 sq. ft.
Draft	3'2"	Hull Type	Modified-V
Weight	30,000#	Deadrise Aft	NA
Clearance	NA	Designer	Ed Monk
Water	150 gals.	Production	1970–79

The 40 Tri Cabin was the first big yacht ever built by Tollycraft. A conservative design, she features a straightforward double-cabin interior with the galley in the salon and wraparound cabin windows. The aft stateroom is particularly spacious for a boat this size and came with twin single berths and direct access to the cockpit (a double berth was optional). Sliding deck doors are located port and starboard in the salon, and the interior woodwork was updated from mahogany to teak in 1977. Typical of West Coast boats, the Tollycraft 40 Tri Cabin has extra-wide sidedecks and a complete lower helm station. Her big 100 sq. ft. aft deck is roomy enough for entertaining or dinghy storage. Construction is solid fiberglass, and the engineering and workmanship are above average throughout. Standard 454-cid gas engines provide a cruising speed of 16–17 knots and a top speed of about 26. The optional 210-hp Cat diesels will average an efficient 1 mpg at a steady 15-knot cruising speed (about 18 knots top). ❏

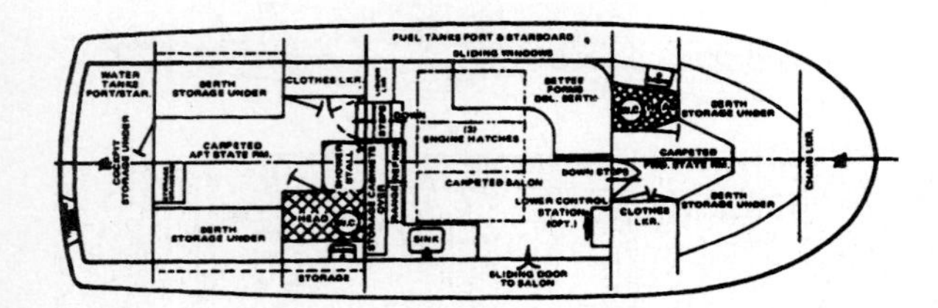

See page 255 for pricing information.

TOLLYCRAFT 40 SUNDECK MY

SPECIFICATIONS

Length	40'2"	Water	140 gals.
LWL	NA	Fuel	300/398 gals.
Beam	14'8"	Hull Type	Modified-V
Draft	3'0"	Deadrise Aft	10°
Weight	26,000#	Designer	Ed Monk
Clearance	12'0"	Production	1985–Current

A popular boat, the Tollycraft 40 Sundeck is a modern double-cabin design with attractive styling and a first-rate interior layout. Aimed at the upscale family cruiser market, the 40's profile has not been subjected to the Euro-style overkill seen in many other boats her size. Like all Tollycraft designs, her modified-V hull handles well, and construction is above average for a production yacht. Inside, the large deckhouse windows combine with a well-crafted teak interior to create an open and very upscale decor. The galley and dinette are down, and there's a deck door in the salon (a lower helm is optional). Note that a queen forward layout (with a stall shower in the head) became available in 1992. Outside, the full-width aft deck provides plenty of entertainment space, and the flybridge has seating for six. Standard 454-cid gas engines cruise at 17–18 knots (26 knots top). Optional 375-hp Cat diesels cruise around 23 knots (27 top), and the 400-hp 6V53s cruise around 24 knots (27 top). The fuel capacity was increased in 1991. ❏

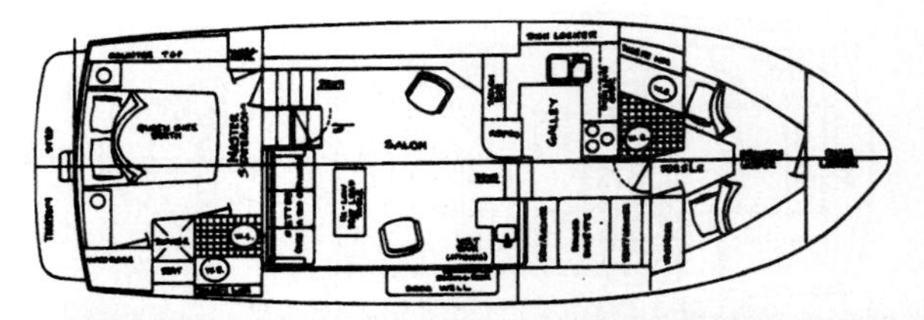

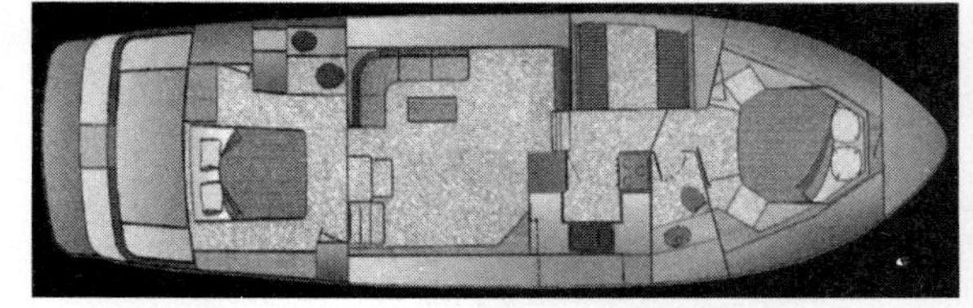

See page 255 for pricing information.

TOLLYCRAFT 43 COCKPIT MY

SPECIFICATIONS

Length	43'4"	Cockpit	NA
LWL	39'5"	Water	140 gals.
Beam	14'2"	Fuel	400 gals.
Draft	3'5"	Hull Type	Semi-Disp.
Weight	30,000#	Designer	Ed Monk
Clearance	13'9"	Production	1980–86

For those who are serious about their cruising and who value seaworthiness over the glitz of some of today's Euro-style bay boats, the Tollycraft 43 Cockpit Motor Yacht should receive strong consideration. Built on a rugged semi-displacement, solid fiberglass hull and bearing a distinct trawler-style profile, the 43 CMY is widely regarded as a superior heavy-weather boat. A sharp entry and a full-length keel provide good handling characteristics, while her flat aftersections allow the 43 to cruise efficiently at 15 knots (15 gph) with the small 210-hp Cat diesels. The accommodations are well-organized with double berths in each stateroom and a serving bar separating the salon from the lower-level galley. An inside helm and deck access door were standard, and the interior is finished with well-crafted teak cabinetry and paneling throughout. Other features include a transom door, wide walkaround sidedecks, radar mast, and top-quality appliances, furnishings, and hardware. A proven design, the Tollycraft 43 Cockpit MY is a classic Pacific Northwest cruiser with considerable eye appeal. ❑

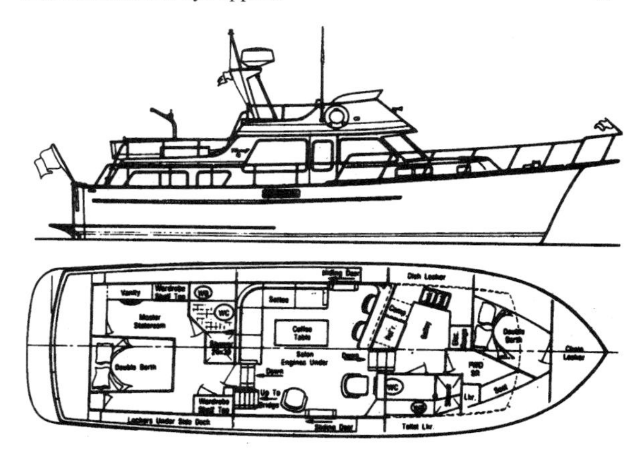

See page 255 for pricing information.

TOLLYCRAFT 44/45 COCKPIT MY

SPECIFICATIONS

Length	44'2"	Water	140 gals.
Beam	14'8"	Fuel	300/398 gals.
Draft	3'0"	Hull Type	Modified-V
Weight	28,000#	Deadrise Aft	10°
Clearance	12'0"	Designer	Ed Monk
Cockpit	42 sq. ft.	Production	1986–93

The addition of a cockpit to the Tollycraft 40 Sundeck created the 44 Cockpit MY, a popular model with an even more attractive profile than her sistership thanks to the extra hull length. The interior accommodations are the same in both boats—a two-stateroom arrangement with the salon wide open to the lower level galley and dinette. V-berths were standard in the forward stateroom until 1992 when a double berth became optional. All of the interior woodwork is teak. The cockpit is too small for any serious fishing, but it certainly makes boarding a lot easier. Standard 454-cid gas engines cruise at 18–19 knots (27 knots top). Optional 375-hp Cat diesels will cruise around 23 knots (about 26 top), and the 400-hp 6V53s cruise around 24 knots (27 knots top). Note that the fuel was increased in 1991. With over 100 built, the 44 CMY was one of Tollycraft's all-time best-selling yachts. Note that the Tollycraft 45 Cockpit MY (1994–current) is basically the same boat with a integral swim platform. ❑

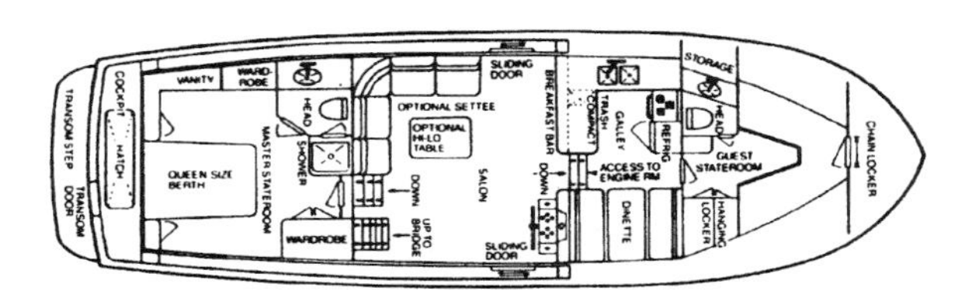

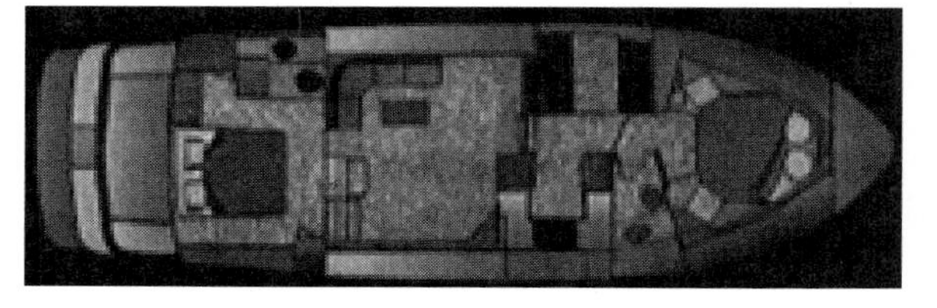

See page 255 for pricing information.

TOLLYCRAFT 48 COCKPIT MY

SPECIFICATIONS

Length	48'2"	Fuel	600 gals.
Beam	15'2"	Cockpit	65 sq. ft.
Draft	3'8"	Hull Type	Semi-Disp.
Weight	42,000#	Designer	Ed Monk
Clearance	17'0"	Production	1976–86
Water	188 gals.		1991–Current

Long regarded as a top-quality West Coast cruiser during the 1970s and 80s, the 48 Cockpit MY was reintroduced into the Tollycraft lineup in 1991. Not much changed from the original—the interior layout is basically the same (the decor has been updated), and the rugged semi-displacement hull with rounded chines and a long keel is retained as well. The efficient two-stateroom layout is arranged with the galley and dinette down from the salon. An inside helm was standard, and there are port and starboard deck doors in the salon. Exterior features include a cockpit transom door, full walkaround sidedecks with protective bulwarks, a functional mast and boom, and space for dinghy storage on the cabintop. Modest styling changes to the flybridge were made in 1985. Early models with 320-hp Cat diesels will cruise at 16 knots (around 20 knots top), and the newer boats with 300-hp Cummins or 3116 Cats cruise at about the same speed. Eighty 48 Cockpit MYs were built before the original production run ended in 1986. ❑

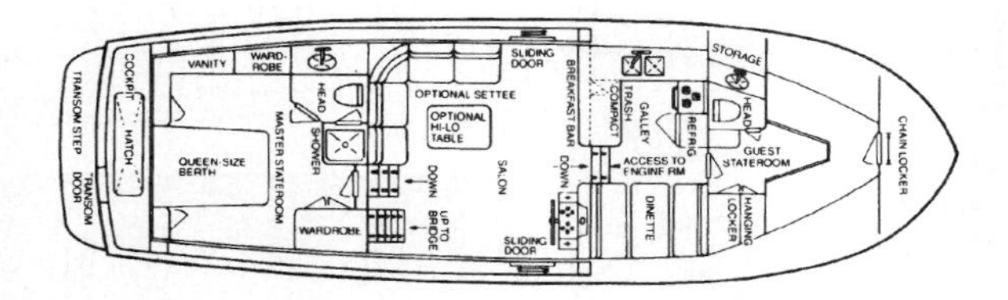

See page 255 for pricing information.

TOLLYCRAFT 53/57 MOTOR YACHT

Tollycraft 53 MY

Tollycraft 57 MY

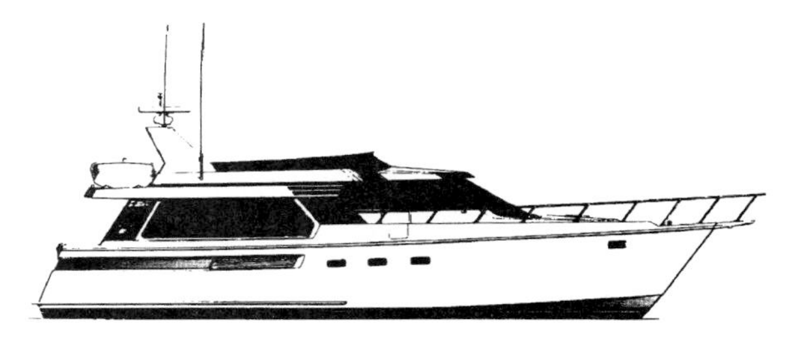

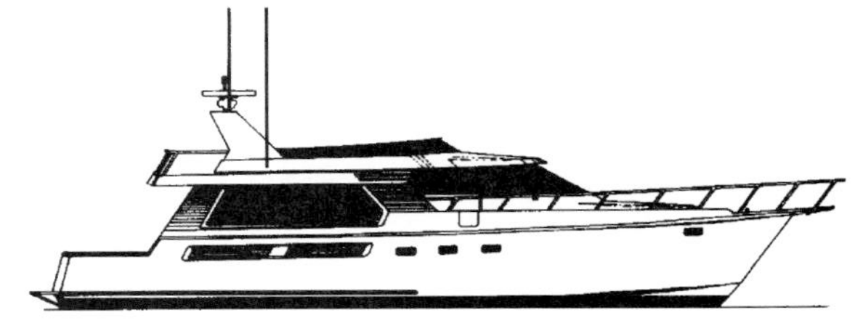

SPECIFICATIONS

Length	52'11"/57'0"	Fuel	800/1,200 gals.
Beam	16'11"	Cockpit	75/122 sq. ft.
Draft	3'6"	Hull Type	Modified-V
Weight	55/58,000#	Deadrise Aft	11°
Clearance	16'3"	Designer	Ed Monk
Water	280 gals.	Production	1989–Current

With the introduction of the 53 and 57 Motor Yachts, Tollycraft made a clean break from their long-standing reputation for producing conservative West Coast designs. The 53 and 57 are basically the same boat with the 57 having a larger cockpit and additional fuel. They're built on Tollycraft's efficient Quadra-Lift hull—a modified-V with 11° of deadrise aft and balsa coring above the waterline. The floorplan is unique: the pilothouse shares space with the galley, and a single portside wing door in the salon complements the starboard sliding door in the pilothouse. The staterooms are forward (the full-width master is opulent), and both guest cabins share a common head. The engine room is entered from the cockpit—not the stand-up variety, but outboard access is excellent. The teak joinerwork in the salon and pilothouse is very good, and the flybridge dimensions are extravagant. Those with 550-hp 6V92 Detroits cruise around 19 knots. Optional 735-hp 8V92s cruise at 23 knots, and 665-hp MTUs cruise at 20–21 knots. A popular model, about thirty 57s have been built. ❑

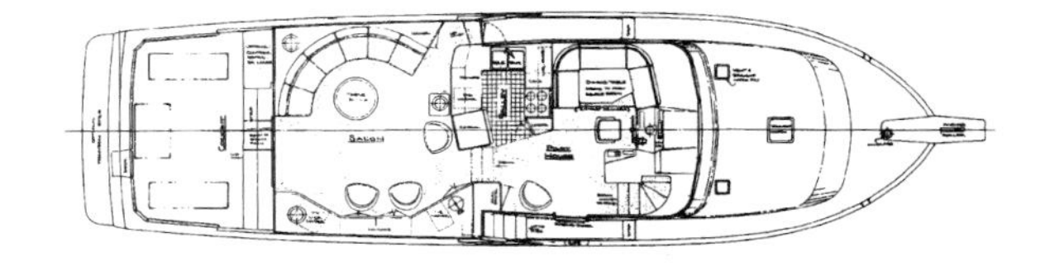

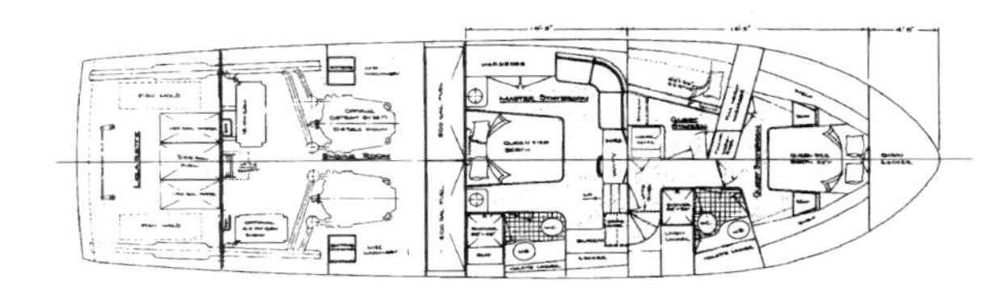

See page 255 for pricing information.

TOLLYCRAFT 61 MOTOR YACHT

SPECIFICATIONS

Length	61'2"	Fuel	1,160 gals.
Beam	17'8"	Cockpit	64 sq. ft.
Draft	4'0"	Hull Type	Modified-V
Weight	65,000#	Deadrise Aft	10°
Clearance	17'6"	Designer	Ed Monk
Water	400 gals.	Production	1983–Current

A popular model, the Tollycraft 61 MY is a handsome pilothouse design with tremendous eye appeal and excellent seakeeping qualities. Unlike conventional double-deck motor yachts her size, the 61's low profile keeps guests and crew at the water level for much of the boat's length. Built on a very efficient hull, her pilothouse layout offers obvious advantages for cruising in cold or wet weather. While the 61 was designed as a Pacific Northwest passagemaker, her graceful profile and economical operation appeal to experienced yachtsmen in virtually every market. Practical features include covered sidedecks, a shallow keel for prop protection, cored hull construction, and a traditional teak interior. Most have been sold with 650-hp 8V92s (20 knots cruise/23 top) or 735-hp 8V92s (22 knots cruise/25 top), although she's still a good performer with the original 485-hp 6-71s (15–16 knots cruise/19 top). Later models with 665-hp MTUs will cruise at 20 knots (23 knots top). Note that the Tollycraft 65 MY is the same boat with a bigger cockpit. About forty have been built. ❏

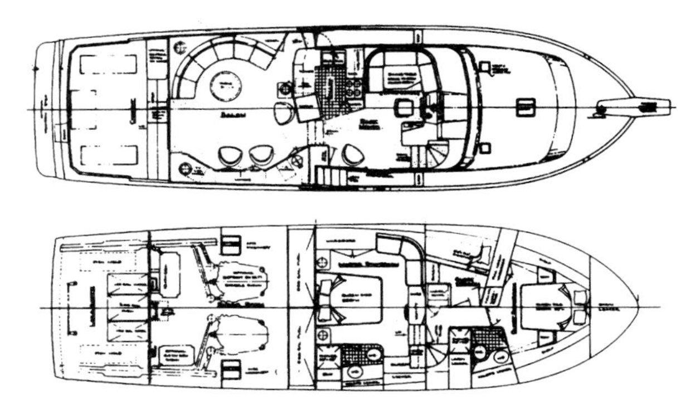

See page 256 for pricing information.

TROJAN 36 TRI CABIN

SPECIFICATIONS

Length	36'0"	Water	66/85 gals.
LWL	NA	Fuel	150/220/300 gals.
Beam	13'0"	Hull Type	Modified-V
Draft	2'11"	Deadrise Aft	9°
Weight	17,500#	Designer	Trojan
Clearance	12'3"	Production	1970–87

Trojan introduced this aft-cabin design in 1970 as the 36 Sea Raider. Built on a fiberglass hull with a wood deckhouse and teak cockpit and decks, the transition to all-fiberglass construction was made in 1972 when she became the 36 Tri Cabin. Her modified-V hull was also used in the production of the Trojan 36 Convertible. The long-running popularity of the Tri Cabin had much to do with an affordable price and the privacy afforded by her double-cabin floorplan. Three galley-up layouts were offered over the years with the latest (introduced in 1986) featuring centerline double berths fore and aft and a U-shaped galley in the salon. Although diesels were an option, the majority were sold with gas power. Note that the Sea Raider and a few of the early model Tri Cabins were fitted with V-drives with the engines located beneath the berths in the owner's stateroom (!). With the 250-hp Chryslers, the Trojan 36 Tri Cabin will cruise at 16 knots and reach about 23 knots wide open. ❏

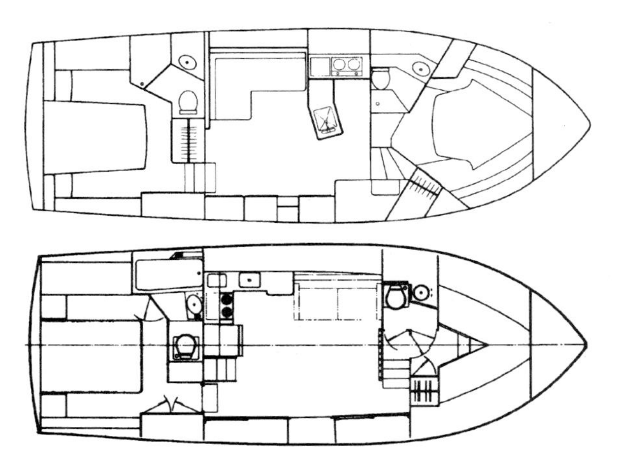

See page 256 for pricing information.

TROJAN 40 MOTOR YACHT

SPECIFICATIONS

Length	40'3"	Water	125 gals.
LWL	NA	Fuel	445 gals.
Beam	14'3"	Hull Type	Modified-V
Draft	3'10"	Deadrise Aft	NA
Weight	29,000#	Designer	Jim Wynne
Clearance	15'6"	Production	1979–84

Designed with a *dual mode* hull capable of economical operation at slower displacement speeds and planing-speed performance with larger engines, the 40 MY was Trojan's response to the fuel uncertainties of the late 1970s. The hull has a slight hump forward on the underbody (which is quite unique) with wide 8" chine flats for lift. The relatively long waterline length results in a fairly expansive interior. The salon measures a full 10'x13' and is separated from the galley by a convenient serving bar. Teak paneling and trim were used throughout the interior. A notable feature of the Trojan 40 MY is the underwater exhaust system—only a few production builders were using them in the early 1980s. Several diesel options were offered in addition to the standard gas engines. The popular 310-hp 6-71N diesels cruise at 14–15 knots (about 17 knots top). At an 8-knot displacement speed, she gets an honest 1 mpg. Note that the Trojan 47 YF (the same boat with a cockpit) runs two knots faster with the same engines. ❑

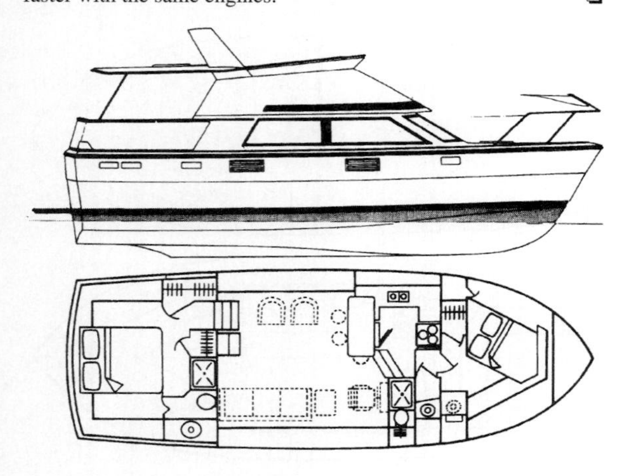

See page 256 for pricing information.

TROJAN 44 MOTOR YACHT

SPECIFICATIONS

Length	44'3"	Water	120 gals.
LWL	38'0"	Fuel	420/520 gals.
Beam	14'11"	Hull Type	Modified-V
Draft	4'0"	Deadrise Aft	NA
Weight	32,000#	Designer	Trojan
Clearance	17'2"	Production	1974–84

The original Trojan 44 MY was introduced in 1974 as an all-fiberglass replacement for the highly successful Trojan 42 mahogany-planked motor yacht. She was extensively redesigned in 1978 with an all-new superstructure, wing doors for the afterdeck, and a revised bridge profile—big styling improvements from the original model. The modified-V hull has a relatively wide beam (for the early 1980s) with a sharp entry, a well-flared bow, hard aft chines, and underwater exhausts. A long, fairly deep keel and deep rudders aid in tracking. The interior accommodations are about average for a 44' motor yacht, with a walkaround queen berth in the aft stateroom, traditional teak paneling in the salon, a U-shaped galley, and two head compartments. (A guest cabin replaced the dinette in the alternate three-stateroom floorplan.) Wide sidedecks allow for secure passage around the house, and there's bench seating on the foredeck. Among several engine options, the popular 310-hp 6-71N diesels will cruise the Trojan 44 MY at 15 knots and reach a top speed of around 18 knots. ❑

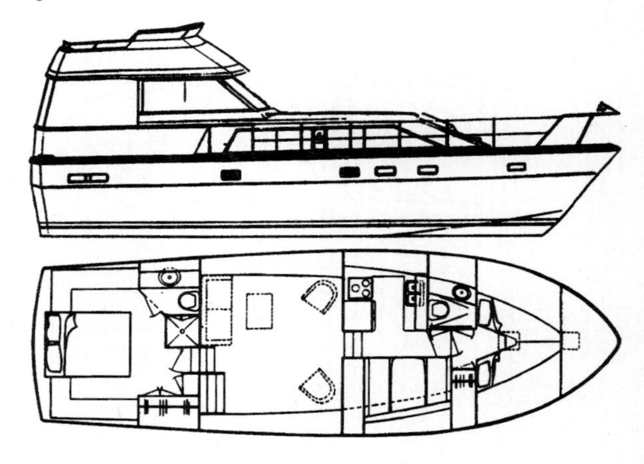

See page 256 for pricing information.

UNIFLITE 36 DOUBLE CABIN

SPECIFICATIONS

Length	36'0"	Fuel, Std.	200 gals.
LWL	32'0"	Fuel, Opt.	300 gals.
Beam	12'4"	Hull Type	Modified-V
Draft	2'8"	Deadrise Aft	NA
Weight	21,000#	Designer	A. Nordtvedt
Clearance	12'3"	Production	1972–84
Water	100 gals.		

The Uniflite 36 Double Cabin was one of the earliest aft-cabin designs to incorporate a full-width master stateroom floorplan. She was a good seller for Uniflite right up to the company's demise in 1984. Her galley-down layout remained relatively unchanged over the years with only minor updates in the galley and master stateroom. Engineering and construction were held to generally high standards, and fire-retardant resin was used in the laminate. Standard 454-cid gas engines will cruise the Uniflite 36 at 19 knots with a top speed of about 28 knots. In 1984, the 36 II version was introduced with a fully cored hull, a new flybridge profile, and a revised interior layout. The weight savings (3,000 lbs.) allowed the smaller 350-cid (270-hp) gas engines to nearly match the performance of the larger 454s used in the earlier models. When Uniflite closed in 1984, Chris Craft picked up the molds and reintroduced her as the Chris 362 Catalina. Note that the Uniflite 41 Yacht Fisherman (1981–84) is a 36 Double Cabin with a cockpit. ❑

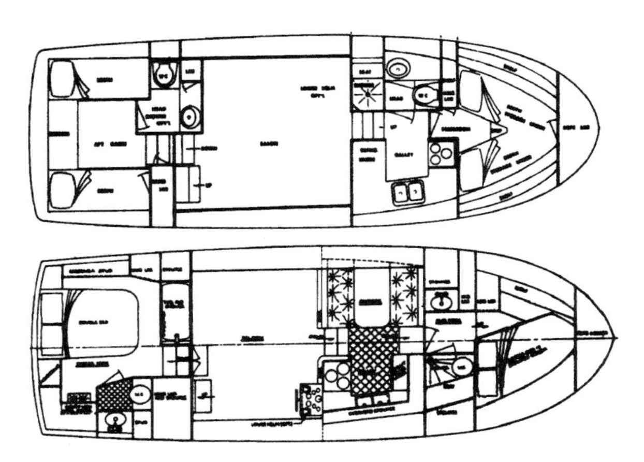

See page 256 for pricing information.

UNIFLITE 42 DOUBLE CABIN

SPECIFICATIONS

Length	42'0"	Water	160 gals.
LWL	37'7"	Fuel	400/500 gals.
Beam	14'9"	Hull Type	Modified-V
Draft	3'9"	Deadrise Aft	NA
Weight	35,000#	Designer	A. Nordtvedt
Clearance	12'10"	Production	1971–84

The Uniflite 42 Double Cabin was one of the earliest production motor yachts to be designed with a raised, full-width aft deck. Used models remain popular due to their sturdy construction and wide open interior layout. (Indeed, the 42's hull and deck molds were later used in the production of the Chris Craft 427 Catalina after Uniflite closed down in 1984.) The standard galley-down floorplan featured twin berths in the aft cabin, a deck access door at the optional lower helm, dinette, and teak woodwork throughout. Those models built after 1978 have a stall shower in the forward head and a rearranged galley/dinette area. Quality components and hardware make used Uniflite 42 DCs a good candidate for refits and interior upgrades. In 1984, the 42 MK II SE version appeared with a revised interior layout which was retained in Chris Craft's 427 model. The popular 310-hp J&T 6-71s will cruise the Uniflite 42 Double Cabin at 18 knots with a top speed of 20–21 knots. Note that the fuel capacity was increased in 1977 to 500 gallons. ❑

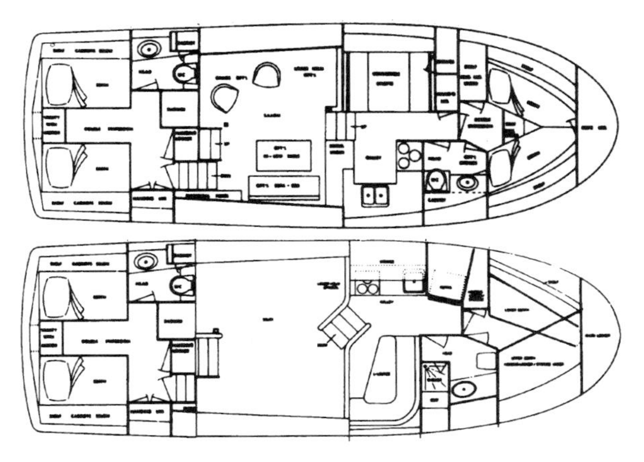

See page 256 for pricing information.

UNIFLITE 48 YACHT FISHERMAN

SPECIFICATIONS

Length	48'0"	Fuel	690 gals.
LWL	43'9"	Cockpit	80 sq. ft.
Beam	14'9"	Hull Type	Modified-V
Draft	3'9"	Deadrise Aft	NA
Weight	39,000#	Designer	A. Nordtvedt
Clearance	12'10"	Production	1980–84
Water	160 gals.		

The Uniflite 48 Yacht Fisherman is basically a Uniflite 42 Double Cabin with a 6-foot cockpit extension. Not only does the increased waterline length add a knot or two to her performance, but the 48's lengthened profile is more pleasing than the 42 DC's ever was. A side benefit of the cockpit is the ability to move the generator from the engine room (where the noise can be heard in the salon above) to beneath the cockpit sole where noise and vibration are kept well away from the interior. The galley-down, two-stateroom layout of the 48 YF is identical to that of the 42 DC—a floorplan notable for its wide-open salon. Built on an efficient hull, the Uniflite 48 YF will cruise at 19 knots (about 22 knots top) with 310-hp 6-71 diesels. With 410-hp 6-71s, the cruising speed jumps to 23 knots (26 top). When the Uniflite factory closed in 1984, the molds were acquired by Chris Craft who reintroduced her the following year as the 480 Catalina—a very popular model. ❑

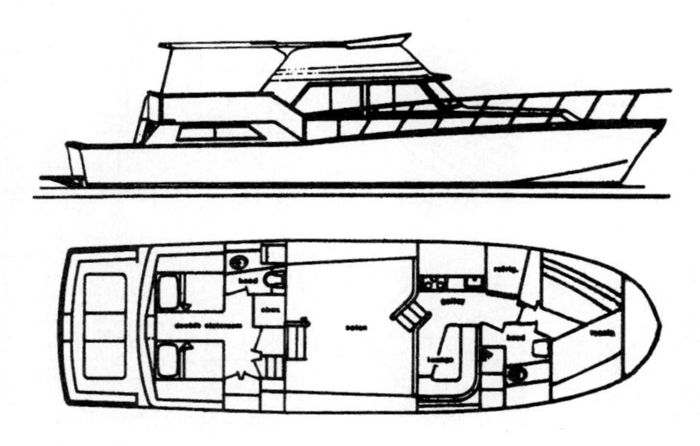

See page 256 for pricing information.

VANTARE 64 COCKPIT MY

SPECIFICATIONS

Length	64'0"	Water	390 gals.
LWL	57'9"	Fuel	1,300 gals.
Beam	17'6"	Hull Type	Modified-V
Draft	3'7"	Deadrise Aft	10°
Weight	70,000#	Designer	Jack Sarin
Clearance	19'6"	Production	1987–92

With her short foredeck and decidedly boxy profile, the Vantare 58 and 64 Cockpit MYs may not have the most graceful profiles in the marina, but both have the interior volume of much larger boats. She was built in Taiwan on a fully cored modified-V hull with a relatively wide beam and a shallow keel. Inside, the four-stateroom floorplan is fairly conventional with two exceptions: the main deck dimensions seem particularly spacious, and one guest stateroom is designed to serve as an office with a built-in desk and bookshelves. The full-width salon is open to the galley, and sliding glass doors lead to the semi-enclosed afterdeck. Light oak or traditional teak cabinetry and woodwork are used throughout the interior. Note that the Vantare 64 was available with the engine room aft (using V-drives) resulting in a roomier lower-level floorplan. Standard 375-hp Cat diesels will cruise at only 13–14 knots. Optional 550-hp 6V92s cruise around 15 knots and 735-hp 8V92s cruise at 18–19 knots. ❑

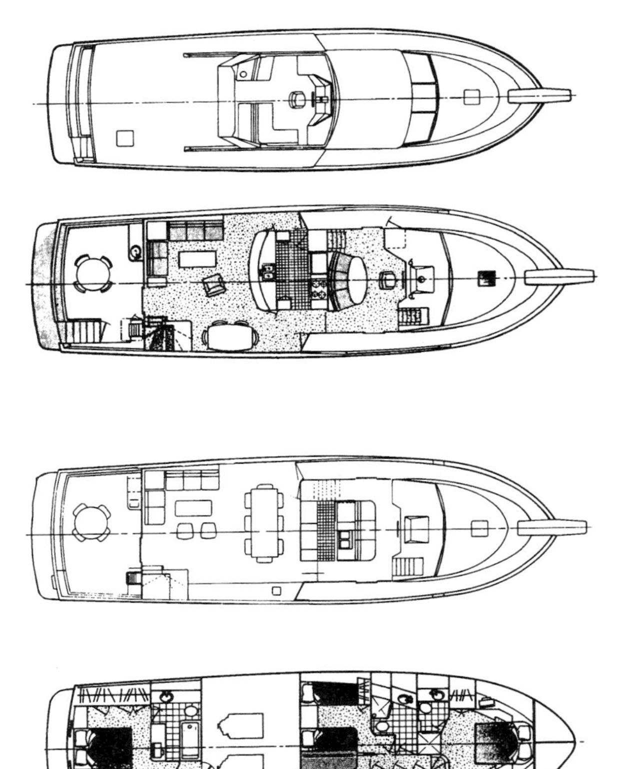

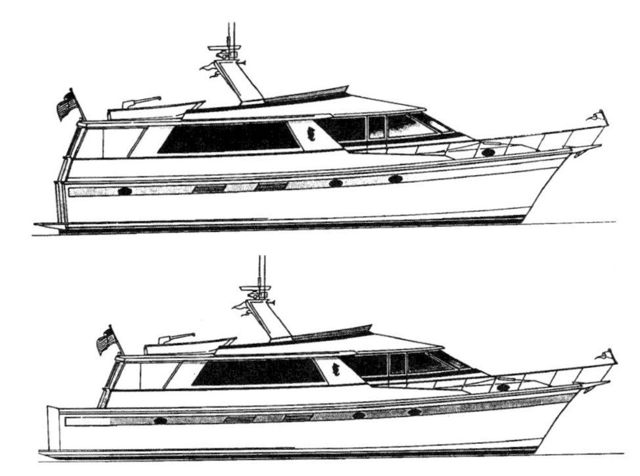

See page 256 for pricing information.

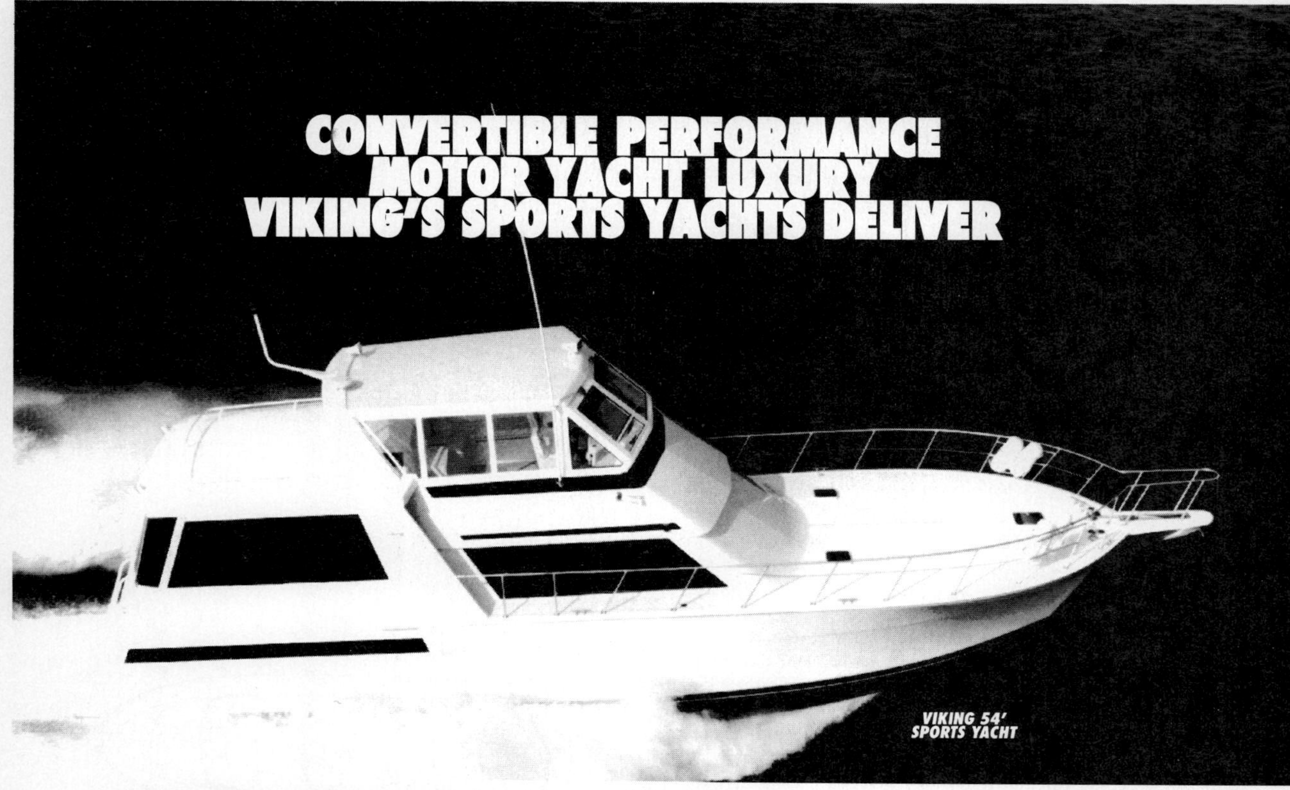

CONVERTIBLE PERFORMANCE
MOTOR YACHT LUXURY
VIKING'S SPORTS YACHTS DELIVER
VIKING 54'
SPORTS YACHT

The all-new Viking Sports Yachts are capable of speeds in excess of 30 knots.

Having built more than 3,300 yachts for more than 30 years, Viking Yachts brings you a whole new class of yacht in two models: the 54' Sports Yacht and the 60' Cockpit Sports Yacht. Experience all the luxurious livability of a motor yacht with all the high performance, handling and speed of a convertible. All built into the Sports Yacht.

The 54' SY and 60' CSY both feature a roomy 17'5" beam, a spacious three stateroom/three head interior, walk-in engine room and 186 square foot enclosable aft deck with integrated swim platform for indoor/outdoor living at its best.

VIKING 60' COCKPIT SPORTS YACHT

At Viking, the definition of luxury includes speed. The Sports Yachts are fast yachts. Exceptional performance comes from Viking's famous modified V-hull, powered by twin turbo-charged diesels which deliver high speeds, economical fuel consumption and outstanding range.

Viking means quality control. Ninety percent of each and every yacht's component parts are produced in-house. The Viking Credo: There's only one way to build a yacht. The right way.

The all-new Sports Yachts join Viking's fleet of 14 models.

- *Motor Yachts 50', 57', 65'.*
- *Convertibles 38', 43', 47', 50', 53', 58', 68'.*
- *Open 43' Sportfish and Open 43' Express/Cruiser.*

"On the Bass River," New Gretna, NJ 08224, (609) 296-6000, Fax: (609) 296-3956

For those who know better.

VIKING 43 DOUBLE CABIN

SPECIFICATIONS

Length..........................42'8"
Beam ....14'9"
Draft...............................3'9"
Weight.....................34,000#
Clearance.....................12'0"
Water....................100 gals.

Fuel........................350 gals.
Hull Type........Modified-V
Deadrise Aft ....17°
Designer..................Viking
Production............1975–82

Viking's first fiberglass motor yacht design, the 43 Double Cabin was built on a stretched Viking 40 Convertible hull and incorporated the same interior floorplan from the salon forward. The additional length was used to provide a full-width aft stateroom together with a large afterdeck. Notably, her wood-grain mica interior was upgraded to teak in 1982—the last year of production. Although some consider the styling a compromise, the Viking 43's quality engineering and muscular construction have made used models popular in most markets. Principal features include excellent storage space, two stall showers, molded steps (not a ladder) leading up to the bridge, and top-quality deck hardware. The helm console is set well aft on the flybridge with bench seating forward. Gas engines were standard, but diesels were the choice of most buyers. With 310-hp GM 6-71Ns, the Viking 43 DC will cruise at 18–19 knots, and the larger 410-hp 6-71s cruise around 22 knots. A popular boat, over 250 Viking 43 Double Cabins were sold during her production run. ❑

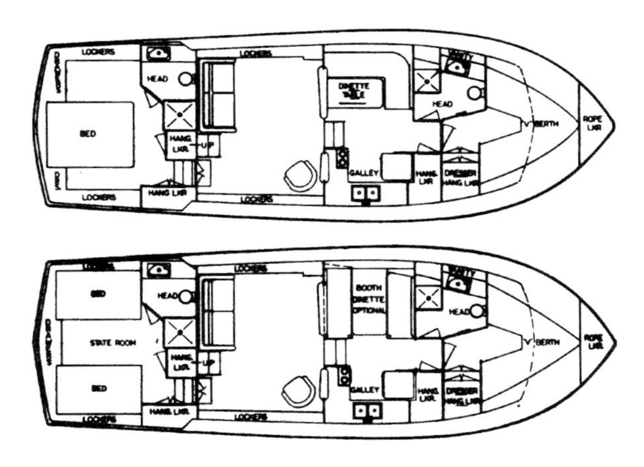

See page 256 for pricing information.

VIKING 44 MOTOR YACHT

Viking 44 MY (1982–88)

Viking 44 MY (1989–91)

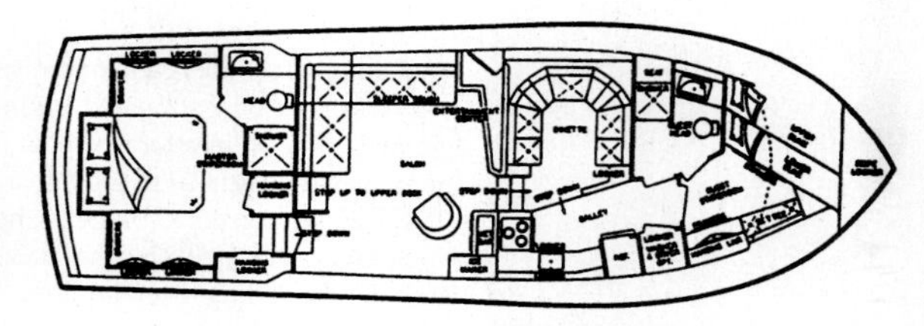

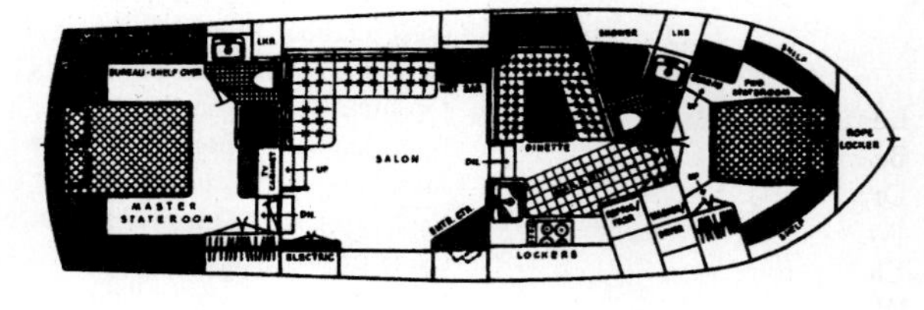

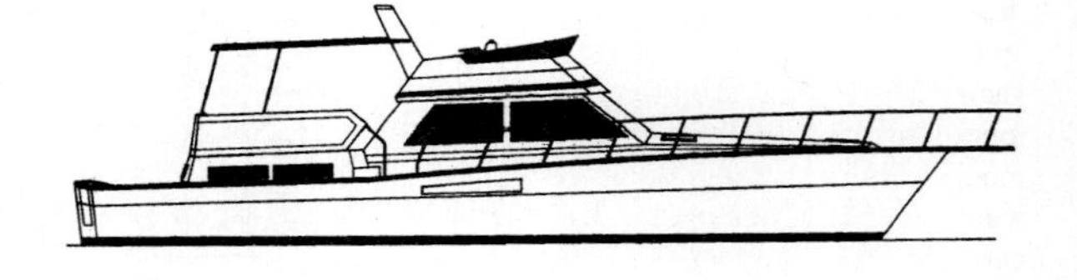

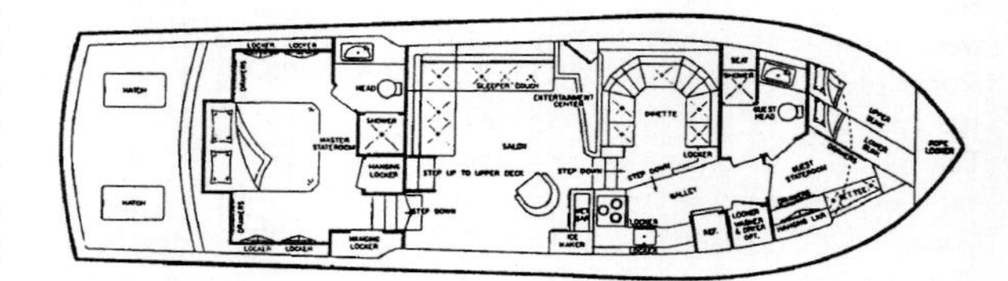

SPECIFICATIONS

Length	44'0"	Water	180 gals.
LWL	37'10"	Fuel	460 gals.
Beam	15'0"	Hull Type	Modified-V
Draft	4'0"	Deadrise Aft	15.5°
Weight	40,000#	Designer	B. Wilson
Clearance	14'6"	Production	1982–91

The Viking 44 MY replaced the popular 43 Double Cabin in the Viking fleet in 1982. Designed for fast and comfortable cruising and showing the improved low-profile styling characteristics of the Viking 46 Convertible, the 44 MY is slightly larger inside than the 43 Double Cabin and carries more fuel and additional water. While the floorplans are similar, the 44 has a larger main salon with the forward companionway offset to starboard. The elegant teak interior is a superb blend of natural woodwork and top-quality fabrics and furnishings. Outside, the 44's flybridge is arranged with passenger seating forward and to starboard of the helm. Additional features include a spacious afterdeck, optional hardtop and arch, balsa coring in the hullsides, and steel engine mounts—a Viking trademark. On the downside, the engine room is rather a tight fit. A good-running boat with optional 485-hp 6-71s, the 44 MY will cruise at a fast 24 knots and turn 28+ knots wide open. Note that the Viking 50 Cockpit MY (1983–87) is a 44 MY with a cockpit. ❑

See page 257 for pricing information.

VIKING 48 MOTOR YACHT

Viking 48 MY

Viking 55 CMY

SPECIFICATIONS

Length	48'7"	Fuel	645 gals.
Beam	16'0"	Hull Type	Modified-V
Draft	4'7"	Deadrise Aft	15.5°
Weight	48,500#	Designer	B. Wilson
Clearance	12'5"	Production	1986–88
Water	200 gals.		

Utilizing the Viking 48 Convertible's hull and deckhouse, the Viking 48 MY combines surprisingly brisk performance with an unusually spacious layout that compares well with many larger yachts. Her elegant three-stateroom interior is finished with the same top-quality teak woodwork, appliances, and fabrics common to all modern Viking interiors. A centerline queen berth is fitted in the forward stateroom, and offset single berths are in the portside guest cabin. Another walkaround queen berth is located in the owner's stateroom aft. Note that the Viking 48's raised aft deck is a fully enclosed *second* salon complete with teak paneling, wing doors, a wet bar, and an L-shaped lounge. Her big flybridge (also borrowed from the 48 Convertible) will seat up to eight. The 48 MY came standard with 565-hp 6V92s for a cruising speed of 21–22 knots. The optional (and more popular) 735-hp 8V92s provide a fast 25-knot cruising speed and about 28 knots top. Note that the Viking 55 Cockpit MY (1988 only) is the same boat with a cockpit (but only a few were built). ❑

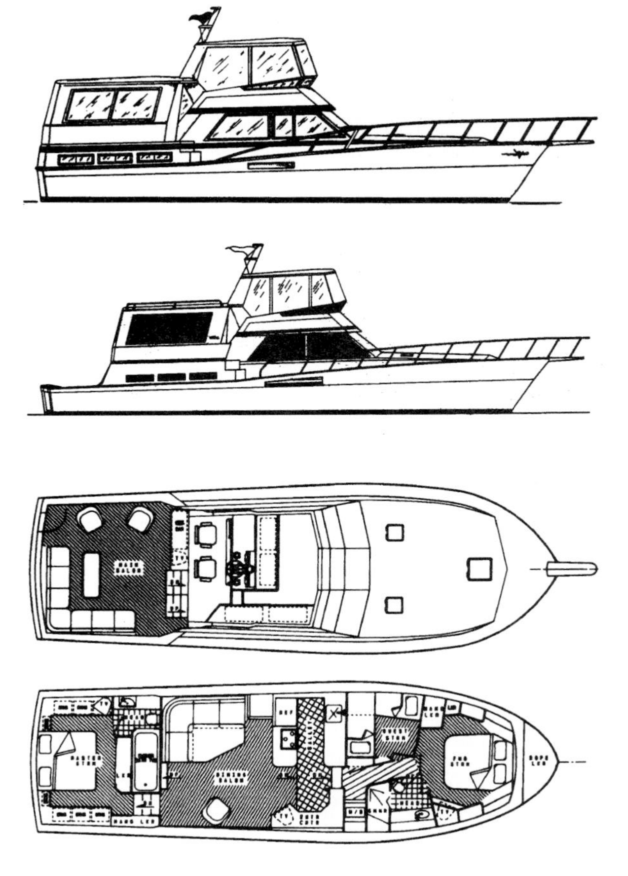

See page 257 for pricing information.

VIKING 50 MOTOR YACHT

SPECIFICATIONS

Length	50'6"	Fuel	770 gals.
Beam	16'4"	Hull Type	Modified-V
Draft	4'3"	Deadrise Aft	14°
Weight	65,000#	Designer	B. Wilson
Clearance	20'0"	Production	1990–Current
Water	250 gals.		

A big boat on the inside with the accommodations of a 55-footer, the look of the Viking 50 MY is stylish but decidedly boxy. Designed to be owner-operated, the beamy modified-V hull is a blend of high-tech fabrics, pre-preg composites, and vacuum-bagging—typical Viking state-of-the art construction. Inside, the truly expansive double-deck floorplan of the Viking 50 MY includes three *double* staterooms on the lower level—each with a private head and stall shower—in addition to a full-width engine room. A lower helm is optional, and the galley/dinette area is open to the full-width salon, where a staircase leads up to the flybridge. The teak interior woodwork is impressive. A bridge overhang protects the aft deck, where an outside ladder provides access to the flybridge. Topside, there's room on the extended flybridge for a Whaler and a dozen guests. No lightweight, standard 735-hp 8V92 diesels will cruise the Viking 50 MY around 18–20 knots and reach a top speed of 23 knots. ❑

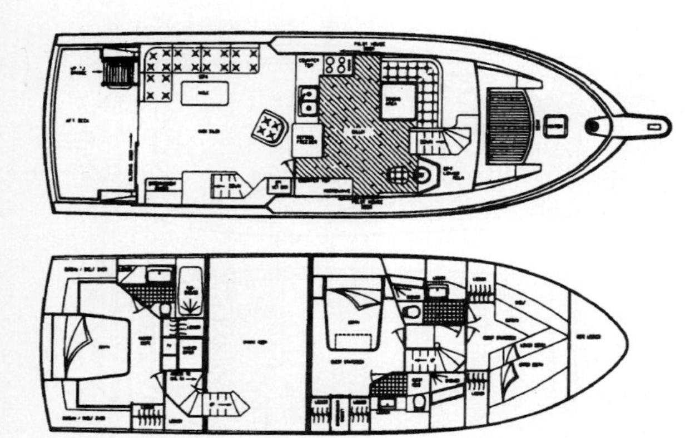

See page 257 for pricing information.

VIKING 54 SPORTS YACHT

SPECIFICATIONS

Length	54'1"	Water	200 gals.
LWL	47'9"	Fuel	900 gals.
Beam	17'5"	Hull Type	Modified-V
Draft	4'10"	Deadrise Aft	15.5°
Weight	75,000#	Designer	B. Wilson
Clearance	NA	Production	1992–Current

The Viking 54 is a great-looking high-performance cruiser with a clean-cut profile and plenty of interior volume. Her good performance is due in part to the fact that she's built on a reworked version of Viking's 53 Convertible hull (with a wider beam) and because her center of gravity is lower than her double-deck counterparts in the mid-50' motor yacht market. Inside, the three-stateroom floorplan includes a spacious and beautifully appointed salon wide open to the mid-level galley. The dinette is forward in the salon and all three heads come with stall showers. The aft stateroom is huge with excellent storage and nearby access to the stand-up engine room. The sprawling aft deck (almost 200 sq. ft.) can be fully enclosed as a factory option. Additional features include a big flybridge with the helm forward and an extremely upscale decor package with some beautiful teak woodwork. A fast boat in spite of her heavy displacement, standard 820-hp MAN diesels will cruise the Viking 54 at 26 knots and deliver 32 knots top. ❑

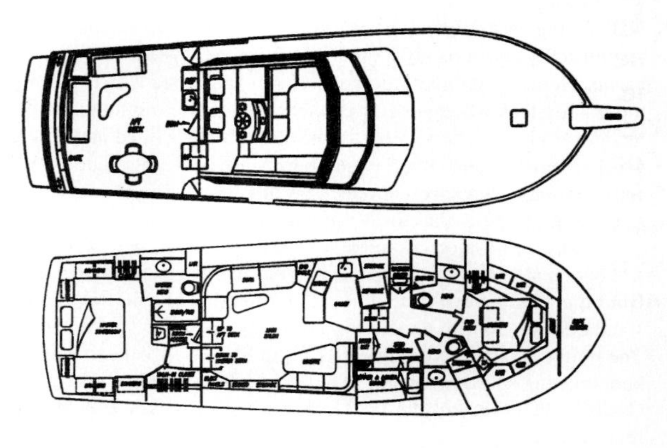

See page 257 for pricing information.

VIKING 55 MOTOR YACHT

Gulfstar 55 MY

Viking 55 MY

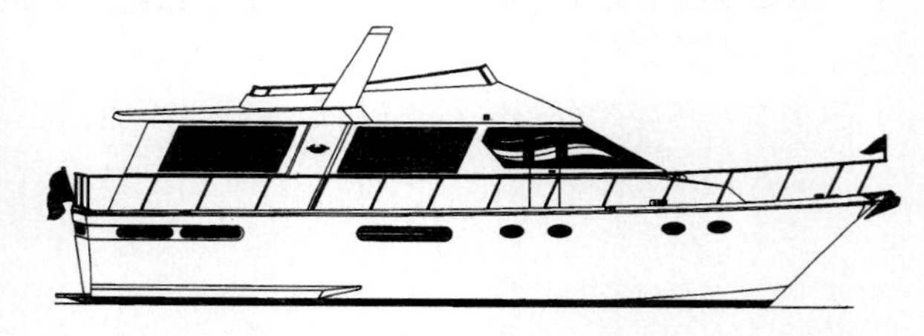

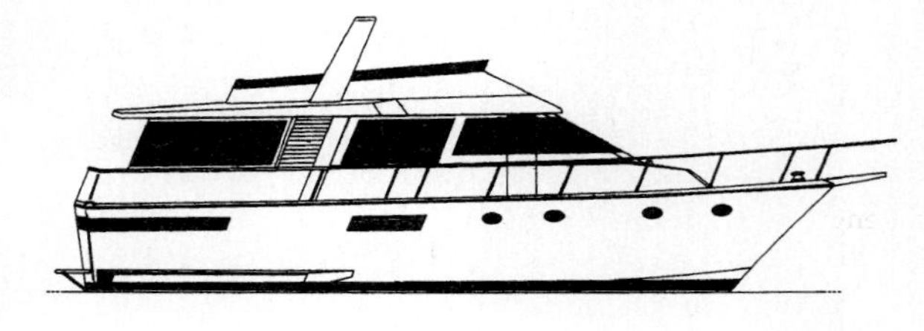

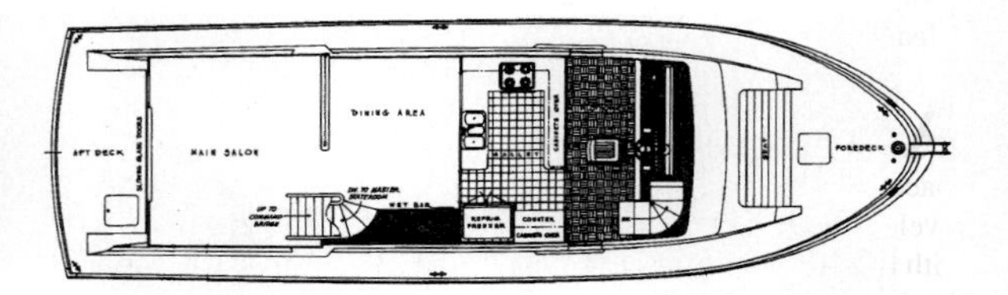

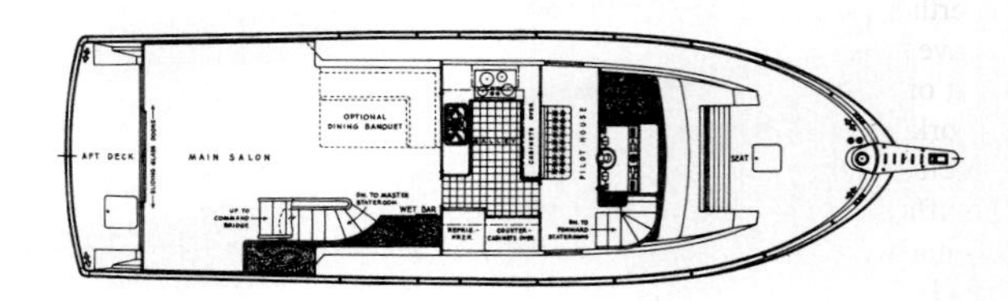

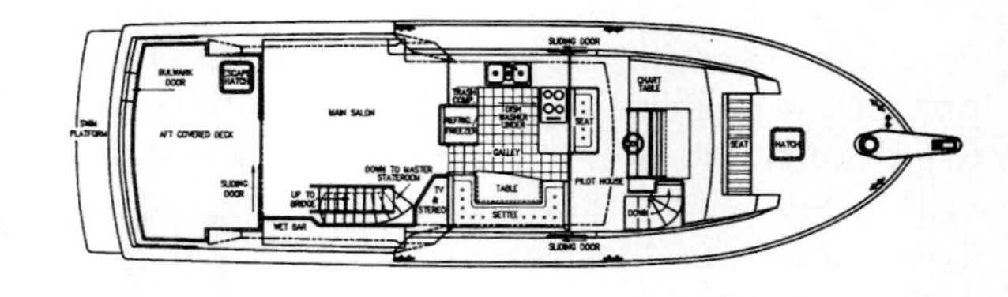

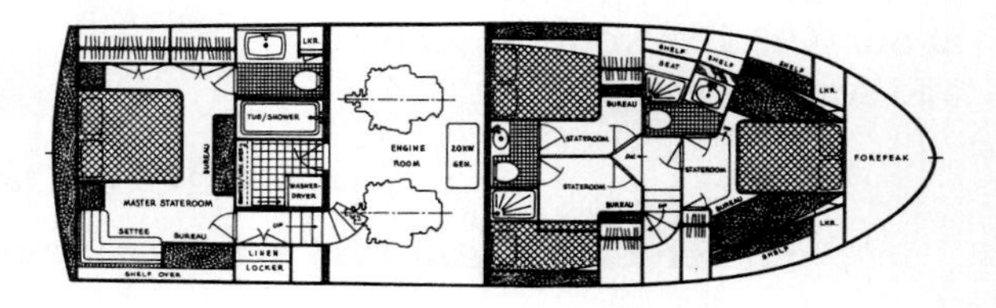

SPECIFICATIONS

Length	55'7"	Water	350 gals.
LWL	47'2"	Fuel	770 gals.
Beam	17'4"	Hull Type	Modified-V
Draft	4'5"	Deadrise Aft	15°
Weight	56,000#	Designer	R. Lazzara
Clearance	21'5"	Production	1987–91

Introduced in 1987 as the Gulfstar 55 MY, the Viking 55 was a break-through design in the late 1980s for her then-modern engineering and lightweight construction. Gulfstar was a technology-driven company when they were acquired by Viking in 1987. Built on a fully cored hull with moderate transom deadrise, the 55 is noted for her good performance, excellent interior space utilization and woodwork, and—unfortunately—a somewhat noisy master stateroom while underway. Her standard four-stateroom layout (three with double berths) has the main salon/dining area separated from the galley and wheelhouse. Notable features include a salon staircase leading up to the flybridge (a big improvement over the normal wheelhouse ladder) and a separate utility room with washer/dryer and workbench. Originally designed with walkaround sidedecks, the full-width salon layout became standard in 1989, and the Extended Aft Deck model (with a much smaller salon) came out in 1990. Standard 735-hp 8V92s will cruise the Gulfstar/Viking 55 Motor Yacht at 19–20 knots and reach 23 knots top. A total of 40 were built. ❑

See page 257 for pricing information.

VIKING 57 MOTOR YACHT

SPECIFICATIONS

Length	57'7"	Water	350 gals.
LWL	49'6"	Fuel	750 gals.
Beam	17'4"	Hull Type	Modified-V
Draft	4'5"	Deadrise Aft	15°
Weight	78,000#	Designer	R. Lazarra
Clearance	21'2"	Production	1991–Current

A big boat for her size, the Viking 57 MY is a spin-off of the original Gulfstar/Viking 55 hull with the extra length used to add space in the engine room for lift mufflers and—on the deckhouse level—a more spacious aft deck. She's built on a fully cored hull with a relatively wide beam, a shallow keel, and moderate transom deadrise. Her galley-up, four stateroom floorplan is arranged with the full-width salon separated from the galley and dinette. A queen berth is located in the amidships VIP guest cabin, and all three heads have stall showers. The engine room is entered from the utility room aft of the salon. The upscale decor of the Viking 57—teak woodwork, Corian countertops, etc.—is impressive. Additional features include narrow service decks around the house and quiet, lift-type mufflers (not the baffled exhausts of the 55 MY) in the engine room. Standard 820-hp MANs will cruise at 19–20 knots (23 knots top), and optional 730-hp 8V92s cruise at 18 knots (21 knots top). ❑

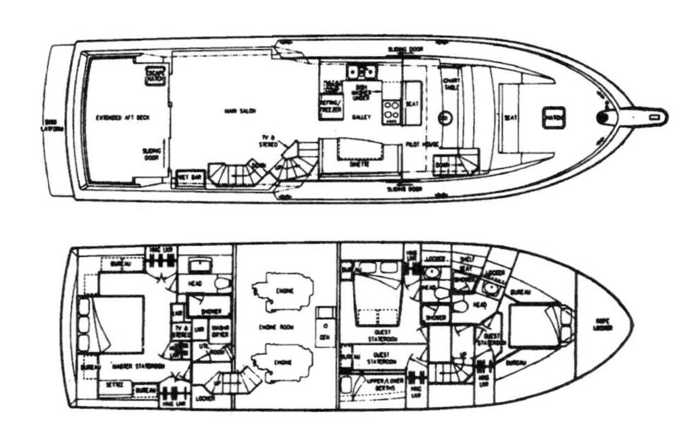

See page 257 for pricing information.

VIKING 60 COCKPIT SPORTS YACHT

SPECIFICATIONS

LOA	65'4"	Water	200 gals.
Hull Length	60'1"	Fuel	1,300 gals.
Beam	17'5"	Hull Type	Modified-V
Draft	4'10"	Deadrise Aft	15.5°
Weight	68,000#	Designer	Viking
Clearance	16'6"	Production	1994–Current

A good-looking yacht with a very aggressive profile, the Viking 60 Cockpit Sport Yacht is basically a Viking 54 Sport Yacht with a small cockpit extension. A cockpit adds great versatility to any motor yacht, and, while the 58 sq. ft. cockpit of the 60 Sport Yacht is modest in size, it nonetheless provides a good swimming and diving platform and also makes boarding much easier. Indeed, the molded steps and storage lockers only add to the cockpit's overall appeal. Inside, the three-stateroom floorplan includes a spacious and beautifully appointed salon wide open to the mid-level galley. The aft stateroom is huge with excellent storage and nearby access to the stand-up engine room. The sprawling aft deck (almost 200 sq. ft.) can be fully enclosed as a factory option. Additional features include a big flybridge with the helm forward and an extremely upscale decor package with some beautiful teak paneling. A good performer for her size, standard 820-hp MAN diesels will cruise the Viking 60 at 26 knots and deliver 32 knots top. ❑

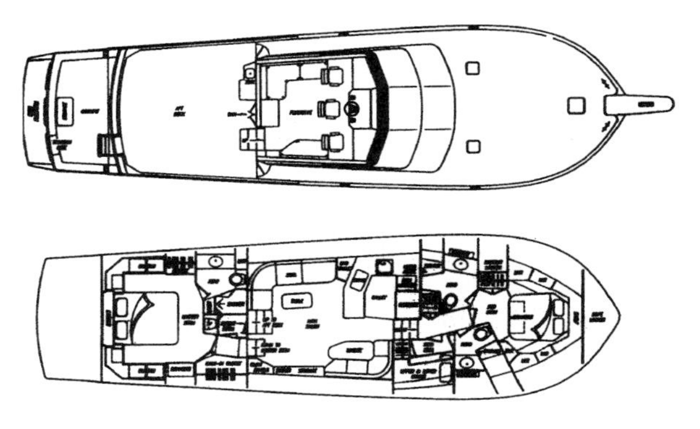

See page 257 for pricing information.

VIKING 63 MOTOR YACHT

Gulfstar 63 MY

Viking 63 MY

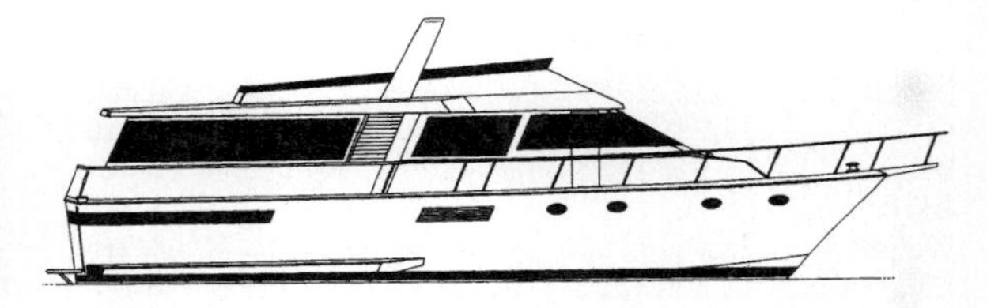

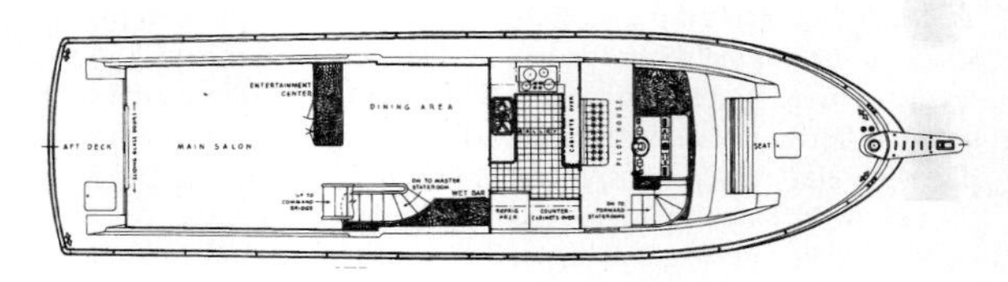

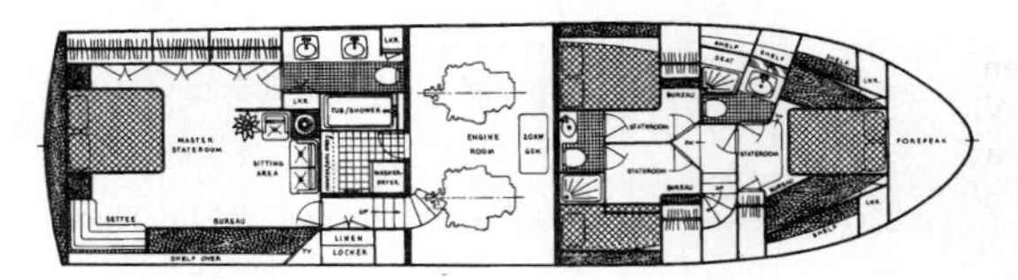

SPECIFICATIONS

Length	62'6"	Water	350 gals.
LWL	54'2"	Fuel	1,080 gals.
Beam	17'4"	Hull Type	Modified-V
Draft	4'9"	Deadrise Aft	15°
Weight	61,500#	Designer	R. Lazzara
Clearance	21'5"	Production	1988–91

The Viking 63 MY began life as the Gulfstar 63 MY—an enlarged version of the original 55 motor yacht that Gulfstar introduced in 1987. She was offered in three configurations: a Walkaround with full sidedecks and a small aft deck; a Widebody with a full-width main salon, a small aft deck, and no sidedecks; and an Extended Aft Deck model with walkaround decks, a smaller salon, and an enlarged afterdeck. While her four-stateroom layout is similar to the original 55 MY's floorplan, the additional length of the 63 is used to extend the main salon and to create a bigger master stateroom. The high-style decor is a blend of traditional teak woodwork, luxurious furnishings, and rich designer fabrics. A utility room contains a workbench as well as access to the engine room. Topside, the flybridge is fitted with enough wraparound lounge seating for a crowd. Standard 735-hp 8V92s will deliver a cruising speed of 17–18 knots (about 20 knots top), and the popular 900-hp 12V71s cruise at 20–21 knots and reach 23 knots top. ❑

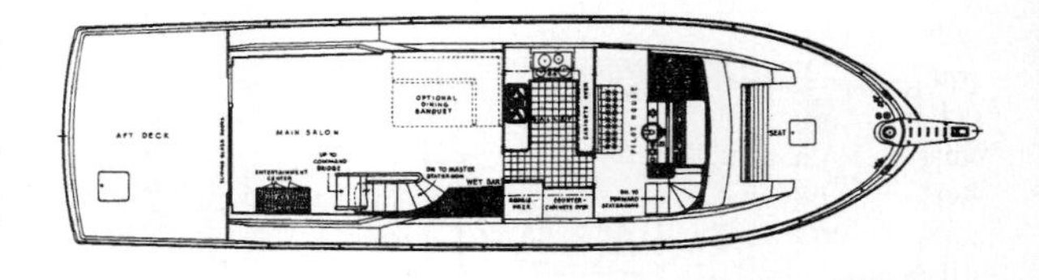

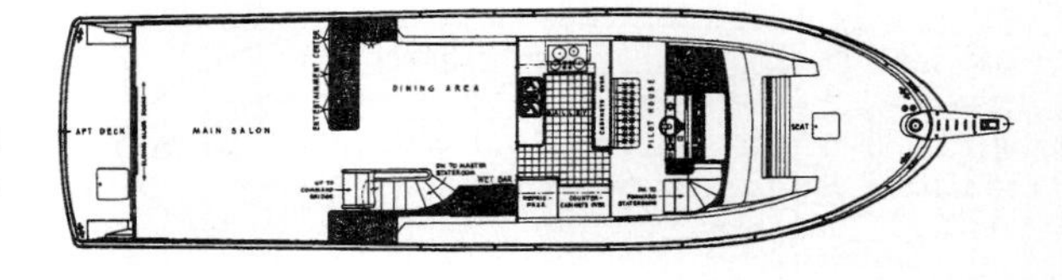

See page 257 for pricing information.

VIKING 63 COCKPIT MY

Gulfstar 63 CMY

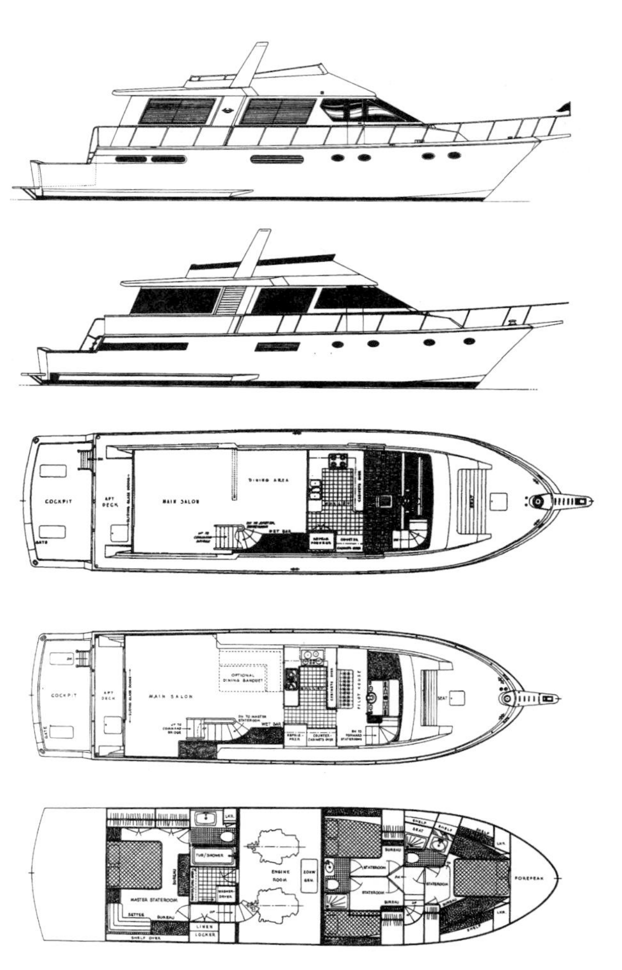

Viking 63 CMY

SPECIFICATIONS

Length	62'6"	Water	350 gals.
LWL	54'4"	Fuel	1,080 gals.
Beam	17'4"	Hull Type	Modified-V
Draft	4'9"	Deadrise Aft	15°
Weight	61,500#	Designer	R. Lazzara
Clearance	21'5"	Production	1987–91

The Viking 63 Cockpit MY began life in 1987 as the Gulfstar 63 CMY—basically a Gulfstar 55 MY with a cockpit extension. Gulfstar built only a couple of these models before the company was acquired by Viking late in 1987. Reintroduced as the Viking 63, the deckhouse was slightly redesigned, but the four-stateroom interior remained unchanged. This is an expansive galley-up floorplan with a separate wheelhouse and two salon staircases—one leading up to the bridge and the other leading down to the master stateroom. Two of the three guest staterooms have queen berths, and a utility room aft contains a washer/dryer and workbench. Viking offered the 63 CMY in Walkaround (with full sidedecks) and Widebody (no sidedecks but with a spacious full-width salon) configurations. The aft deck is small, but the cockpit is large enough for deck chairs or fishing. Additional features include a big engine room, foredeck seating, and a formal dining area in the salon. Standard 735-hp 8V92s cruise at 18 knots, and optional 900-hp 12V71s cruise around 21 knots.❑

See page 257 for pricing information.

VIKING 65 MOTOR YACHT

SPECIFICATIONS

Length ....64'7"	Water.....................300 gals.
LWL ....55'0"	Fuel....................1,030 gals.
Beam ....17'4"	Hull Type........Modified-V
Draft..............................4'9"	Deadrise Aft ....15°
Weight ....94,000#	Designer ....R. Lazarra
Clearance....................20'8"	Production...1991–Current

The Viking 65 MY combines a lavish four-stateroom floorplan with a spacious aft deck and a huge flybridge to create a versatile and extremely good-looking luxury cruiser. The styling is clearly improved over the 63 MY (the boat she replaced in the Viking fleet) with a sweeping flybridge profile and distinctive wraparound bulwarks. There are many highlights in this boat—the full walkaround sidedecks and convenient on-deck day head will be appreciated by experienced owners—but the opulent master suite with its walk-in wardrobe, home-size head, and floor-to-ceiling entertainment center (with refrigerator and wineglass storage, no less) simply dominates the layout. A sliding glass salon door opens to the aft deck, and the sliding (not hinged) wing doors are unique. Topside, the extended flybridge has a wraparound helm and seating for a dozen guests. The large engine room allows space for lift mufflers—much quieter than the baffled exhausts used in earlier Viking 55s and 63 MYs. A good seaboat, standard 1,000-hp MANs cruise at 20 knots and deliver a top speed of about 23 knots. ❑

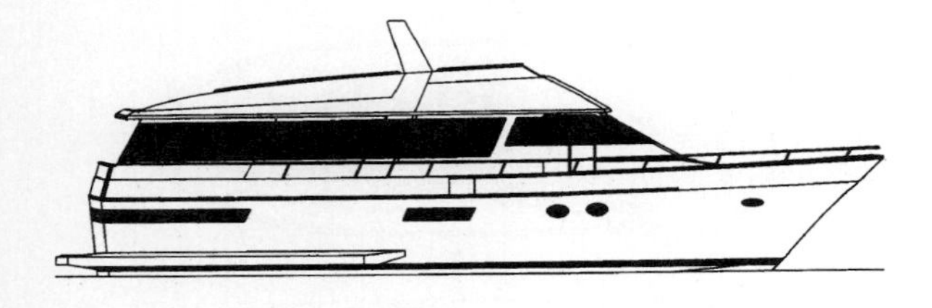

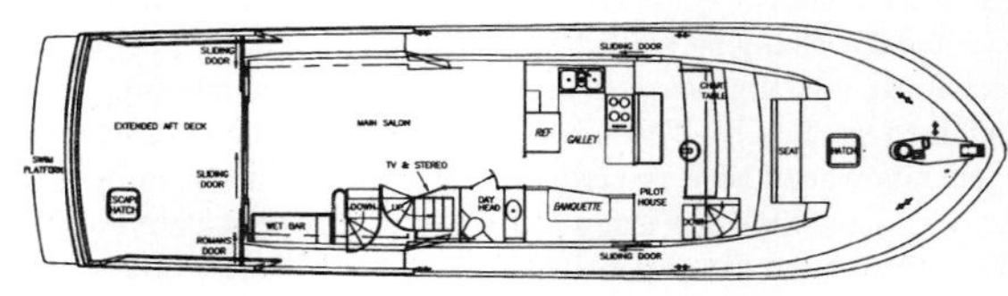

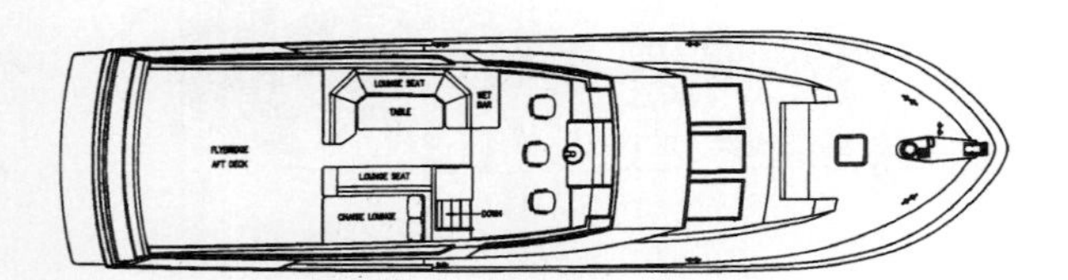

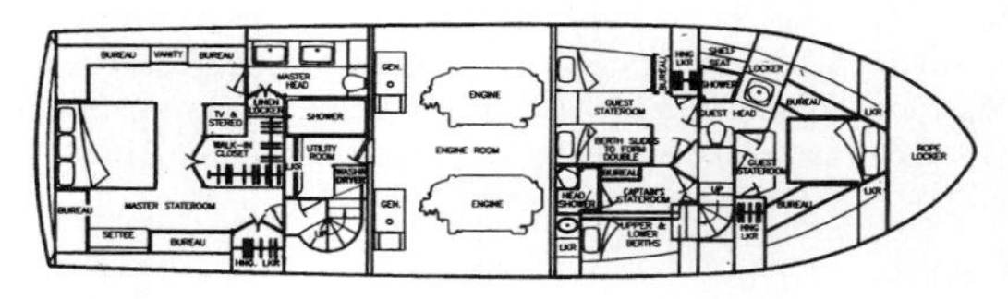

See page 257 for pricing information.

VIKING 65 COCKPIT MY

SPECIFICATIONS

Length	64'7"	Water	300 gals.
LWL	55'0"	Fuel	1,030 gals.
Beam	17'4"	Hull Type	Modified-V
Draft	4'9"	Deadrise Aft	15°
Weight	91,000#	Designer	R. Lazzara
Clearance	20'8"	Production	1991–Current

The Viking 65 is one of the few cockpit motor yachts that isn't simply a cockpit version of a previously-existing motor yacht model. Constructed on a fully cored modified-V hull, she's a good-looking design with a rakish flybridge and a bold Euro-style profile. Inside, the four-stateroom floorplan is arranged with the deckhouse galley separate from the salon. This is a wide-open layout with an impressive array of top-quality joinerwork and plush furnishings throughout. (Except for the smaller master stateroom dimensions, the floorplan is the same as the Viking 65 MY.) Note the day head in the salon. A small aft deck overlooks the cockpit, and there are port and starboard wing doors in the wheelhouse. Additional features include a utility room aft with access to the stand-up engine room, a spacious flybridge with a wraparound helm console, raised bulwarks around the decks, and a transom door and storage lockers in the cockpit. Standard 1,000-hp MAN diesels will cruise the Viking 65 CMY at 21 knots and deliver 23–24 knots wide open. ❑

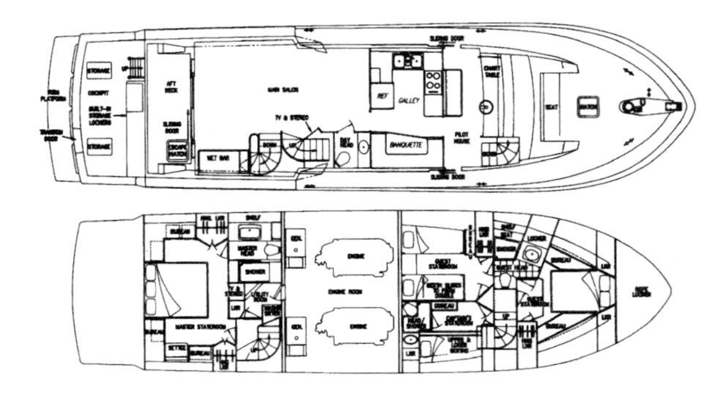

See page 257 for pricing information.

VIKING 72 MOTOR YACHT

SPECIFICATIONS

Length	72'0"	Water	280 gals.
LWL	NA	Fuel	1,470 gals.
Beam	17'5"	Hull Type	Modified-V
Draft	4'10"	Deadrise Aft	15°
Weight	107,000#	Designer	R. Lazarra
Clearance	21'2"	Production	1989–Current

A beautiful motor yacht with modern styling and state-of-the-art construction, the Viking 72 MY is able to combine a surprising degree of performance with her upscale interior accommodations. Note that she's built on a stretched and re-worked version of the fully cored hull first used in the production of the Gulfstar 55 MY. The original four-stateroom floorplan of the 72 MY had two staterooms forward of the engine room and two aft, including a huge master stateroom with a Jacuzzi and separate his-and-hers heads. In 1994, Viking engineers completely rearranged the lower level layout by moving the engine room aft (at the expense of the master stateroom dimensions) and placing all of the guest staterooms forward of the engine room. The deckhouse level was also redesigned in the new floorplan, although the convenient day head and enclosed wheelhouse are retained. Performance is exceptional: she'll cruise at 20 knots with standard 870-hp 12V71s (23 knots top), and optional 1,040-hp 12V92s deliver 22 knots at cruise and 25 knots wide open—very impressive indeed for a boat this size. ❑

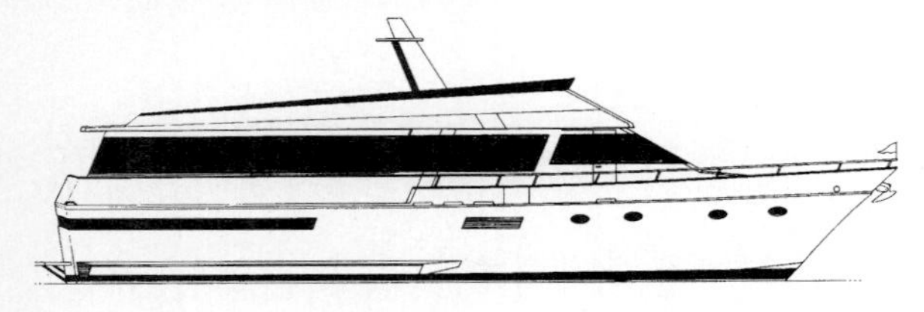

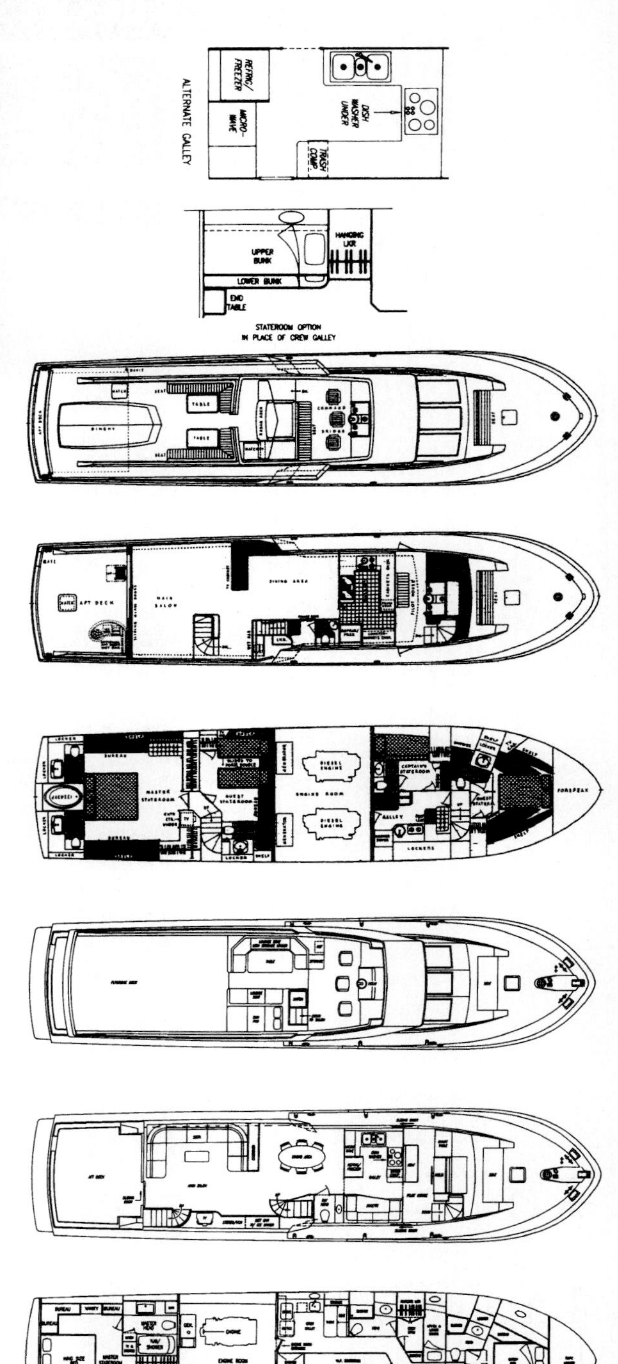

See page 257 for pricing information.

VIKING 72 COCKPIT MY

SPECIFICATIONS

Length..........................72'0"
LWL...............................NA
Beam............................17'5"
Draft.............................4'10"
Weight..................105,000#
Clearance....................21'2"
Water....................350 gals.
Fuel....................1,470 gals.
Cockpit.................92 sq. ft.
Hull Type........Modified-V
Deadrise Aft.................15°
Designer............R. Lazarra
Production...1990–Current

Not a stretched Viking 63 MY, the Viking 72 Cockpit MY is built from the 72 MY mold and features the same distinctive low-profile deckhouse styling and raised bulwarks surrounding the foredeck. She's built on a modern fully cored hull, and her 17'5" beam is modest for a boat this size. The four-stateroom floorplan includes a crew cabin forward in addition to an absolutely extravagant master suite aft that rivals the owner's accommodations found in much larger boats. In 1994, Viking rearranged the deckhouse layout by enlarging the aft deck area (at the expense of the salon dimensions) and adding a dinette opposite the newly redesigned galley. Outside, the large cockpit includes tackle lockers, a transom door, swim platform with ladder, and a fresh-water shower—everything required for fun on the water. A good performer with only 870-hp 12V71 diesels, the Viking 72 CMY has a cruising speed of 20 knots and a top speed of 23 knots. Optional 1,040-hp 12V92s will cruise at 23 knots and reach a top speed of about 26 knots. ❑

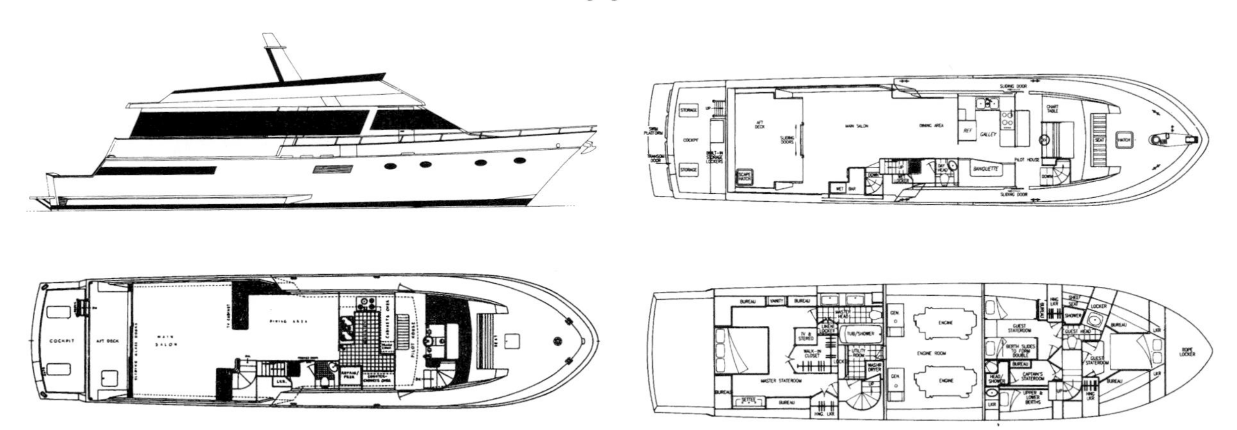

See page 257 for pricing information.

WELLCRAFT 43 SAN REMO

SPECIFICATIONS

Length	42'10"	Water	120 gals.
LWL	NA	Fuel	300/400 gals.
Beam	14'6"	Hull Type	Modified-V
Draft	3'2"	Deadrise Aft	14°
Weight	25,000#	Designer	B. Collier
Clearance	12'6"	Production	1988–90

Notably, the 43 San Remo was the first motor yacht to bear the Wellcraft name since the Californian series was spun off several years before. She uses the Portifino's modified-V hull with prop pockets and rather unattractive exhaust tunnels on the hullsides. The San Remo's lines are low and sleek in spite of her raised deckhouse configuration, and her Eurostyle reverse transom is especially attractive. Below, the interior is quite expansive with emphasis placed in the stylish main salon. There were two floorplans available, the difference being that one has rounded sofas and countertops in the salon, and the other offers a more conventional layout with fewer curves. While the salon dimensions are generous, the aft stateroom is not particularly spacious by motor yacht standards, and the aft deck is correspondingly small. Updates in 1990 included reduced hull graphics, a fresh interior decor package, and a stylish all-white Euro-style helm console. Optional Cat 375-hp diesels will cruise the San Remo around 20 knots with a top speed of 23–24 knots. ❑

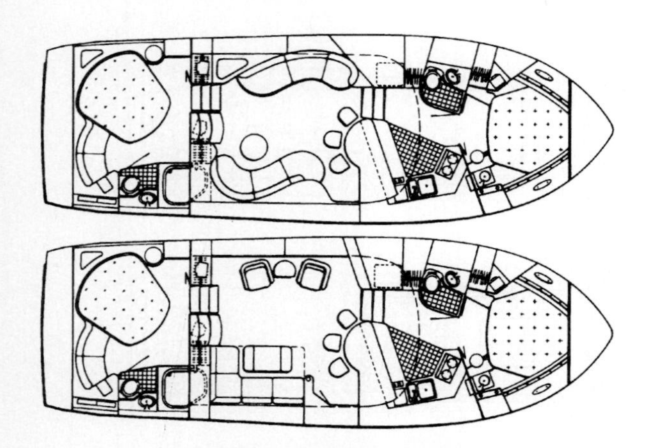

See page 257 for pricing information.

WELLCRAFT 46 COCKPIT MY

SPECIFICATIONS

Length	46'3"	Fuel	400 gals.
Beam	14'6"	Cockpit	46 sq. ft.
Draft	3'2"	Hull Type	Modified-V
Weight	27,000#	Deadrise Aft	14°
Clearance	14'0"	Designer	B. Collier
Water	120 gals.	Production	1990–Current

With the addition of a small cockpit, Wellcraft's 43 San Remo MY was transformed into the 46 Cockpit MY, the largest model in the Wellcraft fleet. Cockpit yachts have become increasingly popular in recent years due to the increased versatility they offer. Boarding is made much easier and it's also possible to do a little fishing or diving from this convenient platform. Equally important, the extra hull length makes the Wellcraft 46 a better-looking yacht than the original San Remo model. The two-stateroom floorplan was available with a choice of salon configurations: a glitzy Euro-style layout with facing S-shaped salon settees and lots of rounded corners, and a more contemporary (and now standard) arrangement with an L-shaped settee in the salon. Note that the 46 eliminated the distracting exhaust tunnels found on the hullsides of the San Remo. Gas engines are standard, although most buyers will surely want diesel power in a boat of this size. Caterpillar 425-hp diesels will cruise the 46 CMY at 19 knots and deliver a top speed of about 24 knots. ❑

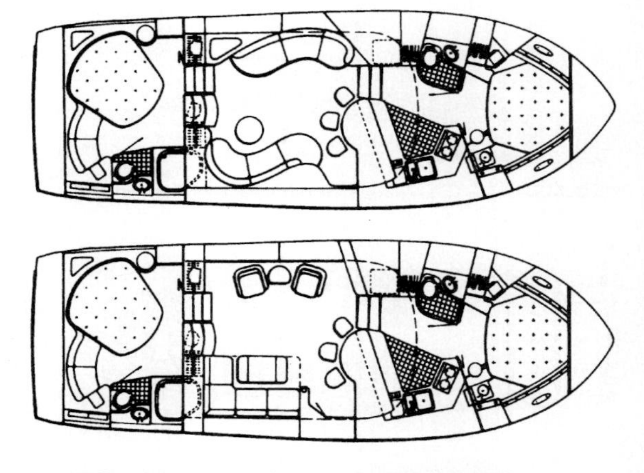

See page 257 for pricing information.

WEST BAY 4500 PILOTHOUSE

SPECIFICATIONS

Length	45'0"	Water	206 gals.
LWL	NA	Fuel	540 gals.
Beam	14'10"	Hull Type	Modified-V
Draft	3'0"	Deadrise Aft	NA
Weight	32,000#	Designer	B. Vermuelen
Clearance	NA	Production	1985–91

For those more interested in seaworthy construction than glitz, the West Bay 4500 PH is a reminder of why many boating enthusiasts favor vessels from the Pacific Northwest. The West Bay 4500 was built in British Columbia on a commercial hull originally designed for Canadian Coast Guard search-and-rescue purposes. Constructed with a watertight bulkhead forward and a watertight engine room, her solid glass hull features a flared bow, moderate beam, and a shallow keel. Inside, the layout is arranged with access to the lower-level staterooms via a salon companionway rather than from the pilothouse. The galley is forward in the salon, and there are three staterooms and two full heads in the standard floorplan. (A two-stateroom layout with an enlarged master was optional.) Additional features include a spacious cockpit with engine room access, walkaround sidedecks, and oak cabinetry and trim throughout. Among several engine options, 300-hp Cummins or 320-hp Cats will cruise around 19–20 knots and 375-hp/425-hp Cats cruise at 22 and 23 knots respectively. Fifteen West Bay 4500s were built. ❑

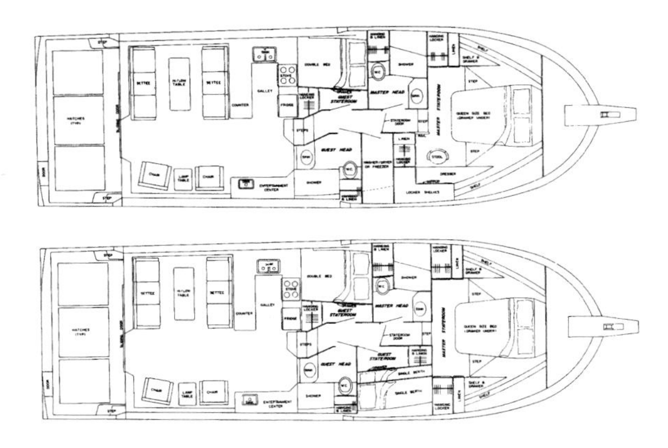

See page 257 for pricing information.

WILLARD 30/4 TRAWLER

SPECIFICATIONS

Length	30'6"	Water	80 gals.
LWL	27'6"	Fuel	150 gals.
Beam	10'6"	Hull Type	Disp.
Draft	3'6"	Designer	Willard
Weight	17,000#	Production	1977–Current

The Willard 30/4 is a salty-looking little trawler with a true displacement hull and a ballasted keel. A little topheavy in appearance, her classic double-ended hull and relatively heavy displacement provide a secure and comfortable ride in a variety of sea conditions. With only 30'6" of LOA, the Willard offers a surprising amount of interior space for her size. There's a compact salon with the galley aft and an L-shaped settee/dinette to port. The lower helm is standard, of course, and V-berths with a small head compartment are forward. Never a full production boat (none were built from 1982–87), the new Willard 30 (1988–current) features a direct-drive engine (replacing the V-drive of previous models) and internal lead ballast rather than external cast iron. With a 50-hp Perkins diesel, the efficiency of the Willard 30/4 is truly phenomenal—at 6 knots, she's actually burning less than 1 gph! With only 150 gallons of fuel capacity her average cruising range is still over 1,000 nautical miles. ❑

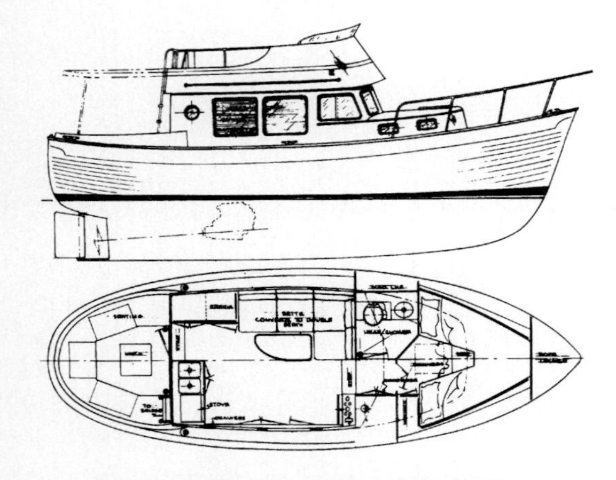

See page 257 for pricing information.

WILLARD 40 TRAWLER

SPECIFICATIONS

Length	39'9"	Water	260 gals.
LWL	36'1"	Fuel	600 gals.
Beam	13'8"	Cockpit	79 sq. ft.
Draft	4'3"	Hull Type	Disp.
Weight	33,000#	Designer	Willard
Clearance	12'0"	Production	1973–Current

The Willard 40 is a full-displacement, ballasted trawler designed for extended offshore cruising. Built on a semi-custom basis in Anaheim, California, she is highly regarded for her seakindly hull and quality construction. The mere fact that the Willard 40 is a real (i.e., displacement) trawler nearly puts her in a class by herself in today's market. Her sedan layout includes two staterooms below with the owner's stateroom and private head to port and V-berths forward. A second head with stall shower is opposite the master stateroom. The salon and galley are on the deckhouse level, and the interior is tastefully finished in teak and rattan woodwork and vinyl wall coverings—not a plush interior, but eminently practical for serious cruising. The flybridge features unique drop-down steps to the foredeck—a good idea since the sidedecks are somewhat narrow. The mast (for a steadying sail) is functional rather than cosmetic. With the standard 130-hp Perkins diesel, the Willard 40 will cruise at 8 knots burning 3 gph for a range of 1,500+ miles. Willard also builds a pilothouse model on this hull. ❑

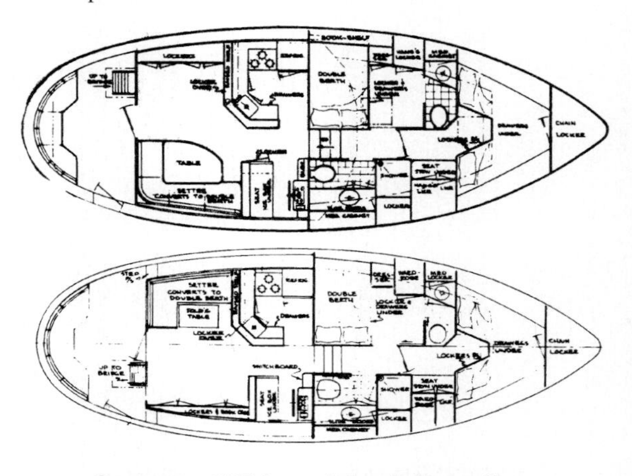

See page 257 for pricing information.

New & Used Boat Prices

Six asterisks (******) indicate insufficient data to render a value for a particular year.
S designates single engine models; T, twins.

Albin 33 Trawler

Year	Power	Retail Low	Retail High
1980	S/Diesel	40,050	44,550
1979	S/Diesel	39,160	43,560

Albin 36 Trawler

Year	Power	Retail Low	Retail High
1994	S/135D	137,060	152,460
1994	S/210D	145,070	161,370
1993	S/135D	126,380	140,580
1993	S/210D	133,500	148,500
1992	S/135D	115,700	128,700
1992	S/210D	121,930	135,630
1991	S/135D	106,800	118,800
1991	S/210D	113,030	125,730
1990	S/135D	97,010	107,910
1990	S/210D	103,240	114,840
1989	S/135D	86,330	96,030
1989	S/210D	91,670	101,970
1988	S/135D	77,430	86,130
1988	S/210D	81,880	91,080
1987	S/135D	69,420	77,220
1986	S/135D	63,190	70,290
1985	S/120D	58,740	65,340
1984	S/120D	55,180	61,380
1983	S/120D	52,510	58,410
1982	S/120D	49,840	55,440
1981	S/120D	47,170	52,470
1980	S/120D	44,500	49,500
1979	S/120D	42,720	47,520
1978	S/120D	40,940	45,540

Albin 40 Trawler

Year	Power	Retail Low	Retail High
1994	T/135D	185,120	205,920
1994	T/210D	195,800	217,800
1993	T/135D	167,320	186,120
1993	T/210D	178,890	198,990
1992	T/135D	150,410	167,310
1992	T/210D	162,870	181,170
1991	T/135D	138,840	154,440
1991	T/210D	148,630	165,330
1990	T/135D	128,160	142,560
1990	T/210D	136,170	151,470
1989	S/135D	107,690	119,790
1989	T/135D	119,260	132,660
1988	S/135D	100,570	111,870
1988	T/135D	111,250	123,750
1987	S/135D	97,010	107,910
1987	T/135D	105,910	117,810

Albin 40 Sundeck

Year	Power	Retail Low	Retail High
1994	T/135D	193,130	214,830
1994	T/210D	203,810	226,710
1993	T/135D	174,440	194,040
1993	T/210D	186,010	206,910
1992	T/135D	156,640	174,240
1992	T/210D	169,100	188,100
1991	T/135D	144,180	160,380
1991	T/210D	153,970	171,270
1990	T/135D	132,610	147,510
1990	T/210D	140,620	156,420
1989	T/135D	111,250	123,750
1989	T/210D	122,820	136,620
1988	T/135D	103,240	114,840
1988	T/210D	113,920	126,720
1987	T/135D	99,680	110,880
1987	T/210D	108,580	120,780

Albin 43 Trawler

Year	Power	Retail Low	Retail High
1994	T/210D	227,840	253,440
1993	T/210D	217,160	241,560
1992	T/210D	202,920	225,720
1991	T/210D	187,790	208,890
1990	T/210D	170,880	190,080
1989	T/135D	147,740	164,340
1989	T/210D	157,530	175,230
1988	T/135D	141,510	157,410
1988	T/210D	143,290	159,390
1987	T/135D	121,930	135,630
1987	T/210D	131,720	146,520
1986	T/135D	113,030	125,730
1986	T/210D	122,820	136,620
1985	T/135D	105,910	117,810
1985	T/210D	115,700	128,700
1984	T/135D	102,350	113,850
1984	T/210D	110,360	122,760

Year	Power	Retail Low	Retail High
1983	T/120D	97,010	107,910
1982	T/120D	92,560	102,960
1981	T/120D	89,000	99,000
1980	T/120D	85,440	95,040
1979	T/120D	80,990	90,090

Albin 43 Sundeck

Year	Power	Retail Low	Retail High
1994	T/210D	238,520	265,320
1993	T/210D	226,950	252,450
1992	T/210D	211,820	235,620
1991	T/210D	195,800	217,800
1990	T/210D	179,780	199,980
1989	T/135D	152,190	169,290
1989	T/210D	161,980	180,180
1988	T/135D	137,950	153,450
1988	T/210D	147,740	164,340
1987	T/135D	128,160	142,560
1987	T/210D	137,060	152,460
1986	T/135D	118,370	131,670
1986	T/210D	127,270	141,570
1985	T/135D	109,470	121,770
1985	T/210D	119,260	132,660
1984	T/135D	105,910	117,810
1984	T/210D	113,920	126,720
1983	T/120D	98,790	109,890
1982	T/120D	92,560	102,960
1981	T/120D	88,110	98,010

Albin 48 Cutter

Year	Power	Retail Low	Retail High
1989	T/307D	194,910	216,810
1989	T/375D	203,810	226,710
1988	T/307D	182,450	202,950
1988	T/375D	191,350	212,850
1987	T/307D	169,990	189,090
1987	T/375D	178,890	198,990
1986	T/255D	159,310	177,210
1985	T/255D	149,520	166,320
1984	T/255D	139,730	155,430
1983	T/255D	129,940	144,540

Albin 49 Cockpit

Year	Power	Retail Low	Retail High
1994	T/250D	293,700	326,700
1993	T/250D	278,570	309,870
1992	T/250D	262,550	292,050
1991	T/250D	244,750	272,250
1990	T/250D	220,720	245,520
1989	T/135D	185,120	205,920
1989	T/210D	195,800	217,800
1988	T/135D	175,330	195,030
1988	T/210D	186,010	206,910
1987	T/135D	167,320	186,120
1987	T/210D	178,000	198,000
1986	T/135D	161,980	180,180
1985	T/120D	150,410	167,310
1984	T/120D	141,510	157,410
1983	T/120D	133,500	148,500
1982	T/120D	127,270	141,570
1981	T/120D	122,820	136,620
1980	T/120D	119,260	132,660
1979	T/120D	115,700	128,700

Albin 49 Tri Cabin

Year	Power	Retail Low	Retail High
1994	T/250D	307,050	341,550
1993	T/250D	289,250	321,750
1992	T/250D	272,340	302,940
1991	T/250D	252,760	281,160
1990	T/250D	227,840	253,440
1989	T/135D	203,810	226,710
1989	T/210D	214,490	238,590
1988	T/135D	191,350	212,850
1988	T/210D	202,030	224,730
1987	T/135D	172,660	192,060
1987	T/210D	183,340	203,940
1986	T/135D	166,430	185,130
1985	T/120D	153,970	171,270
1984	T/120D	145,070	161,370
1983	T/120D	137,060	152,460
1982	T/120D	130,830	145,530
1981	T/120D	126,380	140,580
1980	T/120D	122,820	136,620
1979	T/120D	118,370	131,670

Aquarius 41 MY

Year	Power	Retail Low	Retail High
1993	T/Diesel	******	******
1992	T/Diesel	******	******
1991	T/Diesel	******	******
1990	T/Diesel	******	******
1989	T/Diesel	******	******
1988	T/Diesel	******	******
1987	T/Diesel	******	******

Atlantic 37 DC

Year	Power	Retail Low	Retail High
1992	T/135D	142,400	158,400
1992	T/250D	155,750	173,250
1991	T/135D	136,170	151,470
1991	T/250D	145,960	162,360
1990	T/135D	129,050	143,550
1990	T/250D	137,060	152,460
1989	T/135D	121,930	135,630
1989	T/250D	129,050	143,550
1988	T/Diesel	117,480	130,680
1987	T/Diesel	108,580	120,780
1986	T/Diesel	100,570	111,870
1985	T/Diesel	91,670	101,970
1984	T/Diesel	83,660	93,060
1983	T/Diesel	76,540	85,140
1982	T/Diesel	69,420	77,220

Atlantic 44 MY

Year	Power	Retail Low	Retail High
1992	T/Diesel	245,640	273,240
1991	T/Diesel	234,070	260,370
1990	T/Diesel	223,390	248,490
1989	T/Diesel	214,490	238,590
1988	T/Diesel	204,700	227,700
1987	T/Diesel	193,130	214,830
1986	T/Diesel	182,450	202,950
1985	T/Diesel	171,770	191,070
1984	T/Diesel	160,200	178,200
1983	T/Diesel	146,850	163,350
1982	T/Diesel	133,500	148,500
1981	T/Diesel	118,370	131,670
1980	T/Diesel	106,800	118,800
1979	T/Diesel	97,900	108,900
1978	T/Diesel	90,780	100,980
1977	T/Diesel	87,220	97,020

Atlantic 47 MY

Year	Power	Retail Low	Retail High
1992	T/375D	311,500	346,500
1992	T/450D	328,410	365,310
1991	T/375D	291,920	324,720
1991	T/450D	307,050	341,550
1990	T/375D	276,790	307,890
1990	T/450D	290,140	322,740
1989	T/375D	263,440	293,040
1989	T/450D	275,900	306,900
1988	T/375D	251,870	280,170
1988	T/450D	262,550	292,050
1987	T/375D	238,520	265,320
1987	T/435D	249,200	277,200
1986	T/375D	225,170	250,470
1986	T/435D	235,850	262,350

Year	Power	Retail Low	Retail High
1985	T/375D	210,930	234,630
1985	T/435D	221,610	246,510
1984	T/355D	196,690	218,790
1984	T/410D	207,370	230,670
1983	T/355D	181,560	201,960
1983	T/410D	193,130	214,830
1982	T/355D	168,210	187,110
1982	T/410D	179,780	199,980

Bayliner 4387 Aft Cabin

Year	Power	Retail Low	Retail High
1993	T/Gas	165,540	184,140
1993	T/Diesel	183,340	203,940
1992	T/Gas	154,860	172,260
1992	T/Diesel	169,990	189,090
1991	T/Gas	142,400	158,400
1991	T/Diesel	157,530	175,230
1990	T/Gas	129,050	143,550
1990	T/Diesel	144,180	160,380

Bayliner 4587 MY

Year	Power	Retail Low	Retail High
1994	T/250D	196,690	218,790
1994	T/310D	213,600	237,600

Bayliner 4588 PH MY

Year	Power	Retail Low	Retail High
1993	T/Diesel	213,600	237,600
1992	T/Diesel	196,690	218,790
1991	T/Diesel	181,560	201,960
1990	T/Diesel	167,320	186,120
1989	T/Diesel	159,310	177,210
1988	T/Diesel	150,410	167,310
1987	T/Diesel	137,950	153,450
1986	T/Diesel	127,270	141,570
1985	T/Diesel	118,370	131,670
1984	T/Diesel	112,140	124,740

Bayliner 4788 MY

Year	Power	Retail Low	Retail High
1994	T/250D	264,330	294,030
1994	T/310D	284,800	316,800
1993	T/250D	225,170	250,470
1993	T/250D	244,750	272,250

Bertram 42 MY

Year	Power	Retail Low	Retail High
1987	T/Gas	198,470	220,770
1987	T/Diesel	233,180	259,380
1986	T/Gas	195,800	217,800
1986	T/Diesel	218,050	242,550
1985	T/Gas	182,450	202,950
1985	T/Diesel	204,700	227,700
1984	T/Gas	172,660	192,060
1984	T/Diesel	194,020	215,820
1983	T/Gas	161,980	180,180
1983	T/Diesel	184,230	204,930
1982	T/Gas	153,080	170,280
1982	T/Diesel	175,330	195,030
1981	T/Gas	145,070	161,370
1981	T/Diesel	167,320	186,120
1980	T/Gas	129,940	144,540
1980	T/Diesel	152,190	169,290
1979	T/Gas	119,260	132,660
1979	T/Diesel	142,400	158,400
1978	T/Gas	112,140	124,740
1978	T/Diesel	133,500	148,500
1977	T/Gas	99,680	110,880
1977	T/Diesel	120,150	133,650
1976	T/Gas	87,220	97,020
1976	T/Diesel	105,910	117,810
1975	T/Gas	79,210	88,110
1975	T/Diesel	97,900	108,900

Bertram 46 MY

Year	Power	Retail Low	Retail High
1987	T/Diesel	319,510	355,410
1986	T/Diesel	302,600	336,600
1985	T/Diesel	284,800	316,800
1984	T/Diesel	267,890	297,990
1983	T/Diesel	244,750	272,250
1982	T/Diesel	218,050	242,550
1981	T/Diesel	191,350	212,850
1980	T/Diesel	164,650	183,150
1979	T/Diesel	154,860	172,260
1978	T/Diesel	145,070	161,370
1977	T/Diesel	133,500	148,500
1976	T/Diesel	123,710	137,610
1975	T/Diesel	112,140	124,740

Bertram 58 MY

Year	Power	Retail Low	Retail High
1986	T/Diesel	551,800	613,800
1985	T/Diesel	507,300	564,300
1984	T/Diesel	476,150	529,650
1983	T/Diesel	436,100	485,100
1982	T/Diesel	409,400	455,400
1981	T/Diesel	378,250	420,750
1980	T/Diesel	356,000	396,000
1979	T/Diesel	333,750	371,250
1978	T/Diesel	311,500	346,500
1977	T/Diesel	289,250	321,750
1976	T/Diesel	271,450	301,950

Bluewater 44 Coastal Cruiser

Year	Power	Retail Low	Retail High
1989	T/Gas	146,850	163,350
1989	T/Diesel	165,540	184,140
1988	T/Gas	129,940	144,540
1988	T/Diesel	144,180	160,380
1987	T/Gas	115,700	128,700
1987	T/Diesel	129,050	143,550

Bluewater 462

Year	Power	Retail Low	Retail High
1994	T/IO/Gas	281,240	312,840
1994	T/IO/Diesel	329,300	366,300
1993	T/IO/Gas	321,290	357,390
1993	T/IO/Diesel	275,010	305,910
1992	T/IO/Gas	199,360	221,760
1992	T/IO/Diesel	234,960	261,360

Bluewater 48 Coastal Cruiser

Year	Power	Retail Low	Retail High
1992	T/Gas	200,250	222,750
1992	T/Diesel	222,500	247,500
1991	T/Gas	178,890	198,990
1991	T/Diesel	201,140	223,740
1990	T/Gas	160,200	178,200
1990	T/Diesel	180,670	200,970

Bluewater 51 Coastal Cruiser

Year	Power	Retail Low	Retail High
1989	T/Gas	160,200	178,200
1989	T/Diesel	184,230	204,930
1988	T/Gas	147,740	164,340
1988	T/Diesel	166,430	185,130
1987	T/Gas	132,610	147,510
1987	T/Diesel	150,410	167,310
1986	T/Gas	121,040	134,640
1986	T/Diesel	136,170	151,470
1985	T/Gas	107,690	119,790
1985	T/Diesel	121,930	135,630
1984	T/Gas	94,340	104,940
1984	T/Diesel	116,590	129,690

Bluewater 543

Year	Power	Retail Low	Retail High
1994	T/Gas	329,300	366,300
1994	T/Diesel	373,800	415,800
1993	T/Gas	302,600	336,600
1993	T/Diesel	354,220	394,020
1992	T/Gas	273,230	303,930

Year	Power	Retail Low	Retail High
1992	T/Diesel	309,720	344,520
1991	T/Gas	232,290	258,390
1991	T/Diesel	267,890	297,990

Bluewater 55 Coastal Cruiser

Year	Power	Retail Low	Retail High
1989	T/Gas	187,790	208,890
1989	T/Diesel	228,730	254,430
1988	T/Gas	178,890	198,990
1988	T/Diesel	210,930	234,630
1987	T/Gas	168,210	187,110
1987	T/Diesel	199,360	221,760

Bluewater 55 Yacht

Year	Power	Retail Low	Retail High
1992	T/Diesel	402,280	447,480
1991	T/Diesel	355,110	395,010
1990	T/Diesel	309,720	344,520

Bluewater 622C

Year	Power	Retail Low	Retail High
1994	T/Gas	509,080	566,280
1994	T/Diesel	577,610	642,510
1993	T/Gas	468,140	520,740
1993	T/Diesel	525,990	585,090
1992	T/Gas	406,730	452,430
1992	T/Diesel	461,020	512,820
1991	T/Gas	360,450	400,950
1991	T/Diesel	411,180	457,380
1990	T/Gas	320,400	356,400
1990	T/Diesel	365,790	406,890

Bluewater 623

Year	Power	Retail Low	Retail High
1994	T/Gas	523,320	582,120
1994	T/Diesel	592,740	659,340
1993	T/Gas	480,600	534,600
1993	T/Diesel	539,340	599,940
1992	T/Gas	418,300	465,300
1992	T/Diesel	472,590	525,690
1991	T/Gas	371,130	412,830
1991	T/Diesel	420,970	468,270
1990	T/Gas	330,190	367,290
1990	T/Diesel	375,580	417,780

Californian 34 LRC

Year	Power	Retail Low	Retail High
1982	T/85D	48,950	54,450
1982	T/210D	59,630	66,330
1981	T/85D	47,170	52,470
1981	T/210D	56,960	63,360
1980	T/85D	44,500	49,500
1980	T/210D	54,290	60,390
1979	T/85D	38,270	42,570
1979	T/210D	50,730	56,430

Californian 35 MY

Year	Power	Retail Low	Retail High
1987	T/Gas	78,320	87,120
1987	T/Diesel	89,000	99,000
1986	T/Gas	68,530	76,230
1986	T/Diesel	79,210	88,110
1985	T/Gas	57,850	64,350
1985	T/Diesel	68,530	76,230

Californian 38 LRC

Year	Power	Retail Low	Retail High
1984	T/Diesel	83,660	93,060
1983	T/Diesel	80,990	90,090
1982	T/Diesel	78,320	87,120
1981	T/Diesel	74,760	83,160
1980	T/Diesel	71,200	79,200

Californian 38 MY

Year	Power	Retail Low	Retail High
1987	T/Diesel	112,140	124,740
1986	T/Diesel	106,800	118,800
1985	T/Diesel	101,460	112,860
1984	T/Diesel	96,120	106,920
1983	T/Diesel	90,780	100,980

Californian 42 LRC

Year	Power	Retail Low	Retail High
1984	T/Diesel	111,250	123,750
1983	T/Diesel	107,690	119,790
1982	T/Diesel	104,130	115,830
1981	T/Diesel	100,570	111,870
1980	T/Diesel	94,340	104,940
1979	T/Diesel	89,000	99,000

Californian 43 CMY

Year	Power	Retail Low	Retail High
1987	T/Diesel	153,080	170,280
1986	T/Diesel	144,180	160,380
1985	T/Diesel	134,390	149,490
1984	T/Diesel	125,490	139,590
1983	T/Diesel	118,370	131,670

Californian 45 MY

Year	Power	Retail Low	Retail High
1991	T/375D	289,250	321,750
1991	T/485D	305,270	339,570
1990	T/375D	271,450	301,950
1990	T/485D	284,800	316,800
1989	T/375D	253,650	282,150
1989	T/485D	267,000	297,000
1988	T/375D	231,400	257,400
1988	T/450D	241,190	268,290

Californian 48 MY

Year	Power	Retail Low	Retail High
1991	T/375D	302,600	336,600
1991	T/485D	317,730	353,430
1990	T/375D	282,130	313,830
1990	T/485D	295,480	328,680
1989	T/375D	263,440	293,040
1989	T/485D	276,790	307,890
1988	T/375D	242,080	269,280
1988	T/450D	251,870	280,170
1987	T/375D	226,950	252,450
1987	T/450D	235,850	262,350
1986	T/375D	219,830	244,530

Californian 48 CMY

Year	Power	Retail Low	Retail High
1989	T/Diesel	263,440	293,040
1988	T/Diesel	247,420	275,220
1987	T/Diesel	233,180	259,380
1986	T/Diesel	224,280	249,480

Californian 52 CMY

Year	Power	Retail Low	Retail High
1991	T/375D	328,410	365,310
1991	T/485D	348,880	388,080
1990	T/375D	311,500	346,500
1990	T/485D	333,750	371,250

Californian 55 CMY

Year	Power	Retail Low	Retail High
1991	T/485D	396,050	440,550
1991	T/550D	409,400	455,400
1990	T/485D	444,110	494,010
1990	T/550D	368,460	409,860
1989	T/485D	327,520	364,320
1989	T/550D	340,870	379,170
1988	T/485D	314,170	349,470
1988	T/550D	326,630	363,330

Camargue 48 YF

Year	Power	Retail Low	Retail High
1993	T/Diesel	280,350	311,850
1992	T/Diesel	256,320	285,120
1991	T/Diesel	238,520	265,320
1990	T/Diesel	225,170	250,470
1989	T/Diesel	214,490	238,590
1988	T/Diesel	204,700	227,700
1987	T/Diesel	195,800	217,800

Carver 32 Aft Cabin

Year	Power	Retail Low	Retail High
1990	T/Diesel	64,970	72,270
1989	T/Diesel	59,630	66,330

Year	Power	Retail Low	Retail High
1988	T/Diesel	55,180	61,380
1987	T/Diesel	50,730	56,430
1986	T/Diesel	47,170	52,470
1985	T/Diesel	43,610	48,510
1984	T/Diesel	40,940	45,540
1983	T/Diesel	38,270	42,570

Carver 350 Aft Cabin

Year	Power	Retail Low	Retail High
1994	T/Gas	113,030	125,730
1994	T/Diesel	139,730	155,430
1993	T/Gas	105,020	116,820
1993	T/Diesel	129,050	143,550
1992	T/Gas	99,680	110,880
1992	T/Diesel	120,150	133,650
1991	T/Gas	93,450	103,950
1991	T/Diesel	111,250	123,750

Carver 36 Aft Cabin

Year	Power	Retail Low	Retail High
1989	T/Gas	94,340	104,940
1988	T/Gas	84,550	94,050
1987	T/Gas	77,430	86,130
1986	T/Gas	71,200	79,200
1985	T/Gas	64,970	72,270
1984	T/Gas	58,740	65,340
1983	T/Gas	52,510	58,410
1982	T/Gas	45,390	50,490

Carver 370 Aft Cabin MY

Year	Power	Retail Low	Retail High
1994	T/Gas	142,400	158,400
1994	T/Diesel	173,550	193,050
1993	T/Gas	130,830	145,530
1993	T/Diesel	161,090	179,190
1992	T/Gas	122,820	136,620
1992	T/Diesel	151,300	168,300
1991	T/Gas	113,030	125,730
1991	T/Diesel	137,950	153,450
1990	T/Gas	101,460	112,860
1990	T/Diesel	122,820	136,620

Carver 390 Aft Cabin

Year	Power	Retail Low	Retail High
1994	T/Gas	164,650	183,150
1994	T/Diesel	195,800	217,800
1993	T/Gas	155,750	173,250
1993	T/Diesel	186,900	207,900
1992	T/Gas	146,850	163,350
1992	T/Diesel	177,110	197,010
1991	T/Gas	137,950	153,450
1991	T/Diesel	165,540	184,140
1990	T/Gas	129,050	143,550
1990	T/Diesel	155,750	173,250
1989	T/Gas	118,370	131,670
1989	T/Diesel	143,290	159,390
1988	T/Gas	107,690	119,790
1988	T/Diesel	131,720	146,520
1987	T/Gas	99,680	110,880
1987	T/Diesel	121,040	134,640

Carver 390 CMY

Year	Power	Retail Low	Retail High
1994	T/Gas	128,160	142,560
1994	T/Diesel	151,300	168,300
1993	T/Gas	115,700	128,700
1993	T/Diesel	138,840	154,440

Carver 42 MY

Year	Power	Retail Low	Retail High
1991	T/Gas	164,650	183,150
1991	T/Diesel	192,240	213,840
1990	T/Gas	155,750	173,250
1990	T/Diesel	183,340	203,940
1989	T/Gas	147,740	164,340
1989	T/Diesel	176,220	196,020
1988	T/Gas	139,730	155,430
1988	T/Diesel	166,430	185,130
1987	T/Gas	131,720	146,520
1987	T/Diesel	156,640	174,240
1986	T/Gas	123,710	137,610
1986	T/Diesel	146,850	163,350
1985	T/Gas	115,700	128,700
1985	T/Diesel	134,390	149,490

Carver 430 CMY

Year	Power	Retail Low	Retail High
1994	T/Gas	194,910	216,810
1994	T/Diesel	219,830	244,530
1993	T/Gas	183,340	203,940
1993	T/Diesel	205,590	228,690
1992	T/Gas	172,660	192,060
1992	T/Diesel	195,800	217,800
1991	T/Gas	163,760	182,160
1991	T/Diesel	186,900	207,900

Cheoy Lee 32 Trawler

Year	Power	Retail Low	Retail High
1986	S/Diesel	66,750	74,250
1985	S/Diesel	62,300	69,300
1984	S/Diesel	57,850	64,350
1983	S/Diesel	53,400	59,400
1982	S/Diesel	49,840	55,440
1981	S/Diesel	48,060	53,460
1980	S/Diesel	45,390	50,490
1979	S/Diesel	43,610	48,510
1978	S/Diesel	40,940	45,540
1977	S/Diesel	38,270	42,570

Cheoy Lee 35 Trawler

Year	Power	Retail Low	Retail High
1986	S/Diesel	71,200	79,200
1985	S/Diesel	66,750	74,250
1984	S/Diesel	62,300	69,300
1983	S/Diesel	57,850	64,350
1982	S/Diesel	54,290	60,390
1981	S/Diesel	51,620	57,420
1980	S/Diesel	48,950	54,450
1979	S/Diesel	47,170	52,470

Cheoy Lee 40 LRC

Year	Power	Retail Low	Retail High
1986	T/Diesel	151,300	168,300
1985	T/Diesel	140,620	156,420
1984	T/Diesel	132,610	147,510
1983	T/Diesel	123,710	137,610
1982	T/Diesel	114,810	127,710
1981	T/Diesel	106,800	118,800
1980	T/Diesel	98,790	109,890
1979	T/Diesel	91,670	101,970
1978	T/Diesel	84,550	94,050
1977	T/Diesel	78,320	87,120
1976	T/Diesel	75,650	84,150
1975	T/Diesel	72,980	81,180

Cheoy Lee 46 Trawler

Year	Power	Retail Low	Retail High
1981	T/Diesel	146,850	163,350
1980	T/Diesel	138,840	154,440
1979	T/Diesel	130,830	145,530
1978	T/Diesel	123,710	137,610

Cheoy Lee 50 Trawler

Year	Power	Retail Low	Retail High
1980	T/Diesel	186,900	207,900
1979	T/Diesel	176,220	196,020
1978	T/Diesel	156,640	174,240
1977	T/Diesel	148,630	165,330
1976	T/Diesel	140,620	156,420
1975	T/Diesel	132,610	147,510

Cheoy Lee 52 Efficient

Year	Power	Retail Low	Retail High
1994	T/Diesel	468,140	520,740
1993	T/Diesel	428,980	477,180
1992	T/Diesel	396,940	441,540
1991	T/Diesel	365,790	406,890

Year	Power	Retail Low	Retail High
1990	T/Diesel	338,200	376,200
1989	T/Diesel	312,390	347,490
1988	T/Diesel	289,250	321,750
1987	T/Diesel	271,450	301,950
1986	T/Diesel	255,430	284,130
1985	T/Diesel	241,190	268,290
1984	T/Diesel	231,400	257,400

Cheoy Lee 55 Long Range MY

Year	Power	Retail Low	Retail High
1994	T/Diesel	640,800	712,800
1993	T/Diesel	600,750	668,250
1992	T/Diesel	556,250	618,750
1991	T/Diesel	516,200	574,200
1990	T/Diesel	485,050	539,550
1989	T/Diesel	458,350	509,850
1988	T/Diesel	428,980	477,180
1987	T/Diesel	413,850	460,350
1986	T/Diesel	391,600	435,600
1985	T/Diesel	369,350	410,850
1984	T/Diesel	348,880	388,080
1983	T/Diesel	325,740	362,340
1982	T/Diesel	302,600	336,600
1981	T/Diesel	283,910	315,810

Cheoy Lee 66 Long Range MY

Year	Power	Retail Low	Retail High
1994	T/Diesel	979,000	1,089,000
1993	T/Diesel	890,000	990,000
1992	T/Diesel	841,050	935,550
1991	T/Diesel	778,750	866,250
1990	T/Diesel	741,370	824,670
1989	T/Diesel	712,890	792,990
1988	T/Diesel	685,300	762,300
1987	T/Diesel	666,610	741,510
1986	T/Diesel	644,360	716,760
1985	T/Diesel	600,750	668,250
1984	T/Diesel	569,600	633,600
1983	T/Diesel	535,780	595,980

Cheoy Lee 66 Fast MY

Year	Power	Retail Low	Retail High
1987	T/Diesel	658,600	732,600
1986	T/Diesel	618,550	688,050
1985	T/Diesel	574,050	638,550
1984	T/Diesel	534,000	594,000

Cheoy Lee 83 MY

Year	Power	Retail Low	Retail High
1993	T/Diesel	******	******
1992	T/Diesel	******	******
1991	T/Diesel	******	******
1990	T/Diesel	******	******
1989	T/Diesel	******	******
1988	T/Diesel	******	******
1987	T/Diesel	******	******

Chris 350 Catalina

Year	Power	Retail Low	Retail High
1987	T/Gas	60,520	67,320
1986	T/Gas	56,960	63,360
1985	T/Gas	54,290	60,390
1984	T/Gas	51,620	57,420
1983	T/Gas	48,950	54,450
1982	T/Gas	46,280	51,480
1981	T/Gas	43,610	48,510
1980	T/Gas	41,830	46,530
1979	T/Gas	40,050	44,550
1978	T/Gas	38,270	42,570
1977	T/Gas	36,490	40,590
1976	T/Gas	34,710	38,610
1975	T/Gas	32,930	36,630

Chris 380 Corinthian

Year	Power	Retail Low	Retail High
1986	T/Gas	89,000	99,000
1986	T/Diesel	100,570	111,870
1985	T/Gas	83,660	93,060
1985	T/Diesel	94,340	104,940
1984	T/Gas	77,430	86,130
1984	T/Diesel	87,220	97,020
1983	T/Gas	71,200	79,200
1983	T/Diesel	80,100	89,100
1982	T/Gas	65,860	73,260
1982	T/Diesel	74,760	83,160
1981	T/Gas	60,520	67,320
1981	T/Diesel	68,530	76,230
1980	T/Gas	55,180	61,380
1980	T/Diesel	63,190	70,290
1979	T/Gas	49,840	55,440
1979	T/Diesel	57,850	64,350
1978	T/Gas	45,390	50,490
1978	T/Diesel	53,400	59,400

Chris 381 Catalina

Year	Power	Retail Low	Retail High
1989	T/Gas	105,020	116,820
1989	T/Diesel	121,930	135,630
1988	T/Gas	97,010	107,910
1988	T/Diesel	111,250	123,750
1987	T/Gas	89,890	99,990
1987	T/Diesel	102,350	113,850
1986	T/Gas	83,660	93,060
1986	T/Diesel	95,230	105,930
1985	T/Gas	78,320	87,120
1985	T/Diesel	89,000	99,000
1984	T/Gas	72,980	81,180
1984	T/Diesel	82,770	92,070
1983	T/Gas	66,750	74,250
1983	T/Diesel	75,650	84,150
1982	T/Gas	62,300	69,300
1982	T/Diesel	71,200	79,200
1981	T/Gas	56,960	63,360
1981	T/Diesel	65,860	73,260
1980	T/Gas	53,400	59,400
1980	T/Diesel	62,300	69,300

Chris 410 MY

Year	Power	Retail Low	Retail High
1986	T/Gas	108,580	120,780
1986	T/Diesel	131,720	146,520
1985	T/Gas	103,240	114,840
1985	T/Diesel	122,820	136,620
1984	T/Gas	97,900	108,900
1984	T/Diesel	115,700	128,700
1983	T/Gas	93,450	103,950
1983	T/Diesel	109,470	121,770
1982	T/Gas	89,000	99,000
1982	T/Diesel	102,350	113,850
1981	T/Gas	84,550	94,050
1981	T/Diesel	100,570	111,870
1980	T/Gas	80,100	89,100
1980	T/Diesel	93,450	103,950
1979	T/Gas	76,540	85,140
1979	T/Diesel	88,110	98,010
1978	T/Gas	71,200	79,200
1978	T/Diesel	81,880	91,080
1977	T/Gas	65,860	73,260
1977	T/Diesel	75,650	84,150
1976	T/Gas	61,410	68,310
1976	T/Diesel	70,310	78,210
1975	T/Gas	57,850	64,350
1975	T/Diesel	65,860	73,260

Chris 426/427 Catalina

Year	Power	Retail Low	Retail High
1990	T/Gas	153,970	171,270
1990	T/Diesel	172,660	192,060
1989	T/Gas	144,180	160,380
1989	T/Diesel	162,870	181,170
1988	T/Gas	136,170	151,470

Year	Power	Retail Low	Retail High
1988	T/Diesel	155,750	173,250
1987	T/Gas	129,050	143,550
1987	T/Diesel	147,740	164,340
1986	T/Gas	137,950	153,450
1986	T/Diesel	120,150	133,650
1985	T/Gas	129,050	143,550
1985	T/Diesel	111,250	123,750

Chris 45 Commander MY

Year	Power	Retail Low	Retail High
1981	T/Diesel	137,950	153,450
1980	T/Diesel	128,160	142,560
1979	T/Diesel	119,260	132,660
1978	T/Diesel	111,250	123,750
1977	T/Diesel	103,240	114,840
1976	T/Diesel	95,230	105,930
1975	T/Diesel	87,220	97,020

Chris 47 Commander MY

Year	Power	Retail Low	Retail High
1976	T/Diesel	89,890	99,990
1975	T/Diesel	84,550	94,050

Chris 480 Catalina

Year	Power	Retail Low	Retail High
1989	T/Diesel	196,690	218,790
1988	T/Diesel	183,340	203,940
1987	T/Diesel	170,880	190,080
1986	T/Diesel	161,090	179,190
1985	T/Diesel	152,190	169,290

Chris 500 MY

Year	Power	Retail Low	Retail High
1989	T/Diesel	351,550	391,050
1988	T/Diesel	317,730	353,430
1987	T/Diesel	291,920	324,720
1986	T/Diesel	271,450	301,950
1985	T/Diesel	249,200	277,200

Chris 501 MY

Year	Power	Retail Low	Retail High
1990	T/Diesel	396,050	440,550
1989	T/Diesel	369,350	410,850
1988	T/Diesel	347,100	386,100
1987	T/Diesel	329,300	366,300

Chris 55 Flush Deck

Year	Power	Retail Low	Retail High
1977	T/Diesel	124,600	138,600
1976	T/Diesel	116,590	129,690
1975	T/Diesel	108,580	120,780

Cruisers 3850 Aft Cabin

Year	Power	Retail Low	Retail High
1994	T/Gas	155,750	173,250
1994	T/Diesel	184,230	204,930
1993	T/Gas	146,850	163,350
1993	T/Diesel	170,880	190,080
1992	T/Gas	138,840	154,440
1992	T/Diesel	163,760	182,160
1991	T/Gas	130,830	145,530
1991	T/Diesel	151,300	168,300

Cruisers 4280/4285 Express Bridge

Year	Power	Retail Low	Retail High
1994	265,000	235,850	262,350
1993	T/Diesel	213,600	237,600
1992	T/Gas	155,750	173,250
1992	T/Diesel	195,800	217,800
1991	T/Gas	146,850	163,350
1991	T/Diesel	182,450	202,950
1990	T/Gas	137,950	153,450
1990	T/Diesel	169,100	188,100
1989	T/Gas	129,050	143,550
1989	T/Diesel	160,200	178,200
1988	T/Gas	121,040	134,640
1988	T/Diesel	152,190	169,290

CT 35 Trawler

Year	Power	Retail Low	Retail High
1994	S/Diesel	128,160	142,560
1993	S/Diesel	118,370	131,670
1992	S/Diesel	113,030	125,730
1991	S/Diesel	102,350	113,850
1990	S/Diesel	92,560	102,960
1989	S/Diesel	82,770	92,070
1988	S/Diesel	73,870	82,170
1987	S/Diesel	65,860	73,260
1986	S/Diesel	59,630	66,330
1985	S/Diesel	56,960	63,360
1984	S/Diesel	54,290	60,390
1983	S/Diesel	51,620	57,420
1982	S/Diesel	48,950	54,450
1981	S/Diesel	47,170	52,470
1980	S/Diesel	44,500	49,500
1979	S/Diesel	42,720	47,520
1978	S/Diesel	40,940	45,540
1977	S/Diesel	40,050	44,550

DeFever 40 Offshore Cruiser

Year	Power	Retail Low	Retail High
1991	T/Diesel	124,600	138,600
1990	T/Diesel	114,810	127,710
1989	T/Diesel	105,020	116,820
1988	T/Diesel	95,230	105,930
1987	T/Diesel	89,000	99,000
1986	T/Diesel	82,770	92,070
1985	T/Diesel	78,320	87,120

DeFever 41 Trawler

Year	Power	Retail Low	Retail High
1988	S/Diesel	132,610	147,510
1988	T/Diesel	143,290	159,390
1987	S/Diesel	122,820	136,620
1987	T/Diesel	133,500	148,500
1986	S/Diesel	113,920	126,720
1986	T/Diesel	123,710	137,610
1985	S/Diesel	105,910	117,810
1985	T/Diesel	114,810	127,710
1984	S/Diesel	96,120	106,920
1984	T/Diesel	105,020	116,820
1983	S/Diesel	86,330	96,030
1983	T/Diesel	95,230	105,930

DeFever 43 Trawler

Year	Power	Retail Low	Retail High
1985	T/Diesel	137,950	153,450
1984	T/Diesel	127,270	141,570
1983	T/Diesel	116,590	129,690
1982	T/Diesel	109,470	121,770
1981	T/Diesel	104,130	115,830
1980	T/Diesel	98,790	109,890
1979	T/Diesel	93,450	103,950
1978	T/Diesel	88,110	98,010

DeFever 44 Diesel Cruiser

Year	Power	Retail Low	Retail High
1994	T/Diesel	266,110	296,010
1993	T/Diesel	250,090	278,190
1992	T/Diesel	236,740	263,340
1991	T/Diesel	223,390	248,490
1990	T/Diesel	212,710	236,610
1989	T/Diesel	202,030	224,730
1988	T/Diesel	191,350	212,850
1987	T/Diesel	182,450	202,950
1986	T/Diesel	173,550	193,050
1985	T/Diesel	161,980	180,180
1984	T/Diesel	148,630	165,330
1983	T/Diesel	135,280	150,480
1982	T/Diesel	125,490	139,590
1981	T/Diesel	116,590	129,690
1980	T/Diesel	110,360	122,760

DeFever 48 Diesel Cruiser

Year	Power	Retail Low	Retail High
1994	T/Diesel	324,850	361,350
1993	T/Diesel	302,600	336,600

Year	Power	Retail Low	Retail High
1992	T/Diesel	280,350	311,850
1991	T/Diesel	254,540	283,140
1990	T/Diesel	235,850	262,350
1989	T/Diesel	222,500	247,500
1988	T/Diesel	213,600	237,600
1987	T/Diesel	204,700	227,700
1986	T/Diesel	194,910	216,810
1985	T/Diesel	185,120	205,920
1984	T/Diesel	171,770	191,070
1983	T/Diesel	161,090	179,190
1982	T/Diesel	146,850	163,350
1981	T/Diesel	137,950	153,450
1980	T/Diesel	130,830	145,530
1979	T/Diesel	124,600	138,600
1978	T/Diesel	120,150	133,650

DeFever 49 PH

Year	Power	Retail Low	Retail High
1994	T/Diesel	351,550	391,050
1993	T/Diesel	329,300	366,300
1992	T/Diesel	307,940	342,540
1991	T/Diesel	289,250	321,750
1990	T/Diesel	267,890	297,990
1989	T/Diesel	247,420	275,220
1988	T/Diesel	226,950	252,450
1987	T/Diesel	210,040	233,640
1986	T/Diesel	196,690	218,790
1985	T/Diesel	182,450	202,950
1984	T/Diesel	172,660	192,060
1983	T/Diesel	162,870	181,170
1982	T/Diesel	155,750	173,250
1981	T/Diesel	148,630	165,330
1980	T/Diesel	142,400	158,400
1979	T/Diesel	137,950	153,450
1978	T/Diesel	133,500	148,500
1977	T/Diesel	129,050	143,550

DeFever 52 MY

Year	Power	Retail Low	Retail High
1994	T/Diesel	569,600	633,600
1993	T/Diesel	489,500	544,500
1992	T/Diesel	413,850	460,350
1991	T/Diesel	348,880	388,080
1990	T/Diesel	317,730	353,430
1989	T/Diesel	302,600	336,600
1988	T/Diesel	291,920	324,720
1987	T/Diesel	281,240	312,840
1986	T/Diesel	270,560	300,960
1985	T/Diesel	258,100	287,100
1984	T/Diesel	241,190	268,290
1983	T/Diesel	221,610	246,510
1982	T/Diesel	205,590	228,690
1981	T/Diesel	193,130	214,830
1980	T/Diesel	178,000	198,000

DeFever 53 (POC) MY

Year	Power	Retail Low	Retail High
1992	T/Diesel	409,400	455,400
1991	T/Diesel	383,590	426,690
1990	T/Diesel	353,330	393,030
1989	T/Diesel	324,850	361,350
1988	T/Diesel	293,700	326,700
1987	T/Diesel	275,900	306,900
1986	T/Diesel	261,660	291,060

DeFever 60 Offshore Cruiser

******

Dyna 53 CMY

Year	Power	Retail Low	Retail High
1994	T/Diesel	364,010	404,910
1993	T/Diesel	332,860	370,260
1992	T/Diesel	304,380	338,580
1991	T/Diesel	267,000	297,000
1990	T/Diesel	248,310	276,210

Eagle 32 Trawler

Year	Power	Retail Low	Retail High
1994	S/Diesel	106,800	118,800
1993	S/Diesel	94,340	104,940
1992	S/Diesel	84,550	94,050
1991	S/Diesel	74,760	83,160
1990	S/Diesel	68,530	76,230
1989	S/Diesel	63,190	70,290
1988	S/Diesel	58,740	65,340
1987	S/Diesel	55,180	61,380
1986	S/Diesel	51,620	57,420
1985	S/Diesel	48,950	54,450

Egg Harbor 40 MY

Year	Power	Retail Low	Retail High
1986	T/Gas	124,600	138,600
1986	T/Diesel	146,850	163,350
1985	T/Gas	122,820	136,620
1985	T/Diesel	140,620	156,420
1984	T/Gas	115,700	128,700
1984	T/Diesel	133,500	148,500
1983	T/Gas	111,250	123,750
1983	T/Diesel	124,600	138,600
1982	T/Gas	97,010	107,910
1982	T/Diesel	113,030	125,730

Egg Harbor 41 MY

Year	Power	Retail Low	Retail High
1977	T/Gas	61,410	68,310
1977	T/Diesel	71,200	79,200
1976	T/Gas	56,070	62,370
1976	T/Diesel	65,860	73,260
1975	T/Gas	52,510	58,410
1975	T/Diesel	61,410	68,310

Embassy 444 Sundeck

******

Fleming 55 MY

Year	Power	Retail Low	Retail High
1994	T/Diesel	657,710	731,610
1993	T/Diesel	618,550	688,050
1992	T/Diesel	574,940	639,540
1991	T/Diesel	533,110	593,010
1990	T/Diesel	496,620	552,420
1989	T/Diesel	467,250	519,750
1988	T/Diesel	450,340	500,940
1987	T/Diesel	435,210	484,110

Grand Banks 32

Year	Power	Retail Low	Retail High
1994	S/Diesel	151,300	168,300
1993	S/Diesel	137,060	152,460
1992	S/Diesel	124,600	138,600
1991	S/Diesel	113,920	126,720
1990	S/Diesel	105,020	116,820
1989	S/Diesel	96,120	106,920
1988	S/Diesel	88,110	98,010
1987	S/Diesel	80,990	90,090
1986	S/Diesel	74,760	83,160
1985	S/Diesel	69,420	77,220
1984	S/Diesel	64,970	72,270
1983	S/Diesel	61,410	68,310
1982	S/Diesel	58,740	65,340
1981	S/Diesel	56,070	62,370
1980	S/Diesel	53,400	59,400
1979	S/Diesel	51,620	57,420
1978	S/Diesel	49,840	55,440
1977	S/Diesel	48,060	53,460
1976	S/Diesel	46,280	51,480
1975	S/Diesel	44,500	49,500

Grand Banks 36 (Early)

Year	Power	Retail Low	Retail High
1987	S/Diesel	121,040	134,640
1987	T/Diesel	131,720	146,520
1986	S/Diesel	111,250	123,750

Year	Power	Retail Low	Retail High
1986	T/Diesel	121,040	134,640
1985	S/Diesel	103,240	114,840
1985	T/Diesel	112,140	124,740
1984	S/Diesel	96,120	106,920
1984	T/Diesel	104,130	115,830
1983	S/Diesel	89,890	99,990
1983	T/Diesel	97,010	107,910
1982	S/Diesel	84,550	94,050
1982	T/Diesel	90,780	100,980
1981	S/Diesel	79,210	88,110
1981	T/Diesel	85,440	95,040
1980	S/Diesel	73,870	82,170
1980	T/Diesel	80,100	89,100
1979	S/Diesel	68,530	76,230
1979	T/Diesel	74,760	83,160
1978	S/Diesel	64,080	71,280
1978	T/Diesel	70,310	78,210
1977	S/Diesel	60,520	67,320
1977	T/Diesel	66,750	74,250
1976	S/Diesel	57,850	64,350
1976	T/Diesel	63,190	70,290
1975	S/Diesel	55,180	61,380
1975	T/Diesel	60,520	67,320

Grand Banks 36

Year	Power	Retail Low	Retail High
1994	S/Diesel	209,150	232,650
1994	T/Diesel	236,740	263,340
1993	S/Diesel	195,800	217,800
1993	T/Diesel	218,050	242,550
1992	S/Diesel	183,340	203,940
1992	T/Diesel	203,810	226,710
1991	S/Diesel	171,770	191,070
1991	T/Diesel	186,010	206,910
1990	S/Diesel	160,200	178,200
1990	T/Diesel	172,660	192,060
1989	S/Diesel	149,520	166,320
1989	T/Diesel	160,200	178,200
1988	S/Diesel	138,840	154,440
1988	T/Diesel	148,630	165,330

Grand Banks 36 Europa

Year	Power	Retail Low	Retail High
1994	S/Diesel	207,370	230,670
1994	T/Diesel	234,960	261,360
1993	S/Diesel	200,250	222,750
1993	T/Diesel	222,500	247,500
1992	S/Diesel	186,900	207,900
1992	T/Diesel	207,370	230,670
1991	S/Diesel	174,440	194,040
1991	T/Diesel	188,680	209,880
1990	S/Diesel	161,980	180,180
1990	T/Diesel	174,440	194,040
1989	S/Diesel	150,410	167,310
1989	T/Diesel	161,090	179,190
1988	S/Diesel	138,840	154,440
1988	T/Diesel	148,630	165,330

Grand Banks 42 Classic (Early)

Year	Power	Retail Low	Retail High
1991	T/135D	267,000	297,000
1990	T/135D	247,420	275,220
1989	T/135D	226,950	252,450
1988	T/135D	208,260	231,660
1987	T/135D	190,460	211,860
1986	T/135D	175,330	195,030
1985	T/135D	163,760	182,160
1984	T/135D	153,970	171,270
1983	T/135D	144,180	160,380
1982	T/135D	137,060	152,460
1981	T/135D	129,940	144,540
1980	T/135D	124,600	138,600
1979	T/135D	120,150	133,650
1978	T/135D	115,700	128,700
1977	T/135D	109,470	121,770
1976	T/135D	103,240	114,840
1975	T/135D	97,900	108,900

Grand Banks 42 Classic

Year	Power	Retail Low	Retail High
1994	T/135D	331,080	368,280
1993	T/135D	309,720	344,520
1992	T/135D	291,030	323,730
1991	T/135D	280,350	311,850

Grand Banks 42 Europa

Year	Power	Retail Low	Retail High
1993	T/135D	300,820	334,620
1992	T/135D	278,570	309,870
1991	T/135D	258,100	287,100
1990	T/135D	240,300	267,300
1989	T/135D	224,280	249,480
1988	T/135D	202,920	225,720
1987	T/135D	187,790	208,890
1986	T/135D	170,880	190,080
1985	T/135D	159,310	177,210
1984	T/135D	151,300	168,300
1983	T/135D	145,960	162,360
1982	T/135D	141,510	157,410
1981	T/135D	137,060	152,460
1980	T/135D	132,610	147,510

Grand Banks 42 MY

Year	Power	Retail Low	Retail High
1994	T/135D	347,100	386,100
1993	T/135D	327,520	364,320
1992	T/135D	308,830	343,530
1991	T/135D	293,700	326,700
1990	T/135D	262,550	292,050
1989	T/135D	240,300	267,300
1988	T/135D	218,050	242,550
1987	T/135D	205,590	228,690

Grand Banks 46 Classic

Year	Power	Retail Low	Retail High
1994	T/Diesel	390,710	434,610
1993	T/Diesel	359,560	399,960
1992	T/Diesel	338,200	376,200
1991	T/Diesel	324,850	361,350
1990	T/Diesel	310,610	345,510
1989	T/Diesel	293,700	326,700
1988	T/Diesel	276,790	307,890
1987	T/Diesel	258,100	287,100

Grand Banks 46 Europa

Year	Power	Retail Low	Retail High
1994	T/Diesel	396,050	440,550
1993	T/Diesel	364,900	405,900

Grand Banks 46 MY

Year	Power	Retail Low	Retail High
1994	T/Diesel	407,620	453,420
1993	T/Diesel	375,580	417,780
1992	T/Diesel	356,890	396,990
1991	T/Diesel	339,980	378,180
1990	T/Diesel	325,740	362,340

Grand Banks 49 Classic

Year	Power	Retail Low	Retail High
1994	T/135D	551,800	613,800
1993	T/135D	513,530	571,230
1992	T/135D	467,250	519,750
1991	T/135D	419,190	466,290
1990	T/135D	383,590	426,690
1989	T/135D	352,440	392,040
1988	T/135D	330,190	367,290
1987	T/135D	314,170	349,470
1986	T/135D	303,490	337,590
1985	T/135D	292,810	325,710
1984	T/135D	284,800	316,800
1983	T/135D	276,790	307,890
1982	T/135D	267,890	297,990

Year	Power	Retail Low	Retail High
1981	T/135D	251,870	280,170
1980	T/135D	232,290	258,390

Grand Banks 49 MY

Year	Power	Retail Low	Retail High
1994	T/135D	574,050	638,550
1993	T/135D	533,110	593,010
1992	T/135D	485,050	539,550
1991	T/135D	435,210	484,110
1990	T/135D	396,940	441,540
1989	T/135D	365,790	406,890
1988	T/135D	344,430	383,130
1987	T/135D	327,520	364,320
1986	T/135D	312,390	347,490
1985	T/135D	300,820	334,620

Grand Banks 58 MY

Year	Power	Retail Low	Retail High
1994	T/Diesel	1,032,400	1,148,400
1993	T/Diesel	929,160	1,033,560
1992	T/Diesel	840,160	934,560
1991	T/Diesel	768,960	855,360
1990	T/Diesel	713,780	793,980

Gulfstar 36 Trawler

Year	Power	Retail Low	Retail High
1976	T/Diesel	42,720	47,520
1975	T/Diesel	39,160	43,560

Gulfstar 38 Motor Cruiser

Year	Power	Retail Low	Retail High
1984	T/Diesel	89,890	99,990
1983	T/Diesel	80,990	90,090
1982	T/Diesel	73,870	82,170
1981	T/Diesel	67,640	75,240
1980	T/Diesel	63,190	70,290

Gulfstar 43 Trawler

Year	Power	Retail Low	Retail High
1977	T/Diesel	79,210	88,110
1976	T/Diesel	70,310	78,210
1975	T/Diesel	62,300	69,300

Gulfstar 44 Motor Cruiser

Year	Power	Retail Low	Retail High
1980	T/Diesel	127,270	141,570
1979	T/Diesel	118,370	131,670
1978	T/Diesel	111,250	123,750

Gulfstar 44 MY

Year	Power	Retail Low	Retail High
1986	T/Diesel	171,770	191,070
1985	T/Diesel	165,540	184,140

Gulfstar 44 Widebody MY

Year	Power	Retail Low	Retail High
1988	T/Diesel	213,600	237,600
1987	T/Diesel	200,250	222,750
1986	T/Diesel	189,570	210,870

Gulfstar 48 MY

Year	Power	Retail Low	Retail High
1983	T/Diesel	178,890	198,990
1982	T/Diesel	170,880	190,080
1981	T/Diesel	163,760	182,160

Gulfstar 49 MY

Year	Power	Retail Low	Retail High
1987	T/375D	246,530	274,230
1987	T/435D	256,320	285,120
1986	T/375D	228,730	254,430
1986	T/435D	238,520	265,320
1985	T/350D	211,820	235,620
1985	T/435D	220,720	245,520
1984	T/350D	196,690	218,790

Gulfstar 53 Trawler

Year	Power	Retail Low	Retail High
1976	T/Diesel	111,250	123,750
1975	T/Diesel	101,460	112,860

Hartmann-Palmer 56 MY

Year	Power	Retail Low	Retail High
1987	T/450D	372,910	414,810
1987	T/600D	391,600	435,600
1986	T/450D	339,980	378,180
1986	T/600D	359,560	399,960
1985	T/450D	312,390	347,490
1985	T/600D	331,970	369,270
1984	T/450D	294,590	327,690
1984	T/600D	311,500	346,500

Hatteras 38 DC

Year	Power	Retail Low	Retail High
1978	T/Diesel	89,890	101,000
1977	T/Diesel	80,100	90,000
1976	T/Diesel	71,200	80,000
1975	T/Diesel	65,860	74,000

Hatteras 40 MY

Year	Power	Retail Low	Retail High
1994	T/Diesel	360,450	405,000
1993	T/Diesel	342,650	385,000
1992	T/Diesel	301,710	339,000
1991	T/Diesel	266,110	299,000
1990	T/Gas	204,700	230,000
1990	T/Diesel	240,300	270,000
1989	T/Gas	185,120	208,000
1989	T/Diesel	221,610	249,000
1988	T/Gas	178,000	200,000
1988	T/Diesel	206,480	232,000
1987	T/Gas	161,090	181,000
1987	T/Diesel	188,680	212,000
1986	T/Gas	150,410	169,000
1986	T/Diesel	174,440	196,000

Hatteras 42 LRC

Year	Power	Retail Low	Retail High
1985	T/Diesel	175,330	197,000
1984	T/Diesel	166,430	187,000
1983	T/Diesel	149,520	168,000
1982	T/Diesel	142,400	160,000
1981	T/Diesel	137,060	154,000
1980	T/Diesel	133,500	150,000
1979	T/Diesel	129,940	146,000
1978	T/Diesel	128,160	144,000
1977	T/Diesel	124,600	140,000
1976	T/Diesel	120,150	135,000

Hatteras 42 CMY

Year	Power	Retail Low	Retail High
1994	T/Diesel	413,850	465,000
1993	T/Diesel	391,600	440,000

Hatteras 43 DC

Year	Power	Retail Low	Retail High
1984	T/280D	168,210	189,000
1984	T/390D	181,560	204,000
1983	T/280D	159,310	179,000
1983	T/390D	169,100	190,000
1982	T/280D	149,520	168,000
1982	T/390D	159,310	179,000
1981	T/280D	140,620	158,000
1981	T/390D	151,300	170,000
1980	T/Diesel	145,070	163,000
1979	T/Diesel	137,950	155,000
1978	T/Diesel	129,050	145,000
1977	T/Diesel	120,150	135,000
1976	T/Diesel	110,360	124,000
1975	T/Diesel	104,130	117,000

Hatteras 43 MY

Year	Power	Retail Low	Retail High
1987	T/320D	220,720	248,000
1987	T/340D	231,400	260,000
1986	T/300D	208,260	234,000
1986	T/340D	218,050	245,000
1985	T/300D	199,360	224,000
1985	T/340D	209,150	235,000
1984	T/300D	191,350	215,000
1984	T/340D	200,250	225,000

Hatteras 48 LRC

Year	Power	Retail Low	Retail High
1981	T/Diesel	204,700	230,000

Year	Power	Retail Low	Retail High
1980	T/Diesel	195,800	220,000
1979	T/Diesel	187,790	211,000
1978	T/Diesel	182,450	205,000
1977	T/Diesel	173,550	195,000
1976	T/Diesel	165,540	186,000

Hatteras 48 MY (Early)

Year	Power	Retail Low	Retail High
1984	T/280D	218,050	245,000
1984	T/425D	242,970	273,000
1983	T/280D	200,250	225,000
1983	T/425D	220,720	248,000
1982	T/280D	182,450	205,000
1982	T/425D	204,700	230,000
1981	T/280D	169,100	190,000
1981	T/425D	191,350	215,000

Hatteras 48 CMY (Early)

Year	Power	Retail Low	Retail High
1985	T/280D	226,950	255,000
1985	T/425D	249,200	280,000
1984	T/280D	218,050	245,000
1984	T/425D	241,190	271,000
1983	T/280D	209,150	235,000
1983	T/425D	231,400	260,000
1982	T/280D	198,470	223,000
1982	T/425D	220,720	248,000
1981	T/280D	187,790	211,000
1981	T/425D	210,040	236,000

Hatteras 48 MY

Year	Power	Retail Low	Retail High
1994	T/Diesel	695,980	782,000
1993	T/Diesel	636,350	715,000
1992	T/Diesel	587,400	660,000
1991	T/Diesel	525,100	590,000
1990	T/535D	448,560	504,000
1990	T/720D	480,600	540,000

Hatteras 48 CMY

Year	Power	Retail Low	Retail High
1994	T/Diesel	626,560	704,000
1993	T/Diesel	562,480	632,000

Hatteras 52 CMY

Year	Power	Retail Low	Retail High
1994	T/Diesel	729,800	820,000
1993	T/Diesel	667,500	750,000
1992	T/Diesel	623,000	700,000
1991	T/Diesel	580,280	652,000
1990	T/Diesel	540,230	607,000

Hatteras 52 MY

Year	Power	Retail Low	Retail High
1994	T/Diesel	743,150	835,000
1993	T/Diesel	681,740	766,000

Hatteras 53 MY

Year	Power	Retail Low	Retail High
1988	T/Diesel	427,200	480,000
1987	T/Diesel	400,500	450,000
1986	T/Diesel	367,570	413,000
1985	T/Diesel	339,090	381,000
1984	T/Diesel	320,400	360,000
1983	T/Diesel	302,600	340,000
1982	T/Diesel	284,800	320,000
1981	T/Diesel	274,120	308,000
1980	T/Diesel	263,440	296,000
1979	T/Diesel	254,540	286,000
1978	T/Diesel	244,750	275,000
1977	T/Diesel	228,730	257,000
1976	T/Diesel	210,040	236,000
1975	T/Diesel	187,790	211,000

Hatteras 53 YF

Year	Power	Retail Low	Retail High
1987	T/Diesel	355,110	399,000
1986	T/Diesel	329,300	370,000
1981	T/Diesel	261,660	294,000
1980	T/Diesel	249,200	280,000
1979	T/Diesel	235,850	265,000
1978	T/Diesel	221,610	249,000

Hatteras 53 EDMY

Year	Power	Retail Low	Retail High
1988	T/Diesel	418,300	470,000
1987	T/Diesel	388,930	437,000
1986	T/Diesel	364,010	409,000
1985	T/Diesel	343,540	386,000
1984	T/Diesel	320,400	360,000
1983	T/Diesel	304,380	342,000

Hatteras 54 MY

Year	Power	Retail Low	Retail High
1988	T/Diesel	538,450	605,000
1987	T/Diesel	497,510	559,000
1986	T/Diesel	477,930	537,000
1985	T/Diesel	453,900	510,000

Hatteras 54 EDMY

Year	Power	Retail Low	Retail High
1994	T/Diesel	987,900	1,110,000
1993	T/Diesel	925,600	1,040,000
1992	T/Diesel	854,400	960,000
1991	T/Diesel	787,650	885,000
1990	T/Diesel	707,550	795,000
1989	T/Diesel	629,230	707,000

Hatteras 56 MY

Year	Power	Retail Low	Retail High
1985	T/Diesel	444,110	499,000
1984	T/Diesel	419,190	471,000
1983	T/Diesel	401,390	451,000
1982	T/Diesel	389,820	438,000
1981	T/Diesel	373,800	420,000
1980	T/Diesel	356,000	400,000

Hatteras 58 YF

Year	Power	Retail Low	Retail High
1982	T/Diesel	335,530	377,000
1981	T/Diesel	320,400	360,000
1980	T/Diesel	306,160	344,000
1979	T/Diesel	293,700	330,000
1978	T/Diesel	275,900	310,000
1977	T/Diesel	252,760	284,000
1976	T/Diesel	239,410	269,000
1975	T/Diesel	221,610	249,000

Hatteras 58 LRC

Year	Power	Retail Low	Retail High
1981	T/Diesel	356,890	401,000
1980	T/Diesel	338,200	380,000
1979	T/Diesel	319,510	359,000
1978	T/Diesel	299,930	337,000
1977	T/Diesel	282,130	317,000
1976	T/Diesel	276,790	311,000
1975	T/Diesel	263,440	296,000

Hatteras 58 MY (Early)

Year	Power	Retail Low	Retail High
1981	T/Diesel	310,610	349,000
1980	T/Diesel	297,260	334,000
1979	T/Diesel	283,910	319,000
1978	T/Diesel	267,000	300,000
1977	T/Diesel	246,530	277,000

Hatteras 58 MY

Year	Power	Retail Low	Retail High
1987	T/Diesel	578,500	650,000
1986	T/Diesel	550,910	619,000
1985	T/Diesel	524,210	589,000

Hatteras 58 CMY

Year	Power	Retail Low	Retail High
1981	T/Diesel	315,950	355,000
1980	T/Diesel	307,050	345,000
1979	T/Diesel	298,150	335,000
1978	T/Diesel	275,010	309,000

Hatteras 60 MY

Year	Power	Retail Low	Retail High
1992	T/Diesel	881,100	990,000
1991	T/Diesel	844,610	949,000

Year	Power	Retail Low	Retail High
1990	T/Diesel	776,080	872,000
1989	T/Diesel	729,800	820,000
1988	T/Diesel	676,400	760,000

Hatteras 60 EDMY

Year	Power	Retail Low	Retail High
1994	T/720D	1,227,310	1,379,000
1994	T/870D	1,303,850	1,465,000
1993	T/720D	1,137,420	1,278,000
1993	T/870D	1,237,990	1,391,000
1992	T/720D	1,019,050	1,145,000
1992	T/870D	1,094,700	1,230,000
1991	T/720D	880,210	989,000
1991	T/870D	961,200	1,080,000

Hatteras 61 MY

Year	Power	Retail Low	Retail High
1984	T/Diesel	558,030	627,000
1983	T/Diesel	525,100	590,000
1982	T/Diesel	493,950	555,000
1981	T/Diesel	471,700	530,000

Hatteras 61 CMY

Year	Power	Retail Low	Retail High
1984	T/Diesel	565,150	635,000
1983	T/Diesel	533,110	599,000
1982	T/Diesel	498,400	560,000
1981	T/Diesel	475,260	534,000

Hatteras 63 MY

Year	Power	Retail Low	Retail High
1987	T/Diesel	685,300	770,000
1986	T/Diesel	649,700	730,000

Hatteras 63 CMY

Year	Power	Retail Low	Retail High
1987	T/Diesel	694,200	780,000
1986	T/Diesel	651,480	732,000
1985	T/Diesel	614,100	690,000

Hatteras 64 MY

Year	Power	Retail Low	Retail High
1981	T/Diesel	445,000	500,000
1980	T/Diesel	428,090	481,000
1979	T/Diesel	412,070	463,000
1978	T/Diesel	402,280	452,000
1977	T/Diesel	391,600	440,000
1976	T/Diesel	370,240	416,000
1975	T/Diesel	354,220	398,000

Hatteras 65 LRC

Year	Power	Retail Low	Retail High
1985	T/Diesel	621,220	698,000
1984	T/Diesel	571,380	642,000
1983	T/Diesel	536,670	603,000
1982	T/Diesel	516,200	580,000
1981	T/Diesel	499,290	561,000

Hatteras 65 MY

Year	Power	Retail Low	Retail High
1994	T/Diesel	1,401,750	1,575,000
1993	T/Diesel	1,246,890	1,401,000
1992	T/Diesel	1,143,650	1,285,000
1991	T/Diesel	1,063,550	1,195,000
1990	T/Diesel	979,890	1,101,000
1989	T/Diesel	890,000	1,000,000
1988	T/Diesel	805,450	905,000

Hatteras 67 CMY

Year	Power	Retail Low	Retail High
1994	T/Diesel	1,299,400	1,460,000
1993	T/Diesel	1,201,500	1,350,000
1992	T/Diesel	1,112,500	1,250,000
1991	T/Diesel	1,022,610	1,149,000
1990	T/Diesel	945,180	1,062,000
1989	T/Diesel	898,900	1,010,000
1988	T/Diesel	827,700	930,000

Hatteras 67 EDCMY

Year	Power	Retail Low	Retail High
1994	T/870D	1,367,040	1,536,000
1994	T/1040D	1,406,200	1,580,000
1993	T/870D	1,290,500	1,450,000
1993	T/1040D	1,335,000	1,500,000
1992	T/870D	1,205,950	1,355,000
1992	T/1040D	1,252,230	1,407,000
1991	T/870D	1,121,400	1,260,000
1991	T/1040D	1,157,890	1,301,000

Hatteras 68 CMY

Year	Power	Retail Low	Retail High
1987	T/Diesel	756,500	850,000
1986	T/Diesel	725,350	815,000

Hatteras 70 MY (Early)

Year	Power	Retail Low	Retail High
1981	T/Diesel	591,850	665,000
1980	T/Diesel	544,680	612,000
1979	T/Diesel	516,200	580,000
1978	T/Diesel	495,730	557,000
1977	T/Diesel	468,140	526,000
1976	T/Diesel	447,670	503,000
1975	T/Diesel	425,420	478,000

Hatteras 70 EDMY

Year	Power	Retail Low	Retail High
1983	T/Diesel	701,320	788,000
1982	T/Diesel	663,050	745,000
1981	T/Diesel	614,100	690,000
1980	T/Diesel	566,930	637,000
1979	T/Diesel	537,560	604,000
1978	T/Diesel	516,200	580,000
1977	T/Diesel	489,500	550,000
1976	T/Diesel	474,370	533,000

Hatteras 70 MY

Year	Power	Retail Low	Retail High
1994	T/870D	1,610,900	1,810,000
1994	T/1040D	1,686,550	1,895,000
1993	T/870D	1,499,650	1,685,000
1993	T/1040D	1,535,250	1,725,000
1992	T/870D	1,407,090	1,581,000
1992	T/1040D	1,432,900	1,610,000
1991	T/870D	1,254,900	1,410,000
1991	T/1040D	1,290,500	1,450,000
1990	T/870D	1,146,320	1,288,000
1990	T/1040D	1,177,470	1,323,000
1989	T/870D	1,087,580	1,222,000
1988	T/870D	1,005,700	1,130,000

Hatteras 70 CMY

Year	Power	Retail Low	Retail High
1994	T/870D	1,505,880	1,692,000
1994	T/1040D	1,549,490	1,741,000
1993	T/870D	1,424,000	1,600,000
1993	T/1040D	1,464,050	1,645,000
1992	T/870D	1,326,990	1,491,000
1992	T/1040D	1,352,800	1,520,000
1991	T/870D	1,165,900	1,310,000
1991	T/1040D	1,203,280	1,352,000
1990	T/870D	1,075,120	1,208,000
1990	T/1040D	1,106,270	1,243,000
1989	T/870D	1,043,970	1,173,000
1988	T/870D	970,100	1,090,000

Hatteras 72 MY

Year	Power	Retail Low	Retail High
1992	T/Diesel	1,468,500	1,650,000
1991	T/Diesel	1,370,600	1,540,000
1990	T/Diesel	1,300,290	1,461,000
1989	T/Diesel	1,206,840	1,356,000
1988	T/Diesel	1,103,600	1,240,000
1987	T/Diesel	1,043,970	1,173,000
1986	T/Diesel	979,890	1,101,000
1985	T/Diesel	940,730	1,057,000
1984	T/Diesel	882,880	992,000
1983	T/Diesel	843,720	948,000

Hatteras 77 CMY

Year	Power	Retail Low	Retail High
1985	T/Diesel	1,157,000	1,300,000
1984	T/Diesel	1,076,900	1,210,000
1983	T/Diesel	993,240	1,116,000

Heritage 40 Sundeck

Year	Power	Retail Low	Retail High
1990	T/Diesel	135,280	150,480
1989	T/Diesel	124,600	138,600
1988	T/Diesel	115,700	128,700
1987	T/Diesel	106,800	118,800
1986	T/Diesel	98,790	109,890
1985	T/Diesel	89,890	99,990
1984	T/Diesel	81,880	91,080

Heritage 44 Sundeck

Year	Power	Retail Low	Retail High
1990	T/Diesel	157,530	175,230
1989	T/Diesel	146,850	163,350
1988	T/Diesel	141,510	157,410
1987	T/Diesel	130,830	145,530
1986	T/Diesel	121,040	134,640
1985	T/Diesel	110,360	122,760

Hi-Star 48 MY

Year	Power	Retail Low	Retail High
1992	T/Diesel	276,790	307,890
1991	T/Diesel	258,100	287,100
1990	T/Diesel	240,300	267,300
1989	T/Diesel	226,950	252,450
1988	T/Diesel	217,160	241,560
1987	T/Diesel	204,700	227,700
1986	T/Diesel	194,020	215,820

Hi-Star 55 YF

Year	Power	Retail Low	Retail High
1992	T/Diesel	353,330	393,030
1991	T/Diesel	324,850	361,350
1990	T/Diesel	293,700	326,700
1989	T/Diesel	275,010	305,910
1988	T/Diesel	257,210	286,110
1987	T/Diesel	234,960	261,360
1986	T/Diesel	222,500	247,500

High-Tech 50/55 MY

Year	Power	Retail Low	Retail High
1992	T/Diesel	535,780	595,980
1991	T/Diesel	490,390	545,490
1990	T/Diesel	460,130	511,830
1989	T/Diesel	418,300	465,300
1988	T/Diesel	383,590	426,690

High-Tech 63/65 MY

Year	Power	Retail Low	Retail High
1994	T/Diesel	1,163,230	1,293,930
1993	T/Diesel	1,077,790	1,198,890
1992	T/Diesel	982,560	1,092,960
1991	T/Diesel	906,020	1,007,820
1990	T/Diesel	836,600	930,600

Year	Power	Retail Low	Retail High
1989	T/Diesel	770,740	857,340

Hyatt 40 Sundeck

Year	Power	Retail Low	Retail High
1994	T/300D	252,760	281,160
1994	T/375D	267,890	297,990
1993	T/300D	223,390	248,490
1993	T/375D	236,740	263,340
1992	T/300D	201,140	223,740
1992	T/375D	215,380	239,580
1991	T/300D	182,450	202,950
1991	T/375D	195,800	217,800
1990	T/300D	173,550	193,050
1990	T/375D	184,230	204,930
1989	T/Diesel	170,880	190,080
1988	T/Diesel	160,200	178,200

Hyatt 44 MY

Year	Power	Retail Low	Retail High
1994	T/300D	309,720	344,520
1994	T/375D	324,850	361,350
1993	T/300D	268,780	298,980
1993	T/375D	283,020	314,820
1992	T/300D	251,870	280,170
1992	T/375D	264,330	294,030
1991	T/300D	232,290	258,390
1991	T/375D	244,750	272,250
1990	T/300D	219,830	244,530
1990	T/375D	230,510	256,410
1989	T/Diesel	210,040	233,640
1988	T/Diesel	194,020	215,820

Independence 45 Trawler

Year	Power	Retail Low	Retail High
1994	S/Diesel	342,650	381,150
1994	T/Diesel	361,340	401,940
1993	S/Diesel	274,120	304,920
1993	T/Diesel	293,700	326,700
1992	S/Diesel	250,090	278,190
1992	T/Diesel	265,220	295,020
1991	S/Diesel	225,170	250,470
1991	T/Diesel	239,410	266,310
1990	S/Diesel	205,590	228,690
1990	T/Diesel	218,940	243,540
1989	S/Diesel	193,130	214,830
1989	T/Diesel	205,590	228,690
1988	S/Diesel	174,440	194,040
1988	T/Diesel	185,120	205,920
1987	S/Diesel	164,650	183,150
1987	T/Diesel	174,440	194,040

Island Gypsy 30 Sedan

Year	Power	Retail Low	Retail High
1985	S/Diesel	55,180	61,380
1985	T/Diesel	59,630	66,330
1984	S/Diesel	52,510	58,410
1984	T/Diesel	56,960	63,360
1983	S/Diesel	49,840	55,440
1983	T/Diesel	54,290	60,390
1982	S/Diesel	48,060	53,460
1982	T/Diesel	52,510	58,410
1981	S/Diesel	45,390	50,490
1981	T/Diesel	48,950	54,450
1980	S/Diesel	43,610	48,510
1980	T/Diesel	47,170	52,470
1979	S/Diesel	41,830	46,530
1979	T/Diesel	45,390	50,490
1978	S/Diesel	40,050	44,550
1978	T/Diesel	43,610	48,510
1977	S/Diesel	39,160	43,560
1977	T/Diesel	42,720	47,520
1976	S/Diesel	38,270	42,570
1976	T/Diesel	41,830	46,530
1975	S/Diesel	37,380	41,580
1975	T/Diesel	40,940	45,540

Island Gypsy 32 Sedan

Year	Power	Retail Low	Retail High
1994	S/Diesel	129,050	143,550
1994	T/Diesel	142,400	158,400
1993	S/Diesel	119,260	132,660
1993	T/Diesel	128,160	142,560
1992	S/Diesel	110,360	122,760
1992	T/Diesel	119,260	132,660
1991	S/Diesel	102,350	113,850
1991	T/Diesel	110,360	122,760
1990	S/Diesel	94,340	104,940
1990	T/Diesel	102,350	113,850
1989	S/Diesel	87,220	97,020
1989	T/Diesel	95,230	105,930
1988	S/Diesel	76,540	85,140
1988	T/Diesel	81,880	91,080
1987	S/Diesel	66,750	74,250
1987	T/Diesel	71,200	79,200
1986	S/Diesel	61,410	68,310
1986	T/Diesel	64,970	72,270
1985	S/Diesel	56,960	63,360
1985	T/Diesel	60,520	67,320
1984	S/Diesel	52,510	58,410

Year	Power	Retail Low	Retail High
1984	T/Diesel	54,290	60,390
1983	S/Diesel	46,280	51,480
1983	T/Diesel	49,840	55,440
1982	S/Diesel	42,720	47,520
1982	T/Diesel	46,280	51,480
1981	S/Diesel	40,050	44,550
1981	T/Diesel	43,610	48,510

Island Gypsy 36 Europa

Year	Power	Retail Low	Retail High
1994	S/Diesel	176,220	196,020
1994	T/Diesel	188,680	209,880
1993	S/Diesel	153,080	170,280
1993	T/Diesel	164,650	183,150
1992	S/Diesel	137,950	153,450
1992	T/Diesel	148,630	165,330
1991	S/Diesel	130,830	145,530
1991	T/Diesel	141,510	157,410
1990	S/Diesel	121,930	135,630
1990	T/Diesel	131,720	146,520
1989	S/Diesel	113,030	125,730
1989	T/Diesel	121,930	135,630
1988	S/Diesel	105,910	117,810
1988	T/Diesel	113,030	125,730

Island Gypsy 36 Tri Cabin

Year	Power	Retail Low	Retail High
1994	S/Diesel	181,560	201,960
1994	T/Diesel	194,910	216,810
1993	S/Diesel	157,530	175,230
1993	T/Diesel	169,100	188,100
1992	S/Diesel	145,960	162,360
1992	T/Diesel	156,640	174,240
1991	S/Diesel	137,060	152,460
1991	T/Diesel	146,850	163,350
1990	S/Diesel	128,160	142,560
1990	T/Diesel	137,060	152,460
1989	S/Diesel	121,930	135,630
1989	T/Diesel	129,940	144,540
1988	S/Diesel	116,590	129,690
1988	T/Diesel	121,930	135,630

Island Gypsy 36 Quad Cabin

Year	Power	Retail Low	Retail High
1994	S/Diesel	178,890	198,990
1994	T/Diesel	191,350	212,850
1993	S/Diesel	152,190	169,290
1993	T/Diesel	164,650	183,150
1992	S/Diesel	140,620	156,420
1992	T/Diesel	151,300	168,300

Year	Power	Retail Low	Retail High
1991	S/Diesel	131,720	146,520
1991	T/Diesel	142,400	158,400
1990	S/Diesel	123,710	137,610
1990	T/Diesel	133,500	148,500
1989	S/Diesel	116,590	129,690
1989	T/Diesel	125,490	139,590
1988	S/Diesel	108,580	120,780
1988	T/Diesel	115,700	128,700
1987	S/Diesel	100,570	111,870
1987	T/Diesel	107,690	119,790
1986	S/Diesel	93,450	103,950
1986	T/Diesel	99,680	110,880
1985	S/Diesel	85,440	95,040
1985	T/Diesel	91,670	101,970
1984	S/Diesel	79,210	88,110
1984	T/Diesel	84,550	94,050
1983	S/Diesel	72,980	81,180
1983	T/Diesel	78,320	87,120
1982	S/Diesel	67,640	75,240
1982	T/Diesel	72,980	81,180
1981	S/Diesel	62,300	69,300
1981	T/Diesel	67,640	75,240
1980	S/Diesel	58,740	65,340
1980	T/Diesel	63,190	70,290
1979	S/Diesel	55,180	61,380
1979	T/Diesel	59,630	66,330
1978	S/Diesel	51,620	57,420
1978	T/Diesel	56,070	62,370
1977	S/Diesel	48,060	53,460
1977	T/Diesel	44,500	49,500

Island Gypsy 40 Flush Aft Deck

Year	Power	Retail Low	Retail High
1994	T/135D	233,180	259,380
1994	T/375D	267,890	297,990
1993	T/135D	216,270	240,570
1993	T/375D	250,090	278,190
1992	T/135D	204,700	227,700
1992	T/375D	237,630	264,330
1991	T/135D	200,250	222,750
1991	T/375D	225,170	250,470
1990	T/135D	190,460	211,860
1990	T/375D	213,600	237,600
1989	T/135D	178,890	198,990
1989	T/375D	202,030	224,730
1988	T/135D	165,540	184,140
1988	T/375D	183,340	203,940

Year	Power	Retail Low	Retail High
1987	T/135D	152,190	169,290
1987	T/375D	161,090	179,190
1986	T/135D	140,620	156,420
1986	T/375D	159,310	177,210

Island Gypsy 40 Motor Cruiser

Year	Power	Retail Low	Retail High
1994	T/135D	231,400	257,400
1994	T/375D	258,100	287,100
1993	T/135D	215,380	239,580
1993	T/375D	241,190	268,290
1992	T/135D	202,920	225,720
1992	T/375D	227,840	253,440
1991	T/135D	188,680	209,880
1991	T/375D	213,600	237,600
1990	T/135D	178,000	198,000
1990	T/375D	201,140	223,740
1989	T/135D	166,430	185,130
1989	T/375D	189,570	210,870
1988	T/135D	154,860	172,260
1988	T/375D	175,330	195,030
1987	T/135D	143,290	159,390
1987	T/375D	161,090	179,190
1986	T/135D	131,720	146,520
1986	T/375D	150,410	167,310

Island Gypsy 40 Classic

Year	Power	Retail Low	Retail High
1994	T/135D	242,080	269,280
1994	T/300D	267,890	297,990
1993	T/135D	215,380	239,580
1993	T/300D	238,520	265,320

Island Gypsy 44 Flush Aft Deck

Year	Power	Retail Low	Retail High
1994	T/135D	280,350	311,850
1994	T/300D	311,500	346,500
1993	T/135D	258,100	287,100
1993	T/300D	284,800	316,800
1992	T/135D	244,750	272,250
1992	T/375D	267,890	297,990
1991	T/135D	231,400	257,400
1991	T/375D	256,320	285,120
1990	T/135D	220,720	245,520
1990	T/375D	241,190	268,290
1989	T/135D	204,700	227,700
1989	T/375D	231,400	257,400
1988	T/135D	193,130	214,830
1987	T/135D	174,440	194,040

Year	Power	Retail Low	Retail High
1986	T/135D	157,530	175,230
1985	T/135D	144,180	160,380
1984	T/135D	137,060	152,460
1983	T/135D	128,160	142,560
1982	T/120D	124,600	138,600
1981	T/120D	120,150	133,650
1980	T/120D	112,140	124,740
1979	T/120D	106,800	118,800

Island Gypsy 44 Motor Cruiser

Year	Power	Retail Low	Retail High
1994	T/135D	275,900	306,900
1994	T/300D	308,830	343,530
1993	T/135D	258,100	287,100
1993	T/375D	290,140	322,740
1992	T/135D	246,530	274,230
1992	T/375D	270,560	300,960
1991	T/135D	226,950	252,450
1991	T/375D	256,320	285,120
1990	T/135D	219,830	244,530
1990	T/375D	240,300	267,300
1989	T/135D	195,800	217,800
1989	T/375D	222,500	247,500
1988	T/135D	184,230	204,930
1987	T/135D	165,540	184,140
1986	T/135D	156,640	174,240
1985	T/135D	144,180	160,380
1984	T/135D	138,840	154,440
1983	T/135D	129,940	144,540

Island Gypsy 49 Raised PH

Year	Power	Retail Low	Retail High
1994	T/210D	398,720	443,520
1994	T/375D	445,000	495,000
1993	T/135D	377,360	419,760
1993	T/375D	411,180	457,380
1992	T/135D	347,990	387,090
1992	T/375D	379,140	421,740
1991	T/135D	331,970	369,270
1991	T/375D	357,780	397,980

Island Gypsy 51 MY

Year	Power	Retail Low	Retail High
1994	T/375D	631,900	702,900
1994	T/550D	680,850	757,350
1993	T/375D	544,680	605,880
1993	T/550D	587,400	653,400
1992	T/375D	470,810	523,710
1992	T/550D	510,860	568,260
1991	T/375D	394,270	438,570
1991	T/550D	435,210	484,110
1990	T/375D	342,650	381,150
1990	T/550D	373,800	415,800
1989	T/375D	293,700	326,700
1989	T/550D	329,300	366,300
1988	T/550D	310,610	345,510
1987	T/550D	299,930	333,630
1986	T/550D	289,250	321,750
1985	T/355D	279,460	310,860
1984	T/355D	267,000	297,000
1983	T/355D	253,650	282,150

Jefferson 37 Viscount MY

Year	Power	Retail Low	Retail High
1994	T/300D	191,350	212,850
1994	T/375D	221,610	246,510
1993	T/300D	173,550	193,050
1993	T/375D	196,690	218,790
1992	T/300D	164,650	183,150
1992	T/375D	181,560	201,960
1991	T/300D	157,530	175,230
1991	T/375D	171,770	191,070
1990	T/300D	148,630	165,330
1990	T/375D	160,200	178,200
1989	T/300D	135,280	150,480
1989	T/375D	150,410	167,310
1988	T/300D	128,160	142,560
1988	T/375D	142,400	158,400

Jefferson 42 Sundeck

Year	Power	Retail Low	Retail High
1989	T/210D	159,310	177,210
1989	T/375D	178,000	198,000
1988	T/210D	142,400	158,400
1988	T/320D	159,310	177,210
1987	T/200D	129,050	143,550
1987	T/320D	144,180	160,380
1986	T/200D	120,150	133,650
1986	T/320D	135,280	150,480
1985	T/200D	113,030	125,730
1985	T/300D	129,050	143,550

Jefferson 42 Viscount

Year	Power	Retail Low	Retail High
1994	T/300D	324,850	361,350
1994	T/375D	267,000	297,000
1993	T/300D	234,960	261,360
1993	T/375D	250,090	278,190
1992	T/300D	218,050	242,550
1992	T/375D	233,180	259,380
1991	T/300D	203,810	226,710
1991	T/375D	214,490	238,590
1990	T/300D	187,790	208,890
1990	T/375D	196,690	218,790

Jefferson 43 Marlago

Year	Power	Retail Low	Retail High
1994	T/300D	256,320	285,120
1994	T/425D	301,710	335,610
1993	T/300D	252,760	281,160
1993	T/425D	278,570	309,870
1992	T/300D	230,510	256,410
1992	T/425D	258,100	287,100
1991	T/300D	218,940	243,540
1991	T/425D	241,190	268,290

Jefferson 45 MY

Year	Power	Retail Low	Retail High
1989	T/200D	191,350	212,850
1989	T/320D	210,930	234,630
1988	T/200D	175,330	195,030
1988	T/320D	192,240	213,840
1987	T/200D	162,870	181,170
1987	T/320D	176,220	196,020
1986	T/200D	151,300	168,300
1986	T/320D	164,650	183,150
1985	T/200D	142,400	158,400
1985	T/320D	154,860	172,260
1984	T/200D	131,720	146,520
1984	T/320D	144,180	160,380
1983	T/200D	121,040	134,640
1983	T/320D	134,390	149,490
1982	T/200D	113,030	125,730
1982	T/320D	125,490	139,590

Jefferson 46 Sundeck MY

Year	Power	Retail Low	Retail High
1989	T/200D	200,250	222,750
1989	T/375D	220,720	245,520
1988	T/200D	186,010	206,910
1988	T/375D	203,810	226,710
1987	T/200D	169,100	188,100
1987	T/375D	184,230	204,930
1986	T/200D	156,640	174,240
1986	T/375D	169,990	189,090
1985	T/200D	145,960	162,360
1985	T/375D	159,310	177,210

Jefferson 46 Marlago MY

Year	Power	Retail Low	Retail High
1994	T/300D	286,580	318,780
1994	T/375D	318,620	354,420

Year	Power	Retail Low	Retail High
1993	T/300D	272,340	302,940
1993	T/375D	292,810	325,710
1992	T/300D	250,090	278,190
1992	T/375D	267,000	297,000
1991	T/300D	240,300	267,300
1991	T/375D	252,760	281,160

Jefferson 48 Rivanna MY

Year	Power	Retail Low	Retail High
1994	T/300D	347,100	386,100
1994	T/425D	375,580	417,780
1993	T/300D	323,070	359,370
1993	T/425D	347,100	386,100
1992	T/300D	309,720	344,520
1992	T/425D	327,520	364,320
1991	T/300D	293,700	326,700
1991	T/425D	308,830	343,530
1990	T/300D	277,680	308,880
1990	T/425D	293,700	326,700

Jefferson 52 Monticello

Year	Power	Retail Low	Retail High
1989	T/Diesel	265,220	295,020
1988	T/Diesel	248,310	276,210
1987	T/Diesel	231,400	257,400
1986	T/Diesel	218,050	242,550

Jefferson 52 Marquessa

Year	Power	Retail Low	Retail High
1994	T/Diesel	502,850	559,350
1993	T/Diesel	458,350	509,850
1992	T/Diesel	419,190	466,290
1991	T/Diesel	385,370	428,670
1990	T/Diesel	354,220	394,020
1989	T/Diesel	335,530	373,230

Jefferson 60 MY

Year	Power	Retail Low	Retail High
1994	T/Diesel	818,800	910,800
1993	T/Diesel	738,700	821,700
1992	T/Diesel	667,500	742,500
1991	T/Diesel	618,550	688,050
1990	T/Diesel	580,280	645,480
1989	T/Diesel	540,230	600,930
1988	T/Diesel	500,180	556,380
1987	T/Diesel	466,360	518,760

Kha Shing 40 Sundeck

Year	Power	Retail Low	Retail High
1994	T/Diesel	******	******
1994	T/Diesel	******	******
1993	T/Diesel	******	******
1993	T/Diesel	******	******
1992	T/150D	145,960	162,360
1992	T/250D	163,760	182,160
1991	T/150D	139,730	155,430
1991	T/250D	152,190	169,290
1990	T/150D	133,500	148,500
1990	T/250D	142,400	158,400
1989	T/150D	126,380	140,580
1989	T/250D	134,390	149,490
1988	T/150D	122,820	136,620
1988	T/250D	124,600	138,600
1987	T/150D	107,690	119,790
1987	T/250D	114,810	127,710
1986	T/165D	97,900	108,900
1986	T/260D	105,020	116,820
1985	T/165D	89,000	99,000
1985	T/260D	96,120	106,920
1984	T/165D	81,880	91,080
1984	T/260D	89,000	99,000
1983	T/165D	77,430	86,130
1983	T/260D	84,550	94,050
1982	T/165D	73,870	82,170
1982	T/260D	80,990	90,090
1981	T/165D	71,200	79,200
1981	T/260D	77,430	86,130
1980	T/165D	64,080	71,280
1980	T/260D	71,200	79,200

Krogen Manatee 36

Year	Power	Retail Low	Retail High
1991	S/Diesel	118,370	131,670
1990	S/Diesel	111,250	123,750
1989	S/Diesel	106,800	118,800
1988	S/Diesel	103,240	114,840
1987	S/Diesel	98,790	109,890
1986	S/Diesel	91,670	101,970
1985	S/Diesel	85,440	95,040
1984	S/Diesel	80,100	89,100

Krogen 42 Trawler

Year	Power	Retail Low	Retail High
1994	S/Diesel	253,650	282,150
1993	S/Diesel	233,180	259,380
1992	S/Diesel	213,600	237,600
1991	S/Diesel	202,030	224,730
1990	S/Diesel	192,240	213,840
1989	S/Diesel	181,560	201,960
1988	S/Diesel	173,550	193,050
1987	S/Diesel	165,540	184,140
1986	S/Diesel	156,640	174,240
1985	S/Diesel	150,410	167,310
1984	S/Diesel	133,500	148,500
1983	S/Diesel	125,490	139,590
1982	S/Diesel	117,480	130,680
1981	S/Diesel	111,250	123,750
1980	S/Diesel	105,020	116,820
1979	S/Diesel	98,790	109,890

Krogen Silhouette 42

Year	Power	Retail Low	Retail High
1991	T/Diesel	186,900	207,900
1990	T/Diesel	170,880	190,080
1989	T/Diesel	157,530	175,230
1988	T/Diesel	145,960	162,360
1987	T/Diesel	136,170	151,470

Krogen Whaleback 48

Year	Power	Retail Low	Retail High
1994	S/Diesel	360,450	400,950
1993	S/Diesel	330,190	367,290

Lord Nelson Victory 37

Year	Power	Retail Low	Retail High
1989	S/Diesel	147,740	164,340
1988	S/Diesel	137,950	153,450
1987	S/Diesel	129,050	143,550
1986	S/Diesel	121,040	134,640
1985	S/Diesel	112,140	124,740
1984	S/Diesel	103,240	114,840

Mainship 34 Sedan

Year	Power	Retail Low	Retail High
1982	S/Diesel	40,940	45,540
1981	S/Diesel	38,270	42,570
1980	S/Diesel	34,710	38,610
1979	S/Diesel	32,930	36,630
1978	S/Diesel	31,150	34,650

Mainship 36 DC

Year	Power	Retail Low	Retail High
1989	T/Gas	84,550	94,050
1989	T/Diesel	107,690	119,790
1988	T/Gas	75,650	84,150
1988	T/Diesel	93,450	103,950
1987	T/Gas	64,970	72,270
1987	T/Diesel	81,880	91,080
1986	T/Gas	62,300	69,300
1986	T/Diesel	73,870	82,170
1985	T/Gas	56,960	63,360
1985	T/Diesel	65,860	73,260
1984	T/Gas	52,510	58,410
1984	T/Diesel	61,410	68,310

Mainship 40 DC

Year	Power	Retail Low	Retail High
1988	T/Gas	89,890	99,990
1988	T/Diesel	106,800	118,800
1987	T/Gas	82,770	92,070
1987	T/Diesel	97,900	108,900
1986	T/Gas	77,430	86,130
1986	T/Diesel	90,780	100,980
1985	T/Gas	68,530	76,230
1985	T/Diesel	81,880	91,080
1984	T/Gas	62,300	69,300
1984	T/Diesel	74,760	83,160

Mainship 41 Grand Salon

Year	Power	Retail Low	Retail High
1990	T/Gas	107,690	119,790
1990	T/Diesel	140,620	156,420
1989	T/Gas	104,130	115,830
1989	T/Diesel	126,380	140,580

Mainship 47 MY

Year	Power	Retail Low	Retail High
1994	T/Diesel	413,850	460,350
1993	T/Diesel	357,780	397,980
1991	T/Diesel	242,970	270,270
1990	T/Diesel	226,950	252,450

Marine Trader 34 DC (Early)

Year	Power	Retail Low	Retail High
1992	S/Diesel	93,450	103,950
1991	S/Diesel	89,000	99,000
1990	S/Diesel	81,880	91,080
1989	S/Diesel	76,540	85,140
1988	S/Diesel	71,200	79,200
1987	S/Diesel	66,750	74,250
1986	S/Diesel	63,190	70,290
1985	S/Diesel	58,740	65,340
1984	S/Diesel	55,180	61,380
1983	S/Diesel	52,510	58,410
1982	S/Diesel	49,840	55,440
1981	S/Diesel	47,170	52,470
1980	S/Diesel	44,500	49,500
1979	S/Diesel	41,830	46,530
1978	S/Diesel	40,050	44,550
1977	S/Diesel	38,270	42,570
1976	S/Diesel	36,490	40,590
1975	S/Diesel	34,710	38,610

Marine Trader 34 DC

Year	Power	Retail Low	Retail High
1994	S/Diesel	111,250	123,750
1993	S/Diesel	103,240	114,840
1992	S/Diesel	96,120	106,920

Marine Trader 34 Sedan

Year	Power	Retail Low	Retail High
1994	S/Diesel	114,810	127,710
1993	S/Diesel	105,910	117,810
1992	S/Diesel	97,010	107,910
1991	S/Diesel	89,000	99,000
1990	S/Diesel	80,990	90,090
1989	S/Diesel	75,650	84,150
1988	S/Diesel	70,310	78,210
1987	S/Diesel	65,860	73,260
1986	S/Diesel	62,300	69,300
1985	S/Diesel	57,850	64,350
1984	S/Diesel	55,180	61,380
1983	S/Diesel	52,510	58,410
1982	S/Diesel	49,840	55,440
1981	S/Diesel	47,170	52,470
1980	S/Diesel	44,500	49,500
1979	S/Diesel	41,830	46,530
1978	S/Diesel	40,050	44,550
1977	S/Diesel	38,270	42,570
1976	S/Diesel	36,490	40,590
1975	S/Diesel	34,710	38,610

Marine Trader 36 DC

Year	Power	Retail Low	Retail High
1993	S/Diesel	124,600	138,600
1993	T/Diesel	135,280	150,480
1992	S/Diesel	117,480	130,680
1992	T/Diesel	127,270	141,570
1991	S/Diesel	108,580	120,780
1991	T/Diesel	118,370	131,670
1990	S/Diesel	101,460	112,860
1990	T/Diesel	110,360	122,760
1989	S/Diesel	93,450	103,950
1989	T/Diesel	102,350	113,850
1988	S/Diesel	85,440	95,040
1988	T/Diesel	94,340	104,940
1987	S/Diesel	77,430	86,130
1987	T/Diesel	86,330	96,030
1986	S/Diesel	72,090	80,190
1986	T/Diesel	80,100	89,100
1985	S/Diesel	64,970	72,270
1985	T/Diesel	72,980	81,180
1984	S/Diesel	59,630	66,330
1984	T/Diesel	67,640	75,240
1983	S/Diesel	56,960	63,360
1983	T/Diesel	63,190	70,290
1982	S/Diesel	53,400	59,400
1982	T/Diesel	60,520	67,320
1981	S/Diesel	51,620	57,420
1981	T/Diesel	56,960	63,360
1980	S/Diesel	48,060	53,460
1980	T/Diesel	53,400	59,400
1979	S/Diesel	42,720	47,520
1979	T/Diesel	49,840	55,440
1978	S/Diesel	39,160	43,560
1978	T/Diesel	46,280	51,480
1977	S/Diesel	37,380	41,580
1977	T/Diesel	43,610	48,510
1976	S/Diesel	35,600	39,600
1976	T/Diesel	41,830	46,530
1975	S/Diesel	33,820	37,620
1975	T/Diesel	40,050	44,550

Marine Trader 36 Sedan

Year	Power	Retail Low	Retail High
1993	S/Diesel	122,820	136,620
1993	T/Diesel	133,500	148,500
1992	S/Diesel	115,700	128,700
1992	T/Diesel	125,490	139,590
1991	S/Diesel	106,800	118,800
1991	T/Diesel	115,700	128,700
1990	S/Diesel	100,570	111,870
1990	T/Diesel	109,470	121,770
1989	S/Diesel	92,560	102,960
1989	T/Diesel	101,460	112,860
1988	S/Diesel	84,550	94,050
1988	T/Diesel	93,450	103,950
1987	S/Diesel	77,430	86,130
1987	T/Diesel	86,330	96,030
1986	S/Diesel	72,090	80,190
1986	T/Diesel	80,100	89,100
1985	S/Diesel	64,970	72,270
1985	T/Diesel	72,980	81,180
1984	S/Diesel	59,630	66,330
1984	T/Diesel	67,640	75,240
1983	S/Diesel	56,960	63,360
1983	T/Diesel	63,190	70,290
1982	S/Diesel	53,400	59,400
1982	T/Diesel	60,520	67,320
1981	S/Diesel	51,620	57,420
1981	T/Diesel	56,960	63,360
1980	S/Diesel	48,060	53,460
1980	T/Diesel	53,400	59,400
1979	S/Diesel	42,720	47,520

Year	Power	Retail Low	Retail High
1979	T/Diesel	49,840	55,440
1978	S/Diesel	38,270	42,570
1978	T/Diesel	45,390	50,490
1977	S/Diesel	36,490	40,590
1977	T/Diesel	42,720	47,520
1976	S/Diesel	34,710	38,610
1976	T/Diesel	40,940	45,540
1975	S/Diesel	32,930	36,630
1975	T/Diesel	39,160	43,560

Marine Trader 36 Sundeck

Year	Power	Retail Low	Retail High
1993	S/Diesel	127,270	141,570
1993	T/Diesel	137,950	153,450
1992	S/Diesel	120,150	133,650
1992	T/Diesel	129,940	144,540
1991	S/Diesel	111,250	123,750
1991	T/Diesel	121,040	134,640
1990	S/Diesel	104,130	115,830
1990	T/Diesel	113,030	125,730
1989	S/Diesel	97,010	107,910
1989	T/Diesel	105,910	117,810
1988	S/Diesel	89,000	99,000
1988	T/Diesel	97,900	108,900
1987	S/Diesel	80,990	90,090
1987	T/Diesel	89,890	99,990
1986	S/Diesel	75,650	84,150
1986	T/Diesel	83,660	93,060
1985	S/Diesel	66,750	74,250
1985	T/Diesel	74,760	83,160

Marine Trader 38 DC

Year	Power	Retail Low	Retail High
1994	S/Diesel	153,970	171,270
1994	T/Diesel	167,320	186,120
1993	S/Diesel	145,070	161,370
1993	T/Diesel	155,750	173,250
1992	S/Diesel	137,060	152,460
1992	T/Diesel	146,850	163,350
1991	S/Diesel	128,160	142,560
1991	T/Diesel	138,840	154,440
1990	S/Diesel	120,150	133,650
1990	T/Diesel	129,050	143,550
1989	S/Diesel	113,030	125,730
1989	T/Diesel	121,930	135,630
1988	S/Diesel	104,130	115,830
1988	T/Diesel	113,920	126,720
1987	S/Diesel	96,120	106,920
1987	T/Diesel	105,020	116,820
1986	S/Diesel	89,890	99,990
1986	T/Diesel	97,010	107,910
1985	S/Diesel	80,990	90,090
1985	T/Diesel	87,220	97,020
1984	S/Diesel	73,870	82,170
1984	T/Diesel	80,100	89,100
1983	S/Diesel	67,640	75,240
1983	T/Diesel	74,760	83,160
1982	S/Diesel	63,190	70,290
1982	T/Diesel	71,200	79,200
1981	S/Diesel	58,740	65,340
1981	T/Diesel	66,750	74,250
1980	S/Diesel	54,290	60,390
1980	T/Diesel	62,300	69,300

Tradewinds 39 Sundeck

Year	Power	Retail Low	Retail High
1993	T/135D	178,000	198,000
1992	T/135D	167,320	186,120
1991	T/135D	157,530	175,230
1990	T/135D	147,740	164,340
1989	T/135D	138,840	154,440

Marine Trader 40 DC

Year	Power	Retail Low	Retail High
1986	S/Diesel	93,450	103,950
1986	T/Diesel	103,240	114,840
1985	S/Diesel	86,330	96,030
1985	T/Diesel	95,230	105,930
1984	S/Diesel	80,990	90,090
1984	T/Diesel	89,000	99,000
1983	S/Diesel	65,860	73,260
1983	T/Diesel	83,660	93,060
1982	S/Diesel	63,190	70,290
1982	T/Diesel	80,100	89,100
1981	S/Diesel	58,740	65,340
1981	T/Diesel	76,540	85,140
1980	S/Diesel	64,970	72,270
1980	T/Diesel	72,980	81,180
1979	S/Diesel	60,520	67,320
1979	T/Diesel	68,530	76,230
1978	S/Diesel	56,070	62,370
1978	T/Diesel	64,080	71,280
1977	S/Diesel	52,510	58,410
1977	T/Diesel	59,630	66,330
1976	S/Diesel	48,950	54,450
1976	T/Diesel	56,070	62,370
1975	S/Diesel	45,390	50,490
1975	T/Diesel	51,620	57,420

Marine Trader 40 Sedan

Year	Power	Retail Low	Retail High
1986	S/Diesel	92,560	102,960
1986	T/Diesel	102,350	113,850
1985	S/Diesel	84,550	94,050
1985	T/Diesel	93,450	103,950
1984	S/Diesel	79,210	88,110
1984	T/Diesel	87,220	97,020
1983	S/Diesel	74,760	83,160
1983	T/Diesel	81,880	91,080
1982	S/Diesel	71,200	79,200
1982	T/Diesel	79,210	88,110
1981	S/Diesel	67,640	75,240
1981	T/Diesel	74,760	83,160
1980	S/Diesel	63,190	70,290
1980	T/Diesel	70,310	78,210
1979	S/Diesel	57,850	64,350
1979	T/Diesel	64,970	72,270
1978	S/Diesel	55,180	61,380
1978	T/Diesel	60,520	67,320

Marine Trader 40 Sundeck MY

Year	Power	Retail Low	Retail High
1994	S/Diesel	164,650	183,150
1994	T/Diesel	184,230	204,930
1993	S/Diesel	155,750	173,250
1993	T/Diesel	169,100	188,100
1992	S/Diesel	148,630	165,330
1992	T/Diesel	160,200	178,200
1991	S/Diesel	141,510	157,410
1991	T/Diesel	150,410	167,310
1990	S/Diesel	133,500	148,500
1990	T/Diesel	141,510	157,410
1989	S/Diesel	125,490	139,590
1989	T/Diesel	132,610	147,510
1988	S/Diesel	115,700	128,700
1988	T/Diesel	122,820	136,620
1987	S/Diesel	106,800	118,800
1987	T/Diesel	113,920	126,720
1986	S/Diesel	99,680	110,880
1986	T/Diesel	105,910	117,810

Marine Trader 42 Sedan

Year	Power	Retail Low	Retail High
1994	S/Diesel	169,100	188,100
1994	T/Diesel	182,450	202,950
1993	S/Diesel	161,090	179,190
1993	T/Diesel	172,660	192,060
1992	S/Diesel	153,970	171,270

Year	Power	Retail Low	Retail High
1992	T/Diesel	164,650	183,150
1991	S/Diesel	147,740	164,340
1991	T/Diesel	156,640	174,240
1990	S/Diesel	141,510	157,410
1990	T/Diesel	149,520	166,320
1989	S/Diesel	135,280	150,480
1989	T/Diesel	142,400	158,400
1988	S/Diesel	126,380	140,580
1988	T/Diesel	133,500	148,500
1987	S/Diesel	115,700	128,700
1987	T/Diesel	122,820	136,620

LaBelle 43 MY

Year	Power	Retail Low	Retail High
1988	T/Diesel	149,520	166,320
1987	T/Diesel	138,840	154,440
1986	T/Diesel	129,050	143,550
1985	T/Diesel	120,150	133,650
1984	T/Diesel	112,140	124,740
1983	T/Diesel	105,910	117,810

Tradewinds 43 MY

Year	Power	Retail Low	Retail High
1994	T/Diesel	230,510	256,410
1993	T/Diesel	212,710	236,610
1992	T/Diesel	196,690	218,790
1991	T/Diesel	186,900	207,900
1990	T/Diesel	177,110	197,010
1989	T/Diesel	162,870	181,170
1988	T/Diesel	151,300	168,300
1987	T/Diesel	140,620	156,420
1986	T/Diesel	130,830	145,530

Marine Trader 44 Tri Cabin

Year	Power	Retail Low	Retail High
1988	S/Diesel	137,060	152,460
1988	T/Diesel	146,850	163,350
1987	T/Diesel	137,950	153,450
1986	T/Diesel	129,940	144,540
1985	T/Diesel	121,040	134,640
1984	T/Diesel	112,140	124,740
1983	T/Diesel	104,130	115,830
1982	T/Diesel	98,790	109,890
1981	T/Diesel	91,670	101,970
1980	T/Diesel	85,440	95,040
1979	T/Diesel	80,100	89,100
1978	T/Diesel	74,760	83,160
1977	T/Diesel	70,310	78,210

Marine Trader 46 DC

Year	Power	Retail Low	Retail High
1994	T/Diesel	221,610	246,510
1993	T/Diesel	206,480	229,680
1992	T/Diesel	194,910	216,810
1991	T/Diesel	184,230	204,930
1990	T/Diesel	174,440	194,040

Tradewinds 47 MY

Year	Power	Retail Low	Retail High
1994	T/Diesel	266,110	296,010
1993	T/Diesel	246,530	274,230
1992	T/Diesel	232,290	258,390
1991	T/Diesel	218,050	242,550
1990	T/Diesel	203,810	226,710
1989	T/Diesel	191,350	212,850
1988	T/Diesel	178,890	198,990
1987	T/Diesel	167,320	186,120
1986	T/Diesel	155,750	173,250

Marine Trader 49 PH

Year	Power	Retail Low	Retail High
1990	T/Diesel	214,490	238,590
1989	T/Diesel	197,580	219,780
1988	T/Diesel	185,120	205,920
1987	T/Diesel	173,550	193,050
1986	T/Diesel	163,760	182,160
1985	T/Diesel	157,530	175,230
1984	T/Diesel	152,190	169,290
1983	T/Diesel	146,850	163,350
1982	T/Diesel	141,510	157,410
1981	T/Diesel	136,170	151,470
1980	T/Diesel	131,720	146,520
1979	T/Diesel	127,270	141,570

Marine Trader 50 MY

Year	Power	Retail Low	Retail High
1993	T/Diesel	310,610	345,510
1992	T/Diesel	288,360	320,760
1991	T/Diesel	268,780	298,980
1990	T/Diesel	253,650	282,150
1989	T/Diesel	240,300	267,300
1988	T/Diesel	222,500	247,500
1987	T/Diesel	202,920	225,720
1986	T/Diesel	185,120	205,920
1985	T/Diesel	174,440	194,040
1984	T/Diesel	161,980	180,180
1983	T/Diesel	153,970	171,270
1982	T/Diesel	146,850	163,350
1981	T/Diesel	138,840	154,440
1980	T/Diesel	131,720	146,520
1979	T/Diesel	127,270	141,570

Marine Trader Med 14 Meter

Year	Power	Retail Low	Retail High
1994	T/Diesel	284,800	316,800
1993	T/Diesel	258,100	287,100
1992	T/Diesel	241,190	268,290
1991	T/Diesel	228,730	254,430

Monk 36 Trawler

Year	Power	Retail Low	Retail High
1994	S/Diesel	144,180	160,380
1994	T/Diesel	160,200	178,200
1993	S/Diesel	127,270	141,570
1993	T/Diesel	142,400	158,400
1992	S/Diesel	115,700	128,700
1992	T/Diesel	129,940	144,540
1991	S/Diesel	106,800	118,800
1991	T/Diesel	119,260	132,660
1990	S/Diesel	98,790	109,890
1990	T/Diesel	109,470	121,770
1989	S/Diesel	90,780	100,980
1989	T/Diesel	100,570	111,870
1988	S/Diesel	82,770	92,070
1988	T/Diesel	91,670	101,970
1987	S/Diesel	75,650	84,150
1987	T/Diesel	84,550	94,050
1986	S/Diesel	69,420	77,220
1986	T/Diesel	78,320	87,120
1985	S/Diesel	64,970	72,270
1985	T/Diesel	72,980	81,180
1984	S/Diesel	59,630	66,330
1984	T/Diesel	67,640	75,240
1983	S/Diesel	55,180	61,380
1983	T/Diesel	63,190	70,290
1982	S/Diesel	54,290	60,390
1982	T/Diesel	60,520	67,320

Nordhavn 46

Year	Power	Retail Low	Retail High
1994	S/Diesel	367,570	408,870
1993	S/Diesel	318,620	354,420
1992	S/Diesel	291,030	323,730
1991	S/Diesel	276,790	307,890
1990	S/Diesel	258,100	287,100
1989	S/Diesel	241,190	268,290

Nordic 32 Tug

Year	Power	Retail Low	Retail High
1994	S/Diesel	128,160	142,560
1993	S/Diesel	117,480	130,680
1992	S/Diesel	107,690	119,790

Year	Power	Retail Low	Retail High
1991	S/Diesel	100,570	111,870
1990	S/Diesel	93,450	103,950
1989	S/Diesel	86,330	96,030
1988	S/Diesel	80,990	90,090
1987	S/Diesel	73,870	82,170
1986	S/Diesel	68,530	76,230
1985	S/Diesel	65,860	73,260

Nordic 480 MY

Year	Power	Retail Low	Retail High
1991	T/Diesel	298,150	331,650
1990	T/Diesel	272,340	302,940
1989	T/Diesel	264,330	294,030
1988	T/Diesel	255,430	284,130
1987	T/Diesel	242,970	270,270
1986	T/Diesel	231,400	257,400
1985	T/Diesel	216,270	240,570

Ocean 40+2 Trawler

Year	Power	Retail Low	Retail High
1980	T/Diesel	93,450	103,950
1979	T/Diesel	85,440	95,040
1978	T/Diesel	76,540	85,140

Ocean 42 Sunliner

Year	Power	Retail Low	Retail High
1985	T/Diesel	134,390	149,490
1984	T/Diesel	124,600	138,600
1983	T/Diesel	116,590	129,690
1982	T/Diesel	108,580	120,780
1981	T/Diesel	102,350	113,850

Ocean 44 MY

Year	Power	Retail Low	Retail High
1994	T/435D	378,250	420,750
1994	T/485D	391,600	435,600
1993	T/435D	333,750	371,250
1993	T/485D	347,100	386,100
1992	T/435D	302,600	336,600
1992	T/485D	315,950	351,450

Ocean 46 Sunliner

Year	Power	Retail Low	Retail High
1986	T/Diesel	193,130	214,830
1985	T/Diesel	180,670	200,970
1984	T/Diesel	170,880	190,080
1983	T/Diesel	160,200	178,200

Ocean 48 MY

Year	Power	Retail Low	Retail High
1994	T/Diesel	551,800	613,800
1993	T/Diesel	471,700	524,700
1992	T/Diesel	403,170	448,470
1991	T/Diesel	356,000	396,000
1990	T/Diesel	328,410	365,310
1989	T/Diesel	302,600	336,600

Ocean 53 MY

Year	Power	Retail Low	Retail High
1991	T/Diesel	462,800	514,800
1990	T/Diesel	427,200	475,200
1989	T/Diesel	407,620	453,420
1988	T/Diesel	391,600	435,600

Ocean 55 Sunliner

Year	Power	Retail Low	Retail High
1986	T/Diesel	320,400	356,400
1985	T/Diesel	308,830	343,530
1984	T/Diesel	298,150	331,650
1983	T/Diesel	289,250	321,750

Ocean 56 CMY

Year	Power	Retail Low	Retail High
1994	T/Diesel	596,300	663,300
1993	T/Diesel	538,450	598,950
1992	T/Diesel	495,730	551,430
1991	T/Diesel	465,470	517,770

Ocean Alexander 38 DC

Year	Power	Retail Low	Retail High
1987	T/Diesel	114,810	127,710
1986	T/Diesel	103,240	114,840
1985	T/Diesel	95,230	105,930
1984	T/Diesel	89,000	99,000

Ocean Alexander 390 Sundeck

Year	Power	Retail Low	Retail High
1994	T/250D	226,060	251,460
1994	T/375D	251,870	280,170
1993	T/250D	201,140	223,740
1993	T/375D	225,170	250,470
1992	T/250D	190,460	211,860
1992	T/375D	210,040	233,640
1991	T/250D	178,890	198,990
1991	T/375D	196,690	218,790
1990	T/250D	169,100	188,100
1990	T/375D	185,120	205,920
1989	T/250D	160,200	178,200
1989	T/375D	173,550	193,050
1988	T/250D	152,190	169,290
1988	T/375D	165,540	184,140
1987	T/250D	144,180	160,380
1987	T/375D	157,530	175,230
1986	T/250D	137,950	153,450
1986	T/375D	150,410	167,310

Ocean Alexander 40 DC

Year	Power	Retail Low	Retail High
1985	T/Diesel	116,590	129,690
1984	T/Diesel	106,800	118,800
1983	T/Diesel	102,350	113,850
1982	T/Diesel	94,340	104,940
1981	T/Diesel	87,220	97,020
1980	T/Diesel	81,880	91,080

Ocean Alexander 40 Sedan

Year	Power	Retail Low	Retail High
1985	T/Diesel	115,700	128,700
1984	T/Diesel	104,130	115,830
1983	T/Diesel	97,900	108,900
1982	T/Diesel	89,000	99,000
1981	T/Diesel	80,100	89,100
1980	T/Diesel	72,980	81,180

Ocean Alexander 440 Cockpit

Year	Power	Retail Low	Retail High
1994	T/250D	262,550	292,050
1994	T/375D	284,800	316,800
1993	T/250D	249,200	277,200
1993	T/375D	268,780	298,980
1992	T/250D	234,960	261,360
1992	T/375D	253,650	282,150
1991	T/250D	219,830	244,530
1991	T/375D	236,740	263,340
1990	T/250D	210,040	233,640
1990	T/375D	223,390	248,490
1989	T/250D	195,800	217,800
1989	T/375D	210,040	233,640
1988	T/250D	186,900	207,900
1988	T/375D	199,360	221,760
1987	T/250D	175,330	195,030
1987	T/375D	188,680	209,880

Ocean Alexander 42 Sedan

Year	Power	Retail Low	Retail High
1994	T/250D	240,300	267,300
1994	T/375D	258,100	287,100
1993	T/250D	223,390	248,490
1993	T/375D	240,300	267,300
1992	T/250D	210,040	233,640
1992	T/375D	226,060	251,460
1991	T/250D	199,360	221,760
1991	T/375D	214,490	238,590
1990	T/250D	190,460	211,860
1990	T/375D	203,810	226,710
1989	T/250D	181,560	201,960
1989	T/375D	194,910	216,810
1988	T/250D	174,440	194,040

Year	Power	Retail Low	Retail High
1988	T/375D	186,900	207,900

Ocean Alexander 423 Classico

Year	Power	Retail Low	Retail High
1994	T/210D	267,000	297,000
1993	T/210D	244,750	272,250

Ocean Alexander 43 DC

Year	Power	Retail Low	Retail High
1985	T/Diesel	146,850	163,350
1984	T/Diesel	135,280	150,480
1983	T/Diesel	124,600	138,600
1982	T/Diesel	114,810	127,710
1981	T/Diesel	105,020	116,820
1980	T/Diesel	96,120	106,920

Ocean Alexander 456 Classico

Year	Power	Retail Low	Retail High
1994	T/210D	317,730	353,430
1994	T/375D	345,320	384,120
1993	T/210D	299,930	333,630
1993	T/375D	321,290	357,390
1992	T/210D	283,910	315,810
1992	T/375D	302,600	336,600

Ocean Alexander 48 Sedan

Year	Power	Retail Low	Retail High
1991	T/Diesel	359,560	399,960
1990	T/Diesel	327,520	364,320
1989	T/Diesel	302,600	336,600
1988	T/Diesel	276,790	307,890

Ocean Alexander 486 Classico

Year	Power	Retail Low	Retail High
1994	T/Diesel	458,350	509,850
1993	T/Diesel	418,300	465,300

Ocean Alexander 50 PH

Year	Power	Retail Low	Retail High
1983	T/Diesel	187,790	208,890
1982	T/Diesel	168,210	187,110
1981	T/Diesel	153,080	170,280
1980	T/Diesel	140,620	156,420
1979	T/Diesel	131,720	146,520
1978	T/Diesel	124,600	138,600
1977	T/Diesel	119,260	132,660

Ocean Alexander 50 MK II PH

Year	Power	Retail Low	Retail High
1990	T/Diesel	324,850	361,350
1989	T/Diesel	294,590	327,690
1988	T/Diesel	265,220	295,020
1987	T/Diesel	238,520	265,320
1986	T/Diesel	224,280	249,480
1985	T/Diesel	212,710	236,610
1984	T/Diesel	202,030	224,730

Ocean Alexander 51 Sedan

Year	Power	Retail Low	Retail High
1994	T/400D	436,100	485,100
1994	T/735D	520,650	579,150
1993	T/400D	422,750	470,250
1993	T/735D	493,950	549,450
1992	T/400D	400,500	445,500
1992	T/735D	471,700	524,700
1991	T/400D	382,700	425,700
1991	T/735D	445,000	495,000
1990	T/400D	366,680	407,880
1990	T/735D	428,090	476,190
1989	T/400D	356,890	396,990
1989	T/735D	408,510	454,410

Ocean Alexander 520 PH

Year	Power	Retail Low	Retail High
1994	T/Diesel	489,500	544,500
1993	T/Diesel	453,900	504,900
1992	T/Diesel	427,200	475,200
1991	T/Diesel	403,170	448,470

Ocean Alexander 54 CMY

Year	Power	Retail Low	Retail High
1989	T/Diesel	******	******
1988	T/Diesel	******	******
1987	T/Diesel	******	******
1986	T/Diesel	******	******
1985	T/Diesel	******	******

Ocean Alexander 60 MY (Early)

Year	Power	Retail Low	Retail High
1987	T/Diesel	514,420	572,220
1986	T/Diesel	463,690	515,790
1985	T/Diesel	427,200	475,200
1984	T/Diesel	383,590	426,690

Ocean Alexander 60 MY

Year	Power	Retail Low	Retail High
1994	T/735D	933,610	1,038,510
1993	T/735D	879,320	978,120
1992	T/735D	832,150	925,650
1991	T/735D	773,410	860,310
1990	T/735D	712,890	792,990
1989	T/735D	671,950	747,450
1988	T/735D	618,550	688,050
1987	T/735D	539,340	599,940

Ocean Alexander 63 MY

Year	Power	Retail Low	Retail High
1992	T/735D	907,800	1,009,800
1991	T/735D	841,050	935,550
1990	T/735D	797,440	887,040
1989	T/735D	736,030	818,730
1988	T/735D	663,940	738,540
1987	T/735D	590,070	656,370
1986	T/735D	525,990	585,090

Ocean Alexander 630 MY

Year	Power	Retail Low	Retail High
1994	T/735D	1,045,750	1,163,250
1993	T/735D	962,980	1,071,180
1992	T/735D	898,900	999,900

Ocean Alexander 66 MY

Year	Power	Retail Low	Retail High
1994	T/735D	1,112,500	1,237,500
1993	T/735D	1,055,540	1,174,140
1992	T/735D	994,130	1,105,830
1991	T/735D	954,080	1,061,280
1990	T/735D	898,900	999,900
1989	T/735D	809,900	900,900
1988	T/735D	729,800	811,800
1987	T/735D	625,670	695,970
1986	T/735D	582,060	647,460

Offshore 48 Sedan

Year	Power	Retail Low	Retail High
1994	T/Diesel	364,900	405,900
1993	T/Diesel	327,520	364,320
1992	T/Diesel	302,600	336,600
1991	T/Diesel	279,460	310,860
1990	T/Diesel	258,100	287,100
1989	T/Diesel	239,410	266,310
1988	T/Diesel	226,060	251,460
1987	T/Diesel	213,600	237,600

Offshore 48 YF

Year	Power	Retail Low	Retail High
1994	T/Diesel	378,250	420,750
1993	T/Diesel	338,200	376,200
1992	T/Diesel	311,500	346,500
1991	T/Diesel	288,360	320,760
1990	T/Diesel	268,780	298,980
1989	T/Diesel	249,200	277,200
1988	T/Diesel	235,850	262,350
1987	T/Diesel	222,500	247,500
1986	T/Diesel	213,600	237,600
1985	T/Diesel	205,590	228,690

Offshore 52 Sedan MY

Year	Power	Retail Low	Retail High
1994	T/Diesel	520,650	579,150
1993	T/Diesel	471,700	524,700
1992	T/Diesel	441,440	491,040
1991	T/Diesel	404,950	450,450

Offshore 55 PH

Year	Power	Retail Low	Retail High
1994	T/Diesel	654,150	727,650
1993	T/Diesel	618,550	688,050
1992	T/Diesel	570,490	634,590
1991	T/Diesel	560,700	623,700
1990	T/Diesel	534,000	594,000

PT 35 Sundeck

Year	Power	Retail Low	Retail High
1990	T/Diesel	121,930	135,630
1989	T/Diesel	111,250	123,750
1988	T/Diesel	101,460	112,860
1987	T/Diesel	93,450	103,950
1986	T/Diesel	85,440	95,040
1985	T/Diesel	78,320	87,120
1984	T/Diesel	72,090	80,190

PT 42 CMY

Year	Power	Retail Low	Retail High
1990	T/Diesel	158,420	176,220
1989	T/Diesel	141,510	157,410
1988	T/Diesel	132,610	147,510
1987	T/Diesel	124,600	138,600
1986	T/Diesel	114,810	127,710
1985	T/Diesel	104,130	115,830
1984	T/Diesel	95,230	105,930

PT 52 CMY

Year	Power	Retail Low	Retail High
1990	T/Diesel	284,800	316,800
1989	T/Diesel	258,100	287,100
1988	T/Diesel	234,960	261,360
1987	T/Diesel	222,500	247,500
1986	T/Diesel	213,600	237,600

Pacemaker 40 MY

Year	Power	Retail Low	Retail High
1980	T/Gas	71,200	79,200
1980	T/Diesel	83,660	93,060
1979	T/Gas	66,750	74,250
1979	T/Diesel	79,210	88,110
1978	T/Gas	63,190	70,290
1978	T/Diesel	72,980	81,180
1977	T/Gas	58,740	65,340
1977	T/Diesel	68,530	76,230
1976	T/Gas	52,510	58,410
1976	T/Diesel	62,300	69,300
1975	T/Gas	48,950	54,450
1975	T/Diesel	57,850	64,350

Pacemaker 46 MY

Year	Power	Retail Low	Retail High
1980	T/Diesel	137,950	153,450
1979	T/Diesel	130,830	145,530
1978	T/Diesel	124,600	138,600
1977	T/Diesel	118,370	131,670

Pacemaker 57 MY

Year	Power	Retail Low	Retail High
1980	T/Diesel	271,450	301,950
1979	T/Diesel	253,650	282,150
1978	T/Diesel	238,520	265,320
1977	T/Diesel	222,500	247,500

Pacemaker 62 MY

Year	Power	Retail Low	Retail High
1980	T/Diesel	323,070	359,370
1979	T/Diesel	302,600	336,600
1978	T/Diesel	290,140	322,740
1977	T/Diesel	274,120	304,920
1976	T/Diesel	262,550	292,050
1975	T/Diesel	250,980	279,180

Pacemaker 62 CMY

Year	Power	Retail Low	Retail High
1980	T/Diesel	329,300	366,300
1979	T/Diesel	307,940	342,540
1978	T/Diesel	293,700	326,700
1977	T/Diesel	280,350	311,850
1976	T/Diesel	272,340	302,940
1975	T/Diesel	263,440	293,040

Pacemaker 66 MY

Year	Power	Retail Low	Retail High
1980	T/Diesel	393,380	437,580
1979	T/Diesel	377,360	419,760
1978	T/Diesel	357,780	397,980
1977	T/Diesel	339,090	377,190

Pearson 38 DC

Year	Power	Retail Low	Retail High
1991	T/Gas	131,720	146,520
1991	T/Diesel	155,750	173,250
1990	T/Gas	121,930	135,630
1990	T/Diesel	143,290	159,390
1989	T/Gas	113,920	126,720
1989	T/Diesel	131,720	146,520
1988	T/Gas	103,240	114,840
1988	T/Diesel	121,040	134,640

Pearson 43 MY

Year	Power	Retail Low	Retail High
1987	T/Gas	121,040	134,640
1987	T/Diesel	134,390	149,490
1986	T/Gas	112,140	124,740
1986	T/Diesel	123,710	137,610
1985	T/Gas	104,130	115,830
1985	T/Diesel	114,810	127,710
1984	T/Gas	97,010	107,910
1984	T/Diesel	106,800	118,800
1976	T/Gas	61,410	68,310
1976	T/Diesel	68,530	76,230
1975	T/Gas	58,740	65,340
1975	T/Diesel	65,860	73,260

Pilgrim 40

Year	Power	Retail Low	Retail High
1991	S/Diesel	152,190	169,290
1990	S/Diesel	142,400	158,400
1989	S/Diesel	133,500	148,500
1988	S/Diesel	125,490	139,590
1987	S/Diesel	117,480	130,680
1986	S/Diesel	111,250	123,750
1985	S/Diesel	105,020	116,820
1984	S/Diesel	89,890	99,990

Present 42 Sundeck MY

Year	Power	Retail Low	Retail High
1987	T/135D	115,700	128,700
1987	T/225D	119,260	132,660
1986	T/135D	99,680	110,880
1986	T/225D	110,360	122,760
1985	T/135D	93,450	103,950
1985	T/225D	103,240	114,840
1984	T/135D	89,000	99,000
1984	T/225D	97,900	108,900
1983	T/135D	82,770	92,070
1983	T/225D	90,780	100,980

President 395 DC

Year	Power	Retail Low	Retail High
1994	T/Diesel	229,620	255,420
1993	T/Diesel	206,480	229,680
1992	T/Diesel	189,570	210,870

President 41 DC

Year	Power	Retail Low	Retail High
1988	T/135D	115,700	128,700
1988	T/225D	124,600	138,600
1987	T/135D	107,690	119,790
1987	T/225D	115,700	128,700
1986	T/135D	99,680	110,880

Year	Power	Retail Low	Retail High
1986	T/225D	107,690	119,790
1985	T/135D	91,670	101,970
1985	T/225D	99,680	110,880
1984	T/135D	83,660	93,060
1984	T/225D	91,670	101,970
1983	T/135D	76,540	85,140
1983	T/225D	84,550	94,050
1982	T/135D	72,090	80,190
1982	T/225D	77,430	86,130

President 43 DC

Year	Power	Retail Low	Retail High
1990	T/135D	152,190	169,290
1990	T/275D	162,870	181,170
1989	T/135D	143,290	159,390
1989	T/275D	153,080	170,280
1988	T/135D	133,500	148,500
1988	T/275D	142,400	158,400
1987	T/135D	125,490	139,590
1987	T/275D	133,500	148,500
1986	T/135D	116,590	129,690
1986	T/275D	124,600	138,600
1985	T/135D	107,690	119,790
1985	T/225D	115,700	128,700
1984	T/135D	99,680	110,880
1984	T/225D	107,690	119,790

President 46/485 MY

Year	Power	Retail Low	Retail High
1994	T/Diesel	351,550	391,050
1993	T/Diesel	320,400	356,400

President 52/545 CMY

Year	Power	Retail Low	Retail High
1994	T/Diesel	431,650	480,150
1993	T/Diesel	399,610	444,510

Sabreline 34

Year	Power	Retail Low	Retail High
1994	T/Diesel	191,350	212,850
1993	T/Diesel	178,890	198,990
1992	T/Diesel	168,210	187,110
1991	T/Diesel	158,420	176,220

Sabreline 36

Year	Power	Retail Low	Retail High
1994	T/Diesel	222,500	247,500
1993	T/Diesel	205,590	228,690
1992	T/Diesel	193,130	214,830
1991	T/Diesel	180,670	200,970
1990	T/Diesel	169,990	189,090
1989	T/Diesel	159,310	177,210

Sea Ranger 36 Sundeck

Year	Power	Retail Low	Retail High
1986	T/Diesel	56,070	62,370
1986	T/Diesel	64,970	72,270
1985	T/Diesel	51,620	57,420
1985	T/Diesel	59,630	66,330
1984	T/Diesel	48,950	54,450
1984	T/Diesel	55,180	61,380
1983	T/Diesel	48,060	53,460
1983	T/Diesel	52,510	58,410
1982	T/Diesel	46,280	51,480
1982	T/Diesel	49,840	55,440
1981	T/Diesel	44,500	49,500
1981	T/Diesel	48,060	53,460
1980	T/Diesel	42,720	47,520
1980	T/Diesel	46,280	51,480

Sea Ranger 39 Sedan

Year	Power	Retail Low	Retail High
1989	S/Diesel	80,100	89,100
1989	T/Diesel	89,890	99,990
1988	S/Diesel	71,200	79,200
1988	T/Diesel	80,990	90,090
1987	S/Diesel	64,970	72,270
1987	T/Diesel	73,870	82,170
1986	S/Diesel	58,740	65,340
1986	T/Diesel	67,640	75,240
1985	S/Diesel	54,290	60,390
1985	T/Diesel	62,300	69,300
1984	S/Diesel	50,730	56,430
1984	T/Diesel	57,850	64,350
1983	S/Diesel	48,950	54,450
1983	T/Diesel	55,180	61,380
1982	S/Diesel	49,840	55,440
1982	T/Diesel	53,400	59,400
1981	S/Diesel	48,060	53,460
1981	T/Diesel	51,620	57,420
1980	S/Diesel	46,280	51,480
1980	T/Diesel	49,840	55,440

Sea Ranger 39 Sundeck

Year	Power	Retail Low	Retail High
1989	T/Diesel	93,450	103,950
1988	T/Diesel	85,440	95,040
1987	T/Diesel	78,320	87,120
1986	T/Diesel	72,980	81,180
1985	T/Diesel	67,640	75,240
1984	T/Diesel	64,970	72,270
1983	T/Diesel	61,410	68,310

Sea Ranger 45 Sundeck

Year	Power	Retail Low	Retail High
1989	T/135D	166,430	185,130
1989	T/375D	188,680	209,880
1988	T/135D	149,520	166,320
1988	T/375D	170,880	190,080
1987	T/135D	140,620	156,420
1987	T/375D	160,200	178,200
1986	T/135D	132,610	147,510
1986	T/375D	150,410	167,310
1985	T/135D	115,700	128,700
1985	T/375D	141,510	157,410
1984	T/120D	106,800	118,800
1984	T/355D	131,720	146,520
1983	T/120D	102,350	113,850
1983	T/355D	117,480	130,680

Sea Ranger 47 PH

Year	Power	Retail Low	Retail High
1989	T/135D	163,760	182,160
1989	T/375D	186,900	207,900
1988	T/135D	146,850	163,350
1988	T/375D	169,100	188,100
1987	T/135D	140,620	156,420
1987	T/375D	159,310	177,210
1986	T/135D	131,720	146,520
1986	T/375D	150,410	167,310
1985	T/135D	116,590	129,690
1985	T/375D	141,510	157,410
1984	T/120D	106,800	118,800
1984	T/355D	131,720	146,520
1983	T/120D	106,800	118,800
1983	T/355D	121,930	135,630
1982	T/120D	103,240	114,840
1982	T/265D	113,920	126,720
1981	T/120D	94,340	104,940
1981	T/265D	105,020	116,820
1980	T/120D	87,220	97,020
1980	T/265D	97,010	107,910

Sea Ranger 51 MY

Year	Power	Retail Low	Retail High
1989	T/260D	250,090	278,190
1989	T/650D	284,800	316,800
1988	T/260D	233,180	259,380
1988	T/650D	265,220	295,020
1987	T/260D	221,610	246,510
1987	T/650D	248,310	276,210
1986	T/260D	191,350	212,850
1986	T/650D	218,050	242,550

Year	Power	Retail Low	Retail High
1985	T/260D	187,790	208,890
1985	T/650D	203,810	226,710
1984	T/265D	178,890	198,990
1983	T/265D	170,880	190,080

Sea Ranger 55 PH

Year	Power	Retail Low	Retail High
1989	T/320D	321,290	357,390
1989	T/650D	356,000	396,000
1988	T/320D	306,160	340,560
1988	T/650D	334,640	372,240
1987	T/320D	293,700	326,700
1987	T/650D	320,400	356,400
1986	T/310D	277,680	308,880
1986	T/650D	302,600	336,600
1985	T/310D	262,550	292,050
1985	T/650D	287,470	319,770
1984	T/310D	246,530	274,230
1984	T/650D	267,890	297,990
1983	T/Diesel	235,850	262,350
1982	T/Diesel	220,720	245,520
1981	T/Diesel	205,590	228,690
1980	T/Diesel	188,680	209,880
1979	T/Diesel	172,660	192,060
1978	T/Diesel	160,200	178,200
1977	T/Diesel	151,300	168,300
1976	T/Diesel	144,180	160,380

Sea Ranger 65 MY

Year	Power	Retail Low	Retail High
1989	T/Diesel	559,810	622,710
1988	T/Diesel	514,420	572,220
1987	T/Diesel	473,480	526,680
1986	T/Diesel	439,660	489,060
1985	T/Diesel	411,180	457,380
1984	T/Diesel	388,930	432,630
1983	T/Diesel	368,460	409,860
1982	T/Diesel	348,880	388,080
1981	T/Diesel	330,190	367,290
1980	T/Diesel	307,940	342,540
1979	T/Diesel	290,140	322,740
1978	T/Diesel	275,900	306,900
1977	T/Diesel	261,660	291,060
1976	T/Diesel	247,420	275,220

Sea Ray 360 Aft Cabin

Year	Power	Retail Low	Retail High
1987	T/Gas	71,200	79,200
1987	T/Diesel	81,880	91,080
1986	T/Gas	62,300	69,300
1986	T/Diesel	73,870	82,170
1985	T/Gas	56,960	63,360
1985	T/Diesel	67,640	75,240
1984	T/Gas	53,400	59,400
1984	T/Diesel	62,300	69,300
1983	T/Gas	48,950	54,450
1983	T/Diesel	57,850	64,350

Sea Ray 380 Aft Cabin

Year	Power	Retail Low	Retail High
1991	T/Gas	124,600	138,600
1991	T/Diesel	146,850	163,350
1990	T/Gas	114,810	127,710
1990	T/Diesel	134,390	149,490
1989	T/Gas	105,020	116,820
1989	T/Diesel	123,710	137,610

Sea Ray 440 Aft Cabin

Year	Power	Retail Low	Retail High
1991	T/Gas	161,980	180,180
1991	T/Diesel	196,690	218,790
1990	T/Gas	152,190	169,290
1990	T/Diesel	178,890	198,990
1989	T/Gas	142,400	158,400
1989	T/Diesel	164,650	183,150
1988	T/Gas	121,040	134,640
1988	T/Diesel	141,510	157,410
1987	T/Gas	109,470	121,770
1987	T/Diesel	129,940	144,540
1986	T/Gas	101,460	112,860
1986	T/Diesel	120,150	133,650

Sea Ray 500 Sedan Bridge

Year	Power	Retail Low	Retail High
1994	T/650D	520,650	579,150
1993	T/650D	467,250	519,750
1992	T/650D	427,200	475,200
1991	T/650D	400,500	445,500
1990	T/650D	372,910	414,810
1989	T/650D	349,770	389,070

Sea Ray 550 Sedan Bridge

Year	Power	Retail Low	Retail High
1994	T/650D	578,500	643,500
1993	T/650D	525,100	584,100
1992	T/650D	484,160	538,560

Sea Ray 650 MY

Year	Power	Retail Low	Retail High
1994	T/Diesel	1,415,100	1,574,100
1993	T/Diesel	1,160,560	1,290,960
1992	T/Diesel	929,160	1,033,560

Seamaster 48 MY

Year	Power	Retail Low	Retail High
1989	T/Diesel	223,390	248,490
1988	T/Diesel	204,700	227,700
1987	T/Diesel	190,460	211,860
1986	T/Diesel	180,670	200,970
1985	T/Diesel	169,990	189,090
1984	T/Diesel	160,200	178,200
1983	T/Diesel	149,520	166,320

Shannon 36 Voyager

Year	Power	Retail Low	Retail High
1994	T/Diesel	234,070	260,370
1993	T/Diesel	207,370	230,670
1992	T/Diesel	186,900	207,900
1991	T/Diesel	171,770	191,070

Silverton 34 MY

Year	Power	Retail Low	Retail High
1994	T/Gas	112,140	124,740
1994	T/Diesel	137,950	153,450
1993	T/Gas	105,910	117,810
1993	T/Diesel	123,710	137,610

Silverton 37 MY

Year	Power	Retail Low	Retail High
1989	T/Gas	90,780	100,980
1989	T/Diesel	110,360	122,760
1988	T/Gas	81,880	91,080
1988	T/Diesel	98,790	109,890

Silverton 40 Aft Cabin

Year	Power	Retail Low	Retail High
1990	T/Gas	108,580	120,780
1990	T/Diesel	138,840	154,440
1989	T/Gas	100,570	111,870
1989	T/Diesel	127,270	141,570
1988	T/Gas	93,450	103,950
1988	T/Diesel	117,480	130,680
1987	T/Gas	86,330	96,030
1987	T/Diesel	108,580	120,780
1986	T/Gas	80,100	89,100
1986	T/Diesel	99,680	110,880
1985	T/Gas	75,650	84,150
1985	T/Diesel	91,670	101,970
1984	T/Gas	70,310	78,210
1984	T/Diesel	83,660	93,060
1983	T/Gas	65,860	73,260
1983	T/Diesel	76,540	85,140
1982	T/Gas	59,630	66,330
1982	T/Diesel	69,420	77,220

Year	Power	Retail Low	Retail High
Silverton 41 Aft Cabin			
1994	T/Gas	198,470	220,770
1994	T/Diesel	253,650	282,150
1993	T/Gas	180,670	200,970
1993	T/Diesel	229,620	255,420
1992	T/Gas	167,320	186,120
1992	T/Diesel	207,370	230,670
1991	T/Gas	153,970	171,270
1991	T/Diesel	184,230	204,930
Silverton 46 MY			
1994	T/Diesel	373,800	415,800
1993	T/Diesel	333,750	371,250
1992	T/Diesel	306,160	340,560
1991	T/Diesel	277,680	308,880
1990	T/Diesel	248,310	276,210
1989	T/Diesel	221,610	246,510
Stevens 59 MY			
1992	T/Diesel	******	******
1991	T/Diesel	******	******
1990	T/Diesel	******	******
1989	T/Diesel	******	******
Tollycraft 34 Tri Cabin			
1985	T/Gas	76,540	85,140
1985	T/Diesel	88,110	98,010
1984	T/Gas	72,090	80,190
1984	T/Diesel	81,880	91,080
1983	T/Gas	65,860	73,260
1983	T/Diesel	75,650	84,150
1982	T/Gas	62,300	69,300
1982	T/Diesel	71,200	79,200
1981	T/Gas	58,740	65,340
1981	T/Diesel	65,860	73,260
1980	T/Gas	52,510	58,410
1980	T/Diesel	56,960	63,360
1979	T/Gas	48,950	54,450
1979	T/Diesel	52,510	58,410
1978	T/Gas	45,390	50,490
1977	T/Gas	42,720	47,520
1976	T/Gas	40,050	44,550
1975	T/Gas	37,380	41,580
Tollycraft 34 Sundeck			
1988	T/Gas	111,250	123,750
1988	T/Diesel	125,490	139,590
1987	T/Gas	98,790	109,890
1987	T/Diesel	110,360	122,760
1986	T/Gas	86,330	96,030
1986	T/Diesel	97,900	108,900
Tollycraft 40 Tri Cabin			
1979	T/Gas	80,990	90,090
1979	T/Diesel	95,230	105,930
1978	T/Gas	72,980	81,180
1978	T/Diesel	85,440	95,040
1977	T/Gas	65,860	73,260
1977	T/Diesel	77,430	86,130
1976	T/Gas	59,630	66,330
1976	T/Diesel	69,420	77,220
1975	T/Gas	54,290	60,390
1975	T/Diesel	62,300	69,300
Tollycraft 40 Sundeck			
1994	T/Gas	258,100	287,100
1994	T/Diesel	310,610	345,510
1993	T/Gas	236,740	263,340
1993	T/Diesel	284,800	316,800
1992	T/Gas	213,600	237,600
1992	T/Diesel	258,990	288,090
1991	T/Gas	188,680	209,880
1991	T/Diesel	232,290	258,390
1990	T/Gas	170,880	190,080
1990	T/Diesel	210,040	233,640
1989	T/Gas	152,190	169,290
1989	T/Diesel	187,790	208,890
1988	T/Gas	148,630	165,330
1988	T/Diesel	177,110	197,010
1987	T/Gas	132,610	147,510
1987	T/Diesel	157,530	175,230
1986	T/Gas	123,710	137,610
1986	T/Diesel	150,410	167,310
1985	T/Gas	115,700	128,700
1985	T/Diesel	140,620	156,420
Tollycraft 43 CMY			
1986	T/Diesel	178,000	198,000
1985	T/Diesel	167,320	186,120
1984	T/Diesel	156,640	174,240
1983	T/Diesel	146,850	163,350
1982	T/Diesel	137,950	153,450
1981	T/Diesel	129,940	144,540
1980	T/Diesel	121,930	135,630
Tollycraft 44 CMY			
1993	T/Gas	262,550	292,050
1993	T/Diesel	303,490	337,590
1992	T/Gas	249,200	277,200
1992	T/Diesel	283,910	315,810
1991	T/Gas	226,060	251,460
1991	T/Diesel	258,990	288,090
1990	T/Gas	209,150	232,650
1990	T/Diesel	240,300	267,300
1989	T/Gas	192,240	213,840
1989	T/Diesel	220,720	245,520
1988	T/Gas	177,110	197,010
1988	T/Diesel	202,030	224,730
1987	T/Gas	164,650	183,150
1987	T/Diesel	189,570	210,870
1986	T/Gas	151,300	168,300
1986	T/Diesel	178,000	198,000
Tollycraft 48 CMY			
1993	T/Diesel	337,310	375,210
1992	T/Diesel	312,390	347,490
1991	T/Diesel	297,260	330,660
1986	T/Diesel	228,730	254,430
1985	T/Diesel	214,490	238,590
1984	T/Diesel	202,920	225,720
1983	T/Diesel	192,240	213,840
1982	T/Diesel	183,340	203,940
1981	T/Diesel	175,330	195,030
1980	T/Diesel	167,320	186,120
1979	T/Diesel	156,640	174,240
1978	T/Diesel	146,850	163,350
1977	T/Diesel	137,060	152,460
1976	T/Diesel	128,160	142,560
Tollycraft 53 CMY			
1994	T/Diesel	731,580	813,780
1993	T/Diesel	668,390	743,490
1992	T/Diesel	606,090	674,190
1991	T/Diesel	537,560	597,960
1990	T/Diesel	480,600	534,600
1989	T/Diesel	447,670	497,970
Tollycraft 57 CMY			
1994	T/Diesel	838,380	932,580
1993	T/Diesel	774,300	861,300
1992	T/Diesel	694,200	772,200

Year	Power	Retail Low	Retail High
1991	T/Diesel	626,560	696,960
1990	T/Diesel	560,700	623,700
1989	T/Diesel	518,870	577,170

Tollycraft 61 MY

Year	Power	Retail Low	Retail High
1994	T/665D	979,000	1,089,000
1993	T/665D	900,680	1,001,880
1992	T/735D	840,160	934,560
1991	T/735D	792,990	882,090
1990	T/735D	709,330	789,030
1989	T/735D	659,490	733,590
1988	T/735D	612,320	681,120
1987	T/650D	578,500	643,500
1986	T/550D	549,130	610,830
1985	T/550D	498,400	554,400
1984	T/550D	472,590	525,690
1983	T/550D	440,550	490,050

Trojan 36 Tri Cabin

Year	Power	Retail Low	Retail High
1987	T/Gas	94,340	104,940
1986	T/Gas	85,440	95,040
1985	T/Gas	76,540	85,140
1984	T/Gas	68,530	76,230
1983	T/Gas	62,300	69,300
1982	T/Gas	57,850	64,350
1981	T/Gas	56,070	62,370
1980	T/Gas	54,290	60,390
1979	T/Gas	52,510	58,410
1978	T/Gas	49,840	55,440
1977	T/Gas	45,390	50,490
1976	T/Gas	40,940	45,540
1975	T/Gas	37,380	41,580

Trojan 40 MY

Year	Power	Retail Low	Retail High
1984	T/Gas	106,800	118,800
1984	T/Diesel	134,390	149,490
1983	T/Gas	98,790	109,890
1983	T/Diesel	124,600	138,600
1982	T/Gas	95,230	105,930
1982	T/Diesel	117,480	130,680
1981	T/Gas	89,000	99,000
1981	T/Diesel	109,470	121,770
1980	T/Gas	86,330	96,030
1980	T/Diesel	101,460	112,860
1979	T/Gas	72,980	81,180
1979	T/Diesel	92,560	102,960

Trojan 44 MY

Year	Power	Retail Low	Retail High
1984	T/Gas	135,280	150,480
1984	T/Diesel	163,760	182,160
1983	T/Gas	127,270	141,570
1983	T/Diesel	153,080	170,280
1982	T/Gas	121,930	135,630
1982	T/Diesel	143,290	159,390
1981	T/Gas	115,700	128,700
1981	T/Diesel	136,170	151,470
1980	T/Gas	108,580	120,780
1980	T/Diesel	129,940	144,540
1979	T/Gas	103,240	114,840
1979	T/Diesel	122,820	136,620
1978	T/Gas	97,900	108,900
1978	T/Diesel	115,700	128,700
1977	T/Gas	90,780	100,980
1977	T/Diesel	105,910	117,810
1976	T/Gas	81,880	91,080
1976	T/Diesel	96,120	106,920
1975	T/Gas	75,650	84,150
1975	T/Diesel	88,110	98,010

Uniflite 36 DC

Year	Power	Retail Low	Retail High
1984	T/Gas	80,100	89,100
1984	T/Diesel	90,780	100,980
1983	T/Gas	72,980	81,180
1983	T/Diesel	83,660	93,060
1982	T/Gas	69,420	77,220
1982	T/Diesel	78,320	87,120
1981	T/Gas	65,860	73,260
1981	T/Diesel	74,760	83,160
1980	T/Gas	62,300	69,300
1980	T/Diesel	71,200	79,200
1979	T/Gas	59,630	66,330
1979	T/Diesel	67,640	75,240
1978	T/Gas	64,970	72,270
1978	T/Diesel	64,080	71,280
1977	T/Gas	49,840	55,440
1977	T/Diesel	57,850	64,350
1976	T/Gas	43,610	48,510
1976	T/Diesel	51,620	57,420
1975	T/Gas	38,270	42,570
1975	T/Diesel	45,390	50,490

Uniflite 42 DC

Year	Power	Retail Low	Retail High
1984	T/Gas	114,810	127,710
1984	T/Diesel	138,840	154,440
1983	T/Gas	111,250	123,750
1983	T/Diesel	133,500	148,500
1982	T/Gas	106,800	118,800
1982	T/Diesel	128,160	142,560
1981	T/Gas	104,130	115,830
1981	T/Diesel	123,710	137,610
1980	T/Gas	97,010	107,910
1980	T/Diesel	114,810	127,710
1979	T/Gas	89,890	99,990
1979	T/Diesel	107,690	119,790
1978	T/Gas	85,440	95,040
1978	T/Diesel	101,460	112,860
1977	T/Gas	75,650	84,150
1977	T/Diesel	96,120	106,920
1976	T/Gas	68,530	76,230
1976	T/Diesel	82,770	92,070
1975	T/Gas	62,300	69,300
1975	T/Diesel	75,650	84,150

Uniflite 48 YF

Year	Power	Retail Low	Retail High
1984	T/Diesel	162,870	181,170
1983	T/Diesel	153,970	171,270
1982	T/Diesel	146,850	163,350
1981	T/Diesel	141,510	157,410
1980	T/Diesel	134,390	149,490

Vantare 58/64 CMY

Year	Power	Retail Low	Retail High
1992	T/Diesel	******	******
1991	T/Diesel	******	******
1991	T/Diesel	******	******
1990	T/Diesel	******	******
1989	T/Diesel	******	******

Viking 43 DC

Year	Power	Retail Low	Retail High
1982	T/Gas	126,380	140,580
1982	T/Diesel	151,300	168,300
1981	T/Gas	122,820	136,620
1981	T/Diesel	144,180	160,380
1980	T/Gas	118,370	131,670
1980	T/Diesel	137,060	152,460
1979	T/Gas	110,360	122,760
1979	T/Diesel	129,050	143,550
1978	T/Gas	103,240	114,840
1978	T/Diesel	120,150	133,650
1977	T/Gas	97,010	107,910
1977	T/Diesel	112,140	124,740

Year	Power	Retail Low	Retail High
1976	T/Gas	89,890	99,990
1976	T/Diesel	103,240	114,840
1975	T/Gas	84,550	94,050
1975	T/Diesel	96,120	106,920

Viking 44 MY

Year	Power	Retail Low	Retail High
1991	T/Diesel	344,430	383,130
1990	T/Diesel	307,940	342,540
1989	T/Diesel	278,570	309,870
1988	T/Diesel	250,980	279,180
1987	T/Diesel	233,180	259,380
1986	T/Diesel	215,380	239,580
1985	T/Diesel	199,360	221,760
1984	T/Diesel	191,350	212,850
1983	T/Diesel	183,340	203,940
1982	T/Diesel	175,330	195,030

Viking 48 MY

Year	Power	Retail Low	Retail High
1988	T/735D	331,080	368,280
1987	T/735D	301,710	335,610
1986	T/735D	282,130	313,830

Viking 50 MY

Year	Power	Retail Low	Retail High
1994	T/Diesel	756,500	841,500
1993	T/Diesel	676,400	752,400
1992	T/Diesel	618,550	688,050
1991	T/Diesel	569,600	633,600
1990	T/Diesel	531,330	591,030

Viking 54 Sports Yacht

Year	Power	Retail Low	Retail High
1994	T/Diesel	801,000	891,000
1993	T/Diesel	753,830	838,530
1992	T/Diesel	718,230	798,930

Viking 55 MY

Year	Power	Retail Low	Retail High
1991	T/Diesel	640,800	712,800
1990	T/Diesel	604,310	672,210
1989	T/Diesel	560,700	623,700
1988	T/Diesel	511,750	569,250
1987	T/Diesel	448,560	498,960

Viking 57 MY

Year	Power	Retail Low	Retail High
1994	T/Diesel	907,800	1,009,800
1993	T/Diesel	863,300	960,300
1992	T/Diesel	801,000	891,000
1991	T/Diesel	732,470	814,770

Viking 60 Cockpit Sports Yacht

Year	Power	Retail Low	Retail High
1994	T/Diesel	979,000	1,089,000

Viking 63 MY

Year	Power	Retail Low	Retail High
1991	T/735D	827,700	920,700
1991	T/900D	865,080	962,280
1990	T/735D	736,920	819,720
1990	T/900D	770,740	857,340
1989	T/735D	694,200	772,200
1989	T/900D	724,460	805,860
1988	T/735D	636,350	707,850
1988	T/900D	659,490	733,590

Viking 63 CMY

Year	Power	Retail Low	Retail High
1991	T/Diesel	792,100	881,100
1990	T/Diesel	712,000	792,000
1989	T/Diesel	667,500	742,500
1988	T/Diesel	623,000	693,000
1987	T/Diesel	581,170	646,470

Viking 65 MY

Year	Power	Retail Low	Retail High
1994	T/Diesel	1,299,400	1,445,400
1993	T/Diesel	1,181,030	1,313,730
1992	T/Diesel	1,085,800	1,207,800
1991	T/Diesel	1,025,280	1,140,480

Viking 65 CMY

Year	Power	Retail Low	Retail High
1994	T/Diesel	1,173,910	1,305,810
1993	T/Diesel	1,089,360	1,211,760
1992	T/Diesel	1,005,700	1,118,700
1991	T/Diesel	930,940	1,035,540

Viking 72 MY

Year	Power	Retail Low	Retail High
1994	T/Diesel	1,619,800	1,801,800
1993	T/Diesel	1,513,000	1,683,000
1992	T/Diesel	1,409,760	1,568,160
1991	T/Diesel	1,272,700	1,415,700
1990	T/Diesel	1,173,020	1,304,820
1989	T/Diesel	1,103,600	1,227,600

Viking 72 CMY

Year	Power	Retail Low	Retail High
1994	T/Diesel	1,486,300	1,653,300
1993	T/Diesel	1,424,000	1,584,000
1992	T/Diesel	1,330,550	1,480,050

Year	Power	Retail Low	Retail High
1991	T/Diesel	1,237,100	1,376,100
1990	T/Diesel	1,142,760	1,271,160

Wellcraft 43 San Remo

Year	Power	Retail Low	Retail High
1990	T/Gas	135,280	150,480
1990	T/Diesel	161,090	179,190
1989	T/Gas	127,270	141,570
1989	T/Diesel	151,300	168,300
1988	T/Gas	122,820	136,620
1988	T/Diesel	145,070	161,370

Wellcraft 46 CMY

Year	Power	Retail Low	Retail High
1994	T/Diesel	302,600	336,600
1993	T/Diesel	264,330	294,030
1992	T/Diesel	239,410	266,310
1991	T/Diesel	218,050	242,550
1990	T/Diesel	187,790	208,890

West Bay 4500 PH

Year	Power	Retail Low	Retail High
1991	T/Diesel	******	******
1990	T/Diesel	******	******
1989	T/Diesel	******	******
1988	T/Diesel	******	******
1987	T/Diesel	******	******
1986	T/Diesel	******	******
1985	T/Diesel	******	******

Willard 30 Trawler

Year	Power	Retail Low	Retail High
1994	S/Diesel	89,000	99,000
1986	S/Diesel	66,750	74,250
1977	S/Diesel	49,840	55,440

Willard 40 Trawler

Year	Power	Retail Low	Retail High
1994	S/Diesel	204,700	227,700
1986	S/Diesel	124,600	138,600
1975	S/Diesel	89,000	99,000

Cross-Reference Guide

If you are unable to locate a particular model in the Table of Contents, check this cross-reference to see if information might be found elsewhere. Many models are very simliar in design.

Can't Find:	Go To:
46 Marlago Cockpit MY	Jefferson 43 Marlago
Albin 49 Cockpit	Albin 43 Sundeck
Albin Palm Beach 48	Albin 48 Cutter
Ocean Alexander 50 Sedan	Ocean Alexander 48/50 Sedan
Bluewater 42 CC	Bluewater 44 CC
Bluewater 45 Motor Yacht	Bluewater 462
Bluewater 53 Coastal Cruiser	Bluewater 543
Bluewater 55 Coastal Cruiser	Bluewater 55 Yacht
Bluewater 60 Yacht	Bluewater 622C
C&L 36 Trawler	Sea Ranger 36 Sundeck
C&L 39 Sedan	Sea Ranger 39 Sedan
C&L 51 Motor Yacht	Sea Ranger 51 MY
Californian 38 MY	Californian 43 CMY
Californian 42 MY	Californian 48 CMY
Camargue 42 Sundeck	Camargue 48 YF
Carver 33 Aft Cabin	Carver 350 Aft Cabin
Carver 350 Motor Yacht	Carver 390 Aft Cabin
Carver 36 Aft Cabin MY	Carver 370 Aft Cabin MY
Carver 38 Aft Cabin	Carver 390 CMY
Carver 390 MY	Carver 430 CMY
CHB 34 DC	Marine Trader 34 DC (Early)
CHB 42 MY	Present 42 Sundeck MY
CHB 48 MY	Seamaster 48 MY
Cheoy Lee 47 Efficient MY	Cheoy Lee 52 Efficient MY
Cheoy Lee 61 CMY	Cheoy Lee 55 Long Range MY
Cheoy Lee 77 MY	Cheoy Lee 83 CMY
Chris 362 Catalina	Uniflite 36 Double Cabin
Chris Craft 501 MY	Jefferson 52 Marquessa EDMY
Gulfstar 55 MY	Viking 55 MY
Island Gypsy 36 Aft Cabin	Island Gypsy 36 Tri-Cabin
Island Gypsy 36 Extended FB	Island Gypsy 36 Europa
Jefferson 40 Viscount	Jefferson 37 Viscount MY
Jefferson 52 Cockpit MY	Jefferson 46 Sundeck MY
Jefferson 52 Rivanna CMY	Jefferson 48 Rivanna MY
Jefferson 65 Cockpit MY	Jefferson 60 MY
Johnson 63 MY	High-Tech 63 MY
Marine Trader 38 Sedan	Marine Trader 42 Sedan
Marine Trader 44 YF	Marine Trader 40 Sundeck MY
Marine Trader 49 Sundeck	Marine Trader 49 PH
Nordic 520 MY	Nordic 480 MY
Ocean Alexander 50 MY	Ocean Alexander 54 CMY
Ocean Alexander 600 MY	Ocean Alexander 630 MY
Ocean Alexander 63 YF	Ocean Alexander 63 MY
Ocean Alexander 70 CMY	Ocean Alexander 63 MY
Offshore 58 PH	Offshore 55/58 PH
Positive 42 Trawler	Independence 45 Trawler
Prairie 36 Coastal Cruiser	Atlantic 37 Double Cabin
Prairie 46 LRC	Atlantic 47 MY
President 37 Double Cabin	President 395 Double Cabin
PT 46 Sundeck	PT 52 Cockpit MY
Sea Ranger 50 Cockpit MY	Sea Ranger 45 Sundeck
Southern Star 40 MY	Kha Shing 40 Sundeck
Spindrift 40 MY	Kha Shing 40 Sundeck
Stevens 67 Cockpit MY	Stevens 59 MY
Tradewinds 47 YF	Tradewinds 43 MY
Trojan 47 YF	Trojan 40 MY
Uniflite 41 Yacht Fisherman	Uniflite 36 Double Cabin
Viking 50 Cockpit MY	Viking 44 MY
Viking 55 Cockpit MY	Viking 48 MY

Advertiser's Index

INTERNATIONAL MARINE
A Division of McGraw-Hill, Inc.
Camden, Maine